Lecture Notes in Mathematics Vol. 1189
ISBN 978-3-540-16474-6

Lukeš J., Malý J., Zajíček L.

Fine Topology Methods in Real Analysis and Potential Theory

Errata

page	instead of	should be
2^{6}	$r_{n+1} + k^{-1}$	$r_n + k^{-1}$
6, 7	$\lfloor\ ,\ \rfloor$	$[\ ,\ ]$
42	"	"
9^{12}	of.	cf.
34^{5}	$\beta X = X = \emptyset = ibX$	$\beta X = X \neq \emptyset = ibX$
39_{4}	$A \quad X$	$A \subset X$
51^{5}	$A \quad x \to 0$	$A \ni x \to 0$
64_{3}	that the set	that $f_n \to f$ and the set
72_{3}	from	from $\mathcal{H}$
72_{1}	$\mathcal{F}^{\uparrow}$	the family of all real functions from $\mathcal{F}^{\uparrow}$
73^{1}	"	"
82^{7}	lead	led
97_{3}	staight	straight
144_{3}		(iv)(a) every open connected set is τ-connected, (b) if V and W are disjoint τ–open τ–connected sets, then $V \cap \overline{W} = \emptyset$.
149^{9}	Remark 6.16.a	Remark 6.16
152_{8}	$d^e(x, M)$	$d^e(x, \mathbb{R} \setminus M)$
155_{6}	$x \in F \setminus G$	$x \in F \cap G$
171^{14}	Theorem 8.1	Ch. II, §32
176^{15}	D. Preiss (1981)	D. Preiss (1983)
179_{4}	G–insertion	G_δ–insertion
187_{6}	$\lambda_n K_i)$	$\lambda_n(K_i)$
187_{4}	ohne	one

page	instead of	should be
189^{12}	M. Chlebík (1984)	M. Chlebík
217^{14}	$L(A) = \{x \in X : x \in \mathrm{int}_\tau(A \cup \{x\})\}$	$L(A) = \mathrm{int}_\tau \overline{A}^\tau$
252_{15}	Let (X, ρ) be a metric space.	Let (X, ρ) be a metric space and $(\mathfrak{M}, m)$ be on $P = X$ as in Section 6B, p. 163
252_{14}	on X	on $(X, \mathfrak{M}, m)$
252_{13}	the condition (ω) of Section 6.C	the condition (i) of 6.34 (D)
253_{10}	on $(\mathbb{R}, \mathfrak{M}, \lambda)$	on $\mathbb{R}$
261^{10}	$(0, 1)$	$\mathbb{R}$
261^{12}	"	"
262^{15}	M. Laczkovich and G. Petruska	G. Petruska and M. Laczkovich
263_{10}	Lebesgue point	Lebesgue bounded
263_{8}	"	"
267_{10}	type G	type G_δ
293_{13}	$P = \overline{H}$	$P = M \times [0, +\infty)$
343_{6}	function on	function f on
343_{6}	at f	at
346_{14}	$M \quad y \to x$	$M \ni y \to x$
351_{5}	11.B.1c	11.B.1d
358_{4}	11.D.2b	10.D.2b
372_{2}	locally bounded	locally lower bounded
374^{11}	12.A.6	$\boxed{\text{12.A.6}}$
374_{3}	(e)	
400^{12}	form	from
449_{8}	characterisation	characterization

Lecture Notes in Mathematics

Edited by A. Dold and B. Eckmann

Subseries: Mathematisches Institut der Universität Erlangen-Nürnberg

Adviser: H. Bauer

1189

Jaroslav Lukeš

Jan Malý

Luděk Zajíček

Fine Topology Methods in Real Analysis and Potential Theory

Springer-Verlag

Berlin Heidelberg New York London Paris Tokyo

Authors

Jaroslav Lukeš
Jan Malý
Luděk Zajíček
Matematicko-fyzikáln fakulta University Karlovy
Sokolovská 83, 186 00 Praha 8, ČSSR

Mathematics Subject Classification (1980): 26 A 21, 26 A 24, 26 A 27, 26 B 05, 28 A 15, 28 A 51, 31 C 05, 31 D 05, 54 A 05, 54 C 20, 54 D 05, 54 D 15, 54 E 52, 54 E 55

ISBN 3-540-16474-X Springer-Verlag Berlin Heidelberg New York
ISBN 0-387-16474-X Springer-Verlag New York Berlin Heidelberg

© Springer-Verlag Berlin Heidelberg 1986
Printed in Germany

Printing and binding: Beltz Offsetdruck, Hemsbach/Bergstr.
2146/3140-543210

P r e f a c e

There are two well-established seminars at the Department of Mathematical Analysis of the Charles University Prague. One is the seminar on Potential Theory and the other the seminar on Modern Theory of Real Functions. It has been established that some of the problems subjected to study in these seminars are closely related resulting in the combined investigation of the very foundations and interrelations of the

The first version of the manuscript was prepared in April 1982 in order to serve as a complement to lectures at Erlangen and Eichstätt Universities. Since that time the work has been greatly extended and is now quite voluminous. It is only fair to admit that the manuscript is not free of misprints and shortcomings and does not contain all sources dealing with the subject. Any useful commentary and supplementary contributions would be gratefully received by the authors.

As many persons cooperated in the preparatory work on the manuscript let us mention at least some of those who extended their generous help in support of the authors. Let us quote amongst others our colleagues O.John, D.Preiss, J.Veselý and students M.Chlebík, J.Hrdina, V.Kelar, T.Schütz, V.Šverák and R.Thomas.

At this occasion let us express our gratitude and indebtness to Mrs. Erika Ritter from Erlangen who has devoted the best care to the final re-typing of the text. Insertion of mathematical symbols was done by the authors who themselves are responsible for all the possible mistakes.

In concluding this preface we have to stress our special thanks to those who have tirelessly supported us in our endeavour to bring this work to a succesful finish. We thank, in particular, Dr. Niels Jacob, Erlangen, for his assistance in proof reading. Special thanks are due to Professor Heinz Bauer without whose participation it would have been impossible to carry out our task.

J.L.
J.M.
L.Z.

Byčiarky 1982 - Praha 1985

C O N T E N T

INTRODUCTION

Mathematical analysis has undergone its most outstanding and rapid development in the present century. In many instances the classical penetrating results stemming from the beginning of the century have been newly formulated only recently. Certain new methods and specific approaches have brought more understanding of some classical parts and have demonstrated similarities in disciplines where no connections may be observed prima faciae. The fine topologies discussed in this book serve as an example.

The concept of approximate continuity which plays an important part in the theory of real functions was defined at the beginning of this century. But the concept of the density topology naturally connected with that of approximate continuity has been introduced and investigated (relatively) only recently. The fine topology introduced in the forties plays an important part in potential theory. Its methods in the theory of harmonic functions facilitated many excelent results in this classical part of mathematical analysis.

It has become apparent that in spite of having different structures these topologies have many fundamental properties in common. Besides, certain methods and procedures used in connection with density topology have recently influenced the investigation of fine topology in potential theory.

Both the density topology and the fine topology in potential theory are "fine topologies" on a set which has already been equipped with an original coarser (metrizable) topology. Phenomena where two topologies are defined on a set can be encountered in mathematics quite easily. There is even a discipline of general topology - - theory of "bitopological spaces" - which investigates these situations.

Thus nearly the whole first part of our book, i.e. "Abstract Fine Topology", can be considered as belonging to this part of topology. But we investigate mainly those concepts which are related in a way to two fundamental examples of fine topologies and concentrate chiefly upon that part of general theory which is suitable for applications in the theory of real functions or in the potential theory.

Among the general properties of fine topologies one of the most important qualities and also one of the most suitable qualities from the point of applications is the Lusin-Menchoff property to which we devote greater part of this volume. Though it often happens that fine topologies in which we take interest are not normal, nevertheless, they allow us to separate each closed set and finely closed set by "alternately" open sets. We call this property for which the theory of bitopological spaces uses the word "binormality" the Lusin-Menchoff property as it is usual in the theory of real functions when the density topology is investigated. Among others, the Lusin-Menchoff property enables - as well as normality does - various constructions of finely continuous functions.

The fundamental technical means for constructing functions is the "Abstract inbetween theorem" which is simple and not new but which is - perhaps for the first time - explicitly formulated here and provided with a simple proof. Other important concepts are M-modification of a fine topology - which enables to extend finely continuous functions "with the preservation of continuity and semicontinuity" - and the Lusin-Menchoff property of base operators which finds applications in the following text. New "Approximation Theorem" in I/3.6 and its applications also deserve attention.

One of the other problems investigated in this treatise is the following: under which circumstances a space equipped with a fine topology is a Baire space, a strong Baire space or a Blumberg space (I/4), and when finely continuous functions are in Baire class one (I/2D). We also investigate properties of fine limits in an abstract case (I/2C).

Other interesting and not unimportant results (some of which are new) are contained in I/5 which concerns the connectedness of fine topologies.

The beginning of the first chapter is devoted to the study of base operator spaces. But the idea of this generalization of topological spaces is not new at any rate. Among others, we produce some results already proved in the potential theory for concrete base operators in a completely general version. Besides, the concept of base operator (with the Lusin-Menchoff property) appears useful also in real analysis.

In the second part, i.e. in "Fine Topology Methods in Real Analysis", we investigate in detail some fine topologies that occur in real analysis and employ in high degree the general theorems produced in the first part.

The fundamental example of fine topology is a density topology which we study in various degrees of generality (from the density topology on the real line to the "abstract density topologies "which are investigated in lifting theory). We have taken pains to find simple criteria which make it possible to obtain information about some properties of the density topology under the assumption that we know certain properties of the respective "differentiation basis". In our opinion it is especially interesting (and perhaps new) to find simple conditions under which the density topology has the Lusin-Menchoff property (6.34.B), the internal characterization of abstract density topologies (Th. 6.39) and the possibility to use this characterization in lifting theory.

We investigate (II/7.A.7.B) the concepts of a.e.-modification and r-modification of a given fine topology which are generalizations of topologies defined by R.J. O'Malley. The main new contribution is that our proofs enable us to prove the Lusin-Menchoff property of a.e.-topology and r-topology under very general circumstances. The new theorem about the approximation of r-continuous function by approximately differentiable functions is also worth attention.

Another example of a fine topology with the Lusin-Menchoff property is the "fine boundary topology" (II/7.D) and the contingent topology which is related to it and investigated in "Exercises".

The Lusin-Menchoff property of topologies mentioned above is applied in II/8 to obtain results concerning constructions of various functions, especially approximately continuous functions where many of the results are new. Concerning the theorem about the extension from F_σ-sets in II/8.C we would like to mention that we do not use topological methods here.

In II/9 we concentrate upon the boundary behaviour of functions. The chief new idea is use of the Lusin-Menchoff property of "fine boundary topology". The utility of this method is demonstrated in II/9.C,D,E.

The final third part, i.e. "Fine Topology Methods in Potential Theory", is devoted to fine potential theory. The investigation of fine topology and finely hyperharmonic functions in the framework of abstract harmonic spaces (or even more general standard H-cones) culminates in the study of the Dirichlet problem on finely open sets.

Among the purely topological properties of harmonic spaces the Lusin-Menchoff property of the fine topology is of major importance. It is applied in various connections. In chapters 12-14 we present a selfcontained theory of finely hyperharmonic functions and of fine Dirichlet problem in harmonic spaces; until now this theory has been established only under the assumption of the domination axiom D. The possibility of generalization of the presented results to standard H-cones is suggested in the Appendix.

Developing the theory of finely hyperharmonic functions we derive their basic properties (III/12.A) and take interest in the properties of the cone of all nonnegative finely hyperharmonic functions (III/12.B). Besides, we also study the class of all pointwise hyperharmonic functions on a finely open set and its subclass of all its elements that are lower semicontinuous functions. Among new results let us mention e.g. interesting theorem on removable singularities (Th. 12.20, 14.11).

The Dirichlet problem is solved either by means of Perron type method using fine superfunctions or of the Wiener type method when the given finely open set is exhausted by finely regular sets.

A quite new approach to the solution of the fine Dirichlet problem based on quasitopological concepts is shown in Chapter 14. The quasitopological methods are also used to characterize finely hyperharmonic functions (Th. 14.8). The results contained in Chapter 14 embrace both a part of the preceding chapters as well as the basic results of Fuglede's theory of finely hyperharmonic functions in spaces with axiom D.

In the Appendix we investigate analogous problems within the framework of the abstract theory of standard H-cones. We present an abstract definition of the class of "finely hyperharmonic functions" and demonstrate its connections with the localized cone. Our results make it possible to develop the theory of fine Dirichlet problem also in this very general context.

Note that each section ends with "Exercises" containing not only counterexamples and generalizations of the given theorems but often also further new results (even more general structures are investigated). Some of the results of the theory of bitopological spaces, σ-topologies, even further generalizations of abstract density topologies, the study of Keldych operators on finely open sets, a boundary behaviour of the Perron solution etc. may serve as an example. In the "Exercises" one can also find new results and results of other authors having a close relationship to the investigated problems.

The "Exercises" are followed by "Remarks and Comments" where the authors try to give information about the origin of concepts and results of the respective sections and bring occasionally further quotations of authors dealing with similar problems.

PROLOGUE

In (1915), A. Denjoy in his famous paper generalized the notion of a continuity. A real function f is said to be *approximately continuous* at a point z if there is a measurable set M which is "sufficiently great at z" with

$$\lim_{M \ni x \to z} f(x) = f(z).$$

Greatness of a measurable set M at z is defined by the condition that

$$(*) \qquad \lim_{h \to 0_+} \frac{1}{2h} \lambda(M \cap (z-h, z+h)) = 1,$$

where λ stands for the Lebesgue measure.

The family of all functions approximately continuous at any point of $\mathbb{R}$ has many nice properties. For example:

(a) Every approximately continuous function has the Darboux property and belongs to the first class of Baire.

(b) Any bounded approximately continuous function is a derivative.

Almost half a century later, a new topology on the real line was introduced: The so-called *density topology* consists of all measurable sets M satisfying (*) at each of its points. Let us mention several interesting properties of this topology:

(a) Any countable set of $\mathbb{R}$ is closed in the density topology.

(b) The density topology is not normal, so it is not metrizable.

(c) Only the finite sets are density compact.

It was proved that the "nice" system of all approximately continuous functions is exactly the class of all continuous functions in the "bad" density topology.

By a thorough study of the density topology mathematicians succeeded to discover a lot of its important properties. Even though the density topology was discovered in the fifties, many methods labelled recently as "topological" have been used much earlier. For example, A. J. Ward (1933) used the method similar to the proof of Urysohn's lemma which enables to construct approximately continuous functions (and therefore also derivatives) of various properties. Similarly Z. Zahorski (1950) proved that for any G_δ-set Z of Lebesgue measure zero there is always a non-negative bounded approximately continuous function f such that

$$Z = \{x \in \mathbb{R} : \quad f(x) = 0\}.$$

It is not very difficult to construct on the real line a G_δ - set Z of Lebesgue measure zero which is at the same time everywhere dense. In fact, enumerating all rational numbers by $\{r_n\}$, we can define the desired set as

$$Z = \bigcap_{k=1}^{\infty} \bigcup_{n=1}^{\infty} (r_n - k^{-1}2^{-n}, r_{n+1} + k^{-1}2^{-n}).$$

Using the theorem mentioned above, there is an approximately continuous function f, say $0 \leq f \leq 1$, such that $Z = f^{-1}(0)$. Of course, f cannot be continuous at any point $z \in \mathbb{R} \setminus Z$ (since $f(z) > 0$ and Z is dense). Hence, even though f is a derivative of some function F it is not Riemann integrable on any interval $[a,b]$. By the way, we have just constructed a function F which has the derivative at each point of $\mathbb{R}$ but this derivative F' is discontinuous on a "massive" set $\mathbb{R} \setminus Z$. Moreover, both the sets

$$\{x \in \mathbb{R}: \ F'(x) = 0\}, \{x \in \mathbb{R}: \ F'(x) > 0\}$$

are dense in $\mathbb{R}$. The construction of functions of this type, the so-called Pompeiu functions, goes back to the beginning of our century (for some references and for the constructions of Pompeiu functions see e.g. Blažek, Borák and Malý's paper (1978)).

The introduction of the density topology throws light upon many problems. Starting from it, Zahorski's theorem becomes at once clearer and we can see that some new "modern" methods clear up certain old problems from the theory of real functions. Still more evident becomes an advantage of the use of topological methods in the construction of unbounded derivatives in Section 8.B.

Observe now that even if the density topology is not normal, it has in some sense the property of normality with respect to the Euclidean topology. The use of this so-called "Lusin-Menchoff property" represents almost main topic of this work.

Let us turn our attention to another plentiful part of analysis with a long and interesting history - to the classical theory of harmonic functions. The Dirichlet problem is probably its corner stone. It is historically the oldest problem of existence in potential theory and various attacks on this problem brought many important mathematical discoveries. Given a bounded open set $U \subset \mathbb{R}^n$ and a continuous function f on the boundary ∂U of U, by the *classical solution* of the *Dirichlet problem* for f we understand a continuous function F on the closure $\bar{U}$ of U which is *harmonic* on U (i.e. it satisfies the *Laplace equation*

$$\Delta F = \frac{\partial^2 F}{\partial x_1^2} + \ldots + \frac{\partial^2 F}{\partial x_n^2} = 0$$

in U) and coincides with f on ∂U. A set U is termed *regular* if there exists a classical solution of the Dirichlet problem for each continuous function f on ∂U. Towards the close of the last century, the Dirichlet problem was regarded as always solvable and it was believed that methods of proof powerful enough would be found to confirm this opinion. A new period begins with the recognition that this opinion is not justified. Of course, now it is well known that there are open bounded sets in $\mathbb{R}^n$ which are not regular. On the boundaries of such sets continuous functions can be prescribed for which the Dirichlet problem cannot be solved. Nevertheless, we can assign to those functions something like a solution in a reasonable way. It can be done, for example, in a following manner.

A $\mathcal{C}^2$-function s is called $\mathcal{C}^2$-*superharmonic* on U if $\Delta s \leq 0$ on U; the key role is played by the system of all *superharmonic* functions which arises as a natural generalization from the system of all $\mathcal{C}^2$-superharmonic functions (for example, as the smallest family containing $\mathcal{C}^2$-superharmonic functions closed with respect to the limits of increasing sequences which are finite on a dense set). For a continuous function f on ∂U, if we denote by H_f^U the infimum of the family of all superharmonic functions on U whose lower limit dominates f at each boundary point of U, then H_f^U is a harmonic function on U and it is called the *Perron generalized solution* of the Dirichlet problem for f. A point $z \in \partial U$ is termed a *regular* boundary point if

$$\lim_{U \ni x \to z} H_f^U(x) = f(z)$$

for every continuous function f on ∂U. The remaining points of ∂U are termed *irregular*.

In (1939), the notion of *thinness* of a set A at a point x was introduced by M. Brelot: A set A is thin at $x \in A$ if there is a superharmonic function u defined on a neighborhood of x such that

$$u(x) < \liminf_{A \ni y \to x} u(y).$$

In a letter from 30th December 1940 adressed to M. Brelot, H. Cartan observed that A is thin at x if and only if the complement of A is a neighborhood of x in the coarsest topology on $\mathbb{R}^n$ making all superharmonic functions continuous. This new topology was termed the *fine topology* and in the dimension $n > 1$ it differs from the Euclidean one. Soon the fine topology proved itself to be a natural and important tool in the study of more refined properties in potential theory. The resemblance of its properties with those of the density topology is admirable. In the case $n > 1$, for the fine topology on $\mathbb{R}^n$ the following assertions can be proved:

(a) Each countable set is finely closed, the only compact sets are the finite sets.

(b) The fine topology is neither metrizable nor normal.

(c) Finely continuous functions are of the first class of Baire.

Since the introduction of the fine topology, its methods have served as an effective tool for examinations of many properties in the potential theory. Let us mention, for instance, that irregular points of ∂U can be characterized as those which are isolated points of $\mathbb{R}^n \setminus U$ in the fine topology, or equivalently, in which the set $\mathbb{R}^n \setminus U$ is thin. The series of theorems concerning the fine limits of harmonic or superharmonic functions was proved. Further, the role of stable points of U was clarified; they are those points which lie in the fine closure of $\mathbb{R}^n \setminus U$. In the past years the methods of fine topology originated the theories of finely harmonic and finely holomorphic functions.

1. BASE OPERATOR SPACES

1.A Base and strong base

Both in potential theory and in the density topologies problems certain set operators are closely related to the notion of a fine topology. The structure of spaces equipped with a base operator is slightly richer than their topological structure. Moreover, sometimes we find it useful to introduce some concepts in this more general setting, which enables us to get some new results. The Lusin-Menchoff property of generalized density base operators is a typical example. Similarly, we can introduce such notions as the quasi-Lindelöf property or the semipolar sets.

Recently, various base operators play an important role in the potential theory and its generalizations (e.g. H-cones, balayage spaces). The base operators are used to characterize the set of regular points, the Choquet boundary, the semipolar sets, and many others. Taking advantage of the "Borel type" of a base we will be able to derive certain properties of fine topologies and, in particular, of finely continuous functions.

In what follows, we will summarize general properties of base operator spaces needed in the sequel.

Let X be an abstract set. A mapping $b\colon \exp X \to \exp X$ is called a *base operator* (or briefly a *base*) if

(b_1) $b\emptyset = \emptyset$,

(b_2) $b(A \cup B) = bA \cup bB$ for any $A, B \subset X$.

The couple (X,b) will be called a *base operator space*. Any base preserves the inclusion: If $A \subset B$, then $bB = b(A \cup B) = bA \cup bB$, and therefore $bA \subset bB$. Further, $bB \setminus bA \subset b(B \setminus A)$ whenever $A, B \subset X$. The *b-topology* determined by the base b is defined by saying that a set F is *b-closed* if $bF \subset F$. It is not difficult to verify that the collection of all b-closed sets determines actually a topology on X. In fact, the b-topology is the finest topology τ on X satisfying $bA \subset \bar{A}^{\tau}$ for each $A \subset X$. We denote by $\bar{A}^{b}$ the closure of a set A in the b-topology.

Obviously, any topology is determined by some base operator. We must underline that this base operator is not uniquely determined. Indeed, both the maps $A \mapsto \bar{A}$ and $A \mapsto \operatorname{der} A$ on a topological space have the properties of a base operator, and they determine the same original topology.

1.1 EXAMPLES. (a) Let M be a subset of a topological space (X,τ). Then the map

$$b_M: A \mapsto \mathrm{der}_\tau A \cup (A \cap M)$$

is a base operator on X. Of course, the corresponding b_M-topclogy is equal to τ. The choice of $M = X$ or $M = \emptyset$ yields the examples of typical base cperators: closure and derivative (cf. Theorem 1.2).

(b) Let $\mathcal{S}$ be the set of all nonnegative superharmonic functions on a harmonic space X (see later). Then

$$b: A \mapsto \{x \in X: \hat{R}_s^A(x) = s(x) \text{ for all } s \in \mathcal{S}\}$$

is a base operator (the proof is not obvious!). Similar base cperators can be defined in a more general setting of H-cones or balayage spaces. On the other hand, the operator t of Remark 2.6.b is not, in general, a base.

(c) For any $A \subset \mathbb{R}$ we put

$$bA := \{x \in \mathbb{R}: \limsup_{h \to 0_+} \frac{\lambda^*(A \cap (x-h,x+h))}{2h} > 0\}.$$

Then $(\mathbb{R},b)$ is a base operator space (the proof is not difficult). The corresponding b-topology is called the density topology. In the sequel we prove that a set G is b-open if and cnly if it is Lebesgue measurable and any point of G is its point of density. The notion of a (generalized) density base operator as well as other base operators generated by outer measures will be discussed below (sections 6.B, 6.D, 6.E).

(d) For every $a = (a_1,a_2) \in \mathbb{R}^2$ and for each $\delta > 0$ we denote

$$K(a,\delta) := \{x = (x_1,x_2) \in \mathbb{R}^2: x \in U(a,\delta) \text{ and } x_1 = a_1 \text{ or } x_2 = a_2\}.$$

We put

$$bA := \{z \in \mathbb{R}^2: K(z,\delta) \cap A \neq \emptyset \text{ for each } \delta > 0\}.$$

Then $(\mathbb{R}^2,b)$ is a base operator space. A set G is b-open if and only if for any $z \in G$ there is $\delta > 0$ such that $K(z,\delta) \subset G$. Let us remark that $K(z,\delta)$ is not a b-neighborhood of z ! If $A = [0,1) \times [0,1)$, then

$$\bar{A}^b = bbA = [0,1] \times [0,1], \quad bA = bbA \setminus \{(1,1)\}.$$

Thus bbA is not a subset of bA. The b-topology will be termed the *crosswise topology* on $\mathbb{R}^2$.

(e) Let (X,τ) be a topological space. A set $C \subset X$ is of the *first category at a point* x if there is a neighborhood V of x such that $C \cap V$ is of the first category in X. The mapping

$$A \longmapsto \{x \in X: \ A \text{ is not of the first category at } x\}$$

is a base operator. This example as well as the similar ones will be examined in details in Section 1.C.

REMARK. Let (X,b) be a base operator space. It is not difficult to check that the mapping

$$c: \ A \longmapsto A \cup bA$$

is a *closure operation*, i.e. it satisfies

(c_1) $c\emptyset = \emptyset$,

(c_2) $c(A \cup B) = cA \cup cB$ for any $A,B \subset X$,

(c_3) $A \subset cA$ for any $A \subset X$.

The closure operation c is the smallest closure operation containing b (cf. Ex. 1.A.6a). In particular, $bA \subset cA \subset \bar{A}^b$ for each $A \subset X$. In general, cA is not equal to the closure of A in the b-topology (cf. Example 1.1.d or Example 6.27). The necessary and sufficient conditions for the equality $cA = \bar{A}^b$ are given in the following theorem.

1.2 THEOREM. Let (X,b) be a base operator space, and $M = \{x \in X: \ x \in b\{x\}\}$. Consider the following assertions:

(i) $bbA \subset bA$ for any $A \subset X$,

(ii) c is an idempotent operator,

(iii) $bbA \subset cA$ for any $A \subset X$,

(iv) $\bar{A}^b = cA$ for any $A \subset X$,

(v) $bA = \mathrm{der}_b A \cup (A \cap M)$ for any $A \subset X$.

Then (i) $\Rightarrow$ (ii) $\Leftrightarrow$ (iii) $\Leftrightarrow$ (iv) $\Leftrightarrow$ (v). If one-point sets are b-closed (i.e. if $b\{x\} \subset \{x\}$, or if, equivalently, the b-topology is T_1), then all the assertions are equivalent.

Proof. (i) $\Rightarrow$ (iii) is obvious.

(ii) $\Rightarrow$ (iii): We have $bbA \subset ccA = cA$ for any $A \subset X$.

(iii) $\Rightarrow$ (iv) : Given $A \subset X$,

$$b(A \cup bA) = bA \cup bbA \subset cA.$$

Therefore cA is a b-closed set containing A. On the other hand, $cA \subset \bar{A}^b$.

(iv) $\Rightarrow$ (v): Clearly, $bA \setminus A \subset \bar{A}^b \setminus A \subset der_b A$. If $x \in bA \setminus M$ (notice $bA = b\{x\} \cup b(A \setminus \{x\})$), then

$$x \in b(A \setminus \{x\}) \subset \overline{A \setminus \{x\}}^b, \quad \text{hence } x \in der_b A.$$

On the other hand, $x \in A \cap M$ implies $x \in b\{x\} \subset bA$. Finally, if $x \in der_b A$, then $x \in \overline{A \setminus \{x\}}^b = (A \setminus \{x\}) \cup b(A \setminus \{x\})$, and thus $x \in b(A \setminus \{x\}) \subset bA$.

(v) $\Rightarrow$ (iv): $\bar{A}^b = A \cup der_b A \subset A \cup bA = cA \subset \bar{A}^b$ for every $A \subset X$.

(iv) $\Rightarrow$ (ii) is obvious.

If the b-topology is T_1 (i.e. if singletons are b-closed), we prove (v) $\Rightarrow$ (i). Given $A \subset X$, we wish to show that bA is b-closed. But in every T_1-space the set $der\, A$ is closed. It follows that bA is b-closed since apparently $b(A \cap M) \subset bA$.

A base operator b on X is termed *strong* if $bbA \subset bA$ for any $A \subset X$. In other words, a base b is strong if bA is b-closed for each set $A \subset X$. A base b is *idempotent* if $b^2 = b$. A closure operation is called a *topological closure* (or, a Kuratowski closure operation) if it is idempotent. Of course, a closure operation is a topological closure if and only if it is strong.

EXAMPLES. The base from Example 1.1.a is strong if $M = X$ or if (X,τ) is a T_1-space.

In Example 1.1.b the base is strong, and it is idempotent if and only if the axiom of thinness holds on X (assuming that the points are totally thin).

In the case of the density topology on $\mathbb{R}$ the base bA coincides with the derivative of A in the density topology, and $A \cup bA$ is a measurable cover of A. The Lebesgue density theorem then guarantees that b is even idempotent.

The base b of Example 1.1.e is also idempotent, while the base from Example 1.1.d is not strong.

We will denote by $\bar{b}$ the set operator $A \mapsto \overline{bA}^b$. The following theorem shows that the operator $\bar{b}$ is a certain "strong modification" of b.

1.3 THEOREM. Let (X,b) be a base operator space. Then the operator $\bar{b}$ is a strong base operator which determines the same topology as b. The b-closure of each set $A \subset X$ is given by $\bar{A}^b = A \cup \bar{b}A$.

Proof. Since the composition of two base operators is again a base operator (cf. Ex. 1.A.1), $\bar{b}$ is a base.

Since $bA \subset \bar{b}A \subset \bar{A}^b$, the $\bar{b}$-topology and the b-topology coincide.

The base $\bar{b}$ is strong because $\bar{b}A$ is always a b-closed set, and hence also a $\bar{b}$-closed set.

Finally, in accordance with (i) $\Rightarrow$ (iv) of Theorem 1.2 (applied to $\bar{b}$) we get $\bar{A}^b = \bar{A}^{\bar{b}} = A \cup \bar{b}A$.

REMARKS. (a) The base operator $\bar{b}$ is the smallest strong base operator greater than b (of. Ex. 1.A.6c).

(b) In Section 1.D we associate with every base operator b its "idempotent modification" β which, however, does not preserve a topology.

We conclude this section by defining a concept of *thinness* which is closely related to the notion of a fine neighborhood, being at the same time slightly weaker in the general case (as examples and theorems will show).

Let (X,b) be a base operator space. A subset A of X is called *thin at a point* $x \in X$ if $x \notin bA$.

It is easy to check that the collection of all sets which are thin at x is a nonvoid ideal. Therefore the collection

$$\mathfrak{F}_x := \{M \subset X: \quad x \in CbCM\}$$

is a nonempty filter on X (cf. also Ex. 1.A.2).

If $G \in \mathcal{V}_b(x)$, then $x \in G$ and CG is thin at x. In particular, $\mathcal{V}_b(x) \subset \mathfrak{F}_x$.

On the other hand, if $G \subset X$ is such a set that $x \in G \cap CbCG$, then, in general, G is not in $\mathcal{V}_b(x)$ (cf. Example 1.1.d). But in the case when the base b is strong we have $G \cap CbCG = \operatorname{int}_b G$, and therefore $\mathcal{V}_b(x)$ is formed by the family of all sets containing x whose complements are thin at x. Notice that this observation is analogous to that contained in Cartan's theorem (cf. Theorem 2.5).

EXERCISES

1.A.1 if b_1, b_2 are base operators on X, then

$$b_1 * b_2 : A \mapsto b_1(b_2A)$$

is also a base operator.

1.A.2 (a) Let (X,b) be a base operator space. For every $x \in X$ we put

$$\mathfrak{F}_x := \{M \subset X: \quad x \notin b(X \setminus M)\}.$$

Then $\mathfrak{F}_x$ is a nonempty filter on X.

(b) Conversely, assume that to each $x \in X$ corresponds a nonempty filter $\mathfrak{F}_x$ on X. Then the operator

$$b: \quad M \longmapsto \{x \in X: \quad F \cap M \neq \emptyset \text{ provided } F \in \mathfrak{F}_x\}$$

is a base on X.

(c) The described mappings are (mutually) inverse and this natural correspondence is a bijection between base operators on X and systems $\{\mathfrak{F}_x\}_{x \in X}$ of nonempty filters on X.

(d) If b corresponds to $\{\mathfrak{F}_x\}_{x \in X}$, then $G \subset X$ is b-open if and only if for every $x \in G$ there is $F \in \mathfrak{F}_x$, $F \subset G$.

1.A.3 Let (X,b) be a base operator space and let $\{\mathfrak{F}_x\}_{x \in X}$ correspond to b as in Ex. 1.A.2. Then

(a) b is a closure operation if an only if $x \in F$ whenever $F \in \mathfrak{F}_x$,

(b) b is strong if and only if for each $x \in X$ and every $F \in \mathfrak{F}_x$ there is $B \in \mathfrak{F}_x$, $B \subset F$ such that for every $y \in B$ there is $E \in \mathfrak{F}_y$, $E \subset B$ (in other words, any $\mathfrak{F}_x$ has a filter basis which contains b-open sets only),

(c) b is a topological closure if and only if for each $x \in X$ the filter $\mathfrak{F}_x$ has a filter basis which contains b-open sets only and $x \in F$ whenever $F \in \mathfrak{F}_x$.

In case of (c) (and only in this case) $\mathfrak{F}_x$ is the neighborhood system at x in the b-topology.

1.A.4 Determine an *order relation* on the set $\mathcal{B}$ of all base operators on a set X as follows:

$$b \prec b' \quad \text{if} \quad bA \subset b'A \quad \text{for every} \quad A \subset X.$$

We say also that b' *is greater than* b.

(a) The ordered set $\mathscr{B}$ is order complete and has a smallest and a greatest element. If $\{b_\alpha\}$ is a nonvoid family in $\mathscr{B}$, then:

(a_1) $\sup b_\alpha$ is the base $A \mapsto \bigcup_\alpha b_\alpha A$.

(a_2) $(\inf b_\alpha)(A) \subset \bigcap_\alpha b_\alpha A$.

Hint. Since $\mathscr{B}$ possesses the greatest and the smallest element, to prove that $\mathscr{B}$ is order complete it suffices to show (a_1).

(a_3) If b_1 and b_2 are bases, then the map

$$A \mapsto b_1 A \cap b_2 A$$

need not be a base operator.

(a_4) If $\{b_\alpha\}$ is lower directed, then

$$(\inf b_\alpha)(A) = \bigcap_\alpha b_\alpha A.$$

(a_5) Put $mA := \bigcap_\alpha b_\alpha A$ for each $A \subset X$. We say that $x \in bA$ if for every partition $\{A_1,\dots,A_k\}$ of A there is $j \in \{1,\dots,k\}$ such that $x \in mA_j$. Prove that b is a base operator on X and $bA \subset mA$ for each $A \subset X$. Moreover, $b = \inf b_\alpha$. (Cf. also Ex. 1.A.5.b.)

(b) Prove that the set $\mathscr{B}_c \subset \mathscr{B}$ of all closure operations on X is also order complete. In addition, for $\{b_\alpha\} \subset \mathscr{B}_c$,

$$\sup_{\mathscr{B}} b_\alpha = \sup_{\mathscr{B}_c} b_\alpha, \quad \inf_{\mathscr{B}} b_\alpha = \inf_{\mathscr{B}_c} b_\alpha .$$

Hint. Use the fact that $\inf b_\alpha$ is the supremum of all minorants of $\{b_\alpha\}$.

(c) Let $\mathscr{B}_s \subset \mathscr{B}$ and $\mathscr{B}_t \subset \mathscr{B}$ be the families of all strong bases and all topological closures. Put

$$s := \sup_{\mathscr{B}} b_\alpha, \quad i := \inf_{\mathscr{B}} b_\alpha .$$

Prove that:

(c_1) s need not be strong even in the case all b_α's are,

(c_2) if $\{b_\alpha\} \subset \mathscr{B}_s$, then

$$\sup_{\mathscr{B}_s} b_\alpha = \bar{s}, \quad \inf_{\mathscr{B}_s} b_\alpha = \bar{i},$$

(c_3) if $\{b_\alpha\} \subset \mathcal{B}_t$, then

$$\sup_{\mathcal{B}_t} b_\alpha = t_s, \quad \inf_{\mathcal{B}_t} b_\alpha = t_i$$

(where t_b denotes the closure operator $A \mapsto \bar{A}^b$).

1.A.5 Keeping in mind notations of Ex. 1.A.2 and Ex. 1.A.4, let $\{b_\alpha\} \subset \mathcal{B}$ and let $\{\mathcal{F}_x^\alpha\}_{x \in X}$ be the corresponding collections of filters.

(a) Denote $\mathcal{F}_x^s := \bigcap_\alpha \mathcal{F}_x^\alpha$ and let $\mathcal{F}_x^i$ be a filter on X having as filter subbase $\bigcup_\alpha \mathcal{F}_x^\alpha$. Prove that $\{\mathcal{F}_x^s\}$ corresponds to $s = \sup b_\alpha$ and $\{\mathcal{F}_x^i\}$ corresponds to $i = \inf b_\alpha$.

(b) Using (a) deduce that $x \in iA$ if and only if for each partition $\{A_1,\dots,A_k\}$ of A and for each choice $b_1,\dots,b_k$ of $\{b_\alpha\}$ there is $j \in \{1,\dots,k\}$ such that $x \in b_j A_j$. (Cf. also Ex. 1.A.4.a_5 and [Čech], p. 861.)

1.A.6 Order the set $\mathcal{B}$ as in Ex. 1.A.4. Then the following assertions hold:

(a) The operator c_b: $A \longmapsto A \cup bA$ is the smallest closure operator greater than b.

(b) The operator t_b: $A \longmapsto \bar{A}^b$ is the smallest topological closure greater than b.

(c) The operator $\bar{b}$: $A \longmapsto \overline{bA}^b$ is the smallest strong base operator greater than b.

Hint. Let $\hat{b}$ be a strong base, $b \prec \hat{b}$. Show that $\hat{b}A$ is a b-closed set containing bA.

1.A.7 For any base operator b on X let f(b) be one of the operators c_b, t_b or $\bar{b}$ of the preceding exercise. If $b_1 \prec b_2 \prec b_3$ and $f(b_1) = f(b_3)$, then $f(b_2) = f(b_1) = f(b_3)$.

In particular, c_b and $\bar{b}$ (and, of course, also t_b) determine the same topology as b.

1.A.8 Let (X,b) be a base operator space and let $A \subset X$. Prove that $b\bar{A}^b = \overline{bA}^b$, i.e. the operator b and the b-closure commute.

Hint. Show that bF is always b-closed whenever F is. Hence $\overline{bA}^b \subset b\bar{A}^b$. For the converse inclusion use the identity $\bar{A}^b = A \cup \overline{bA}^b$ (Theorem 1.3).

1.A.9 Let (X,b) be a base operator space and let $A \subset X$. Then $\bar{A}^b$ equals the greatest element of the smallest family $\mathfrak{M}$ of sets having the following properties:

(a) $A \in \mathfrak{M}$.

(b) $B \in \mathfrak{M}$ implies $B \cup bB \in \mathfrak{M}$,

(c) $\bigcup\{M:\ M \in \mathfrak{N}\} \in \mathfrak{M}$ for every $\mathfrak{N} \subset \mathfrak{M}$.

1.A.10 Let (X,b) be a base operator space. For $A \subset X$ define

$$c_0(A) := c_b(A) = A \cup bA.$$

If $\alpha \geq 1$ is an ordinal number and c_β is defined for all $\beta < \alpha$, put

$$c_\alpha = \sup_{\beta < \alpha} c_\beta \quad \text{if } \alpha \text{ is a limit number,}$$

$$c_\alpha = c_b * c_{\alpha-1} \quad \text{if } \alpha \text{ is an isolated number.}$$

(a) Let X be an infinite set and let λ be the first ordinal number of cardinality $> |X|$. Then

$$c_\lambda(A) = \bar{A}^b.$$

(b) What can be asserted in case of X being finite?

(c) Using the transfinite induction prove that

$$\bar{A}^b = A \cup \overline{bA}^b.$$

Hint. Show that

$$c_{\alpha+1}(A) = A \cup c_\alpha(bA) \quad \text{for each finite ordinal } \alpha,$$

and

$$c_\alpha(A) = A \cup c_\alpha(bA) \quad \text{for each infinite ordinal } \alpha.$$

1.A.11 Let (X,b) be a base operator space.

(a) Show that

$$bA \subset \operatorname{der}_b A \cup (A \cap \{x \in X:\ x \in b\{x\}\})$$

for all $A \subset X$.

(b) Show that the implication (ii) $\Rightarrow$ (i) of Theorem 1.2 does not hold in general.

Hint. Consider $X = \{x,y\}$, $b\{x\} = \{y\}$, $b\{y\} = \{x\}$.

(c) If $t_b = c_b$ (cf. Ex. 1.A.6), then $b = \mathrm{der}_b$ if and only if $x \notin b\{x\}$ for any $x \in X$.

1.A.12 Let (X,b) be a base operator space.

(a) Show that the conditions (ii) - (v) of Theorem 1.2 are equivalent to the condition

(o) $\mathrm{der}_b \prec b.$

If (o) holds and the b-topology is T_1, then b is strong.

Hint. Show that (o) $\Rightarrow$ (iv).

(b) If $A \subset X$, then

$$\bar{b}A = \mathrm{der}_b A \cup (A \cap \{x \in X:\ x \in \bar{b}\{x\}\}).$$

Hint. Use (i) $\Rightarrow$ (v) of Theorem 1.2 and the fact that $\mathrm{der}_b = \mathrm{der}_{\bar{b}}$.

(c) The following assertions are equivalent:

(i) b is strong,

(ii) $\bar{b} = b$,

(iii) $bA = \mathrm{der}_b A \cup (A \cap \{x \in X:\ x \in \overline{b\{x\}}^b\})$ for any $A \subset X$,

(iv) $c_b = t_b$ and $b\{x\}$ is b-closed for each $x \in X$,

(v) $\mathrm{der}_b \prec b$ and $b\{x\}$ is b-closed for each $x \in X$.

Hint. Use (b) for (ii) $\Leftrightarrow$ (iii) and Theorem 1.2 ((iv) $\Rightarrow$ (v)) for (iv) $\Rightarrow$ (iii).

1.A.13 If the b-topology is Hausdorff and first countable, then b is strong.

Hint. Consider $A \subset X$ and $x \in \mathrm{der}_b A$. Put $S := \{x_n\}$ where $x_n \in A \setminus \{x\}$, $x_n \to x$. Since S is not b-closed, we get $x \in bS \subset bA$. Now use Ex. 1.A.12.a.

1.A.14 Let X be a topological space. Define for $A \subset X$

$$bA := \overline{\mathrm{int}(\bar{A})}.$$

Prove that b is an idempotent base operator.

Examine also the additivity of the operators

$$A \longmapsto \overline{\operatorname{int} A}, \quad A \longmapsto \operatorname{int}(\bar{A}).$$

1.A.15 Let μ be a monotone subadditive set function on a metric space X, $\mu\emptyset = \emptyset$, and let $\{a_n\}$, $\{\lambda_n\}$ be sequences of positive reals. Define

$$b_1A := \{x \in X: \limsup \lambda_n \cdot \mu(A \cap U(x,a_n)) > 0\}$$

$$b_2A := \{x \in X: \sum_{n=1}^{\infty} \lambda_n \cdot \mu(A \cap U(x,a_n)) = +\infty\}.$$

(a) Prove that b_1 and b_2 are base operators on X.

(b) If $a_n \to 0$, then the b_i-topology $(i = 1,2)$ is finer than the given metric topology.

(c) Find a necessary and sufficient condition in order that $x \in b_i\{x\}$.

(d) In general, the base b_i does not coincide with der_{b_i}.

(e) Show that b_1 need not be strong.

Hint. Consider the superdensity topology from Section 6.D.

REMARK. Compare the definitions of b_1 and b_2 with the definition of a density base operator in Sections 6.B and 6.D and with the Wiener characterization of thinness.

If μ is the Newtonian capacity in $\mathbb{R}^3$ and $a_n = 2^{-n} = \frac{1}{\lambda_n}$, then the base operator b_2 coincides with that of Example 1.1.b.

1.A.16 The collection $\{[x,x+\varepsilon) : x \in \mathbb{R}, \varepsilon > 0\}$ serves as a base for the *Sorgenfrey topology* e^+ on $\mathbb{R}$. (Sometimes $\mathbb{R}$ endowed with the topology e^+ will be called the Sorgenfrey line.)

(a) If $b := \operatorname{der}_{e^+}$, then

$$bA = \{x \in \mathbb{R}: x \in \overline{A \cap (x,+\infty)}\}$$

for each $A \subset X$.

(b) Prove that the base b is strong but not idempotent.

1.A.17 Let X be a set. If for every $x \in X$ there is a nonempty ideal $\mathcal{I}_x$ of subsets of X, then there is a base operator b on X such that A is thin at x if and only if $A \in \mathcal{I}_x$.

1.A.18 Let (X,b) be a base operator space, let $Y \subset X$. Define the operator b_Y on Y by $b_Y A := Y \cap bA$ for $A \subset Y$.

(a) Show that b_Y is a base on Y.

(b) If b is strong, then b_Y is strong as well.

(c) Denote by τ the restriction of the b-topology to Y and let σ be the b_Y-topology on Y. Prove that $\tau = \sigma$ provided b is strong.

(d) Prove that the assertion of (c) is no longer valid if the assumption that b is strong is omitted.

Hint. Let b be the base of Example 1.1.d on $X := [0,1] \times [0,1]$. Put $A := (0,1] \times (0,1]$ and $Y := A \cup \{(0,0)\}$. Then

$$\bar{A}^{\tau} = Y \neq A = \bar{A}^{\sigma}.$$

1.A.19 Let (X,b) be the base operator space of Example 1.1.d . Define

$$b'A := bA \cap (\mathbb{R} \times \{0\}).$$

Then b' is a base. Prove that b' is not strong but

$$b'b'b'A \subset b'b'A$$

for each $A \subset \mathbb{R}^2$.

1.A.20 Let (X,b) be a base operator space. If $bX = X$, then

$$\text{int}_b A \subset bA \subset \bar{A}^b$$

for every $A \subset X$.

Hint. Show that $CbCA \subset bA$. Hence

$$\text{int}_b A \subset C\overline{CA}^b \subset CbCA \subset bA \subset \bar{A}^b.$$

REMARKS AND COMMENTS

The notion of the base of a set was introduced and studied in the potential theory by M. Brelot in (1945). It represents the period when the application of topological methods started to be widely used in this field to obtain new important results.

Being influenced by this we have introduced the notion of the abstract base operator spaces. They include topological spaces as well, but their structure is a little finer.

Naturally various generalizations of the notion of a topological space can be found in literature. Let us mention the study of general closure spaces introduced by E. Čech in (1937) (cf. E. Čech's Topological spaces (1966)) which are distinguished by the fact that the closure does not have to be always a closed set. Certain isotone and additive mappings in the context of extended topology were studied in M.M. Day (1944) and P.C. Hammer (1964). It is interesting to notice that base operators frequently occur also in various branches of mathematics. For example, M. Vlach (1982) and (1984) establishes some "operators which resemble closure and interior operators except that they are neither enlarging nor shrinking" which are used in mathematical programming and in optimization theory.

It must be that in our work we have not got new results in this class of spaces. Base operator spaces in our point of view seem to be useful and advantageous tools not only in the potential theory. Naturally, the base operators can be defined in real analysis as well. As we shall see, we can get new results, for instance, in investigation of the Lusin-Menchoff property.

The notion of thinness was introduced in the classical potential theory by M. Brelot (1939).

Related papers: D. Higgs (1983).

1.8 Quasi-Lindelöf property

Let $\mathcal{Z}$ be a σ-ideal in a topological space (X,τ). We say that X has the *$\mathcal{Z}$-quasi-Lindelöf property* if any family of τ-open sets contains a countable subfamily whose union differs from the union of the whole family by a set belonging to $\mathcal{Z}$.

REMARKS. (a) If $\mathcal{Z} = \{\emptyset\}$, then the $\mathcal{Z}$-quasi-Lindelöf property means the same as the fact that X is *hereditarily Lindelöf*.

(b) Let (X,b) be a base operator space. A set $N \subset X$ is termed (abstract) *totally thin* if $bN = \emptyset$. Obviously, the family of all totally thin sets forms an ideal. Let

$\mathcal{Z}$ be the σ-ideal of all countable unions of totally thin sets. The members of $\mathcal{Z}$ are called (abstract) *semipolar sets*. If X has the $\mathcal{Z}$-quasi-Lindelöf property (X being considered with the b-topology), we say simply that X has the *quasi-Lindelöf property*.

(c) We will see that both the density topology and the fine topology of harmonic spaces have the quasi-Lindelöf property w.r.t. the σ-ideals of null sets or semipolar sets, respectively.

(d) Of course, each hereditarily Lindelöf space (in particular, every space with a countable base) has the $\mathcal{Z}$-quasi-Lindelöf property for any σ-ideal $\mathcal{Z}$ of its subsets.

1.4 THEOREM. Let $\mathcal{Z}$ be a σ-ideal in a topological space (X, τ). Then X has the $\mathcal{Z}$-quasi-Lindelöf property if and only if for every open set $U \subset X$ and for each upper directed set $\mathcal{F}$ of lower semicontinuous functions on U there is an increasing sequence $\{f_n\} \subset \mathcal{F}$ such that

$$\{x \in U: \sup_n f_n(x) < \sup_{f \in \mathcal{F}} f(x)\} \in \mathcal{Z}.$$

Proof. For each $r \in \mathbb{R}$ and each $g \in \mathcal{F}$ we put

$$G_g^r := \{x \in U: g(x) > r\}.$$

For every $r \in \mathbb{R}$, the family $\{G_g^r: g \in \mathcal{F}\}$ covers the set $U_r := \{x \in U: \sup_{f \in \mathcal{F}} f(x) > r\}$. Therefore there is a countable subcollection $\mathcal{G}_r$ of $\mathcal{F}$ such that $\{G_g^r: g \in \mathcal{G}_r\}$ covers U_r up to a set $S_r \in \mathcal{Z}$. Let $\mathcal{G}$ be the union of $\mathcal{G}_r$ over the rationals. Making use of the upper directedness of $\mathcal{F}$, there is an increasing sequence $\{f_n\} \subset \mathcal{F}$ such that $\sup_n f_n \geq \sup \mathcal{G}$. Let $x \in U \setminus \bigcup_{r \in \mathbb{Q}} S_r$ and choose $r \in \mathbb{Q}$, $r < \sup_{f \in \mathcal{F}} f(x)$. Then $x \in U_r \setminus S_r$, hence there is $g \in \mathcal{G}_r$ for which $x \in G_g^r$. We have

$$\sup_n f_n(x) \geq (\sup \mathcal{G})(x) \geq g(x) > r,$$

and by virtue of arbitrary choice of r we get

$$\sup_n f_n(x) = \sup_{f \in \mathcal{F}} f(x).$$

Conversely, given a family $\{G_\alpha\}$ of open sets, we can take for $\mathcal{F}$ the family of the characteristic functions of the unions of all countable subfamilies of $\{G_\alpha\}$.

EXERCISES

1.B.1 The countable chain condition.

A topological space X satisfies the *countable chain condition*, abbreviated (CCC), if every family of disjoint open subsets of X is countable.

(a) Let $\mathcal{C}_I$ denote the σ-ideal of all first category sets in X. If X satisfies (CCC), then X has the $\mathcal{C}_I$-quasi-Lindelöf property.

Hint. Let $\{G_\alpha\}_{\alpha\in A}$ be a family of open sets in X. Say that the family $\{H_b : b\in B \subset A\}$ is admissible if $\{H_b\}_{b\in B}$ are pairwise disjoint nonempty open sets such that $H_b \subset G_b$. By Zorn's lemma there is a maximal admissible family $\{H_b : b\in S\}$. The set S is countable in light of (CCC). Now, it is easy to prove that the set $\bigcup_{\alpha\in A} G_\alpha \setminus \bigcup_{\alpha\in S} G_\alpha$ is even nowhere dense.

(b) Let $\mathcal{Z}$ be a σ-ideal in X not containing any nonempty open set. If X has the $\mathcal{Z}$-quasi-Lindelöf property, then X satisfies (CCC).

(c) If X satisfies (CCC) and if f is a mapping of X into a metric space Y continuous at all points of a set $M \subset X$, then $f(M)$ is separable.

Hint. Assume $T \subset f(M)$ is an uncountable set consisting of isolated points only. If $\{U_t\}_{t\in T}$ is a family of pairwise disjoint neighborhoods of all $t\in T$, then $\{f^{-1}(U_t)\}_{t\in T}$ is a pairwise disjoint collection of sets each of them containing an open nonempty set.

(d) (Cf. K. Janssen and Hadi Ben Saad (*).) Show that the following assertions are equivalent for a topological space X:

(i) X satisfies (CCC),

(ii) for any collection $\{G_\alpha\}_{\alpha\in A}$ of open subsets of X there is a countable set $S\subset A$ such that

$$\overline{\bigcup_{\alpha\in A} G_\alpha} = \overline{\bigcup_{\alpha\in S} G_\alpha},$$

(iii) for any collection $\{G_\alpha\}_{\alpha\in A}$ of open subsets of X there is a countable set $S\subset A$ such that the set $\bigcup_{\alpha\in A} G_\alpha \setminus \bigcup_{\alpha\in S} G_\alpha$ is nowhere dense,

(iv) for any family $\{F_\alpha\}_{\alpha\in A}$ of closed subsets of X there is a countable set $S\subset A$ such that the set $\bigcap_{\alpha\in S} F_\alpha \setminus \bigcap_{\alpha\in A} F_\alpha$ is nowhere dense,

(v) for any family $\{f_\alpha\}_{\alpha\in A}$ of upper semicontinuous functions on X there is a countable set $S\subset A$ such that

$$\widehat{\inf_{\alpha\in A} f_\alpha} = \widehat{\inf_{\alpha\in S} f_\alpha}$$

(where $\hat{f}$ denotes the greatest lower semicontinuous minorant of f),

(vi) for any family $\{f_\alpha\}_{\alpha\in A}$ of upper semicontinuous functions on X there is a countable set $S\subset A$ such that the set $\{x\in X: \inf_{\alpha\in A} f_\alpha(x) < c \leq \inf_{\alpha\in S} f_\alpha(x)\}$ is nowhere dense for any $c\in R$.

Hint. For (i) ⇒ (iii) use Ex. 1.B.1a (the end of the hint).

(e) If X satisfies (CCC) and Y is an open or dense subset of X, then Y satisfies (CCC).

(f) If a topological space X has the property that each subset of X, all of whose points are isolated, is countable, then X satisfies (CCC).

1.B.2 Property (K).

We say that a topological space has *property (K)* if every uncountable collection of open sets has an uncountable subcollection such that each pair meets.

Obviously, the property (K) implies (CCC); if both the Martin axiom and the negation of continuum hypothesis is assumed, then (CCC) implies property (K) (see, e.g. W.W. Comfort and S. Negrepontis (1982), p. 201).

1.B.3 (Doob (1966)) Let X be a Hausdorff topological space and let f be a function on X. We denote

$$\hat{f}(x) := \begin{cases} f(x) & \text{if } x \text{ is an isolated point of } X \\ \liminf\limits_{y\to x,\, y\neq x} f(y) & \text{in the opposite case.} \end{cases}$$

(a) A set $A \subset X$ is said to be

- *weakly quasinull* if there is a lower semicontinuous function f on X such that $A \subset \lfloor f < \hat{f} \rfloor$,
- *quasinull* if it is weakly quasinull and if, moreover, the function $\hat{f}$ in the definition can be chosen to be continuous on X.

(b) The families $\mathcal{Z}_0$ of all quasinull and $\mathcal{Z}_0^w$ of all weakly quasinull sets form σ-ideals.

(c) A set is weakly quasinull (quasinull, resp.) if and only if it contains no isolated points of X and is a countable union of discrete (isolated, resp.) sets.

REMARK. Doob's definition says that X has quasi-Lindelöf property (weak QL-property, resp.) if it has $\mathcal{Z}_0$-quasi-Lindelöf property ($\mathcal{Z}_0^w$-QL property) in our sense.

REMARKS AND COMMENTS

Quasi-Lindelöf property was introduced by J.L.Doob (1966) for the class of all Hausdorff spaces in the exactly same way as described in Ex. 1.B.3. Doob himself proved that both the metric density topology in Euclidean spaces (formed by a Radon measure with respect to certain differentiation basis) and the fine topology on harmonic spaces have this property.

As we shall see further, they are just these topologies in which quasi-Lindelöf property plays an important role. Note that B.Fuglede (1972) used the quasi-Lindelöf property of fine topology in a substantial way proving the fine minimum principle.

The notion of totally thin and semipolar sets appeared in the potential theory where it plays (together with polar sets) the role of "small sets". The transposition of this notion into the base operator spaces is evident and makes no difficulties.

It is easy to see that the property (CCC) in the category of metric spaces is equivalent to separability. In general topological spaces, however, the (CCC) property is nothing but the consequence of separability.

It was Knaster (1945) who introduced the property (K). Both the (CCC) and (K) properties are studied from the set-theoretical point of view. Let us emphasize the fact that it is the density topology on the real line which is often employed for the construction of some counterexamples, and, last but not least, for the constructions of topological spaces with interesting properties as well.

1.C Ideal base operator

Let $\mathcal{Z}$ be an ideal on a base operator space (X,b). Inspired with Example 1.1.e, we investigate the base operators associated with "ideal topologies". For each $A \subset X$ we put

$$iA := \{x \in X: \text{ for every b-neighborhood } U \text{ of } x,\ U \cap A \notin \mathcal{Z}\}.$$

(In fact, this definition depends on the topological structure of (X,b) only.) Of course, $iZ = \emptyset$ for every $Z \in \mathcal{Z}$.

Since $\mathcal{Z}$ is an ideal, i is a base on X. Moreover, by the same reason, it is easy to see that the b-neighborhood U in the definition of iA can be taken to be b-open. We call i a *$\mathcal{Z}$-ideal base* and the corresponding topology is termed a *$\mathcal{Z}$-ideal topology*.

It is obvious that

$$X \setminus iA = \bigcup \{G: G \text{ is b-open},\ G \cap A \in \mathcal{Z}\}.$$

Hence

$$iA = \bigcap \{F: bF \subset F,\ A \setminus F \in \mathcal{Z}\},$$

in particular, iA is a b-closed set. Further, by definition, $iA \subset \bar{A}^b$ for each $A \subset X$. We see that the i-topology is finer than the b-topology. This observation implies that the base i is strong (since iA is always b-closed, it is also i-closed).

An ideal $\mathcal{Z}$ on (X,b) is said to satisfy (*) if $A \in \mathcal{Z}$ whenever

$$\bigcup \{V: V \text{ is b-open},\ V \cap A \in \mathcal{Z}\} \supset A.$$

The condition (*) means nothing else than the fact that each set which is locally in $\mathcal{Z}$ is globally in $\mathcal{Z}$.

Let us notice that (*) is fulfilled if $\mathcal{Z}$ is a σ-ideal and X has the $\mathcal{Z}$-quasi-Lindelöf property (in particular, if X is a second countable topological space).

1.5 EXAMPLES. Many examples of ideals which determine interesting ideal topologies can be constructed. For example, on a base operator space (X,b) we can choose:

(a) $\mathcal{Z} = \{\emptyset\}$. Then $iA = \bar{A}^b$.

(b) $\mathcal{Z} =$ the ideal of all finite subsets of X. If the b-topology is T_1, then $iA = der_b A$ (in general, the condition (*) is violated).

(c) $\mathcal{Z} =$ the σ-ideal of all countable subsets of X. Then iA consists of all condensation points of A. If the b-topology is, for instance, second countable, the condition (*) holds.

(d) $\mathcal{Z} =$ the σ-ideal of all sets of the first category in X. Then i coincides with the base operator from Example 1.1.e (the condition (*) is always satisfied, see [Kur], § 10, III.1).

(e) $\mathcal{Z} =$ the σ-ideal of all Lebesgue measure zero sets on $\mathbb{R}$, $b =$ the Euclidean closure.

(f) $\mathcal{Z} =$ the σ-ideal of all semipolar (polar) sets in a harmonic space with the initial or with the fine topology (and some associated base operator).

(g) $\mathcal{Z} =$ the ideal of all (abstract) totally thin sets (in general, even in harmonic spaces, the condition (*) is violated).

1.6 THEOREM. Let $\mathcal{Z}$ be an ideal on (X,b). Then the following properties are equivalent:

(i) $\mathcal{Z}$ satisfies (*),

(ii) $A \in \mathcal{Z}$ whenever $A \cap iA = \emptyset$,

(iii) $A \setminus iA \in \mathcal{Z}$ for every $A \subset X$,

(iv) $A \in \mathcal{Z}$ whenever $A \cap iA \in \mathcal{Z}$.

Proof. (i) $\Leftrightarrow$ (ii) is trivial because

$$\bigcup \{V: V \text{ is b-open}, V \cap A \in \mathcal{Z}\} = X \setminus iA.$$

(ii) $\Rightarrow$ (iii): We have $(A\setminus iA)\cap i(A\setminus iA) \subset (A\setminus iA)\cap iA = \emptyset$. Thus (ii) applied to $A\setminus iA$ gives (iii).

(iii) $\Rightarrow$ (iv): It is obvious since $A = (A\cap iA)\cup(A\setminus iA)$ and $\mathcal{Z}$ is an ideal.

(iv) $\Rightarrow$ (ii) is trivial.

The main properties of a $\mathcal{Z}$-ideal base operator are summarized now in the following theorem.

1.7 THEOREM. Assume $\mathcal{Z}$ is an ideal on a strong base operator space (X,b) satisfying (*). Then:

(a) for each $A\subset X$, iA is b-closed and $A\setminus iA\in\mathcal{Z}$,

(b) the i-topology is finer than the b-topology,

(c) the operator i is idempotent,

(d) for a set $A\subset X$, the following assertions are equivalent:

(i) $iA\subset bA$,

(ii) $A\setminus bA\in\mathcal{Z}$,

(e) $\mathcal{Z} = \{A\subset X:\ iA = \emptyset\}$.

Proof. (a) and (b) are already proved.

(c) We know that i is strong. On the other hand, $iA\setminus iiA\subset i(A\setminus iA) = \emptyset$ for every $A\subset X$.

(d) The implication (ii) $\Rightarrow$ (i) follows from the expression

$$iA = \bigcap\{F:\ bF\subset F,\ A\setminus F\in\mathcal{Z}\},$$

since $bbA\subset bA$. If (i) is assumed, then $A\setminus bA\subset A\setminus iA$, and we use (a).

(e) If $Z\in\mathcal{Z}$, then obviously $iZ = \emptyset$. If $iA = \emptyset$, then $A = A\setminus iA\in\mathcal{Z}$.

1.8 PROPOSITION. Let $\mathcal{Z}$ be an ideal on (X,b), $G\subset X$. Consider the following assertions:

(i) $G = U\setminus Z$ where U is b-open and $Z\in\mathcal{Z}$,

(ii) G is an i-open set,

(iii) for each $x \in G$ there is a b-open b-neighborhood U of x such that $U \setminus G \in \mathcal{Z}$,

(iv) for each $x \in G$ there is a b-open b-neighborhood U of x and $Z \in \mathcal{Z}$ with $x \in U \setminus Z \subset G$.

Then (i) $\Rightarrow$ (ii) $\Leftrightarrow$ (iii) $\Leftrightarrow$ (iv). If $\mathcal{Z}$ satisfies (*), then all the assertions are equivalent.

Proof. (i) $\Rightarrow$ (ii): Let U be b-open, Z $\mathcal{Z}$. Since $iZ = \emptyset$, Z is i-closed. Hence the set $U \setminus Z$ is i-open (the i-topology is finer than the b-topology).

(ii) $\Rightarrow$ (iii): Let G be i-open. Then $G \subset CiCG$, hence for each $x \in G$ there is a b-open b-neighborhood U of x such that $U \cap CG \in \mathcal{Z}$, i.e. $U \setminus G \in \mathcal{Z}$.

(iii) $\Rightarrow$ (iv): Putting $Z = U \setminus G$ we have $x \in U \setminus Z = U \cap G \subset G$.

(iv) $\Rightarrow$ (ii): Obvious, since $U \setminus Z$ is an i-open set.

If $\mathcal{Z}$ satisfies (*), we prove (iv) $\Rightarrow$ (i). For each $x \in G$ there is a b-open b-neighborhood U_x of x and $Z_x \in \mathcal{Z}$ such that $x \in U_x \setminus Z_x \subset G$. Put

$$U = \bigcup_{x \in G} U_x, \quad Z = U \setminus G.$$

Then U is b-open, $G \subset U$, and $U \setminus Z = G$. Since

$$U_x \cap Z = U_x \setminus G \subset Z_x \in \mathcal{Z}$$

for every $x \in G$, we have

$$Z \subset U \subset \bigcup \{V: \ V \text{ is b-open}, \ V \cap Z \in \mathcal{Z}\},$$

and the condition (*) implies that $Z \in \mathcal{Z}$.

Let $\mathcal{Z}$ be an ideal on a strong base operator space (X,b) satisfying (*). We will say that a strong base t on X is *$\mathcal{Z}$-admissible* in (X,b) if

(a) the t-topology is finer than the b-topology,

(b) $A \in \mathcal{Z}$ implies $tA = \emptyset$,

(c) $A \setminus tA \in \mathcal{Z}$ for every $A \subset X$.

REMARKS. (a) By Theorem 1.7, i is always $\mathcal{Z}$-admissible.

(b) If t is $\mathcal{Z}$-admissible in (X,b) and a strong base a determines a coarser topology on X than b, then t is $\mathcal{Z}$-admissible in (X,a).

(c) It is easy to see that for each $\mathcal{Z}$-admissible base t we have $A \in \mathcal{Z}$ if and only if $A \cap tA \in \mathcal{Z}$ (since $A = (A \setminus tA) \cup (A \cap tA)$), and

$$\mathcal{Z} = \{A \subset X:\ tA = \emptyset\}.$$

1.9 THEOREM. Let t be a $\mathcal{Z}$-admissible base in a strong base operator space (X,b). Then the following assertions are equivalent:

(i) tA is b-closed for any $A \subset X$,

(ii) if s is a $\mathcal{Z}$-admissible base in (X,b), then $sA \subset tA$ for any $A \subset X$,

(iii) t determines the coarsest topology among all admissible bases in (X,b),

(iv) the t-topology and the i-topology coincide,

(v) $t = i$.

Proof. (i) $\Rightarrow$ (ii): Let s be a $\mathcal{Z}$-admissible base, $A \subset X$. Since tA is b-closed, it is also s-closed. Hence

$$sA = s(A \cap tA) \cup s(A \setminus tA) = s(A \cap tA) \subset stA \subset tA.$$

(ii) $\Rightarrow$ (iii) is obvious.

(iii) $\Rightarrow$ (iv): Since i is $\mathcal{Z}$-admissible, the i-topology is finer than the t-topology. Let G be i-open. Then $G = U \setminus Z$ where U is b-open and $Z \in \mathcal{Z}$ (Proposition 1.8). Of course, U is also t-open (t is $\mathcal{Z}$-admissible) and Z is t-closed ($tZ = \emptyset$). Thus G is t-open, and we are through.

(iv) $\Rightarrow$ (v): If s_1, s_2 are $\mathcal{Z}$-admissible and the s_1-topology coincides with the s_2-topology, then $s_1 = s_2$. Indeed, let $A \subset X$. Since $A \subset (A \setminus s_1A) \cup s_1A$, we have

$$s_2A \subset s_2(A \setminus s_1A) \cup s_2s_1A = s_2s_1A \subset s_1A$$

(using the fact that s_1A is s_1-closed, and thus also s_2-closed).

Now we apply this for $s_1 = i$ and $s_2 = t$.

(v) $\Rightarrow$ (i): See Theorem 1.7.a.

REMARKS. (a) A strong base b is $\mathcal{Z}$-admissible in (X,b) if and only if $b = i$.

(b) Although the base operator i is characterized as the greatest of all $\mathcal{Z}$-admissible bases, i is the "smallest base" in the following sense: If t is an arbitrary base with the property that for each $A \subset X$

(a) $A \setminus tA \in \mathcal{Z}$,

(b) tA is b-closed,

then $iA \subset tA$ for every $A \subset X$.

1.10 PROPOSITION. Let a and b be bases on X such that the a-topology is finer than the b-topology. If $aX = X$ and aA is b-closed for each $A \subset X$, then

(a) for every $A \subset X$ there is a b-closed set D such that

$$\operatorname{int}_a A \subset D \subset \bar{A}^a,$$

(b) $\mathcal{C}(X,b) = \mathcal{C}(X,a)$,

(c) if the a-topology is strictly finer than the b-topology, then the a-topology is not regular.

Proof. (a) Put $D = aA$ and use Ex. 1.A.20.

(b) It follows from (a) and from Theorem 2.13 on general measurability.

(c) There is $A \subset X$ such that $A = \bar{A}^a \subsetneq \bar{A}^b$. Choose $x \in \bar{A}^b \setminus A$. The assumed regularity of (X,a) implies the existence of an a-open set V such that $A \subset V \subset \bar{V}^a \subset X \setminus \{x\}$. According to (a), we find a b-closed set D for which $V \subset D \subset \bar{V}^a$. We have

$$\bar{A}^b \subset \bar{D}^b = D \subset X \setminus \{x\} .$$

which is a contradiction with the choice of x.

1.11 COROLLARY. Let $\mathcal{Z}$ be an ideal on (X,b). If $iX = X$, then the set of all i-continuous functions coincides with the set of all b-continuous functions. The i-topology is not regular provided it is strictly finer than the b-topology.

REMARK. The condition $iX = X$ holds if and only if the ideal $\mathcal{Z}$ does not contain any nonempty b-open set.

EXERCISES.

In the following exercises, (X,b) signifies a base operator space and $\mathcal{Z}$ is an ideal on X. Denote by i the corresponding ideal base operator.

1.C.1 (a) If $A, B \subset X$ and $A \setminus B \in \mathcal{Z}$, then $iA \subset iB$.

(b) Let j be the ideal base operator determined by i and $\mathcal{Z}$. Then $j = i$. (So to speak, "i" of "i" is again "i".)

1.C.2 (a) If $\mathrm{der}_b \prec b$ (in particular, if b is strong), then $i \prec b$ if and only if $x \in b\{x\}$ whenever $\mathcal{Z}$ does not contain $\{x\}$.

(b) $i \prec \mathrm{der}_b$ if and only if $\{x\} \in \mathcal{Z}$ for each $x \in X$.

(c) $\mathrm{der}_b \prec i$ if and only if every set in $\mathcal{Z}$ is b-isolated.

(d) $i = \mathrm{der}_b$ if and only if all singletons are in $\mathcal{Z}$ and every set in $\mathcal{Z}$ is b-isolated.

1.C.3 (Semadeni (1963)). We say that an ideal $\mathcal{Z}$ satisfies (**) if $A \in \mathcal{Z}$ whenever $A \subset X$ and $iA = \emptyset$. Prove that:

(a) $\mathcal{Z}$ satisfies (**) provided it satisfies (*),

(b) the condition (**) is essentially weaker than (*).

Hint. Consider the σ-ideal of all countable subsets of the set of all ordinals $\leq \omega_1$.

1.C.4 Let (X, ρ) be a T_1 topological space and let $\mathcal{Z}$ be the σ-ideal of all semipolar sets determined by the derivative base operator. Prove that (**) is satisfied provided X is paracompact. As we remarked, the condition (*) holds if X is hereditarily Lindelöf. In the case when X is a separable metric space, the σ-ideal $\mathcal{Z}$ coincides with the family of all countable subsets of X and iA is exactly the set of all condensation points of A for each $A \subset X$. (Example 1.5.c).

1.C.5 Let der be the Euclidean derivative operator on $X = [0,1]$. Define a base operator b on X by $bA = \mathbb{Q} \cap \operatorname{der} A$. If the ideal base operator i is determined by the ideal of all semipolar sets, then:

(a) $b = \operatorname{der}_b$

(b) each set which is totally thin w.r.t. b is nowhere dense in the Euclidean topology,

(c) $X \setminus bX$ is not semipolar,

(d) $X \setminus iX$ is not semipolar (and thus (*) is not satisfied for the σ-ideal of all semipolar sets).

1.C.6 Prove the assertions stated in Examples 1.5b,c,d.

1.C.7 Let $X = [0,1] \times Y$ where Y is an uncountable discrete space, and let $\mathcal{Z}$ be the σ-ideal of all countable subsets of X. Denote by Φ the collection of all differences $U \setminus Z$ where $U \subset X$ is open and $Z \in \mathcal{Z}$. Then:

(a) Φ does not form a topology,

(b) there are i-open sets which are not in Φ,

(c) $\mathcal{Z}$ does not satisfy (*)

(compare with Proposition 1.8).

Let further $\mathcal{Z}_1$ be the σ-ideal of all locally countable sets of X, and let $\mathcal{Z}_2$ be the ideal of all sets

$$\bigcup_{n=1}^{\infty} (A_n \times \{y_n\})$$

where $A_n \subset [0,1]$ and $y_n \in Y$. Put $\mathcal{Z}_0 = \{A \cup Z\colon\ A \in \mathcal{Z}_2,\ Z \in \mathcal{Z}_1\}$. Then $\mathcal{Z}_0$ does not satisfy (*) even though $A \setminus \operatorname{der} A \in \mathcal{Z}_0$ for every $A \subset X$ (the base operator der is that of topology on X). Compare this result with Theorem 1.6.iii and Ex. 1.C.2b from which it follows that $A \setminus \operatorname{der} A \in \mathcal{Z}$ provided $\mathcal{Z}$ satisfies (*) and singletons are in $\mathcal{Z}$.

1.C.8 Let $\mathcal{Z}$ be an ideal on a strong base operator space (X,b) satisfying (*).

(a) If t is a $\mathcal{Z}$-admissible base in (X,b), and $c_b : A \to A \cup bA$ then

$i = c_b * t$ (i.e. $iA = \overline{tA}^b$).

Hint. Obviously $i \prec c_b * t$. On the other hand,

$$\overline{tA}^b = \overline{t(A \setminus iA)}^b \cup \overline{tiA}^b = \overline{tiA}^b \subset iA.$$

(b) Show that all assertions of Theorem 1.9 are equivalent provided t is a strong base such that $i = c_b * t$.

(c) Let $\mathcal{Z}$ be the ideal of Lebesgue null sets on $\mathbb{R}$ and let b be a density base operator (i.e. bA is a derivative of a set A in the density topology). Further, let $M \subset [0,1]$ for which $\lambda_*(M) = 0$, $\lambda^*(M) = 1$. Put $tA = (bA) \cap M$. Show that $i = c_b * t$ although t is not $\mathcal{Z}$-admissible.

1.C.9 Let i be the ideal topology on $\mathbb{R}^n$ determined by the σ-ideal of all Lebesgue null sets. Prove the following assertions (cf. also S.Scheinberg (1971)):

(a) M is an i-Borel subset of $\mathbb{R}^n$ if and only if M is measurable,

(b) $(\mathbb{R}^n, i)$ is first category in itself,

(c) $\mathbb{R}^n$ is i-connected.

Hint. For (b), there are closed nowhere dense sets F_k and a set of measure zero N so that $\mathbb{R}^n = N \cup \bigcup_{k=1}^{\infty} F_k$.

REMARKS AND COMMENTS

The localization of properties in topological spaces with respect to the various ideals of sets has been studied for a long time. There is an extensive bibliography concerning these topics. For example, reader can be referred to § 7.IV of the book of Kuratowski.

Our task is to investigate the global properties of sets taking in account their local properties only. It is obvious that in spaces having countable topological bases the passage from local to global properties is often trivial and efforts are made to extend this to more general spaces (cf. Preface of the article by Semadeni (1963)). Stress important typical cases only:

- Banach's proof (1930) of the fact that the σ-ideal of first category sets satisfies (*),
- Brelot's result (1957) on locally polar sets in abstract harmonic spaces without countable base,

- the properties of locally negligible sets in Bourbaki's integration theory.

Naturally, many authors on this occasion have studied various properties of ideal topologies (e.g. G. Freud (1958), H. Hashimoto (1952), (1954), (1976), E. Hayashi (1964), (1979), N. F. G. Martin (1961), P. Samuels (1975).

Many authors have investigated the validity of the condition (*) as well.

The notion of admissible topologies was introduced in the work by J. B. Walsh (1971) for the case of Euclidean topology on $\mathbb{R}$ and the σ-ideal of Lebesgue null sets. In his work Theorem 1.9 is also proved in this special case. In general, in this interesting paper the essential fine topology in probabilistic potential theory which is connected with penetration times is also studied.

It is interesting to observe that lately the ideal base operator was studied in the potential theory as well. The original paper by U.Bauermann (1977) containing the investigation of this operator in detail (where iA is also called the quasi-base of A) was followed by the series of works (J.Bliedtner and W. Hansen (1980), W. Hansen (1981), R. Wittmann (1982), J. Bliedtner and W. Hansen (1984). Remark that in their last paper the authors have changed the term using essential base instead of quasi-base.

1.D Essential base operator

The essential base plays a significant role in the potential theory. Among others, this concept is used to characterize the Choquet boundaries of cones of superharmonic functions and to derive the so called essential balayage (J. Bliedtner and W. Hansen (1975)). We will examine this "idempotent modification" of a base operator in a general setting.

Let (X,b) be a base operator space. Given $A\subset X$, by the *essential base* of A we understand the set

$$\beta A = \bigcup \{M \subset \bar{A}:\ M \subset bM\}.$$

(In this section the symbol $\bar{A}$ is reserved for $\bar{A}^b$.) In Theorem 1.13 we will prove that β is actually a base operator. It is called the β-modification of b.

The sets $M\subset X$ satisfying $M\subset bM$ are called *subbasic*.

1.12 LEMMA. (a) Let $V\subset X$ be a b-open set, $A\subset X$, and $x \in V$. Then $x \in bA$ if and only if $x \in b(A\cap V)$ (i.e., the "thinness" is a b-local concept).

(b) If $M\subset X$ is subbasic and $B\subset X$ is b-closed, then $M\setminus B$ is subbasic.

Proof. (a) Obviously $b(A \cap V) \subset bA$. On the other hand, $bA \setminus b(A \cap V) \subset b(A \setminus V) \subset bCV \subset CV$, and hence $V \cap bA \subset b(A \cap V)$.

(b) It is an easy consequence of (a) since the set CB is b-open.

1.13 THEOREM. (a) For every $A \subset X$, βA is a b-closed subbasic set.

(b) β is an idempotent base operator.

(c) If b is strong, then $\beta A \subset bA$ for every $A \subset X$.

Proof. (a) Let $A \subset X$. Then

$$\beta A = \bigcup \{M \subset \bar{A}: \; M \subset bM\} \subset \bigcup \{bM: \; M \subset \bar{A}, \; M \subset bM\} \subset b\beta A.$$

It follows $b\beta A \subset bb\beta A$, so

$$b\beta A \in \{M \subset \bar{A}: \; M \subset bM\}.$$

Thus $b\beta A = \beta A$.

(b) Denote $A' = \beta(A \cup B) \setminus \bar{B}$, $B' = \beta(A \cup B) \setminus \beta A$. By Lemma 1.12.b, A' and B' are subbasic sets. From $A' \subset \overline{A \cup B} \setminus \bar{B} \subset \bar{A}$ we obtain $A' \subset \beta A$, which implies

$$B' \subset (A' \cup \bar{B}) \setminus \beta A \subset (\beta A \cup \bar{B}) \setminus \beta A \subset \bar{B}.$$

But it follows $B' \subset \beta B$, and hence

$$\beta(A \cup B) \subset B' \cup \beta A \subset \beta A \cup \beta B.$$

The converse inclusion and the relation $\beta\emptyset = \emptyset$ are trivial. We have shown that β is a base operator. Since $b\beta A = \beta A = \overline{\beta A}$, we obtain $\beta A = \beta\beta A$ for every $A \subset X$.

(c) If $M \subset \bar{A}$, $M \subset bM$, then

$$M \subset bM \subset b\bar{A} = b(A \cup bA) = bA.$$

Hence $\beta A \subset bA$.

1.14 THEOREM. Let $\mathcal{Z}$ be an ideal on a strong base operator space (X,b) which satisfies (*), and let i be the $\mathcal{Z}$-ideal base. Then the following assertions are equivalent:

(i) $A \setminus bA \in \mathcal{Z}$ for every $A \subset X$,

(ii) $iA \subset bA$ for every $A \subset X$,

(iii) $iA \subset \beta A$ for every $A \subset X$.

Proof. (i) $\Leftrightarrow$ (ii) follows from Theorem 1.7.d.

(iii) $\Rightarrow$ (ii) follows from Theorem 1.13.c.

(ii) $\Rightarrow$ (iii): Let $A \subset X$. According to Theorem 1.7.c and (ii), $iA = iiA \subset biA$. Hence $iA \in \{M \subset \bar{A}: M \subset bM\}$, so $iA \subset \beta A$.

1.15 COROLLARY. If the assumptions of Theorem 1.14 are satisfied and $A \setminus bA \in \mathcal{Z}$ for every $A \subset X$, then $ibA \subset \beta A$ for any $A \subset X$.

Proof. We have $ibA \subset \beta bA \subset \beta\bar{A} = \beta A$.

Recall that a topological space X is called *strong Baire* if every closed subspace of X is a Baire space. Remind also that the family $\mathcal{S}$ of all *semipolar sets* on a base operator space (X,b) is the smallest σ-ideal containing the collection $\{A \subset X: bA = \emptyset\}$ of all *totally thin* sets, i.e. $S \in \mathcal{S}$ if and only if $S = \bigcup_{n=1}^{\infty} A_n$ with $bA_n = \emptyset$.

1.16 THEOREM. Let (X,b) be a base operator space, and let X equipped with the b-topology be a strong Baire space. Let $M \subset X$ be b-closed and subbasic, $B \subset X$ b-closed. Then the set $M \setminus B$ is empty provided it is semipolar.

Proof. Assume $M \setminus B \neq \emptyset$. The space $M \setminus B$ equipped with the induced b-topology is a Baire space. Suppose $T \subset M \setminus B$ is totally thin. Then T is obviously b-closed. Since $M \setminus B$ is subbasic (Lemma 1.12.b), we get

$$M \setminus B \subset b(M \setminus B) = b((M \setminus B) \setminus T) \subset \overline{(M \setminus B) \setminus T}.$$

We see that the complement of T is b-dense in $M \setminus B$, so T is b-nowhere dense in $M \setminus B$. If $M \setminus B$ is semipolar, then $M \setminus B$ is of the first category in $M \setminus B$ equipped with the induced b-topology, which is a contradiction.

1.17 COROLLARY. Let (X,b) be a base operator space, $bX = X$, and let X be strong Baire in the b-topology. Then

(a) If $U \subset X$ is b-open and semipolar, then $U = \emptyset$.

(b) If, moreover, b is strong and i is the ideal base operator determined by the σ-ideal $\mathcal{S}$ of all semipolar sets, if $A \setminus bA \in \mathcal{S}$ for every $A \subset X$, and if $\mathcal{S}$ satisfies (*), then $\beta A = ibA$ for every $A \subset X$.

Proof. (a) Take $M = X$ and $B = CU$ in Theorem 1.16.

(b) By Corollary 1.15, $ibA \subset \beta A$. On the other hand,

$$\beta A \setminus ibA \subset bA \setminus ibA \in \mathcal{S}.$$

Applying Theorem 1.16 to $M = \beta A$ and $B = ibA$, we obtain $\beta A \setminus ibA = \emptyset$.

1.18 EXAMPLE. Let X be the set of all rationals on $\mathbb{R}$, and let bA stand for the set of all rational accumulation points of $A \subset X$. Then $\beta X = X = \emptyset = ibX$. Of course, here (X,b) is not a strong Baire space.

EXERCISES

In what follows, let β be the β-modification of a base operator b on X.

1.D.1 (a) If $A \subset X$ is a subbasic set, then $A \subset \beta A$.

(b) The β-modification of β equals β.

(c) The β-modification of $\bar{b}$ equals β. (Hence, considering β-modifications we can restrict ourselves to the case of strong bases only.)

(d) $\mathcal{C}(X,b) = \mathcal{C}(X,\beta)$ (cf. Proposition 1.10).

1.D.2 Let $A \subset X$. Then βA equals the smallest element of the minimal family $\mathfrak{M}_A$ of sets having the properties:

(a) $\bar{A}^b \in \mathfrak{M}_A$,

(b) $B \in \mathfrak{M}_A$ implies $bB \in \mathfrak{M}_A$,

(c) $\bigcap \mathfrak{N} \in \mathfrak{M}_A$ for every $\mathfrak{N} \subset \mathfrak{M}_A$

(of. with Ex. 1.A.9). If b is strong, the condition (a) can be replaced by

(a´) $bA \in \mathfrak{M}_A$

1.D.3 We put

$$\beta_0(A) = \bar{A}^b$$

for $A \subset X$. If $\alpha \geqq 1$ is an ordinal number and β_γ is defined for all $\gamma < \alpha$, let

$$\beta_\alpha(A) = \beta_{\alpha-1}(A) \cap b(\beta_{\alpha-1}(A)) \quad \text{if } \alpha \text{ is isolated,}$$

$$\beta_\alpha(A) = \bigcap_{\gamma<\alpha} \beta_\gamma(A) \quad \text{if } \alpha \text{ is limit.}$$

If X is infinite and λ is the first ordinal number of cardinality $>|X|$, then $\beta = \beta_\lambda$.

As we can see in the following example, the process of the formation of ß-modification that we have just described is very similar to the Cantor's procedure used in the construction of the kernel of a set.

1.D.4 Let τ be a topology on a set X. For $A \subset X$, let $K(A)$ denote the *kernel* of A. (Recall that $K(A)$ is the greatest subset of A which is dense in itself.)

(a) If β is the ß-modification of der_τ then $\beta A = K(\bar{A}^\tau)$.

(b) Using (a), prove that $\beta X = \emptyset$ if and only if X is scattered (i.e. if every subset of X has an isolated point).

(c) Show that the operator $A \to K(A)$ is not, in general, additive. On the other hand, $A \to K(\bar{A}^\tau)$ is additive (cf. also M. Zarycki (1930)).

1.D.5 Using (c) of the previous exercise, we show that the ß-modification of an arbitrary base operator b is additive.

(a) Show that in the proof of this assertion we can limit ourselves to strong base operators.

Hint. In the proof of Ex. 1.D.1.c the additivity of β is not involved.

(b) Let b be strong and let M be as in Theorem 1.2. Show that

$$\beta A = \overline{\bar{A}^b \cap M}^b \cup K(\bar{A}^b).$$

Hence the additivity of β follows at once from the additivity of $A \to K(\bar{A}^b)$.

1.D.6 Prove that the ß-modification is additive by

(a) using Ex. 1.D.2,

(b) using Ex. 1.D.3.

Hint. (a) Consider the family

$$\mathfrak{M}' := \{A' \in \mathfrak{M}_A : \text{ there is } C \in \mathfrak{M}_{A \cup B} \text{ such that } C \subset A' \cup \bar{B}^b\}$$

and show that $\beta A \in \mathfrak{M}'$. Further consider the family

$$\mathfrak{M}'' := \{B' \in \mathfrak{M}_B : \text{ there is } C \in \mathfrak{M}_{A \cup B} \text{ such that } C \subset B' \cup \beta A\}.$$

(b) First show that

$$\beta(A \cup B) \subset \beta_\alpha(A) \cup \bar{B}^b$$

for each ordinal α. Further prove that

$$\beta\ (A \cup B) \subset \beta A \cup \beta_\alpha(A)$$

for each α.

1.D.7 Let b be the base operator of Example 1.1.d and let β be its β-modification.

(a) If $A = (0,1) \times (0,1)$, then $\beta A \not\subset bA$.

(b) If $F = \{x \in \mathbf{R}^2 : 1/|x| \in \mathbf{N}\}$, then $\beta F = \emptyset$. Hence the β-topology is not locally connected (cf. with Ex. 5.2.b).

1.D.8 (J. Bliedtner, and W. Hansen (1980)).

Let (X,b) be a base operator space. An *abstract base* γ on X(w.r.t. b) is defined to be a base on X which satisfies

$$b(\gamma A) \subset \gamma A \subset bA \quad \text{for all } A \subset X.$$

Prove that:

(a) β-modification of a strong base b is an abstract base,

(b) if $\gamma X = X$, then $\mathcal{C}(X,\gamma) = \mathcal{C}(X,b)$ (use Proposition 1.10 and the fact that γA is always a b-closed set).

1.D.9 Let b be a strong base. Then the β-modification of b is the greatest idempotent base operator smaller than b (cf.Ex. 1.A.6c).

1.D.10 Let b be a strong base operator on X, bX = X, and let $\mathcal{S}$ be the σ-ideal of all semipolar sets. Assuming that X is strong Baire in the b-topology, $\mathcal{S}$ satisfies (*) and $A \setminus bA \in \mathcal{S}$ for each $A \subset X$, prove the following assertions:

(a) (Cf. J. Bliedtner and W. Hansen (1975), Cor.2.9). For every subset A of X, ßA is the smallest b-closed set B such that $\bar{A}^b \setminus B$ is semipolar.

Hint. Use Corollary 1.17.

(b) (An analogy of the Cantor-Bendixson theorem.) If F is a b-closed subset of X, then X contains a b-closed and subbasic subset P and a semipolar set S such that $F = P \cup S$.

Hint. Put P = ßF.

REMARKS AND COMMENTS

The term "essential base" was introduced in the theory of harmonic spaces by J. Bliedtner and W. Hansen (1975) in connection with the investigation of the simpliciality of certain cones of continuous functions in the potential theory.

The importance of this base we appreciate solving various problems. Apart from the works enlisted in the end of the remarks to 1.C quote at least D. Feyel (1981), A. Cornea and H. Hölein (1980), N. Boboc, Gh. Bucur and A. Cornea (1981).

The proof of the fact that ß-modification is a base operator is given in Bliedtner-Hansen (1975) by means of Ex. 1.D.2. Another proof of the same assertion using the methods of the potential theory is in Boboc-Bucur-Cornea (1981).

The proof of the equality ß = ib using substantially the methods of the potential theory can be found in substance in the article of J. Bliedtner and W. Hansen (1975). See also Gh. Bucur and W. Hansen (1984).

Related papers: J.C. Oxtoby (1975).

2. GENERAL PROPERTIES OF FINE TOPOLOGIES

A. General properties

B. Fine topologies induced by a family of functions

C. Fine limits

D. The G_δ -insertion property and Baire one functions

Let (X,ρ) be a topological space. Any topology τ on X finer than ρ is called an (abstract) *fine topology*. Let us remark that in the most important cases the underlying space is always equipped with a metric structure ρ, so we restrict ourselves usually to this special case. Further, by a *fine base operator* b on (X,b) we understand such a base b for which the b-topology is finer than ρ.

In what follows, topological notions referring to the fine topology τ will be qualified by the prefix τ to distinguish them from those pertaining to the initial (metric) topology ρ. For example, $\mathrm{int}_\tau A$ and $\bar{A}^\tau$ denote the τ-interior and the τ-closure of a set A while ∂G will be the ρ-boundary of G.

2.A General properties

We observed in the Prologue that the density topology and the fine topology in potential theory have many joint properties. In general, these properties are far from the metric structure. It can be expected in the case when the fine topology τ is essentially finer than the initial metric topology ρ that any countable set would be finely closed. If it is the case, then the fine topology is seldom metrizable; it is seldom normal, too. Of course, in both cases of the density topology and of the fine topology of potential theory (for $n \geq 2$), any countable set is finely closed. The following theorem describes "bad" properties of fine topologies.

2.1 THEOREM. Let τ be a topology on X. Assume that each countable subset of X is τ-closed and that no countable set is τ-open.
Then:

(a) the set X is uncountable,

(b) there are no τ-isolated points,

(c) the τ-compact subsets of X are exactly the finite sets,

(d) if $\tau\text{-}\lim x_n = x$, then the set $\{n \in \mathbb{N}: x_n \neq x\}$ is finite,

(e) the space (X,τ) is not separable,

(f) no point of X has a countable fundamental system of τ-neighborhoods,

(g) the space (X, τ) is not metrizable,

(h) if, moreover, τ is a fine topology on a separable Baire metric space (P,ϱ) and if τ-continuous functions on P are in Baire class one, then (P, τ) is not normal.

Proof. The assertions (a), (b) and (e) are easy consequences of the assumptions, (g) follows from (f).

(c) Let K be an infinite set. Then there is an infinite countable set $S \subset K$. The family of τ-open sets $\{\{x\} \cup (X \setminus S): x \in S\}$ forms a covering of K from which no finite subcovering can be found.

(d) If the set $\{n \in \mathbb{N}: x_n \neq x\}$ is infinite, then x_n lies outside the τ-neighborhood $X \setminus \{x_n: x_n \neq x\}$ of x for infinitely many $n \in \mathbb{N}$. Therefore $\tau\text{-}\lim x_n = x$ is impossible.

(f) Let $\{V_n\}$ be a countable fundamental system of τ-neighborhoods of a point x. Since $\{x\}$ is not τ-isolated, there are $x_n \in V_n \setminus \{x\}$. Then $X \setminus \{x_n: n \in \mathbb{N}\}$ is a τ-neighborhood of x which does not contain any V_n.

(h) Let $A,B \subset P$ be disjoint countable dense sets. Assuming that (P,τ) is normal, there is a finely continuous function f on P such that $f = 0$ on A, $f = 1$ on B. Since f is in Baire class one, the sets

$$\{f \leq \tfrac{1}{3}\}, \qquad \{f \geq \tfrac{2}{3}\}$$

are disjoint residual sets. But this cannot occur in any Baire space.

Now, we are going to exhibit another criterion of non-normality of fine topologies. First, however, let us agree on some terminology. Let τ be a topology on X, and let $\mathcal{G}$ be a collection of subsets of X. We will say that τ has the $\mathcal{G}$-*insertion property* if for each set A X there ist a set $G \in \mathcal{G}$ such that

$$\operatorname{int}_\tau A \subset G \subset \bar{A}^\tau.$$

It is easy to see that the following two conditions are equivalent:

(i) τ has the $\mathcal{G}$-insertion property,

(ii) for every τ-closed set $F \subset X$ and for every τ-open set U contained in F there is $G \in \mathcal{G}$ such that

$$U \subset G \subset F.$$

2.2 THEOREM. Let τ be a fine topology on a topological space (X, ρ). If there is a collection $\mathcal{A}$ of subsets of X such that:

(a) $\mathcal{A}$ has cardinality α,

(b) τ has the $\mathcal{A}$-insertion property,

and if X contains a τ-isolated set of cardinality α, then τ is not normal.

Proof. Let $\mathcal{S} = \{A \cap B: A \in \mathcal{A}\}$, where B is a τ-isolated subset of X of cardinality α. Choose $T \subset B$ for which $T \notin \mathcal{S}$. Assuming the normality of τ, there is a τ-open set G such that $T \subset G$ and $\bar{G}^{\tau} \cap (B \setminus T) = \emptyset$. Find $A \in \mathcal{A}$ for which $G \subset A \subset \bar{G}^{\tau}$. Then

$$T \subset A \cap B \subset (X \setminus (B \setminus T)) \cap B = T.$$

Therefore $T = A \cap B$, which contradicts to $T \notin \mathcal{S}$.

EXERCISES

2.A.1 Let τ be a fine topology on a Hausdorff topological space (X, ρ). Then a set $A \subset X$ is τ-compact if and only if A is compact and the topologies induced by τ and ρ coincide on A.

2.A.2 The Sorgenfrey line is perfectly normal, hereditarily Lindelöf and paracompact.

Hint. It can be easily checked that the Sorgenfrey topology is T_1, regular and hereditarily Lindelöf. Hence, it is perfectly normal and paracompact.

2.A.3 Prove that $\mathbb{R}^2$ endowed with the crosswise topology τ of Example 1.1.d:

(a) is not regular,

(b) is separable, but not first countable,

(c) is neither paracompact nor Lindelöf.

Hint. (a) Since for every τ-open set U containing

$$A := \{(\frac{p}{q}, \frac{1}{q}) : p,q \in \mathbb{N}\}$$

the set $\{x \in \mathbb{R}: (x,0) \in \bar{U}^{\tau}\}$ is a dense G_δ-subset of $\mathbb{R}$, the point $(0,0)$ and the set A cannot be separated. (Cf. also Ex. 4.A.8.)

(b) The set $\mathbb{Q} \times \mathbb{Q}$ is τ-dense in $\mathbb{R}^2$. Let $\{W_n\}$ be a sequence of τ-neighborhoods of $z \in \mathbb{R}^2$. Find $z^n \in W_n$ such that

$$z_1^n \neq z_1 \ , \ z_2^n \neq z_2 \quad \text{and} \quad |z^n - z| < 1/n$$

and consider the set $W := \mathbb{R}^2 \setminus \{z^n : n \in \mathbb{N}\}$ which is a τ-neighborhood of z. Obviously, W_n is not a subset of W for any n.

(c) The diagonal Δ is an uncountable τ-isolated set. Using this fact prove that the τ-open cover $\{\Gamma_z\}_{z \in \Delta}$ of $\mathbb{R}^2$ (where $\Gamma_z = (\mathbb{R}^2 \setminus \Delta) \cup \{z\}$) has not even a τ-open locally countable refinement.

2.A.4 (a) Show that a separable space containing a set of cardinality c whose derived set is empty cannot be normal ([Eng], p.71).

(b) Show that no space containing a dense subset of cardinality $\aleph_0$ and containing a closed discrete subspace of cardinality c is normal ([Eng], Ex. 2.1.10).

(c) Let τ be a fine topology on a separable complete metric space P and let $\mathcal{B}$ be the Borel field in P. If $\mathcal{B}$ contains an uncountable τ-isolated set and if τ has the $\mathcal{B}$-insertion property, then τ is not normal (Ch. Berg (1971) for the fine topology of harmonic spaces).

Hint. Use Theorem 2.2 and compare with it!

REMARKS AND COMMENTS

In this section we summarize some properties of fine topologies needed in the following. Asking when the finely continuous functions belong to the Baire class one we introduced the concept of $\mathcal{G}$-insertion property (J. Lukeš and L. Zajíček (1977a), (1977b)). A sufficient condition ensuring that the fine topology is not normal is given in Theorem 2.2. A similar theorem with more restrictive assumptions was used by C. Berg (1971). He formulated necessary and sufficient conditions guaranteeing the normality of the fine topology on harmonic spaces (cf. Theorem 10.32). These ideas of proofs of non-normality are not entirely new. F.B.Jones (1937) used similar cardinality arguments for the proof of the non-normality of the Niemytzki plane and these ones can be essentially employed in Ex. 2.A.4a. The cardinality reasonings concerning families of continuous functions which can be applied in Ex. 2.A.4b, were used in M.Katětov (1950). Notice also that the assertion of Theorem 2.1.h uses already methods of fine topologies for non-normality arguments.

2.B. Fine topologies induced by a family of functions

There are several methods how to construct fine topologies. We introduce one of them which is frequently used in potential theory.

Let X be a set, Y a topological space, and let $\mathcal{F}$ be a set of functions from X to Y. The *weak topology* induced on X by $\mathcal{F}$ is the smallest topology on X making each function from $\mathcal{F}$ continuous. It is evident that the sets $f^{-1}(G)$, where $f \in \mathcal{F}$ and G is open in Y, form its subbase. It is easy to see that the weak topology is Hausdorff if and only if $\mathcal{F}$ separates the points of X.

We shall denote by $\mathcal{W}(\mathcal{F})$ the smallest min-stable convex cone containing $\mathcal{F}$ which includes the constant functions. Let τ be the weak topology induced on X by $\mathcal{F}$. As any function of $\mathcal{W}(\mathcal{F})$ is τ-continuous it follows that τ is also the weak topology generated by $\mathcal{W}(\mathcal{F})$.

It is not true that each topology on X can be induced by a family of real functions as the next theorem shows.

2.3 THEOREM. Let τ be the weak topology induced on X by a family $\mathcal{F}$ of functions. If V is a τ-neighborhood of a point $x \in X$, then there are $s,t \in \mathcal{W}(\mathcal{F})$, $s \leqq t$ on X such that

$$x \in [s < t] \subset V.$$

In particular, τ is completely regular.

Proof. We know that the collection of sets $\{[s < a], [t > b] : s, t \in \mathcal{F}, a,b \in \mathbb{R}\}$ is a subbase for τ. Then, of course, the family $\{[s < t] : s,t \in \mathcal{W}(\mathcal{F})\}$ is also a subbase for τ. As $[s < t] = [\min(s,t) < t]$, the collection $\Phi := \{[s < t] : s,t \in \mathcal{W}(\mathcal{F}), s \leqq t \text{ on } X\}$ is again a subbase for τ. Since the collection of all finite intersections of elements form Φ forms a base for τ, and since

$$[s_1 < t_1] \cap [s_2 < t_2] = [s_1+s_2 < \min(s_1+t_2, s_2+t_1)] \in \Phi,$$

Φ is even a base for τ and the assertion easily follows. Now, the function $f := t - s$ is τ-continuous on X, $f(x) > 0$ and $f = 0$ on $X \setminus V$.

2.4 REMARKS. (a) Conversely, if X is completely regular, then the topology of X is the weak topology induced on X by the family $\mathcal{C}(X)$ of all continuous real unctions on X.

(b) If two topologies ρ and τ on a set X are completely regular, then $\rho = \tau$ f and only if $\mathcal{C}(X,\rho) = \mathcal{C}(X,\tau)$.

roof. (a) Obviously, the topology of X is finer than the weak topology genera-ed by $\mathcal{C}(X)$. If V is a neighborhood of z, then there is $f \in \mathcal{C}(X)$ such that $f = 1$ n CV, $f(z) = 0$. Clearly, $[f < 1]$ is a weak neighborhood of z contained in V.

(b) follows immediately from (a).

The important case of a weak topology is contained in the detailed study of M. Bre-ot (1971) where a convex cone Φ of nonnegative lower semicontinuous functions on a opological space (X,ρ) is considered. Following Brelot, we define the *fine topology* f X as the coarsest topology τ on X finer than ρ which makes all functions rom Φ continuous. It is clear that a subbase for τ is formed by all ρ-open sets ogether with all sets of the form $[f < a]$ where $a \in \mathbb{R}$, $f \in \Phi$.

We say that a set A is *thin* at a point $z \notin A$ if there is a neighborhood V of and $f \in \Phi$ such that

$$f(z) < \inf \{f(x) : x \in V \cap A\}.$$

n an equivalent form, a set A is thin at $z \notin A$ if and only if there is a function $g \in \Phi$ such that

$$g(z) < \liminf_{A \ni x \to z} g(x).$$

Of course, for $z \notin \bar{A}$, the set A is always thin at z (it is sufficient to take $g = 0$ - in this case $\liminf_{A \ni x \to z} g(x) = +\infty$).

.5 CARTAN'S THEOREM. Let Φ be a convex cone of nonnegative lower semicontinu-ous functions on a topological space X, and let $z \in X$. Then the fine neighborhoods of z are exactly the complements of sets which do not contain z and which are hin at z.

roof. Let $A \subset X \setminus \{z\}$ be thin at z. There is $g \in \Phi$, a neighborhood U of z nd $\lambda \in \mathbb{R}$ such that $g(z) < \lambda < \inf \{g(x) : x \in U \cap A\}$. Then $[g < \lambda] \cap U$ is a fine neigh-orhood of z contained in CA.

onversely, if V is a fine neighborhood of z, then by a straightforward use of heorem 2.3 it can be shown that

$$s(z) < t(z) \leq \liminf_{CV \ni y \to z} t(y) = \liminf_{CV \ni y \to z} s(y)$$

for some $s, t \in \mathcal{W}(\Phi)$, $s \leq t$ on X (it can be immediately checked that the elements of $\mathcal{W}(\Phi)$ are lower semicontinuous). Since it can be easily shown that each nonconstant function $s \in \mathcal{W}(\Phi)$ is of the form $\min(s_1,\ldots,s_n,c)$ where $s_i \in \Phi$ and c is a constant function, there must be a function $s_i \in \Phi$ with

$$s_i(z) < \liminf_{CV \ni y \to z} s_i(y).$$

2.6 REMARKS. The following assertions are taken from Brelot (1971) (the fine topology τ is considered as above):

(a) A point y is a finely isolated point of a set M if and only if $M \setminus \{y\}$ is thin at y.

(b) For $A \subset X$, define $t(A)$ as the set of all points from CA where A is not thin. Then the fine closure of A is equal to $A \cup t(A)$.

(c) If ρ is completely regular, then τ is completely regular.

(d) If (X,ρ) is locally compact, then (X,τ) is a Baire space.

Let us remark that both the density topology and the fine topology in potential theory are examples of fine topologies which can be derived from certain convex cones of lower semicontinuous functions. More generally, any topology having the Lusin-Menchoff property is the fine topology associated with a suitable cone Φ (see Remark 3.15).

EXERCISES

2.B.1 Let τ be a fine topology on a topological space (X,ρ) and let $\mathcal{B}$ be the collection of all lower semicontinuous and τ-continuous functions on X.

Prove that $\tau \supset \tau_a \supset \tau_b$ provided

(a) τ_a is the topology generated by τ-open sets of the type F_σ,

(b) τ_b is the weak topology induced on X by $\mathcal{B}$.

In particular, examine these topologies in case that ρ is the Euclidean topology on $\mathbb{R}^n$ and τ is the ideal topology determined by the σ-ideal

(α) of all Lebesgue null sets,

(β) of all first category sets.

Hint. In case (α) show that $\tau_a \neq \tau_b$. In (β), the topologies τ, τ_a, τ_b are different.

2.B.2 Describe the weak topology induced on $\mathbb{R}$ by the convex cone of all (lower finite) lower semicontinuous functions being, in addition,

(a) decreasing,

(b) monotone,

(c) derivatives.

Hint. (c) Use the following result of Z.Zahorski (1950): A function f is approximately continuous on $\mathbb{R}$ provided f' is both lower finite and lower semicontinuous there.

2.B.3 Let e and t be the Euclidean topology and the crosswise topology on $\mathbb{R}^2$, respectively. If τ is the weak topology on $\mathbb{R}^2$ induced by the family of all separately continuous functions, then

$$e \subsetneqq \tau \subsetneqq t.$$

2.B.4 If τ is the weak topology induced on a set X by a family $\mathfrak{F}$, then τ is Hausdorff if and only if $\mathfrak{F}$ separates the points of X. (Thus τ is Hausdorff if and only if it is completely regular.)

2.B.5 (H. Bauer). Let Φ be the family of all nonnegative real lower semicontinuous concave functions on a convex subset K of a Hausdorff topological vector space. Show that $K \setminus \{z\}$ is thin at z if and only if z is an extreme point of K.

REMARKS AND COMMENTS

Weak topologies generated by convex cones of nonnegative lower semicontinuous functions were examined in detail by M. Brelot (1971). Both the associated notions of thinness and fine topology in the potential theory were introduced, as it has been already mentioned in Prologue, in the forties by M. Brelot and H. Cartan.

2.C Fine limits

It is known that a real function f is approximately continuous at a point z if and only if f is continuous at z in the density topology (cf Theorem 6.6). Actually it is possible to state more general versions of this theorem which bring out both the character of fine limits and the nature of usual limits on fine neighborhoods. To do this, we need a simple preliminary lemma.

2.7 LEMMA. Let $f: X \to Y$ be a mapping, where (X,τ) and (Y,σ) are topological spaces. The following conditions on $x \in X$ and $y \in Y$ are equivalent:

(i) $$x \in \bigcap_{U \in \mathcal{V}_\sigma(y)} \overline{f^{-1}(U) \setminus \{x\}}^{\tau},$$

(ii) $$y \in \bigcap_{V \in \mathcal{V}_\tau(x)} \overline{f(V \setminus \{x\})}^{\sigma}.$$

Proof. It is not too hard to check that non(i) $\Leftrightarrow$ non(ii).

We say that $y \in Y$ is a *cluster point* of a mapping $f: X \to Y$ at a point $x \in X$ if (i) or (ii) of the previous lemma is satisfied. The set of all cluster points of f at x will be denoted by $Cl(f,x)$. Notice that the set $Cl(f,x)$ is always closed.

Specifying Y to be the set $\bar{\mathbb{R}}$ of extended reals considered with the usual topology, and taking profit of compactness of $\bar{\mathbb{R}}$, we see that $Cl(f,x)$ is nonempty provided x is not isolated. Moreover,

$$\max Cl(f,x) = \limsup_{t \to x} f(t), \quad \min Cl(f,x) = \liminf_{t \to x} f(t).$$

2.8 THEOREM. Let τ be a fine topology on a topological space (X,ρ). The following assertions for $x \in X$ are equivalent:

(i) for each sequence $\{V_n\} \subset \mathcal{V}_\tau(x)$ there is a sequence $\{W_n\} \subset \mathcal{V}_\rho(x)$ such that

$$X \setminus \bigcup_{n=1}^{\infty} (W_n \setminus V_n) \in \mathcal{V}_\tau(x),$$

(ii) for each function f on X there is $V \in \mathcal{V}_\tau(x)$ such that

$$Cl_\tau(f,x) = Cl_\rho(f/V,x),$$

(iii) for each function f on X which is lower τ-semicontinuous at x there is $V \in \mathcal{V}_\tau(x)$ such that

$$f(x) \leq \liminf_{V \ni y \to x} f(y),$$

(iv) for each function f on X which is τ-continuous at x there is $V \in \mathcal{V}_\tau(x)$ such that f/V is continuous at x,

(v) for each function f on X for which $\tau\text{-}\lim_{y\to x} f(y) = A$ there is $V \in \mathcal{V}_\tau(x)$ such that $\lim_{V\ni y\to x} f(y) = A$.

<u>REMARK.</u> The sequence $\{W_n\}$ of (i) can be chosen to be decreasing.

<u>Proof.</u> We may assume that x is not a τ-isolated point of X.

(i) $\Rightarrow$ (ii): Let $\{G_n\}$ be a decreasing sequence of open subsets of $\bar{\mathbb{R}}$ satisfying

$$Cl_\tau(f,x) = \bigcap_{n=1}^{\infty} G_n = \bigcap_{n=1}^{\infty} \bar{G}_n$$

(closures are taken in $\bar{\mathbb{R}}$). Put $V_n = f^{-1}(G_n) \cup \{x\}$. According to (i), there is a sequence $\{W_n\} \subset \mathcal{V}_\rho(x)$ such that

$$V := X \setminus \bigcup_{n=1}^{\infty} (W_n \setminus V_n) = \bigcap_{n=1}^{\infty} (V_n \cup CW_n) \in \mathcal{V}_\tau(x).$$

Let $y \in (V \cap W_n) \setminus \{x\}$. Then $y \in V_n \setminus \{x\}$, and thus $f(y) \in G_n$. It follows that $Cl_\rho(f \wedge V, x) \subset \bar{G}_n$, which yields that $Cl_\rho(f \wedge V, x) \subset Cl_\tau(f,x)$. The converse inclusion is obvious.

(ii) $\Rightarrow$ (iii) $\Rightarrow$ (v) $\Rightarrow$ (iv): Obvious.

(iv) $\Rightarrow$ (i): Let $\{V_n\} \subset \mathcal{V}_\tau(x)$. We may assume that $V_1 \supset V_2 \supset V_3 \supset \dots$. Put

$$f(y) := \inf\{\tfrac{1}{n} : y \in V_n\}, \quad y \in X.$$

We can see that f is τ-continuous at x. By (iv) there is $U \in \mathcal{V}_\tau(x)$ such that $f \wedge U$ is continuous at x. Hence for each $n \in \mathbb{N}$ there is $W_n \in \mathcal{V}_\rho(x)$ such that $f < \frac{1}{n}$ on $W_n \cap U$. Put

$$V := X \setminus \bigcup_{n=1}^{\infty} (W_n \setminus V_n).$$

To conclude the proof it suffices to check that $U \subset V$. If this is not the case, then there is $y \in U \cap (W_n \setminus V_n)$ for some $n \in \mathbb{N}$. Since $y \in W_n \cap U$, we have $f(y) < \frac{1}{n}$. On the other hand, $y \notin V_n$ implies $f(y) > \frac{1}{n}$. But this is inconsistent, thereby completing the proof.

We say that a fine topology τ on a topological space (X,ρ) satisfies the *condition (tFL)* at a point $x \in X$ if (i) of Theorem 2.8 holds.

REMARK. Theorem 2.8 is a special case of the so-called *fine limit principle* only. For more details see Ex. 2.C.10.

2.9 THEOREM. Let τ be a fine topology on a topological space (X,ρ). Pick a point $x \in X$ and assume that $\{x\}$ is a G_δ-set.

If $\mathcal{M} := \{A \subset X: x \in \bar{A}^{\tau} \setminus A\}$, then the following are equivalent:

(i) for each sequence $\{A_n\} \subset \mathcal{M}$ there is a (decreasing) sequence $\{W_n\} \subset \mathcal{V}_\rho(x)$ such that

$$\bigcup_{n=1}^{\infty} (A_n \setminus W_n) \in \mathcal{M},$$

(ii) for each function f on X and for any $c \in \mathrm{Cl}_\tau (f,x)$ there is $M \in \mathcal{M}$ such that

$$c = \lim_{M \ni y \to x} f(y).$$

Proof. (i) $\Rightarrow$ (ii): Let $G_n \in \mathcal{V}_\rho(c)$, $G_1 \supset G_2 \supset \ldots$ and $\bigcap_{n=1}^{\infty} \bar{G}_n = \{c\}$.

Put $A_n := f^{-1}(G_n) \setminus \{x\}$. Obviously $A_n \in \mathcal{M}$, so that there is a decreasing sequence $\{W_n\} \subset \mathcal{V}_\rho(x)$ satisfying

$$M := \bigcup_{n=1}^{\infty} (A_n \setminus W_n) \in \mathcal{M}.$$

If $y \in W_n \cap M$, then $y \in A_j$ for some $j > n$, so that $f(y) \in G_j \subset G_n$. It follows that $\mathrm{Cl}_\rho (f \wedge M,x) \subset \bar{G}_n$, and therefore $c = \lim_{M \ni y \to x} f(y)$.

(ii) $\Rightarrow$ (i): Let $\{A_n\} \subset \mathcal{M}$. Put

$$f(y) := \inf\{\tfrac{1}{n} : y \in A_n\}.$$

By (ii) there is $M \in \mathcal{M}$ for which $\lim_{M \ni y \to x} f(y) = 0$. We can find a decreasing sequence $\{W_n\} \subset \mathcal{V}_\rho(x)$ such that

$$\bigcap_{n=1}^{\infty} W_n = \{x\} \quad \text{and} \quad f < \frac{1}{n} \text{ on } W_n \cap M.$$

We put $A := \bigcup_{n=1}^{\infty} (A_n \setminus W_n)$. We need only to show that $M \cap W_1 \subset A$ (then $A \in \mathcal{M}$). Let $y \in M \cap W_1$. If $f(y) = 0$, then there is $n \in \mathbb{N}$ such that $y \notin W_n$. Further, there is $k > n$ for which $y \in A_k$. Thus $y \in A_k \setminus W_n \subset A_k \setminus W_k \subset A$. If $f(y) \neq 0$, then $f(y) = \frac{1}{n}$ for some $n \in \mathbb{N}$, hence $y \in A_n \setminus W_n \subset A$.

We can now employ the ideas just accomplished to provide one of the important criteria of validity of the condition (tFL), which will be useful in Section 6.C.

Let b be a fine base on a topological space (X,ρ). We say that the *condition* (*bFL*) holds at a point $z \in X$ if for each sequence $\{A_n\}$ of subsets of X which are thin at z there is a (decreasing) sequence $\{U_n\}$ of neighborhoods of z such that the set $\bigcup_{n=1}^{\infty} (A_n \cap U_n)$ is again thin at z.

If b is a strong base and z is a point of X, then (bFL) holds at z if and only if the b-topology satisfies (tFL) at z.

If b is a fine base on a regular topological space and if $\{z\}$ is a G_δ-set, then (bFL) implies (tFL) again for the case of the b-topology. This assertion is the promised criterion. As we shall see, its proof is not an immediate consequence of the definitions.

2.10 PROPOSITION. Let b be a fine base on a regular topological space (X,ρ). Assume that z is a point of X and $\{z\}$ is a G_δ-set. If (bFL) holds at z, then b-topology satisfies (tFL) at z.

Proof. Let $\{V_n\}$ be a sequence of b-neighborhoods of z. We can assume that V_n's are b-open. We are going to exhibit a sequence $\{W_n\} \subset \mathcal{V}_\rho(z)$ such that

$$X \setminus \bigcup_{n=1}^{\infty} (W_n \setminus V_n) \in \mathcal{V}_b(z).$$

According to (bFL) there is a sequence $\{W_n^*\} \subset \mathcal{V}_\rho(z)$ such that the set

$$\bigcup_{n=1}^{\infty} (W_n^* \cap \complement V_n) = \bigcup_{n=1}^{\infty} (W_n^* \setminus V_n)$$

is thin at z.

By our hypotheses, there is a sequence $\{W_n\}$ of open sets such that

$$\bigcap_{n=1}^{\infty} \bar{W}_n = \{z\}, \quad W_1 \supset \bar{W}_2 \supset W_2 \supset \bar{W}_3 \supset \dots$$

and $\bar{W}_i \subset W_i^*$ for each $i \in \mathbb{N}$. The set

$$A := \bigcup_{n=1}^{\infty} (\bar{W}_n \setminus V_n)$$

is obviously thin at z.

We can finish by proving that A is b-closed. Then, of course, the set $X \setminus \bigcup_{n=1}^{\infty} (W_n \setminus V_n) \supset X \setminus A$ will be a b-neighborhood of z.

Let us concentrate on $x \in bA$. Since $x \neq z$, there is $n \in \mathbb{N}$ such that $x \notin \bar{W}_n$.

The set $A \setminus W_n$, being a finite union of b-closed sets, is b-closed. Therefore

$$x \in bA \setminus \bar{W}_n \subset bA \setminus bW_n \subset b(A \setminus W_n) \subset A \setminus W_n \subset A$$

and the proof is complete.

2.11 COROLLARY. Let b be a fine base on a metric space P. If (bFL) holds at any point of a b-open set $U \subset P$, then a function f is b-continuous on U if and only if for each $x \in U$ there is a set $V \in \mathcal{V}_b(x)$ such that

$$\lim_{V \ni y \to x} f(y) = f(x).$$

This corollary says, in fact, that under (bFL), b-continuous functions are exactly the "approximately continuous" functions.

EXERCISES

2.C.1 Let b be a fine base on a topological space X and let $\mathcal{F}_x$ $(x \in X)$ be the corresponding filters (cf.Ex.1.A.2). If x is not a b-isolated point of X then (bFL) holds at x if and only if the following implication is true:

If $A = \lim_{\mathcal{F}_x} f$, then there is a set $F \in \mathcal{F}_x$ such that $A = \lim_{F \ni y \to x} f(y)$.

Hint. Use Theorem 2.8 for the b*-topology. where $b^*(M) = bM \cap \{x\}$.

2.C.2 (a) Prove that the condition (tFL) holds in the following examples:

(a1) X is the real line and τ is the Sorgenfrey topology,

(a2) τ is an ideal topology determined by some σ-ideal on a topological space,

(a3) X is $\mathbb{R}^n$ and τ is the ideal topology determined by the ideal of nowhere dense subsets of $\mathbb{R}^n$, or by the ideal of sets having finite $(n-1)$-dimensional Hausdorff measure.

(b) On the other hand, if X is $\mathbb{R}^2$ and τ is the ideal topology determined by the ideal of all subsets of $\mathbb{R}^2$ which can be covered by a finite system of segments, then (tFL) does not hold.

Show that in this case every τ-continuous function on a τ-open set is continuous (compare with Corollary 2.11).

2.C.3 (a) Let τ be the ordinary density topology on $\mathbb{R}^2$. Put

$$f(x) = x_1 x_2 (x_1^2 + x_2^2)^{-1}$$

for $x = (x_1;x_2) \neq 0$. Then $0 \in Cl_\tau(f,0)$ and there is no set A of positive upper density at 0 for which $\lim_{\substack{A\ x\to 0}} f(x) = 0$. (Since for the density topology (tFL) holds, we see that (i) of Theorem 2.9 does not follow from (tFL). Illustrate this phenomenon also in the case of the density topology on $\mathbb{R}$ for $x = 0$ and $f(x) = \sin \frac{1}{x}$.)

(b) Show that the condition (i) of Theorem 2.9 does not imply (tFL).

Hint. Consider Ex. 2.C.2.b.

2.C.4 Let $\{A_n\}_{n=1}^{\infty}$ be a family of pairwise disjoint dense subsets of $\mathbb{R}$ and $\mathbb{R} = \bigcup_{n=1}^{\infty} A_n$. Define a base b on $\mathbb{R}$ as follows: A point $x \neq 0$ belongs to bA if x is an accumulation point of A. Further, $0 \in bA$ if for every $r > 0$ the set

$$\{n \in \mathbb{N} : (-r,r) \cap A \cap A_n \text{ is not nowhere dense in } \mathbb{R}\}$$

is infinite. Prove that (tFL) holds for the b-topology but (bFL) is violated.

2.C.5 Let b be a fine base on a topological space X. Prove that (bFL) holds at $z \in X$ if and only if for each sequence $\{B_n\}$, $B_n \subset X$, $z \notin B_n$, B_n thin at z, there is a sequence $\{U_n\}$, $U_n \in \mathcal{V}_\rho(z)$ such that $\bigcup_{n=1}^{\infty} (B_n \cap U_n)$ is thin at z.

2.C.6 Let τ be a fine topology on a topological space X. Then the following assertions are equivalent for $z \in X$:

(i) τ satisfies (tFL) at z,

(ii) (bFL) holds for the base operator $b_1 : A \mapsto \bar{A}^{\tau}$,

(iii) (bFL) holds for the base $b_2 : A \mapsto \mathrm{der}_\tau(A)$.

2.C.7 Let b be a strong fine base on a topological space. The following assertions are equivalent:

(i) the b-topology satisfies (tFL),

(ii) (bFL) holds for b.

Hint. Use Ex. 2.C.5 and Ex. 2.C.6.ii.

2.C.8 Put the order topology ρ on the set Ω of all ordinals $\leq \omega_1$ (the first uncountable ordinal) and let L be the set of all limit ordinals. Let J denote the ideal generated by the sets $B_n = \{y+n: y \in L\}$ and let i be the corresponding ideal topology on Ω. Finally, define the base operator b as follows:

a) $\omega_1 \in bM$ if $\omega_1 \in \overline{L \cap M}^{\rho}$,

b) a countable limit ordinal x belongs to bM if $x \in \bar{M}^i$,

c) no nonlimit ordinal lies in bM

(for $M \subset \Omega$). Prove that b is a base operator on Ω. Show that (bFL) holds at ω_1, while (tFL) is violated at ω_1 (for the b-topology).

Hint. Consider the sequence $\Omega \setminus B_n$ of neighborhoods of ω_1.

REMARK. A similar example with (bFL) holding at all points of Ω is more complicated, see the following exercise.

2.C.9 Let (Ω,ρ), L and B_n have the same meaning as in Ex. 2.C.8. Put $S_n = L \cup \bigcup_{n=1}^{\infty} B_n$. For every $x \in \Omega \setminus \{\omega_1\}$, $\{V_n(x)\}$ denotes a local base at x formed by a decreasing sequence of neighborhoods of x. A set $M \subset \Omega$ is said to be thin at ω_1 if there is a neighborhood V of ω_1 such that $M \cap L \cap V = \emptyset$, and thin at $x \in \Omega \setminus \{\omega_1\}$ if there is $m \in \mathbb{N}$ such that $M \cap S_n \cap V_m(x) \subset V_n(x)$ for every $n \in \mathbb{N}$.

Show that the corresponding base operator b (of. Ex. 1.A.17) has the property that (bFL) holds at all points of Ω and (tFL) does not hold at ω_1 for the b-topology.

2.C.10 Fine limit principle. (Cf. also J. Jedrzejewski (1974).) Let x be a point of a topological space (X,ρ), and let $\mathcal{A}$ be a collection of subsets of $X \setminus \{x\}$. Consider the following assertions:

(A_1) If $A \in \mathcal{A}$, $A \subset B \subset X \setminus \{x\}$, then $B \in \mathcal{A}$.

(A_2) If $A \in \mathcal{A}$ and $V \in \mathcal{V}_\rho(x)$ then $A \cap V \in \mathcal{A}$ and $\{x\}$ is a G_δ-set.

(i) If $\{M_n\}$ is a decreasing sequence of sets from $\mathcal{A}$, then there is a decreasing sequence $\{U_n\}$ of (open) neighborhoods of x such that

$$\bigcup_{n=1}^{\infty} (M_n \setminus U_n) \in \mathcal{A}.$$

(ii) If $\{M_n\}$ is a decreasing sequence of sets from $\mathcal{A}$, then there is a decreasing sequence $\{W_n\}$ of (open) neighborhoods of x such that

$$X \setminus \bigcup_{n=1}^{\infty} (W_n \setminus M_n) \in \mathcal{A}.$$

(iii) If f is a function on X for which there is a sequence $\{M_n\} \subset \mathcal{A}$ such that $f(M_n) \subset (-1/n, 1/n)$, then there is $M \in \mathcal{A}$ with $\lim_{M \ni y \to x} f(y) = 0$.

If (A_1) holds, then (i) $\Rightarrow$(ii)$\Leftrightarrow$(iii). If, in addition, (A_2) holds, then (i), (ii) and (iii) are equivalent.

Hint. The proof of (ii) $\Leftrightarrow$ (iii) is almost the same as the proof of Theorem 2.8. For the implication (i) $\Rightarrow$(ii) use the relations

$$(\text{oo}) \qquad X \setminus \bigcup_{n=1}^{\infty} (W_n \setminus M_n) = \bigcap_{n=1}^{\infty} M_n \cup (X \setminus W_1) \cup \bigcup_{n=1}^{\infty} (M_n \setminus W_{n+1}) \supset \bigcup_{n=1}^{\infty} (M_n \setminus W_n)$$

which hold for each decreasing sequences $\{W_n\}$ and $\{M_n\}$ of subsets of X. By (A_2) we can assume in (i) that $M_n \subset V_n$, where $V_n \in \mathcal{V}_\varrho(x)$, $\bigcap_{n=1}^{\infty} V_n = \{x\}$. Hence $\bigcap_{n=1}^{\infty} M_n = \emptyset$, and by (oo)

$$\bigcup_{n=1}^{\infty} (M_n \setminus W_{n+1}) \supset (X \setminus \bigcup_{n=1}^{\infty} (W_n \setminus M_n)) \cap W_1 .$$

Therefore (ii) $\Rightarrow$(i).

2.C.11 Show that the assertions of Theorems 2.8 and 2.9 are only special cases of Fine limit principle.

Hint. Put $\mathcal{A} = \{V \setminus \{x\} : V \in \mathcal{V}_\tau(x)\}$ or $\mathcal{A} = \mathcal{M}$.

2.C.12 Let τ be a fine topology on a topological space (X, ϱ), and let (i*) be the following condition:

(i*) For each sequence $\{V_n\} \subset \mathcal{V}_\tau(x)$ there is a sequence $\{\tilde{W}_n\} \subset \mathcal{V}_\varrho(x)$ such that $\{x\} \cup \bigcup_{n=1}^{\infty} (V_n \setminus \tilde{W}_n) \in \mathcal{V}_\tau(x)$.

Show that:

(a) (i*) $\Rightarrow$ (tFL),

(b) in general, the implication (tFL) $\Rightarrow$ (i*) does not hold,

(c) if $\{x\}$ is a G_δ -set (even if $\{x\}$ is a τ-G_δ -set), then (tFL) $\Rightarrow$ (i*).

Hints. (a) Use Fine limit principle for $\mathcal{A} = \{V \setminus \{x\} : V \in \mathcal{V}_\tau(x)\}$.
(b) Let Ω and ρ have the same meaning as in Ex. 2.C.8. Put $X = \Omega$, $\tau = \rho$ and $x = \omega_1$.
(c) If $\{x\}$ is a τ-G_δ -set only, we can use (oo) in the same way as in the proof of the second part of Fine limit principle.

REMARKS AND COMMENTS

The first theorem on fine limits was proved in the case of approximate continuity already by A. Denjoy himself (1915). Since then a number of papers dealing with this problem have appeared. Let us mention for example the paper by C.V. Smallwood (1972) which proves the equivalence of two different definitions for upper approximate limits, or the general fine limit theorem for topologies on $\mathbb{R}$ given in the paper by J. Jedrzejewski (1974) (cf.W.Poreda, E. Wagner-Bojakowska and W. Wilczyński (1983)). One of the recent proofs of the fine limit theorem for approximately continuous functions in $\mathbb{R}^n$ can be found in M. Vuorinen (1982).

Applications of general theorems of this section will be given in Section 6.C for the case of density topologies on metric spaces. A rather surprising fact is that in metric spaces the condition (bFL) always implies (tFL) and that allows us to prove simply the fine limit theorem for various density and "superdensity" topologies even if the corresponding base operator is not strong. Thus, the validity of (tFL) for the density topologies does not depend on the density theorem.

On the other hand, we shall prove Theorem 6.6 independently without using the general theorem in order to keep the section dealing with the density topology on $\mathbb{R}$ self-contained.

The fine limit theorems in the potential theory originate in the papers of H. Cartan and J.L. Doob for the classical case and in Brelot's papers in axiomatic theory and are summarized by M. Brelot (1971) in an abstract setting. Ex. 10.D.1 represents the application of the theorems given in this section in case of the fine topology in potential theory.

2.D The G_δ-insertion property and Baire one functions

At about the end of the nineteenth century, the important class of Baire one functions consisting of all limits of sequences of real continuous functions was introduced. It seems quite natural that many properties of continuous functions are conserved by functions in the first Baire class, especially if they have in addition the

Darboux property. It is known that various families of functions are subsets of Baire class one. The derivatives, most sorts of generalized derivatives(e.g. approximate or preponderant derivatives) and some systems of finely continuous functions are typical examples. Although these functions can enjoy very bad behavior, they are often somehow "controled" just owing to the fact that they belong to the Baire class one.

Nowadays, necessary and sufficient conditions guaranteeing that a given function is in the Baire class one form a mathematical folklor. We collect some of them in the following theorem.

2.12 THEOREM. Consider the following properties that a function f on a metric space P may possess:

(B_1) f is a Baire one function,

(G_δ) for each $a \in \mathbb{R}$, the sets $[f \leq a]$, $[f \geq a]$ are G_δ,

($In\text{-}G_\delta$) for any couple $a < b$ of reals there are G_δ-sets H_1, H_2 such that $[f \leq a] \subset H_1 \subset [f \leq b]$ and $[f \geq b] \subset H_2 \subset [f \geq a]$,

(Ba) each nonempty closed subset F of P contains a point x such that f/F is continuous at x,

(D-P) for each nonempty closed set $F \subset P$ and for every reals $a < b$, the sets $[f \leq a]$ and $[f \geq b]$ cannot be dense in F simultaneously.

Then (Ba) $\Rightarrow$ (D-P) $\Rightarrow$ (B_1) $\Leftrightarrow$ (G_δ) $\Leftrightarrow$ ($In\text{-}G_\delta$), and all these properties are equivalent if and only if P is a strong Baire space.

Of course, the condition (Ba) goes back to Baire himself. The interesting condition (D-P) appeared in the Denjoy paper (1915), p.184, in the connection with his proof of the assertion that any preponderantly continuous function belongs to the first Baire class. It was rediscovered recently by D. Preiss in (1971).

It is not a hard work to prove the implication (Ba) $\Rightarrow$ (D-P). Indeed, the definition of continuity should be employed only.

The condition (G_δ) goes back to Lebesgue's paper (1905) and expresses nothing but the fact that the function f is "G_δ-measurable". The proof that (B_1) implies (G_δ) is standard and easy. Obviously, (G_δ) implies immediately ($In\text{-}G_\delta$). The other way round ($In\text{-}G_\delta$) leads directly to (G_δ) in view of

$$[f \geq a] = \bigcap_{n=1}^{\infty} [f \geq a - \frac{1}{n}].$$

The *insertion of G_δ- sets method* ($In\text{-}G_\delta$) was used in (1977a) and (1977b) by

J. Lukeš and L. Zajíček. It seems to be a relatively simple way how to prove that certain functions are in the Baire class one, rather important especially from the methodical aspects. Though (In-G_δ) is an almost trivial reformulation of (G_δ), we utilize it in an essential way as a joint in the proof of the implication (D-P) $\Rightarrow$ (B_1).

As we shall see further, the application of (In-G_δ) represents in this case a nice illustration of our methods. Observe that no completeness of P is required, meanwhile it is often necessary if (D-P) is employed. So (In-G_δ) fits well to be applied to arbitrary subsets of a given metric space.

The proof of the implication (B_1) $\Rightarrow$ (Ba) in case of P being a strong Baire space uses classical Baire category arguments applied to a closed set F. For the "converse" assertion see Ex. 2.D.2 and Ex. 2.D.3.

To complete the proof of Theorem 2.12, it suffices to prove (D-P) $\Rightarrow$ (In-G_δ) and (G_δ) $\Rightarrow$ (B_1). The proof of (G_δ) $\Rightarrow$ (B_1) is relatively difficult, it can be found in [Kur] ; later on in Section 3.A (Corollary 3.8) we indicate a different proof.

The proof of the implication (D-P) $\Rightarrow$ (In-G_δ) can be done in the following way: Assume $a < b$ are the reals. Let $\mathcal{G}$ be the family of all open subsets G of P for which there is a G_δ-set H_G satisfying

$$[f \leq a] \cap G \subset H_G \subset [f \leq b] \cap G.$$

Notice that obviously $\mathcal{G}$ is a hereditary family: If $G \in \mathcal{G}$ and $G' \subset G$ is open, then $G' \in \mathcal{G}$. Put $\Gamma = \bigcup_{G \in \mathcal{G}} G$.

In light of paracompactness of metric spaces, the open covering $\mathcal{G}$ of Γ has an open locally finite refinement $\mathcal{U}$.

Then

$$[f \leq a] \cap \Gamma \subset \bigcup_{G \in \mathcal{U}} H_G \subset [f \leq b] \cap \Gamma \quad .$$

Since every locally finite union of G_δ-sets is again a G_δ-set (see e.g. [Čech] , § 30.E or [Kur], § 30.X) we can see that $\Gamma \in \mathcal{G}$. If $\Gamma = P$, we are through. If $P \setminus \Gamma \neq \emptyset$, both the sets $[f \leq a]$ and $[f \geq b]$ cannot be dense in $P \setminus \Gamma$. Therefore there is an open set G with the properties $G \setminus \Gamma \neq \emptyset$ and either $(G \setminus \Gamma) \cap [f \leq a] = \emptyset$ or $(G \setminus \Gamma) \cap [f \geq b] = \emptyset$. In both cases $G \in \mathcal{G}$ which contradicts the maximality of Γ (in the first case we can put $H_G = G \cap H_\Gamma$, and in the second one $H_G = (G \cap H_\Gamma) \cup (G \setminus \Gamma)$).

We can now employ observations just accomplished to provide one of the important and interesting methods of proofs that finely continuous functions are in the Baire class one. The gist of the next result is inspired by the G_δ-insertion method. We introduce a concept obviously related to G_δ-insertion, but slightly more general.

Before moving on to the main theorem and to some examples which typify its importance, it is convenient to introduce some notions.

Let $\mathfrak{M}$ be a family of subsets of a set X. A function f on X is said to be *$\mathfrak{M}$-measurable* if the sets $[f \geq a]$, $[f \leq a]$ belong to $\mathfrak{M}$ for every $a \in \mathbb{R}$.

Notice that in the case $\mathfrak{M}$ is not a σ-algebra it does not hold that the function f is $\mathfrak{M}$-measurable if and only if $[f > a]$, $[f < a] \in \mathfrak{M}$. Some authors define the concept of $\mathfrak{M}$-measurability just in this last way. In this text we hold our original definition because it is more convenient for us.

2.13 THEOREM. Let $\mathfrak{M}$ be a family of subsets of a topological space X closed under the formation of countable intersections. If the topology τ of X has the $\mathfrak{M}$-insertion property, then all τ-continuous functions on X are $\mathfrak{M}$-measurable.

Proof. Let $a \in \mathbb{R}$ and let f be τ-continuous on X. Then for each $n \in \mathbb{N}$ there is a set $A_n \in \mathfrak{M}$ such that

$$[f \geq a] \subset [f > a - 1/n] \subset A_n \subset [f \geq a - 1/n] .$$

Hence

$$[f \geq a] = \bigcap_{n=1}^{\infty} A_n \in \mathfrak{M}.$$

Since $[f \leq a] = [-f \geq -a]$, this set is in $\mathfrak{M}$ too. Therefore f is $\mathfrak{M}$-measurable.

2.14 COROLLARY. If a topology τ has the G_δ-insertion property on a metric space P, then any τ-continuous function on P is in the Baire class one.

REMARK. Notice that the G_δ-insertion property of a topology in the previous corollary can be replaced by the *F_σ-insertion condition.* In fact, it is easy to see that both these conditions are equivalent. A simple "essential radius condition" guaranteeing the G_δ-insertion property which can be used in some concrete cases is contained in Ex. 2.D.16.

Now we are going to introduce a typical application of the G_δ-insertion method whose usefulness we hope to amply illustrate in subsequent sections.

There is also another approach closely related to our investigations, namely that one we will use in Section 9.A. We introduce there sufficient conditions guaranteeing that certain "limit functions" belong to the Baire clase one.

2.15 EXAMPLE. Let P_1 and P_2 be metric spaces. We introduce a fine topology τ on

$P_1 \times P_2$ which is entirely similar to that given in Example 1.1.d. For $a = (a_1, a_2) \in P_1 \times P_2$ and $\delta > 0$ put

$$K(a, \delta) = \{x = (x_1, x_2) \in P_1 \times P_2 : x \in U(a, \delta) \text{ and } x_1 = a_1 \text{ or } x_2 = a_2\}.$$

A set $G \subset P_1 \times P_2$ is called to be τ-open if for each $z \in G$ there is $\delta > 0$ such that $K(z, \delta) \subset G$.

It is not difficult to realize that a function f on $P_1 \times P_2$ is τ-continuous if and only if f is separately continuous.

We now wish to show that τ has the G_δ-insertion property. Let $G \subset P_1 \times P_2$ be τ-open. For each $z \in G$ there is $\delta_z > 0$ such that $K(z, \delta_z) \subset G$. For $n \in \mathbb{N}$ and $z \in G$ put $\varepsilon_z^n = \min(\delta_z, 1/n)$.
Denote

$$G_n := \bigcup_{(z_1, z_2) \in G} U_1(z_1, \varepsilon_z^n) \times U_2(z_2, \varepsilon_z^n)$$

(neighborhoods are taken in P_1 and P_2, respectively). Then G_n is an open subset of $P_1 \times P_2$. It is easy to check that

$$G \subset \bigcap_{n=1}^{\infty} G_n \subset \bar{G}^\tau ,$$

which implies that all separately continuous functions on $P_1 \times P_2$ are in the Baire class one.

REMARK. A direct proof of the last assertion in the case of $\mathbb{R}^2$ is not difficult and it goes back to Baire's thesis (1899). In fact, Baire used in his proof the so-called strip method. Another proof for $P \times Q$ can be found in [Kur]. The variants of the previous example are noted in Ex.2.D.5. See also Section 9.D.1 and Ex. 2.D.18.

Further nontrivial applications are given later in the cases of a metric density topology (Theorem 6.23) and of a fine topology in harmonic spaces (Theorem 11.9.h). We use also the G_δ-insertion method in Section 9.A (especially in the proof of Theorem 9.2).

We close this section by introducing a concept of honorary Baire two functions. We say that a function f on a topological space X is a *honorary Baire two function* if there is a Baire one function g on X and a countable subset S of X such that $f = g$ on $X \setminus S$.

If X is a metric space, then any Baire one function defined on a G_δ-subset of X can be extended to a Baire one function over X (cf. Corollary 3.8.b).

It easily follows that f is a honorary Baire two function on X if and only if there is a countable set $S \subset X$ such that f is a Baire one function on $X \setminus S$.

Before giving a further application of the G_δ-insertion method it will be convenient to have the following lemma.

2.16 LEMMA. For a set $M \subset \mathbb{R}$ and $x \in \mathbb{R}$, let

$$d^{i,+}(x,M) = \limsup_{h \to 0_+} \frac{1}{h} \lambda_*(M \cap (x,x+h))$$

denote the upper right inner density of M at x. Define $d^{i,-}(x,M)$, $d_{i,+}(x,M)$, $d_{i,-}(x,M)$ similarly.

If $A(M)$ is the set of all $x \in \mathbb{R}$ for which either $d_{i,+}(x,M) = 1 > d^{i,-}(x,M)$ or $d^{i,+}(x,M) < 1 = d_{i,-}(x,M)$, then $A(M)$ is countable.

Proof. It follows from the fact that all points of $A(M)$ are *angle points* of the function $x \mapsto \lambda_*(M \cap (0,x))$ (cf. [Saks], p. 261, or [Bru], Chap. IV, §4, Th. 4.1).

2.17 PROPOSITION. If a function f on $\mathbb{R}$ is either left or right approximately continuous at any point of a set $A \subset \mathbb{R}$, then f is a honorary Baire two function on A.

Proof. Put

$$S := \bigcup_{r \in \mathbb{Q}} (A([f > r]) \cup A([f < r]))$$

By Lemma 2.16, S is countable. We claim that $f \in B_1(A \setminus S)$. It only needs to be observed that if $a < b$, then there is a G_δ - subset G of $A \setminus S$ such that

$$A \cap ([f \geq b] \setminus S) \subset G \subset ([f \geq a] \setminus S) \cap A.$$

Indeed, pick $r \in (a,b) \cap \mathbb{Q}$ and put

$$G := \{x \in A \setminus S: \text{ for each } m \in \mathbb{N} \text{ there is } h \in (0,\tfrac{1}{m}) \text{ such that}$$

$$\frac{1}{2h} \lambda_*([f > r] \cap (x-h,x+h)) > \frac{3}{4}\}.$$

Then G is a G_δ - subset of $A \setminus S$. If $x \in [f \geq b] \setminus S$ and f is, say, right approximately continuous at x, then $d_{i,+}(x,[f > r]) = 1$. By the definition of S, $d^{i,-}(x,[f > r]) = 1$, and thus $x \in G$. On the other hand, if $x \in G$ and $f(x) < a$, then either $d_{i,+}(x,[f \leq r]) = 1$ or $d_{i,-}(x,[f \leq r]) = 1$, which contradicts the definition of G.

EXERCISES

2.D.1 Let f be an upper semicontinuous function on a topological space X. Then f satisfies (D-P) of Theorem 2.12.

Hint. Let $\emptyset \neq F \subset X$ be closed and let $a < b$. If $[f \geq b] \cap F$ is dense in F, then $[f \geq b] \cap F = F$.

2.D.2 Let X be a topological space. The following two assertions are equivalent

(i) X is strong Baire,

(ii) (D-P) $\Rightarrow$ (Ba).

Hint. (i) $\Rightarrow$ (ii): Assume that $F \neq \emptyset$ is a closed set on which a function $f \wedge F$ has no point of continuity. Put

$$A^{+}_{r,s} = \{x \in F\colon f(x) < r < s < \limsup_{F \ni t \to x} f(t)\},$$

$$A^{-}_{r,s} = \{x \in F\colon f(x) > s > r > \liminf_{F \ni t \to x} f(t)\}$$

($r,s \in \mathbb{Q}$, $r < s$). Show that there is a nonempty set $U \subset X$ open in F and a couple $r,s \in \mathbb{Q}$ such that either $A^{+}_{r,s}$ or $A^{-}_{r,s}$ is dense in U. Put $F^* = F \cap \bar{U}$ and seek a contradiction with (D-P).

(ii) $\Rightarrow$ (i). Assume that $\emptyset \neq F = \bigcup_{n=1}^{\infty} F_n$, where F is closed and F_n are closed and nowhere dense in F put.

$$f(x) = \begin{cases} \max\ \{1/n\colon x \in F_n\} & \text{if } x \in F, \\ 0 & \text{if } x \in X \setminus F. \end{cases}$$

Then f is upper semicontinuous on X and there is no point of continuity of f on F. Now, use Ex. 2.D.1.

2.D.3 Let X be a topological space. Consider the following assertions:

(i) X is strong Baire,

(ii) given a closed set $F \neq \emptyset$, there are no disjoint G_δ-sets C_1, C_2

which are both dense in F,

(iii) given a closed set $F \neq \emptyset$, there are no disjoint Coz_δ -sets C_1, C_2 which are both dense in F,

(iv) $(B_1) \Rightarrow (\Pi\text{-}P)$.

Then (i) $\Rightarrow$ (ii) $\Rightarrow$ (iii) $\Leftrightarrow$ (iv); if P is a metric space, then all these properties are equivalent.

Hint. (i) $\Rightarrow$ (ii) $\Rightarrow$ (iii) $\Rightarrow$ (iv) is obvious.
(iv) $\Rightarrow$ (iii) Use Ex. 3.A.2.
If P is a metric space, then (ii) and (iii) are apparently equivalent; for the proof of (ii) $\Rightarrow$ (i) use Ex. 4.A.14.

2.D.4 Denote by $\mathbb{D}$ the set of all dyadic rationals. Let τ be a fine topology on $\mathbb{R}$ determined as follows: A set $U \subset \mathbb{R}$ is τ-open provided for every $x \in U$ there is a Euclidean neighborhood V of x such that

$$V \cap \mathbb{D} \subset U \qquad \text{if } x \in \mathbb{D},$$

$$V \subset U \qquad \text{if } x \in \mathbb{Q} \setminus \mathbb{D},$$

$$V \setminus \mathbb{Q} \subset U \qquad \text{if } x \in \mathbb{R} \setminus \mathbb{Q}.$$

Then every τ-continuous function is continuous even though τ has not the G_δ intersection property.

Hint. The τ-closure of the τ-open set $\mathbb{D}$ equals $\mathbb{Q}$.

2.D.5 a) Let Ω be the set of all ordinals $\leq \omega_1$ endowed with the usual order topology. Define a function f on $\Omega \times \Omega$ as follows:

$$f(\alpha,\beta) = \begin{cases} 1 & \text{if } \alpha = \beta \text{ and } \alpha \text{ is not a limit ordinal,} \\ 0 & \text{otherwise.} \end{cases}$$

Then f is separately continuous, but f is not a Baire function.

Hint. Each Baire function on $\Omega \times \Omega$ is constant on a neighborhood of (ω_1, ω_1).

(b) The fact that a function f on $\mathbb{R}^2$ is a Baire one function in each variable spearately does not imply the Borel measurability of f (W. Sierpiński (1920), cf. [Kur], § 31, V).

c) Let f be a function on $\mathbb{R}^3$ which is continuous in each variable separately. Then f is in the second Baire class.

d) There is a function g on $\mathbb{R}^2$ which is right continuous in each variable separately but it is not in Baire class one.

Hint. Let F be a perfect nowhere dense subset of $[0,1]$ containing 0 and 1. Then $[0,1] \setminus F$ is a disjoint union of open intervals (a_n, b_n) $(n \in \mathbb{N})$. Put

$$g(x,y) = \begin{cases} (b_n - x)(b_n - y)(b_n - a_n)^{-2} & \text{if } x,y \in [a_n, b_n], \\ 0 & \text{otherwise.} \end{cases}$$

In the set $M := \{(x,x) : x \in F\}$ there is no point of continuity of $g \restriction M$.

(e) If e^+ is the Sorgenfrey topology on $\mathbb{R}$ and if f is separately e^+-continuous (i.e. if f is right continuous in each variable), then f belongs to $B_1(e^+ \times e^+)$.

Hint. Construct directly a sequence $\{f_n\}$ of $e^+ \times e^+$-continuous functions which tends to f.

(f) Let f be a function on $\mathbb{R}^2$. If $f(x,.)$ is continuous for each $x \in \mathbb{R}$ and if $f(.,y)$ is right continuous for each $y \in \mathbb{R}$, then f is a Baire one function.

<u>REMARK.</u> If f is separately approximately continuous on $\mathbb{R}^2$ then it is in the second Baire class, but it need not be in Baire class one (R.O.Davies(1973)).

2.D.6 (a) Let e^+ be the Sorgenfrey topology of Ex. 1.A.16. Show that e^+ has the G_δ-insertion property (in fact, any e^+-open set is an F_σ-subset of $\mathbb{R}$) and consequently any right continuous function is in $B_1(\mathbb{R})$.

Hint. Use the fact that any union of closed (non-degenerate) intervals is an F_σ-set.

(b) If a function f is either right or left continuous at every point of a set $M \subset \mathbb{R}$, then $f \in B_1(M)$.

Hint. (i) Show that $[f > a]$ is a union of closed non-degenerate intervals.
(ii) Show that there is a topology τ_f on M similar to e^+ such that f is τ_f-continuous and, moreover, τ_f has the G_δ-insertion property (in fact, then any

τ_f-open set is an F_σ-set).

2.D.7 A function f on $\mathbb{R}$ which is approximately continuous at each point either from the right or from the left need not be in Baire class one (cf. also Ex. 2.D.6.b and Proposition 2.17).

Hint. (M.Chlebík). Let F be a perfect nowhere dense subset of $\mathbb{R}$, $\mathbb{R}\setminus F = \bigcup_n (a_n, b_n)$ (disjoint union). Consider the characteristic function of the set

$$\bigcup_n [a_n, a_n + \frac{1}{n}(b_n - a_n)] .$$

2.D.8 If every point of a metric space P has a neighborhood on which a function f is in Baire class one, then f is a Baire one function on P.

Hint. Use $(B_1) \Leftrightarrow (G_\delta)$ (cf. Corollary 3.8) and Th. 1 in § 30.X of [Kur].

2.D.9 If a function f on a metric space P is continuous except a countable set, then f is a Baire one function on P.

Hint. Let D be the set of all points of discontinuity of f, let $a \in \mathbb{R}$. Show that $[f > a] = \operatorname{int}[f > a] \cup \{x \in D: f(x) > a\}$ is an F_σ-set (cf. [Haus], p. 250).

2.D.10 (a) Define a subset of $\mathbb{R}^2$ to be a *trioda* with center $s \in \mathbb{R}^2$ if it is a union of three closed arcs issueing from s such that none two of them intersect outside s. Let T be a trioda in $\mathbb{R}^2$ with center 0, let $z \in \mathbb{R}^2$ and denote

$$T_z = \{z + t : t \in T\} .$$

If a function f on $\mathbb{R}^2$ is continuous with respect to every trioda T_z $(z \in \mathbb{R}^2)$, then f is a Baire one function (cf. also Section 9.D.1).

(b) Deduce from (a) that a function which is continuous with respect to any unit circle is in Baire class one.

2.D.11 (Baire (1899)). Let f be a Baire one function on a topological space X. Then the set of all discontinuity points of f is a first category set in X.

Hint. Show that

$$D \subset \bigcup_{r \in \mathbb{Q}} [([f > r] \setminus \operatorname{int}[f > r]) \cup ([f < r] \setminus \operatorname{int}[f < r])]$$

and that $A \setminus \text{int}\, A$ is a first category set provided A is an F_σ-subset of X. (Cf. [Kur], § 31.X.See also J.C.Oxtoby (1971), Th. 73.)

2.D.12 If a fine topology τ has the G_δ-insertion property on a metric space (X, ϱ), and if a function f on X is τ-continuous at each point of a set $M \subset X$, then $f \restriction M$ is in $B_1(M)$.

Hint. Pick $a \in \mathbb{R}$ and find G_δ-sets G_n,

$$\text{int}_\tau [f < a + \frac{1}{n}] \subset G_n \subset \overline{[f < a + \frac{1}{n}]}^{\tau}.$$

Show that $M \cap [f \leq a] = \bigcap_{n=1}^{\infty} (G_n \cap M)$

2.D.13 (W. Sierpiński (1921a)) Let f be a real function on a perfectly normal space X. Show that the following assertions are equivalent.

(i) $f = f_1 - f_2$ where f_1 and f_2 are real lower semicontinuous functions,

(ii) $f = g_1 - g_2$ where g_1, g_2 are nonnegative real lower semicontinuos functions,

(iii) f is the sum of an absolutely convergent series of real continuous functions.

Hint. (i) $\Rightarrow$ (iii). Use Ex. 3.B.8.c and the fact that $h = h_1 + \sum_{n=1}^{\infty} (h_{n+1} - h_n)$ provided $h_n \nearrow h$ are real functions.

(iii) $\Rightarrow$ (ii). Use the equality $f = \sum_{n=1}^{\infty} h_n^+ - \sum_{n=1}^{\infty} h_n^-$ provided the sum of the absolutely convergent series $\sum_{n=1}^{\infty} h_n$ equals f.

REMARK. As stated in Remarks and Comments to 3.G, there are Baire one functions lacking the property (i).

2.D.14 Subclasses of B_1. Define the following classes of real functions on a topological space X. A function f belongs to:

$B_1^d(X)$ if there is a sequence f_n of real continuous functions on X such that the set $\{n \in \mathbb{N}: f_n(x) \neq f(x)\}$ is finite for every $x \in X$,

$B_1^*(X)$ if for every closed set $F \subset X$, the set of all discontinuity points of $f \restriction F$ is nowhere dense,

$CG(X)$ if there are closed sets $F_n \subset X$ such that $X = \bigcup_{n=1}^{\infty} F_n$ and $f \restriction F_n$ is continuous for each $n \in \mathbb{N}$.

The elements of $CG(X)$ are termed *continuous functions in generalized sense*, while those in $B_1^d(X)$ are *discrete limits* of continuous functions.

Prove the following assertions:

(a) $f \in B_1^*(X)$ if and only if for every nonempty closed set F there is a portion P of F (= nonempty open set in F) such that $f \restriction P$ is continuous,

(b) $B_1^d(X) \subset CG(X)$,

(c) if X is normal, then $B_1^d(X) = CG(X)$,

(d) if X is perfect and paracompact, then $B_1^*(X) \subset CG(X)$,

(e) if X is strong Baire, then $CG(X) \subset B_1^*(X)$,

(f) if X is strong Baire and paracompact, then $B_1^d(X) = CG(X) = B_1^*(X)$.

Hint. (b) Put $F_n = \{x \in X: f_m(x) = f_n(x)$ for each $m \geq n\}$.

(c) Put $Z_n = \bigcup_{i=1}^{n} F_i$ and extend $f \restriction Z_n$ continuously to X.

(d) Let $f \in B_1^*(X)$. If $\mathcal{G} = \{H:$ H is open in X and $f \restriction H \in CG(H)\}$, then $\mathcal{G}$ is a hereditary class and $G := \bigcup_{H \in \mathcal{G}} H \in \mathcal{G}$. Suppose $G \neq X$ and seek a contradiction.

2.D.15 Recall that the family $\mathcal{A}$ of subsets of a topological space X which are both F_σ and G_δ forms an algebra of sets. The elements of $\mathcal{A}$ are termed *ambivalent* sets. $\mathcal{A}$-measurable functions are called also *ambivalent*.

Prove the following assertions:

(a) Any function of $CG(X)$ is a difference of nonnegative real lower semicontinuos functions (cf.Ex.2.D.13).

(b) If X is perfect, then any function in $CG(X)$ is ambivalent.

(c) Any function of $B_1^d(X)$ is ambivalent and is the sum of an absolutely convergent series of real continuous functions.

(d) The characteristic function of an ambivalent set or, more generally, any ambivalent function admitting only of countably many values belongs to $CG(X)$. Consequently, it is a difference of nonnegative real lower semicontinuous functions.

(e) There is an ambivalent function which is not in $CG(X)$.

Hint. (a) Let $f \in CG(X)$. Find closed sets $Z_1 \subset Z_2 \subset \ldots$, $\bigcup_{n=1}^{\infty} Z_n = X$ such that $f \restriction Z_n$ is continuous and $|f| \leq n$ on Z_n. Put

$$u(x) = f^+(x) + \inf\{n \in \mathbb{N}: x \in Z_n\},$$

$$v(x) = f^-(x) + \inf\{n \in \mathbb{N}: x \in Z_n\}$$

and show that u and v are lower semicontinuous.

(c) Use the equality $[f > a] = \bigcap_{n=1}^{\infty} \bigcup_{k=n}^{\infty} [f_k > a]$.

(e) Consider the sum of the Riemann function on $[0,1]$ and the identity function.

REMARK. It follows by (e) that the sum of a continuous function and of an ambivalent function need not be ambivalent.

2.D.16 A topology τ on a metric space (P, ρ) is said to satisfy the *essential radius condition* if for each $x \in P$ and each τ-neighborhood U of x there is an "essential radius" $r(x,U) > 0$ such that

$$\rho(x,y) \leq \min(r(x,U_x), r(y,U_y)) \Rightarrow U_x \cap U_y \neq \emptyset$$

for every τ-neighborhood U_x, U_y of x and y.

Show that the essential radius condition implies the G_δ-insertion property of τ.

Hint. Choose $G \subset P$ τ-open and set

$$A_n := \{x \in G: r(x,G) \geq 1/n\}.$$

Prove that the F_σ-set $\bigcup_{n=1}^{\infty} \bar{A}_n$ is inserted between G and $\bar{G}^\tau$. For contrast, let $y \in \bar{A}_n \setminus \bar{G}^\tau$ for some $n \in \mathbb{N}$. Find $x \in A_n$ such that $\rho(x,y) \leq \min(r(y, P \setminus \bar{G}^\tau), 1/n)$ and deduce that $G \cap (P \setminus \bar{G}^\tau) \neq \emptyset$. Now, use Remark following Corollary 2.14.

2.D.17 Let $\omega \in \Omega_0$ (cf. Section 7.D) and assume that τ is a fine topology on (P,ρ) satisfying the essential radius condition. If f is a real τ-ω-Lipschitzian function at any point $z \in P$ (i.e. there is a τ-neighborhood V_z of z such that $|f(z) - f(y)| \leq \omega(\rho(z,y))$ for each $y \in V_z$), then $f \in CG(P)$. In particular, f is an ambivalent function.

Hint. Set $A_n := \{x \in P: r(x,V_x) > 1/n\}$ and show that f is continuous on $\bar{A}_n$. (Pick $x \in \bar{A}_n$ and $\varepsilon > 0$. Find $d > 0$ such that $\omega(d) < \varepsilon/2$ and put

$$\delta := \min\ (d/2,\ 1/n,\ r(x,\ V_x \cap U(x,\ d/2)).$$

Show that $|f(x) - f(y)| < \varepsilon$ for each $y \in A_n \cap U(x,\delta)$.)

2.D.18 (a) Show that the (crosswise) topology of Example 2.15 satisfies the essential radius condition. Hence, any separately continuous function is in the Baire class one.

(b) Show that the Sorgenfrey topology on $\mathbb{R}$ satisfies the essential radius condition. Consequently, any right continuous function is in $B_1(\mathbb{R})$. (Cf. with Ex. 2.D.6.)

(c) If a function f has at each point of $\mathbb{R}^2$ finite partial derivatives, then $f \in CG(\mathbb{R}^2)$.

Hint. Show that f is τ-Lipschitzian at every point of $\mathbb{R}^2$ where τ is the topology of (a). Then use Ex. 2.D.17 (e.g., for $\omega(t) = \sqrt{t}$).

(d) If a function f has at any point of $\mathbb{R}$ finite right derivative, then $f \in CG(\mathbb{R})$.

Hint. Use the idea of (c) for the Sorgenfrey topology.

2.D.19 We say that a topology τ on a metric space (P,ρ) satisfies the *strong essential radius condition* on a set M if for each $x \in M$ and for each its τ-neighborhood U there is a "strong essential radius" $r(x,U) > 0$ and $K(x,U) > 0$ such that

$$\rho(x,y) \leq \min\ (r(x,U_x), r\ (y,U_y)) \Rightarrow U_x \cap U_y \cap U(x, K(x,U_x)\,\rho(x,y)) \cap U(y, K(y,U_y)\rho(x,y)) \neq \emptyset$$

whenever U_x and U_y are τ-neigborhoods of x and y.

Assume that a topology τ satisfies the strong essential radius condition on a subset M of a separable metric space P. If f is a τ-locally Lipschitzian function on M, then $M = \bigcup_{n=1}^{\infty} M_n$ where M_n are closed in M and f is Lipschitzian on each M_n.

Hint. Given $x \in M$, find its τ-neighborhood U_x and $L_x \geq 0$ such that $|f(x)-f(y)| \leq L_x \cdot \rho(x,y)$ for each $y \in U_x$. Put $B_k := \{x \in M: L_x \leq k,\ K(x,U_x) \leq k$ and $r(x,U_x) \geq 1/k\}$. Let $B_k = \bigcup_{i=1}^{\infty} B_{k,i}$ where diam $B_{k,i} \leq 1/k$. The sequence $\{\bar{B}_{k,i}\}$ has all properties desired. (Cf. Hint of Ex. 2.D.17.)

2.D.20 Show that both the Sorgenfrey topology and the (crosswise) topology of Example 2.15 satisfy the strong essential radius condition. In light of the previous exercise improve the assertions (c) and (d) of Ex. 2.D.18.

2.D.21 Let f be a function on a perfectly normal topological space X. Show that the following assertions are equivalent:

(i) f is a honorary Baire two function,

(ii) there is a countable set $S \subset X$ such that $f \in B_1(X \setminus S)$,

(iii) f is the limit of a sequence of functions, each of which has at most a finite number of points of discontinuity, each of which is removable,

(iv) f is the limit of a sequence of functions, each of which has at most countably many points of discontinuity.

Hint. The implications (i) $\Rightarrow$ (ii), (iii) $\Rightarrow$ (iv) and (iv) $\Rightarrow$ (ii) hold trivially in any topological space. For (i) $\Rightarrow$ (iii), let $\{g_n\}$ be a sequence of continuous functions on X, $g_n \to g$ and $f = g$ on $X \setminus \{a_1, a_2, \ldots\}$. Define $f_n(a_k)$)$=f(a_k)$ for $k \leq n$ and $f_n = g_n$ otherwise. Next prove (ii) $\Rightarrow$ (i): To extend f from $X \setminus S$ to a Baire one function on X use either Ex. 3.A.3 and Ex. 3.A.1, or Ex. 3.A.2 and the fact that G_δ-sets coincide with Coz_δ-sets in perfectly normal spaces.

REMARK. Ch. Tucker (1968) proved the assertion of this exercise using a general "approximation theorem" (cf. Ex. 3.G.3).

REMARKS AND COMMENTS

The class of functions which are the limits of convergent sequences of continuous functions was introduced by R. Baire and their properties were extensively studied in his thesis (1899). Since there have been a number of papers investigating the methods how to prove that certain families of functions belong to this

Baire class one. We already mentioned some of them in the course of the proof of Theorem 2.12. In our approach which is closely related to applications of fine topologies method we draw the attention to a simple G_δ-insertion property of J. Lukeš and L. Zajíček (1977a) and (1977b).

The notion of honorary Baire two functions appeared in E. Bagemihl and G. Piranian (1961) where a characterization given in Ex.2.D.21 was proved. Later on the same assertion was done by Ch. Tucker (1968).

On the other hand, in the past decades several subclasses of Baire one functions appeared. In Ex. 2.D.13 we mentioned *Sierpiński's class* formed by sums of absolutely convergent series of real continuous functions.

It was a paper by S. Verblunsky (1930) where a real function on an interval I is said to have the *property* R if "its points of discontinuity with respect to any perfect set $P \subset I$ are non-dense in P". But some properties of this class of B_1^*-*functions* were studied even formerly. For example, E.W. Chittenden (1919) showed that a limit of each relatively uniformly convergent sequence of continuous functions is a B_1^*-function. Notice also that the property R was generalized by S. Verblunsky (1932) saying that f has the *property* R^* on an interval I if every closed subset of I contains a portion on which f is upper semicontinuous. An interesting characterization of this class of functions appeared in S. Saks (1932). Finally, note that G.Tolstov (1939) showed that approximately derivable functions are in B_1^* (cf. Ex. 2.D.17 and Ex. 6.C.6; cf. also Section 7.B).

The class of functions which we denote $CG(X)$ appeared in H.W.Ellis (1951), who labelled these functions as [CG]-functions.

The *class* $B_1^d(X)$ of all discrete limits of continuous functions is examined by A. Császár and M. Laczkovich (1975), (1979). As mentioned in Ex. 2.D.14, all these classes of functions coincide on intervals of $\mathbb{R}$ (i.e. $B_1^* = CG = B_1^d$) and this important family of functions is studied by R.J. O'Malley in a series of papers (1976a), (1976b), (1977), (1979b) and by M. Laczkovich (1983).

Other additional material on Baire one functions on metric or topological spaces can be found in Section 3.A (especially, in Ex. 3.A.1-3.A.3).

<u>Related papers:</u> H.W. Pu and H.H. Pu (1982).

3. THE LUSIN-MENCHOFF PROPERTY OF FINE TOPOLOGIES

3.A Abstract in-between theorems

As noticed at the end of this section, there is an extensive literature devoted to various insertion problems. The gist of these questions is: Let the families of functions $\mathcal{S}$, $\mathcal{T}$ and $\mathfrak{F}$ be given. It is required to find $f \in \mathfrak{F}$ such that $t \leqq f \leqq s$ whenever $s \in \mathcal{S}$ and $t \in \mathcal{T}$ are given and $t \leqq s$.

Perhaps, the most famous result in this direction goes back to H. Hahn (1917) who proved that for every pair s, t of real functions defined on a metric space X where t is upper semicontinuous, s is lower semicontinuous and $t \leqq s$, there is a continuous function f such that $t \leqq f \leqq s$ on X.

In this section we are going first to exhibit a quite general in-between theorem with a very simple proof. The question we deal with is: Given bounded functions $t \leqq s$, find necessary and sufficient conditions under which there exists a function f belonging to a prescribed family $\mathfrak{F}$ such that $t \leqq f \leqq s$. The main result can be stated in a well-arranged form and as examples will show it provides one of the important and interesting methodical ways in miscellaneous applications.

Remark also that in-between theorems often become the basis for the extension theorems of Tietze's type.

First, however, let us agree on some terminology. In what follows, by an *L-cone* we understand a family $\mathfrak{F}$ of real functions on a set X having the properties:

($\mathfrak{F}_1$) $\mathfrak{F}$ is a lattice cone (i.e. $f, g \in \mathfrak{F}$, $\alpha > 0$ implies that $f+g$, αf, $\max(f,g)$, $\min(f,g) \in \mathfrak{F}$),

($\mathfrak{F}_2$) $\mathfrak{F}$ contains the constant functions.

We say that a family $\mathfrak{F}$ is *closed* if $\mathfrak{F}$ is closed under uniform convergence.

Let $\mathfrak{F}$ be an L-cone. We say that a set A is $\mathfrak{F}$*-separated* from B if there is $f \in \mathfrak{F}$ such that $f(A) = 0$ and $f(B) = 1$.

Notice that, in general, this relation is not symmetric with respect to A and B. It is easy to see that A is $\mathfrak{F}$-separated from B if and only if for every couple $a < b$ of reals there is $g \in \mathcal{F}$ such that $g(A) = a$, $g(B) = b$ and $a \leqq g \leqq b$.

3.1 LEMMA. Let $\mathcal{F}$ be a closed L-cone, and let $t \leqq s$ be real functions on X. If for every $\varepsilon > 0$ there is $f_\varepsilon \in \mathcal{F}$ such that

$$t - \varepsilon \leqq f_\varepsilon \leqq s + \varepsilon \quad \text{on } X,$$

then there is $f \in \mathcal{F}$ for which

$$t \leqq f \leqq s \quad \text{on } X.$$

Proof. There is a sequence $\{g_n\} \subset \mathcal{F}$ such that

$$t - 2^{-n} \leq g_n \leq s + 2^{-1}.$$

Put

$$h_1 = g_1 \text{ and } h_n = \max(h_{n-1} - 2^{-n}, \min(h_{n-1} + 2^{-n}, g_n)).$$

It is easily seen that

$$t - 2^{-n} \leqq h_n \leqq s + 2^{-n}$$

and $|h_n - h_{n-1}| \leqq 2^{-n}$. We can finish by putting $f = \lim h_n$.

3.2 ABSTRACT IN-BETWEEN THEOREM.

Let $\mathcal{F}$ be a closed L-cone, and let $t \leqq s$ be bounded functions on X. Then the following two assertions are equivalent:

(i) there is $f \in \mathcal{F}$ such that $t \leqq f \leqq s$,

(ii) $a < b$ implies that $[s \leqq a]$ is $\mathcal{F}$-separated from $[t \geqq b]$.

Proof. (i)$\Longrightarrow$(ii): Let $t \leqq f \leqq s$ be as in (i) and let $a < b$. Put

$$h(x) = \max(a, \min(f(x), b))$$

for $x \in X$. Then $h \in \mathcal{F}$. Obviously $h = a$ on $[s \leqq a]$ and $h = b$ on $[t \geqq b]$.

(ii)$\Longrightarrow$(i): Adding an appropriate constant function and multiplying by positive reals we get that $0 \leqq t \leqq s \leqq 1$. So we can assume it without loss of generality.

Let $\varepsilon > 0$. Take $p \in \mathbb{N}$ for which $\frac{1}{p} \leqq \varepsilon$. By (ii) we exhibit $f_k \in \mathcal{F}$ ($k = 1,2,\ldots,p$) such that:

$$0 \leqq f_k \leqq \frac{k}{p} \quad \text{on } X,$$

$$f_k = 0 \quad \text{on the set } [s \leqq \frac{k-1}{p}],$$

$$f_k = \frac{k}{p} \quad \text{on the set } [t \geqq \frac{k}{p}].$$

Put $f = \max(f_1,\ldots,f_p)$. Then $f \in \mathcal{F}$ and

$$t(x) - \varepsilon \leqq t(x) - \frac{1}{p} \leqq f(x) \leqq s(x) + \frac{1}{p} \leqq s(x) + \varepsilon$$

for each $x \in X$.

(Indeed, let us concentrate on $x \in X$. There are $i,j \in \{1,\ldots,p\}$ such that

$$s(x) \in [\frac{i-1}{p}, \frac{i}{p}], t(x) \in [\frac{j-1}{p}, \frac{j}{p}].$$

Then

$$t(x) \leqq \frac{j}{p} \leqq f_{j-1}(x) + \frac{1}{p} \leqq f(x) + \frac{1}{p}$$

and

$$s(x) \geqq \frac{m-1}{p} \geqq f_m(x) - \frac{1}{p} \qquad \text{for every } m \in \{1,\ldots,i\}.$$

Since $f_m(x) = 0$ for each $m \in \{i+1,\ldots,p\}$, we get

$$s(x) \geqq f(x) - \frac{1}{p}.)$$

Now, we are going to give several applications of the previous abstract in-between theorem in case of functions on metric spaces. Of course, these results are far from being new. But they illustrate our ideas and, perhaps, they bring new views of some classical theorems. Nontrivial applications of the in-between theorem as well as more general results on topological spaces will be stated in subsequent sections and in exercises.

Let $\mathcal{H}$ be a family of functions on a set X. Denote by $\mathcal{H}^{\uparrow}$ ($\mathcal{H}^{\downarrow}$) the family of all pointwise limits of increasing (decreasing) sequences from .

We start with a simple lemma.

3.3 LEMMA. Let $\mathcal{F}$ be an L-cone. Then $\mathcal{F}^{\uparrow}$ is a closed L-cone.

Proof. Clearly, $\mathcal{F}^{\uparrow}$ is an L-cone. If a sequence $\{f_n\} \subset \mathcal{F}^{\uparrow}$ converges uniformly to f, we can find its subsequence $\{f_{n_k}\}$ such that

$$|f_{n_k} - f| < \frac{1}{k}$$

for each $k \in \mathbb{N}$. Put

$$h_k = \max\left(f_{n_1} - 1,\ f_{n_2} - \frac{1}{2},\ \dots,\ f_{n_k} - \frac{1}{k}\right).$$

Then $h_k \in \mathcal{F}^{\uparrow}$ and $h_k \nearrow f$. A routine argument gives $f \in \mathcal{F}^{\uparrow}$.

(a) Lower semicontinuous functions on metric spaces.

Let F be a closed subset of a metric space P. Put

$$f(x) = \min(1, \operatorname{dist}(x,F))$$

for $x \in P$. It is easy to see that f is continuous on P, $0 \leqq f \leqq 1$ and $F = [f = 0]$.

3.4 PROPOSITION. Any lower semicontinuous lower finite function h on a metric space P belongs to $\mathcal{C}^{\uparrow}(P)$.

Proof. Let $G \subset P$ be an open set. Let $g \geqq 0$ be a continuous function on P such that $G = [g > 0]$. Then

$$\min(1, ng) \nearrow \chi_G,$$

hence $\chi_G \in \mathcal{C}^{\uparrow}(P)$.

At first assume that h is a bounded lower semicontinuous function on P. We apply Theorem 3.2 considering $\mathcal{F} = \mathcal{C}^{\uparrow}(P)$ and $s = t = h$. In view of Lemma 3.3, $\mathcal{C}^{\uparrow}(P)$ is a closed L-cone. Let $a < b$. Put $f = \chi_{[h>a]}$. Then $f \in \mathcal{C}^{\uparrow}(P)$, $f = 0$ on $[h \leqq a]$ and $f = 1$ on $[h \geqq b]$. Hence the sets $[h \leqq a]$ and $[h \geqq b]$ are $\mathcal{C}^{\uparrow}(P)$-separated, and therefore $h \in \mathcal{C}^{\uparrow}(P)$. Now, mapping $\bar{\mathbb{R}}$ homeomorphically onto $[0,1]$ we see that it suffices to find a sequence of continuous functions $g_n: P \to (0,1)$ such that $g_n \nearrow g$ whenever $g: P \to (0,1]$ is lower semicontinuous. This can be realized as follows: Find $p_n \in \mathcal{C}(P)$, $p_n \nearrow g$ and put $s_n = \max(p_n - \frac{1}{n}, 0)$, $g_n = \max(s_n, \sum_{k=1}^{\infty} 2^{-k} s_k)$.

(b) G_δ-measurability and B_1-functions.

We know (cf. Theorem 2.12) that any Baire one function is G_δ-measurable. Now, we propose a proof of the converse assertion in metric spaces. Before doing so we summarize the following lemmas.

3.5 LEMMA. The family $B_1(P)$ of all Baire one functions is closed under uniform convergence.

Proof. Obviously,

$$B_1(P) \subset \mathcal{C}^{\uparrow\downarrow}(P) \cap \mathcal{C}^{\downarrow\uparrow}(P).$$

Let now $s \in \mathcal{C}^{\uparrow}(P)$, $t \in \mathcal{C}^{\downarrow}(P)$ be bounded, $t \leqq s$. It can be easily proved that the set $[s \leqq \lambda]$ is closed for each $\lambda \in \mathbb{R}$. Thus, if $a < b$, the sets $[s \leqq a]$, $[t \geqq b]$ are disjoint and closed, hence they are $\mathcal{C}(P)$-separated. The abstract in-between theorem guarantees the existence of $f \in \mathcal{C}(P)$ such that $t \leqq f \leqq s$. It follows easily that

$$B_1(P) = \mathcal{C}^{\uparrow\downarrow}(P) \cap \mathcal{C}^{\downarrow\uparrow}(P).$$

(Indeed, let $h \in \mathcal{C}^{\uparrow\downarrow}(P) \cap \mathcal{C}^{\downarrow\uparrow}(P)$. There are $g_n \in \mathcal{C}^{\downarrow}(P)$, $f_n \in \mathcal{C}^{\uparrow}(P)$ such that $g_n \nearrow h$, $f_n \searrow h$. For any $n \in \mathbb{N}$ choose $h_n \in \mathcal{C}(P)$ for which $g_n \leqq h_n \leqq f_n$ as above. Then, obviously, $h_n \to h$.) Applying Lemma 3.3 we can see that $B_1(P)$ is closed under uniform convergence.

REMARK. The "classical" proof of the previous lemma can be found in Ex. 3.G.2 (cf. [Haus] , p.237).

3.6 LEMMA. Let G be a G_δ-subset of a metric space P. Then there exists $f \in \mathcal{C}^{\downarrow}(X)$ such that $0 \leqq f \leqq 1$ and $G = [f = 0]$.

Proof. Let $G = \bigcap_{n=1}^{\infty} G_n$ where G_n are open. Put

$$f = 1 - \sum_{n=1}^{\infty} 2^{-n} \chi_{G_n}.$$

In the course of the proof of Proposition 3.4 we proved that $\chi_{G_n} \in \mathcal{C}^{\uparrow}(P)$. According to Lemma 3.3 we have $f \in \mathcal{C}^{\downarrow}(P)$.

3.7 PROPOSITION. Let $t \leqq s$ be functions on a metric space P. The following assertions are equivalent:

(i) there is a Baire one function f such that $t \leqq f \leqq s$ on P,

(ii) given $a<b$, the sets $[s \leqq a]$, $[t \geqq b]$ can be separated by G_δ-sets.

Proof. Obviously, (i)$\Longrightarrow$(ii). For (ii)$\Longrightarrow$(i) we use again the abstract in-between theorem. We may assume that $0 \leqq t \leqq s \leqq 1$ on P. In light of Lemma 3.5, the family $B_1(P)$ is a closed L-cone. Let $a < b$. There are disjoint G_δ-sets A, B such that $[s \leqq a] \subset A$, $[t \geqq b] \subset B$ (examine also the trivial cases when a, b are not in $[0,1]$!). Lemma 3.6 then guarantees the existence of $f, g \in \mathcal{C}^{\downarrow}(P)$ such that $0 \leqq f, g \leqq 1$ and $A = [f = 0]$, $B = [g = 0]$. Put $\varphi = \frac{f}{f+g}$. Obviously, $\varphi \in B_1(P)$ and $\varphi = 0$ on $[s \leqq a]$, $\varphi = 1$ on $[t \geqq b]$, which completes the proof.

3.8 COROLLARY. (a) Every G_δ-measurable function on a metric space belongs to the Baire class one.

(b) Let g be a function on a subset Y of a metric space P. Then there is a Baire one function f on P such that $g = f \wedge Y$ if and only if the sets $\{x \in Y: g(x) \geqq b\}$, $\{x \in Y: g(x) \leqq a\}$ are separated by G_δ-sets for every couple $a < b$ of real numbers.

In particular, any Baire one function defined on a G_δ-subset of a metric space can be extended to a Baire one function on the whole space.

Proof. (a) Let f be G_δ-measurable. Put $s = t = f$ in Proposition 3.7.

(b) Use Proposition 3.7 where $s = t = g$ on Y, $s = +\infty$ and $t = -\infty$ on $P \setminus Y$.

(c) Uniform approximation of approximately continuous functions.

It is known that a real function f on $\mathbb{R}$ is a Baire one function if and only if f is a uniform limit of differences of lower semicontinuous functions (cf. Remarks and comments to 3.G). Our aim is to show that an analogous result holds for the class of all (bounded) approximately continuous functions. Remark that a more general result (which avoids, among others, the condition of boundedness) will be considered in Section 3.G.

We need the following simple fact which we derive in later sections: Given a G_δ-set $A \subset \mathbb{R}$ which is density closed, there is an approximately continuous and upper semicontinuous function h on $\mathbb{R}$ such that $0 \leqq h \leqq 1$ and $A = [h = 0]$ (cf. Zahorski property in Corollary 3.14).

3.9 PROPOSITION. Let f be a bounded approximately continuous function on $\mathbb{R}$. Then f can be uniformly approximated by differences of non-negative, approximately continuous and lower semicontinuous functions.

Proof. As usual, we apply the abstract in-between theorem putting

$\mathcal{L} = \{h - g:$ h, g are non-negative, approximately continuous and lower semicontinuous functions on $\mathbb{R}\}$,

$\mathcal{F}$ = uniform limits of sequences from

and $s = t = f$.

Since $\mathcal{L}$ is an algebra, $\mathcal{F}$ is a closed vector lattice (cf. [Will], § 44). Let $a < b$. Then the sets $[f \leqq a]$, $[f \geqq b]$ are density closed G_δ-sets. As remarked above, there are approximately continuous and upper semicontinuous functions φ, ψ such that

$$0 \leqq \varphi, \psi \leqq 1, \quad [f \leqq a] = Z(\varphi), \quad [f \geqq b] = Z(\psi).$$

Put

$$h = \frac{1}{\varphi+\psi}, \quad g = \frac{1-\varphi}{\varphi+\psi}.$$

Then $h - g \in \mathcal{L}$, $[f \leqq a] = Z(h - g)$ and $h - g = 1$ on $[f \geqq b]$. Hence the function $h - g$ is the one we looked for. Now we apply Theorem 3.2.

We close this section by giving another in-between theorem the proof of which is based on Tietze's type procedure. Now, instead of a lattice of functions we consider a vector space of functions.

3.10 THEOREM. Let $\mathcal{F}$ be a closed vector space of functions on a set X containing the constants. Let s, $-t$ be lower bounded functions on X, $t \leqq s$. If for each $\varphi \in \mathcal{F}$ and for each $a < b$ $(a, b \in \mathbb{R})$ the sets $[s - \varphi \leqq a]$ and $[t - \varphi \geqq b]$ are $\mathcal{F}$-separated, then there is $g \in \mathcal{F}$ such that $t \leqq g \leqq s$.

Proof. It may be assumed without loss of generality that $s \geqq -1$ and $t \leqq 1$. Using the extension procedure of Tietze's type we can define inductively a sequence of functions $\{g_n\} \subset \mathcal{F}$ such that

$$|g_n| \leqq \frac{1}{3}\left(\frac{2}{3}\right)^{n-1} \text{ on } X,$$

$$s - \sum_{j=1}^{n} g_j \geqq -\left(\frac{2}{3}\right)^n, \quad t - \sum_{j=1}^{n} g_j \leqq \left(\frac{2}{3}\right)^n.$$

Indeed, let $g_1 \in \mathcal{F}$, $|g_1| \leqq \frac{1}{3}$ and

$$g_1 = \frac{-1}{3} \text{ on } [s \leqq \tfrac{-1}{3}], \quad g_1 = \frac{1}{3} \text{ on } [t \geqq \tfrac{1}{3}].$$

Then, obviously,

$$s - g_1 \geqq -\frac{2}{3}, \; t - g_1 \leqq \frac{2}{3}.$$

Further, there is $g_2 \in \mathcal{F}$ such that $|g_2| \leqq \frac{2}{9}$ and

$$g_2 = -\frac{2}{9} \text{ on } [s + g_1 \leqq -\frac{2}{9}], \quad g_2 = \frac{2}{9} \text{ on } [t + g_1 \geqq \frac{2}{9}].$$

Then again

$$s - g_1 - g_2 \geqq -\frac{4}{9}, \quad t - g_1 - g_2 \leqq \frac{4}{9}$$

and we can (inductively) continue.

The function

$$g = \sum_{j=1}^{\infty} g_j$$

has all the properties prescribed to it.

REMARK. It is not difficult to realize that Propositions 3.7 and 3.9 can also be derived with the help of the preceding Theorem 3.10.

EXERCISES

3.A.1 Let f be a function on a topological space X. Consider the following conditions:

(B_1) f is a Baire one function,

(Coz_δ) f is Coz_δ-measurable,

(G_δ) f is G_δ-measurable.

Then $(B_1) \Leftrightarrow (\mathrm{Coz}_\delta) \Rightarrow (G_\delta)$; if X is normal then all these properties are equivalent.

Hint. The proof of $(B_1) \Rightarrow (\mathrm{Coz}_\delta)$ is routine, and (Coz_δ) implies (G_δ) obviously. To prove $(\mathrm{Coz}_\delta) \Rightarrow (B_1)$ proceed as above performing the following steps:

(a) Prove that $B_1(X)$ is closed (cf. Lemma 3.5 or Ex. 3.G.2).

(b) Any Coz_δ-subset of X is a zero set of a Baire one function (cf. Lemma 3.6).

(c) Show that A is a $B_1(X)$ - separated from B is and only if A and B are separated by Coz_δ -sets.

(d) Apply the abstract in-between theorem to the case of f being bounded putting $s = t = f$ (cf. Proposition 3.7).

Assume now that X is normal and let f be G_δ-measurable on X. Since

$$[f \leqq a] = \bigcap_{n=1}^{\infty} [f < a + \tfrac{1}{n}] = \bigcap_{n=1}^{\infty} [f \leqq a + \tfrac{1}{n}],$$

it suffices to find a Coz_δ-set G_n such that

$$[f < a + \tfrac{1}{n}] \subset G_n \subset [f \leqq a + \tfrac{1}{n}].$$

It can be easily realized as follows: Given disjoint F_σ-sets A, B in a normal space X there is always a Coz_δ-set G such that $A \subset G \subset X \setminus B$. Indeed if $A = \bigcup_{n=1}^{\infty} F_n$, F_n and B being closed, then obviously

$$A \subset [\sum 2^{-n} f_n > 0] \subset X \setminus B,$$

where $f_n \in \mathcal{C}(X)$, $0 \leqq f_n \leqq 1$, $f_n = 0$ on B and $f_n = 1$ on F_n.

3.A.2 Prove an analogy to Proposition 3.7: If $t \leqq s$ are functions on a topological space X then there is a Baire one function f such that $t \leqq f \leqq s$ on X if and only if the sets $[s \leqq a]$, $[t \geqq b]$ can be separated by Coz_δ -sets whenever $a < b$.

Hint. Cf. with the preceding exercise, (c).

Similarly, if g is a function on $Y \subset X$, then g is the restriction of a Baire one function on X if and only if the sets $\{x \in Y: g(x) \leqq a\}$, $\{x \in Y: g(x) \geqq b\}$ are Coz_δ-separated for any $a, b \in \mathbb{R}$, $a < b$.

3.A.3 Let X be a perfect topological space and let g be a function on a set $Y \subset X$. The following assertions are equivalent:

(i) g can be extended to a G_δ-measurable function on X,

(ii) the sets $\{x \in Y: g(x) \leqq a\}$, $\{x \in Y: g(x) \geqq b\}$ are separated by G_δ-sets for any $a, b \in \mathbb{R}$, $a < b$.

Hint. Use the abstract in-between theorem combined with Lemma 3.26 (for $\mathcal{T}_1$, $\mathcal{T}_2$ discrete) or with the fact that any disjoint G_δ-sets can be separated by an ambivalent set (cf. W. Sierpiński (1924) or [Kur] for metric spaces; see also Ex. 3.B.7b) considering its characteristic function.

3.A.4 If $t \leqq s$ are functions on a metric space P, then there is a Baire two function f on P such that $t \leqq f \leqq s$ if and only if the sets $[s \leqq a]$, $[t \geqq b]$ can be separated by $F_{\sigma\delta}$ -sets for all $a < b$.

Deduce the analogy of Corollaries 3.8.a and 3.8.b to this case as well as the analogy to Ex. 3.A.1, 3.A.2 and 3.A.3.

3.A.5 In-between theorem for closed lattices. Let $\mathcal{F}$ be a closed family of bounded functions on a set X which satisfies the following conditions:

($\mathcal{L}$1) $\mathcal{F}$ is a lattice,

($\mathcal{L}$2) $\mathcal{F}$ contains the constant functions,

($\mathcal{L}$3) if $f \in \mathcal{F}$ and k is a constant function on X, then $f + k \in \mathcal{F}$.

(a) Prove that $\mathcal{F}$ is a convex set.

(b) Let s, $-t$ be lower bounded, $t \leqq s$. If for any couple $a < b$ $(a,b \in \mathbb{R})$ there is $f_a^b \in \mathcal{F}$ such that $a \leqq f_a^b \leqq b$, $f_a^b = a$ on $[s \leqq a]$ and $f_a^b = b$ on $[t \geqq b]$, then there is $f \in \mathcal{F}$ such that $t \leqq f \leqq s$.

Hint. (a) Let $f, g \in \mathcal{F}$. It suffices to prove that $\frac{1}{2}(f + g) \in \mathcal{F}$. But

$$\frac{1}{2}(f + g) = \sup_{c \in \mathbb{R}} (\min(f+c, g-c))$$

and this sup can be uniformly approximated by finite maxima.

(b) Assume that s,t are bounded, $s(x)$, $t(x) \in [-p,p]$ for each $x \in X$. Let $\varepsilon > 0$ and let E be a finite subset of $[-p,p]$ with $\sup\{\operatorname{dist}(c,E) : c \in [-p,p]\} < \varepsilon/2$.

Put

$$f_E = \min_{a \in E} \left(\max_{b \in E,\, b > a} f_a^b\right).$$

Then $f_E \in \mathcal{F}$ and

$t-\varepsilon \leqq f_E \leqq s+\varepsilon$ on X.

Now, the proof of Lemma 3.1 works and yields the assertion.

3.A.6 A family $\mathfrak{F}$ of bounded functions is a closed L-cone if and only if satisfies $(\mathcal{L}1)$, $(\mathcal{L}2)$, $(\mathcal{L}3)$ of the preceding exercise and

$(\mathcal{L}4)$ $\mathfrak{F}$ is closed under multiplication by non-negative reals.

Hint. Use Ex. 3.A.5.a.

3.A.7 Let $\mathfrak{F}$ be a closed L-cone on X. Let $f \in \mathfrak{F}$ be bounded and let g be an increasing continuous function on a closed interval containing $f(X)$. Then $g*f \in \mathfrak{F}$.

Hint. Use the abstract in-between theorem for $s = t = g*f$.

3.A.8 Let $\mathfrak{F}$ be a lattice of functions on a set X containing the constants.

(a) If $t \leqq s$ is a couple of bounded functions on X such that $[s \leqq a]$ is $\mathfrak{F}$-separated from $[t \geqq b]$ for each a,b, $a < b$, and if $\varepsilon > 0$ is given, then there is $f \in \mathfrak{F}$ such that

$$t-\varepsilon \leqq f \leqq s+\varepsilon$$

on X.

Hint. Repeat the proof of Theorem 3.2 using $(\mathcal{L}1)$ and $(\mathcal{L}2)$ only

(b) Show that the conclusion (ii) $\Rightarrow$ (i) of the abstract in-between theorem in general does not hold.

Hint. Put $\mathfrak{F}$ to be the family of all uniformly continuous functions f on $(0,2)$ such that:

α) $f \leqq 0$ provided $\lim\limits_{y\to 0_+} f(y) \leqq 0$,

β) $f(1) = \sup f$,

γ) $f(x) \leqq x-1$ for each $x \geqq 1 + \lim\limits_{y\to 0_+} f(y)$.

Consider now $t(x) = x-1$, $s(x) = x$.

3.A.9 Let $\mathcal{F}$ be the family of all functions f on a metric space (P,ρ) such that $|f(x)-f(y)| \leqq \rho(x,y)$ for each $x,y \in P$. Show that $\mathcal{F}$ satisfies $(\mathcal{L}1)$, $(\mathcal{L}2)$ and $(\mathcal{L}3)$ of Ex. 3.A.3. On the other hand, $(\mathcal{L}4)$ of Ex. 3.A.4 is not satisfied.

Prove that, given functions $t \leqq s$, there is $f \in \mathcal{F}$ such that $t \leqq f \leqq s$ if and only if $\operatorname{dist}([s \leqq a], [t \geqq b]) \geqq b-a$ whenever $a < b$.

Hint. Put $f(x) = \inf\{s(y) + \rho(x,y) : y \in P\}$. You can use also Ex. 3.A.5.b.

REMARKS AND COMMENTS

As pointed out in the introductory remarks of this section, one of the oldest and most important in-between theorems was proved by H.Hahn (1917) for the case of metric spaces consisting in the insertion of a continuous function between two functions which are semicontinuous. This theorem was later generalized by J. Dieudonné (1944) for paracompact spaces, by H. Tong (1948), (1952) and M. Katětov (1951), (1953) for the case of normal spaces, and nowadays is known as the Tong-Katětov characterization of normal spaces.

Up to now, a number of papers on in-between theorems have been published in which the original Hahn's version is generalized in various ways. Recall for instance the characterization of countably paracompact normal spaces (M. Katětov (1951), C.H. Dowker (1951)), of perfectly normal spaces (E. Michael (1956)), or extremaly disconnected spaces (M.H. Stone (1949)). The general in-between theorems for continuous functions and the characterizations of different variants of normal spaces have been dealt with by E.P. Lane in a series of papers (1971), (1975), (1976), (1979), (1981), (1983) and by a number of other authors.

Among papers published lately and related to our results let us mention the following ones: J. Blatter and G.L.Seever (1975), (1976) (very general in-between theorems for closed L-cones of functions, also immediately yielding our in-between theorem though provided with a more complicated proof), D. Preiss and J. Vilímovský (1980) (mostly insertion of uniformly continuous functions), or J. Hoffmann-Jørgensen (1982) (insertion of functions from a very general function system though under relatively complicated assumptions).

In our text, much thought is given to applications of in-between theorems to extension theorem, especially to the extension theorems of derivatives. For this purpose we first derive the so-called B_1-extension theorem 3.29. The proof of this theorem (or, more precisely, of its consequence concerning the extension of Baire one functions from G_δ - sets of measure zero resulting in approximately continuous functions) given by D. Preiss (Conference held in Dědinky, 1972), M. Laczkovich and G. Petruska (1973) or by J. Lukeš (1977) in a more general context used Tietze's method.

Tietze's approach is still sufficient for extension theorems with control (cf. e.g. Theorem 3.10). The use thereof for extension preserving the semicontinuity was no more possible as the differences of two semicontinuous functions need not be semicontinuous any longer. Thus we were lead, in fact, to our abstract in-between Theorem 3.2 and to its consequence in the form of the asymmetrical in-between Theorem 3.27.

As already mentioned, our abstract in-between theorem is essentially a special case of the general theorems from J. Blatter and G. L. Seever (1976); our proof is quite elementary, indeed, and we think it to be applicable, together with Corollaries 3.3-3.9, to basic courses of analysis.

The result in Ex. 3.A.1 on characterization of Baire one functions in normal topological spaces as those which are G_δ-measurable was proved by M. Laczkovich in his thesis and was mentioned without proof in M. Laczkovich (1983).

Related papers: M. Powderly (1981), J. M. Boyle and E. P. Lane (1975).

3.B The Lusin-Menchoff-property

Even though the normality of fine topologies is missing in the most important examples, it can be replaced in many considerations by the so called Lusin-Menchoff property. We shall see that the Lusin-Menchoff property leads to interesting applications not only in the density topology or in the fine topology in potential theory. It can be said without exaggeration that the Lusin-Menchoff property plays the key role in our study of fine topologies.

3.11 THEOREM. Let τ be a fine topology on a topological space (X,ρ). Then the following assertions are equivalent:

(i) For each pair of disjoint subsets F and F_τ of X, F closed, F_τ τ-closed, there are $G, G_\tau \subset X$, G open, G_τ τ-open such that

$$F_\tau \subset G, \quad F \subset G_\tau, \quad G \cap G_\tau = \emptyset.$$

(ii) For any couple $F_\tau \subset U$ of X, F_τ τ-closed, U open, there is an open set G such that

$$F_\tau \subset G \subset \bar{G}^\tau \subset U$$

(iii) For each couple $F \subset U_\tau$, F closed, U_τ τ-open, there is a τ-open set G_τ such that

$$F \subset G_\tau \subset \overline{G}_\tau \subset U_\tau .$$

(iv) Given any pair of disjoint subsets F, F_τ of X, F closed, F_τ τ-closed, there is a τ-continuous and upper semicontinuous function h on X such that

$$0 \leqq h \leqq 1, \quad h = 0 \text{ on } F_\tau, \quad h = 1 \text{ on } F.$$

(v) Given any pair of functions t, s, $t \leqq s$, t upper semicontinuous and s lower τ-semicontinuous on X, there is a τ-continuous and upper semicontinuous function f on X such that $t \leqq f \leqq s$.

(vi) Given a τ-continuous and upper semicontinuous function h on a τ-closed set $A \subset X$, there is a τ-continuous and upper semicontinuous function h^* on X such that $h = h^*$ on A (moreover, h^* can be found in such a way that

$$\sup_A h = \sup_X h^*, \quad \inf_A h = \inf_X h^*).$$

(vii) Given sets $F, F_\tau, A, A_\tau \subset X$, F closed, F_τ τ-closed, A of type G_δ, A_τ of type G_δ in τ, $F_\tau \subset A$, $F \subset A_\tau$, there are τ-continuous and upper semicontinuous functions f, g on X such that $f \geqq 0$, $g \leqq 0$,

$$F_\tau \subset Z(f) \subset A, \quad F \subset Z(g) \subset A_\tau$$

(recall that $Z(h) = [h = 0]$).

(viii) for each pair of sets P, Q, P of type F_σ in ρ, Q of type F_σ in τ, fulfilling $\overline{P} \cap Q = \emptyset = P \cap \overline{Q}^\tau$, there are sets G, U, G open, U τ-open, such that

$$P \subset U, \quad Q \subset G, \quad G \cap U = \emptyset .$$

Proof. The conditions (i), (ii) and (iii) are obviously equivalent.

(ii) $\Rightarrow$ (iv): Let F_τ be τ-closed, U be open and $F_\tau \subset U$. Imitating the proof of well known Urysohn's lemma ([Kel] , Chap. IV, Lemma 4), we associate by induction with any dyadic rational number $r \in [0,1]$ an open set $W(r)$ such that $W(1) = U$, $W(0)$ is an open set containing F_τ whose τ-closure is in U, and if r, s are dyadic rationals of $[0,1]$, $r < s$, then

$$W(r) \subset \overline{W(r)}^{\tau} \subset W(s).$$

Put

$$h(x) = \inf \{r \in [0,1], x \in W(r)\} .$$

Since

$$[h < c] = \bigcup \{W(r): r < c\} ,$$

$$[h \leqq c] = \bigcap \{W(r): r > c\} ,$$

h is upper semicontinuous and τ-continuous on X. Obviously, $h = 0$ on F_{τ} , $h = 1$ on $X \setminus U$ and $0 \leqq h \leqq 1$ on X.

(iv)$\Longrightarrow$(v): Without loss of generality we can assume that the values of t and s are contained in $[0,1]$. Taking $\mathfrak{F}$ to be the family of all τ-continuous and upper semicontinuous functions on X it is easy to check that we can apply Theorem 3.2.

(v)$\Longrightarrow$(vi): If h is non-constant, we reduce again to the case where

$$\inf_A h = 0 , \quad \sup_A h = 1.$$

Put

$$s = \begin{cases} h & \text{on } A \\ 1 & \text{on } X \setminus A, \end{cases}$$

$$t(x) = \begin{cases} h(x) & \text{for } x \in A \\ \limsup\limits_{y \to x, y \in A} h(y) & \text{for } x \in \bar{A} \setminus A \\ 0 & \text{for } x \in X \setminus \bar{A}. \end{cases}$$

Then s is lower τ-semicontinuous, t is upper semicontinuous on X and $t \leqq s$. In light of (v) there is a τ-continuous and upper semicontinuous function h^* on X, $t \leqq h^* \leqq s$. Obviously, h^* is the desired extension of h.

(vi)$\Longrightarrow$(vii): The set $X \setminus A$ is ot the type F_{σ} , i.e. there are closed sets F_n such that $X \setminus A = \bigcup\limits_{n=1}^{\infty} F_n$. Let f_n be a τ-continuous and upper semicontinuous function,

$$0 \leqq f_n \leqq 1, \; f_n = 0 \text{ on } F, \; f_n = 1 \text{ on } F_n.$$

Put

$$f = \sum_{n=1}^{\infty} 2^{-n} f_n .$$

Then f is τ-continuous and upper semicontinuous, $0 \leqq f \leqq 1$ and $F \subset Z(f) \subset A$. Analogously we can construct the function g.

(vii) $\Rightarrow$ (viii): Using (vii), there are τ-continuous and upper semicontinuous functions f,g, $f \geqq 0$, $g \leqq 0$ such that

$$Q \subset Z(f) \subset X \setminus P, \quad P \subset Z(g) \subset X \setminus Q.$$

It is sufficient to put

$$G = [f + g < 0], \qquad U = [f + g > 0].$$

(viii) $\Rightarrow$ (i). Obvious.

REMARK. The question of when, under the assumption $t < s$ in (v), even a function f can be chosen with $t < f < s$, or of whether a real function h in (vi) has even a real extension will be discussed in exercises. As for negative results, see Ex. 3.B.10 and 3.B.11. Positive results appear in Ex. 3.E.9 and 3.E.10.

We say that the fine topology τ on (X,ρ) has the *Lusin-Menchoff property* (with respect to ρ) if the condition (i) of Theorem 3.11 is satisfied. Of course, if $\tau = \rho$ then the Lusin-Menchoff property is equivalent to the normality of the space (X,ρ).

Generally, a topology with the Lusin-Menchoff property, even in the case ρ being the Euclidean topology, need not be normal.

If a set X is equipped with two topologies τ and ρ (not necessarily comparable) then the "Lusin-Menchoff property" of the topology τ with respect to the topology ρ means nothing else as the pairwise normality (or binormality) of the bitopological space (X,ρ,τ) introduced by J. C. Kelly (1963) (cf. Ex. 3.B.5).

3.12 REMARK. If (X,ρ) is a metric space, then each closed set is a zero set of a continuous function and therefore the Lusin-Menchoff property of τ turns to be equivalent with the condition:

(vii*): For any τ-closed set F_τ and for each set A of type G_δ, $F_\tau \subset A$, there is a τ-continuous and upper semicontinuous function f, $f \geqq 0$, such that $F_\tau \subset Z(f) \subset A$.

3.13 COROLLARY. Every fine topology with the Lusin-Menchoff property on a T_1-space is completely regular.

3.14 COROLLARY (*Zahorski property* of τ). If τ has the Lusin-Menchoff property, then every τ-closed set of the type G_δ is a zero set of a τ-continuous and upper semicontinuous function on X.

3.15 REMARK. Assume that the fine topology τ on X has the Lusin-Menchoff property with respect to a T_1- topology ρ. Putting

$$\Phi = \{f: f \geqq 0, \ f \text{ is } \tau\text{-continuous and lower semicontinuous on } X\},$$

we can easily check that Φ is a convex cone on X. The coarsest topology τ_Φ on X finer than ρ making all functions from Φ continuous is exactly τ. Indeed, let U be τ-open, $x \in U$. There is a function f, $0 \leqq f \leqq 1$, $1-f \in \Phi$, such that $f(x) = 1$, $X \setminus U \subset f^{-1}(0)$. Then

$$x \in [1 - f < \tfrac{1}{2}] \subset U.$$

On the other hand, τ is obviously finer than τ_Φ. Thus, we can apply the results of Section 2.B. In particular, we obtain immediately the following proposition.

3.16 PROPOSITION. If τ has the Lusin-Menchoff property with respect to a locally compact T_1 space (X,ρ), then (X,τ) is a Baire space.

It is well-known that a subspace of a normal topological space need not be normal. If every subspace of a space X is normal, X is said to be *completely normal* (or *hereditarily normal*).

Analogously, we say that a fine topology has the *complete Lusin-Menchoff property* with respect to a topological space (X,ρ) if it has the Lusin-Menchoff property on every subspace of X.

3.17 REMARK. Notice that the discrete topology has the complete Lusin-Menchoff property with respect to any topology on X.

3.18 THEOREM. The following conditions are equivalent:

(i) τ has the complete Lusin-Menchoff property with respect to (X,ρ),

(ii) τ has the Lusin-Menchoff property on every τ-open subset of X,

(iii) whenever A and B are biseparated subsets of X (i.e. $\bar{A} \cap B = \emptyset = A \cap \cap \bar{B}^{\tau}$), then there are disjoint τ-open set $U \supset A$ and open set $G \supset B$.

Proof. (i)$\Rightarrow$(ii) : Obvious.

(ii)$\Rightarrow$(iii) : Let $M := X \setminus (\bar{A} \cap \bar{B}^{\tau})$. Then M is τ-open in X and $A, B \subset M$. Since $\bar{A} \cap \bar{B}^{\tau} \cap M = \emptyset$, according to the Lusin-Menchoff property of τ on M, there exist a τ-open set U (in M) and an open set G^* (in M) such that

$$U \cap G^* = \emptyset , \; A \subset U \subset M, \quad B \subset G^* \subset M.$$

Clearly, U is τ-open in X, and there is a set G open in X such that $G \cap M = G^*$. Then $U \cap G = \emptyset$ and $B \subset G$, $A \subset U$.

(iii)$\Rightarrow$(i) : Let M be a subset of X and $A, B \subset M$ be disjoint, A closed in M, B τ-closed in M. Of course, $\bar{A} \cap B = \emptyset = A \cap \bar{B}^{\tau}$.
Therefore there are disjoint τ-open set $U \supset A$ and open set $G \supset B$. Now, $M \cap U$ is τ-open in M, $M \cap G$ is open in M, $M \cap U \supset A$, $M \cap G \supset B$ and $(M \cap U) \cap (M \cap G) = \emptyset$.

3.19 THEOREM. Assume that a fine topology τ has the Lusin-Menchoff property with respect to a metrizable topological space (P, ρ). If for each set M there is a set $b(M)$ of the type G_δ such that $\bar{M}^{\tau} = M \cup b(M)$, then τ has the complete Lusin-Menchoff property.

Proof. Let $A \subset P$ be an arbitrary set. Let F_τ be a τ-closed set in P, F closed, $F_\tau \cap F \cap A = \emptyset$. Then $M := F \setminus b(F_\tau \setminus F)$ is a F_σ-set, and

$$F \setminus F_\tau \subset F \setminus \overline{F_\tau \setminus F}^{\tau} \subset F \setminus b(F_\tau \setminus F) = M.$$

Since

$$\overline{F_\tau \setminus F}^{\tau} = (F_\tau \setminus F) \cup b(F_\tau \setminus F) \subset P \setminus M,$$

there is a τ-continuous and upper semicontinuous function g on P such that

$$0 \leqq g \leqq 1 , \quad \overline{F_\tau \setminus F}^{\tau} \subset Z(g) \subset P \setminus M$$

(cf. Remark 3.12). Let $f \geqq 0$ be a continuous function on P, $F = Z(f)$. Then for $x \in F_\tau \cap A \subset F_\tau \setminus F$ we have $g(x) - f(x) < 0$ and for $y \in F \cap A \subset F \setminus F_\tau \subset M$ we have $g(y) - f(y) > 0$. It is enough to put

$$U := [g - f > 0], \quad G := [g - f < 0] .$$

Obviously, U is τ-open, G is open, $U \cap A \cap G = \emptyset$, $F_\tau \cap A \subset G \cap A$, $F \cap A \subset U \cap A$.

EXERCISES

3.B.1 Heredity of the Lusin-Menchoff property.

(a) Prove that the Lusin-Menchoff property is hereditary with respect to F_σ -subsets or with respect to finely closed sets.

Hint. Use Theorem 3.11.viii.

(b) Show that the following assertions are equivalent:

(i) τ has the complete Lusin-Menchoff property on (X,ρ),

(ii) if F is closed and F_τ τ-closed, then there are a closed set F^* and a τ-closed set F_τ^* such that $F^* \cup F_\tau^* = X$, $F^* \cap (F \cup F_\tau) = F$, $F_\tau^* \cap (F \cup F_\tau) = F_\tau$ (cf. [Kur], § 14. V, Th. 2).

(iii) given a τ-open set W and $A \subset W$, there is a τ-open set U such that $A \subset U \subset \bar{U} \subset W \cup \bar{A}$.

Hint. (i)$\Rightarrow$(ii). The sets $F \setminus F_\tau$ and $F_\tau \setminus F$ are biseparated. By Theorem 3.18.iii there are disjoint sets G_τ and G, $G_\tau \supset F \setminus F_\tau$, $G \supset F_\tau \setminus F$. Put $F^* = F \cup (X \setminus G)$, $F_\tau^* = F_\tau \cup (X \setminus G_\tau)$.

(ii)$\Rightarrow$(i). For $A, B \subset X$ biseparated, apply (ii) in case of $F = \bar{A}$, $F_\tau = \bar{B}^\tau$.

(c) If the assumptions of Theorem 3.19 are satisfied and if F is closed, F_τ τ-closed, then there is a τ-continuous and upper semicontinuous function f such that

$$f > 0 \text{ on } F \setminus (F \cap F_\tau), \quad f < 0 \text{ on } F_\tau \setminus (F \cap F_\tau).$$

Hint. Examine the proof of Theorem 3.19.

(d) Show that the conclusion of (c) is not true in case of the density topology d on $\mathbb{R}$ although d has the complete Lusin-Menchoff property (cf. 6.34.B).

Hint. Let F be the Cantor discontinuum and let

$$[0,1] \setminus F = \bigcup_{i=1}^{\infty} (a_i, b_i).$$

Making use of Lemma 2 of J. Malý (1979a) we can show that there is a d-closed set F_τ such that

$$F_\tau \subset \bigcup_{i=1}^{\infty} [a_i, b_i] \quad \text{and} \quad \mathrm{der}_d F_\tau \supset \bigcup_{i=1}^{\infty} \{a_i, b_i\}.$$

Assuming that a function f has the required properties, we get that

$$F \setminus \bigcup_{i=1}^{\infty} \{a_i, b_i\} = \{x \in F: \ f(x) > 0\} \text{ is an } F_\sigma\text{-set.}$$

(e) *Šverák's example.* Arbitrary subspaces of normal spaces need not be normal. Hence, there are fine topologies having the Lusin-Menchoff property but lacking the complete one. The question of whether or not there is a fine topology having this property with respect to a metric space has been answered by V. Šverák (unpublished).

By a *Lusin space* we mean an uncountable space such that every nowhere dense set is countable. It can be proved that the continuum hypothesis entails the existence of Lusin spaces of the real line (cf. [Kur], § 40.VII). A simple reasoning yields even a Lusin subspace L of $\mathbb{R}$ such that $I \cap L$ is uncountable for each interval $I \subset \mathbb{R}$. Analogously, under CH there is also a Lusin subspace M of the Cantor set C (cf. also E.K. van Douwen, F.D.Tall, and W.A.R. Weiss (1977)). Let further $S \subset C \setminus M$ be a countable dense subset of C. There is an (uncountable) family $\{N_x\}_{x \in M}$ of pairwise disjoint dense subsets of $L \setminus C$, $\bigcup_{x \in M} N_x = L \setminus C$.

Put $T_x := \{a_n^x: \ n \in \mathbb{N}\}$ where $\{a_n^x\}$ is a sequence in N_x converging to x.

Define the š-topology on $\mathbb{R}$ as follows: A set G is š-open if any point $x \in G \cap (S \cup M)$ has a (Euclidean) neighborhood V such that

$$V \subset G \ \text{ provided } \ x \in S,$$

$$V \cap T_x \subset G \ \text{ provided } \ x \in M.$$

(α) Show that š has the Lusin-Menchoff property (w.r.t. Euclidean topology).

Hint. Assume $F \subset G$ where F is closed and G is š-open. Then $F \setminus \mathrm{int}\, G$ is a closed set disjoint with S. Hence $F \setminus \mathrm{int}\, G$ is nowhere dense in C. It follows that $(F \setminus \mathrm{int}\, G) \cap M$ is countable, say $(F \setminus \mathrm{int}\, G) \cap M = \{x_n\}$. Find a sequence $\{V_n\}$ where V_n are neighborhoods of x_n for which diam $V_n \to 0$ and $V_n \cap T_{x_n} \subset G$. Since $\mathbb{R} \setminus (F \setminus \mathrm{int}\, G)$ is normal, there is an open set H such that $F \setminus \mathrm{int}\, G \subset \overline{H} \subset H \subset G$. Put $G' = H \cup F \cup \bigcup_{n=1}^{\infty} (V_n \cap T_{x_n})$ and prove that G' is š-open. Further $\overline{G'} \subset G' \cup \overline{H} \subset G$.

(β) Show that the $\check{s}$-topology has not the complete Lusin-Menchoff property.

Hint. Consider in the space $\mathbb{R}\setminus S$ a closed set $F = C\setminus S$ and a $\check{s}$-open subset $G = F\cup(L\setminus C)$. Assume, in order to obtain a contradiction, that there is a $\check{s}$-open set G' such that $F\subset G'\subset \overline{G'}\subset G$. But then $\overline{G'}\cap L$ is nowhere dense in L and uncountable.

3.B.2 Let τ have the Lusin-Menchoff property with respect to a completely normal topological space (X,ρ). If F is τ-closed and G open, $F\subset G$, then there is an open set $H\subset X$ such that

$$F\subset H\subset \bar{H}^{\tau}\subset G \quad\text{and}\quad \bar{H}\setminus G\subset \bar{F}\setminus G.$$

Hint. Put $Z:=\bar{F}\setminus G$ and find an open set U such that

$$\bar{F}\setminus Z\subset U\subset \bar{U}\setminus Z\subset G.$$

Then use the Lusin-Menchoff property for a couple $F\subset U$.

3.B.3 Let τ be a fine topology on a metric space (P,ρ). Show that τ has the Lusin-Menchoff property if and only if

(a) τ has the Zahorski property,

and

(b) for every τ-closed set F and every open set $G\supset F$ there is a τ-closed G_δ-set M such that $F\subset M\subset G$.

Hint. Use Remark 3.12.

3.B.4 Show that the Zahorski property does not imply the Lusin-Menchoff property, in general.

Hint. Let $N\subset[0,1]$, $\lambda_* N = 0$, $\lambda^* N = 1$. Let $\mathcal{I}$ be the σ-ideal of all subsets of N. Consider the ideal topology i corresponding to the density topology and to $\mathcal{I}$. Show that any i-closed G_δ-set is also density closed and therefore i has the Zahorski property. On the other hand, i has not the Lusin-Menchoff property since it is not regular (cf. Corollary 1.11).

3.B.5 <u>Binormality of σ-topologies.</u> A collection τ of subsets of a set X is called a <u>σ-topology</u> if $\emptyset$ and X belong to τ, and if all countable unions and finite intersections of elements of τ belong to τ. The couple (X,τ) is said to be a *σ-topological space.*

(a) Every topology is a σ-topology.

(b) The collections of all F_σ-subsets or, more generally, the families of all sets of additive class α (α ordinal) on a topological space form σ-topologies.

(c) Let Φ be a uniformly closed vector lattice of functions containing the constants (e.g. the family of all Baire one functions, the set of all continuous functions on a topological space or the set of all finely continuous Baire one functions). Then the collections $\mathrm{Coz}(\Phi)$ and $Z_\sigma(\Phi)$ of all Φ-cozero sets and of all Φ-zero$_\sigma$-sets, respectively, form σ-topologies.

(d) The collection of all quasi-open sets in a harmonic space forms a σ-topology, see ex. 14.A.1.

Subsets of X belonging to a σ-topology τ are called (τ-)open. The notions of closed sets, continuity, semicontinuity and a subspace with usual conventions are introduced as in topological spaces.

(e) Let a set X be equipped with two (generally non-comparable) σ-topologies τ and ϱ. Prove that the following conditions are equivalent.

Binormality condition: Whenever A and B are disjoint sets in X, A τ-closed, B ϱ-closed, there are disjoint sets G_A, G_B, G_A ϱ-open, G_B τ-open with $A\subset G_A$, $B\subset G_B$.

Urysohn's condition: Given A and B as in the binormality condition, there is a ϱ-upper semicontinuous and τ-lower semicontinuous function $f: X\to[0,1]$ with $f(A)= 0$ and $f(B) = 1$.

In-between condition: Given a couple $t\leqq s$ of functions, t ϱ-upper semicontinuous, s τ-lower semicontinuous on X, there is a ϱ-upper semicontinuous and τ-lower semicontinuous function h with $t\leqq h\leqq s$ on X.

Couple-in-between condition: Given a couple $t\leqq s$ of functions, t ϱ-upper semicontinuous, s τ-lower semicontinuous, there is a couple s^*, t^* of functions, s^* τ-lower semicontinuous, t^* ϱ-upper semicontinuous such that $t\leqq s^*\leqq t^*\leqq s$.

Hint. Imitate the proofs of Theorem 3.11. For the proof of the existence of Urysohn's functions from the binormality condition proceed as follows: By induction define chains $\{G^r_\tau\}$ and $\{G^r_\varrho\}$ of τ-open and ϱ-open subsets of X indexed by all dyadic rational numbers in $[0,1]$ such that

$$G^0_\varrho = G^1_\tau = \emptyset,\quad G^0_\tau = G^1_\varrho = X,\quad G^r_\tau\cap G^r_\varrho = \emptyset \text{ and}$$

$$G_\tau^r \cup G_\rho^s = X \quad \text{for} \quad s > r,$$

and look at the function $f(x) = \inf\{r: \quad x \in G_\rho^r\}$.

(f) If σ-topologies τ and ρ satisfy the binormality condition on X, and if $t \leqq s$ on X, then the following conditions are equivalent:

(i) there is a τ-upper semicontinuous and ρ-lower semicontinuous function f such that $t \leqq f \leqq s$,

(ii) for any couple $a < b$ there are a τ-closed set F and a ρ-closed set H such that $[t \geqq b] \subset F$, $[s \leqq a] \subset H$, $F \cap H = \emptyset$.

Hint. Use the abstract in-between theorem. Cf. also Gh. Bucur (1969).

REMARK. A space X on which two topologies τ and ρ are defined is called a *bitopological space*. The triple (X,τ,ρ) is said to be a *binormal* bitopological space if τ and ρ satisfy the binormality condition.

(g) Let τ be the σ-topology determined by the family of all F_σ-subsets of a metric space (P,ρ). Prove that τ and ρ satisfy the binormality condition.

(h) Prove that the σ-topological spaces of (c) (whose σ-topology τ is determined by the families $\mathrm{Coz}(\Phi)$ or by $Z_\delta(\Phi)$) are "normal" (i.e. τ and $\rho = \tau$ satisfy the binormality condition).

(j) Assume that σ-topologies τ and ρ satisfy the binormality condition on X. If $A \subset X$ is such a set that there are a ρ-closed set F_ρ and a τ-closed set F_τ with $A = F_\tau \cap F_\rho$, then any τ-lower semicontinuous and ρ-upper semicontinuous function f on A has an extension to all of X which is of the same type (*Tietze's condition*).

Hint. Use (f).

(k) If σ-topologies τ and ρ satisfy the binormality condition on X, then all τ-closed ρ-G_δ-sets are zero sets of a τ-lower semicontinuous and ρ-upper semicontinuous function (*Zahorski property*).

3.B.6 Density and Sorgenfrey's topologies on $\mathbb{R}$. We write e, e^+, d and d^+, respectively, for the Euclidean topology on $\mathbb{R}$, the Sorgenfrey topology on $\mathbb{R}$ (Ex.1.A.16), the density topology and the right density topology on $\mathbb{R}$. (Remark that the right density topology is determined by the family of all measurable subsets of $\mathbb{R}$ having each point as a right density point). The topologies e^- and d^- are defined analogously in an obvious way.

(a) The density topology d has the complete Lusin-Menchoff property with respect to e (cf. Theorem 6.9.g).

(b) Show that d^+ and d^- have the complete Lusin-Menchoff property w.r.t. e.

Hint. Suppose $\overline{A}^e \cap B = \emptyset = A \cap \overline{B}^{d^+}$. Taking into account Ex. 6.A.7 or 6.34.E there is an open set G such that $B \subset G \subset \mathbb{R} \setminus \bar{A}$ and $d^e(x, G \setminus B) = 0$ for each $x \in \bar{A}$. Obviously $A \cap \overline{G}^{d^+} = \emptyset$.

(c) Show that each couple of topologies e, e^+, e^- satisfies the binormality condition.

Hint. Consider e^+ and e^-. Pick an e^+-closed set F^+ and an e^--closed set $F^- \subset \mathbb{R} \setminus F^+$. Given $x \in F^-$, there is $r_x > 0$ such that $[x, x+r_x) \cap F^+ = \emptyset$. Analogously for $x \in F^+$.

Put

$$G^+ = \bigcup_{x \in F^-} [x, x+r_x/2), \quad G^- = \bigcup_{x \in F^+} (x-r_x/2, x]$$

and show that $G^+ \cap G^- = \emptyset$. The proof for the couples (e^+, e) and (e^-, e) is analogous (cf. also Examples of Section 10.B and Corollary 10.26).

(d) Show that (e^+, d) and (e^+, d^+) are not binormal.

Hint. Let $C \subset [0,1]$ be the Cantor set, and let $(a_n, b_n) \subset (0,1)$ be the component intervals of $[0,1] \setminus C$. Put

$$F^+ = C \setminus \bigcup_{n=1}^{\infty} \{a_n\} \quad \text{and} \quad F^d = \bigcup_{n=1}^{\infty} \{a_n\}.$$

Given a sequence $\varepsilon_n > 0$ show that the set

$$\overline{\bigcup_{n=1}^{\infty} [a_n, a_n + \varepsilon_n)}^{d^+}$$

is residual in C, and hence it is uncountable.

(e) Show that $(\mathbb{R}, e^+, d^-)$ is a binormal bitopological space.

Hint. Let F_e be e^+-closed, F_d d^--closed, $F_e \cap F_d = \emptyset$. Since $\mathbb{R} \setminus F_e$ is e^+-open, there is a countable family $\{(a_n, b_n)\}_{n \in I}$ of disjoint open intervals and $I_0 \subset I$ such that

$$\mathbb{R}\setminus F_e = \bigcup_{n\in I}(a_n,b_n) \cup \bigcup_{n\in I_0}\{a_n\}.$$

Put $H = \bigcup_{n\in I}(a_n,b_n)$ and find an open set T such that

$$F_d \cap H \subset T \subset H \quad \text{and} \quad \bar{T}^d \cap F_e = \emptyset .$$

Given $x \in F_d \setminus H$, we have $x = a_i$ for some $i \in I_0$. Denote $A_x := [a_i, a_i+(b_i-a_i)/i)$. Put

$$G = T \cup \bigcup_{x\in F_d\setminus H} A_x .$$

Then G is e^+-open, $F_d \subset G \subset \mathbb{R}\setminus F_e$, and it remains to show that $\bar{G}^{d^-} \cap F_e = \emptyset$.

3.B.7 Let $\mathcal{G}$ be a family of subsets of a set X closed under finite unions and intersections. Denote $\mathcal{F} = \{X\setminus G: G\in\mathcal{G}\}$. We say that $\mathcal{G}$ is *perfect* if $\mathcal{G}\subset\mathcal{F}_\sigma$ (or, equivalently, if $\mathcal{F}\subset\mathcal{G}_\delta$).

(a) Show that the family $\mathcal{F}_\sigma$ forms a σ-topology on X.

(b) If $\mathcal{G}$ is perfect, then $\mathcal{F}_\sigma$ is a perfect and normal σ-topology.

Hint. Assume that $A,B\in\mathcal{G}_\delta$, $A\cap B=\emptyset$. Let $X\setminus A = \bigcup F_n$, $X\setminus B = \bigcup F_n^*$, $F_n, F_n^* \in \mathcal{F}$. Let $\{Z_n\}$ be a sequence formed by all sets F_n and F_n^* and put

$$W_1 = Z_1,\quad W_n = Z_n \setminus \bigcup_{i=1}^{n-1} Z_i .$$

Denote $I = \{n\in\mathbb{N}: W_n\cap A \neq \emptyset\}$ and put $H_A = \bigcup_{n\in I} W_n$, $H_B = \bigcup_{n\notin I} W_n$. Show that $A\subset H_A$, $B\subset H_B$, $H_A\cap H_B = \emptyset$, $H_A\cup H_B = X$, $H_A, H_B \in \mathcal{F}_\sigma\cap\mathcal{G}_\delta$.

In what follows, (X,τ) will be a perfect topological space and $\alpha\geq 1$ will be a fixed countable ordinal. We denote by $\mathcal{G}_\alpha$ ($\mathcal{F}_\alpha$) the additive (multiplicative, resp.) family of all Borel sets of class α.

(c) Show that $\mathcal{G}_\alpha$ forms a perfect and normal σ-topology. The set of all functions continuous in this σ-topology is exactly the class of all Borel α-measurable functions.

Hint. Use transfinite induction and (b). For α being a limit ordinal, consider $\mathcal{G} = \bigcup_{\beta<\alpha}\mathcal{G}_\beta$.

(d) If $t \leq s$ are functions on X, then there is a Borel α-measurable function f such that $t \leq f \leq s$ if and only if for each couple $a < b$, $a,b \in \mathbb{R}$ there are $A,B \in \mathfrak{F}_\alpha$ such that

$$[s \leq a] \subset A, \qquad [t \geq b] \subset B, \qquad A \cap B = \emptyset.$$

Hint. Use (c) and Ex. 3.B.5.f. You can use also the Abstract in-between theorem directly together with the separation theorem of (b) (disjoint sets $A,B \in \mathfrak{F}_\alpha$ can be separated by the characteristic function of a set H_B).

(e) A function f on a set $C \subset X$ can be extended to the whole space X resulting in a Borel α-measurable function if and only if for each $a < b$, the sets $\{x \in C: f(x) \geq b\}$ and $\{x \in C: f(x) \leq a\}$ can be separated by sets of $\mathfrak{F}_\alpha$.

Hint. Use (d).

(f) If $C \in \mathfrak{F}_\alpha$ and if f is a Borel α-measurable set on C, then f can be extended to a Borel α-measurable function F on X. The function F can be chosen real provided f is real.

Hint. Use (e), or Tietze's condition of Ex. 3.B.5.j for $\tau = \varrho = \mathcal{U}_\alpha$. If f is real, put $F^* = \max(-G, \min(F,G))$ where $G \geq 0$ is such a function that $C = [G = +\infty]$ (cf. Zahorski property of Ex. 3.B.5.k).

3.B.8 Perfectly binormal spaces. Let σ_1 and σ_{-1} be σ-topologies on a set X. A triple $(X,\sigma_1, \sigma_{-1})$ is called *perfectly binormal* if σ_1 and σ_{-1} satisfy the binormality condition of Ex. 3.B.5.e and if each σ_i-closed set is a σ_{-i}-G_δ-set $(i = 1, -1)$.

(a) If $\tau = \sigma_1 = \sigma_{-1}$ is a topology, then $(X, \sigma_1, \sigma_{-1})$ is perfectly binormal if and only if the topological space (X,τ) is perfectly normal.

(b) Keep the notation of Ex. 3.B.6. Show that:

(b1) $(\mathbb{R},e,e^+)$ is perfectly binormal,

(b2) $(\mathbb{R},e^+e^-)$ is perfectly binormal,

(b3) $(\mathbb{R},e,d)$ is not perfectly binormal.

(c) The following assertions for σ-topologies σ_1 and σ_{-1} on X are equivalent:

(i) $(X, \sigma_1, \sigma_{-1})$ is perfectly binormal,

(ii) each σ_i-closed set is a zero set of a σ_i-lower semicontinuous and σ_{-i}-upper semicontinuous function $f \geq 0$ (N.B. Vedenisov (1936) and (1940) for topological spaces),

(iii) for every σ_i-lower semicontinuous function g on X, $0 \leq g \leq 1$, there exists a sequence $0 \leq g_n \leq 1$ of σ_i-lower semicontinuous and σ_{-i}-upper semicontinuous functions such that $g_n \nearrow g$ (H. Tong (1952) for topological spaces),

(iv) the same as (iii) and, moreover, g_n can be chosen such that $0 < g_n(x) < g(x)$ whenever $g(x) > 0$,

(v) for every lower finite, σ_i-lower semicontinuous function g on X there is a sequence $\{g_n\}$ of real σ_i-lower semicontinuous and σ_{-i}-upper semicontinuous functions such that $g_n < g$ and $g_n \nearrow g$,

(vi) if $t \leq s$ where s is a σ_i-lower semicontinuous function and t is a σ_{-i}-upper semicontinuous function, then there is a σ_i-lower semicontinuous and σ_{-i}-upper semicontinuous function f such that $t \leq f \leq s$ and $t(x) < f(x) < s(x)$ whenever $t(x) < s(x)$ (E.A. Michael (1956) for topological spaces).

Hint. For the proof of (i)$\Rightarrow$(ii) use Ex. 3.B.5.e and methods yielding the assertion of Corollary 3.14. The implication (ii)$\Rightarrow$(i) is obvious. To prove (ii)$\Rightarrow$(iii)$\Rightarrow$(iv) $\Rightarrow$(v) use the procedure as in Proposition 3.4.

(v)$\Rightarrow$(vi). It can be assumed that $t \leq s$ are bounded. Let $\varphi_n \nearrow s$, $\varphi_n < s$, $\psi_n \searrow t$, $\psi_n > t$, where φ_n, ψ_n are real functions with the desired continuity properties. Using the Hausdorff method of Ex. 3.E.4a construct the corresponding ω_n and put $f = \omega$.

(vi)$\Rightarrow$(ii). Use (vi) for $t = 0$ and s being the characteristic function of a set in consideration.

3.B.9 The perfect Lusin-Menchoff property. We say that a fine topology τ on a topological space (X, ρ) possesses the perfect Lusin-Menchoff property, whenever the space (X, ρ, τ) is perfectly binormal; equivalently whenever τ has the Lusin-Menchoff property and every τ-closed set is a G_δ-set. State some further equivalent conditions as in (c) of the previous exercise. In a similar way define the notion of the perfect Lusin-Menchoff property of a "fine σ-topology". (Cf. also Ex. 3.B.20.)

In applying general results one example has become of some interest: Let $\mathcal{K}$ be the σ-topology of all open F_σ-subsets of a normal T_1 topological space (X,ϱ). Show that:

(a) $\mathcal{K}$ is perfect and normal (cf. Ex. 3.B.7).

Hint. Use that fact that $\mathcal{K} = \operatorname{Coz} \mathbf{C}(X)$, or insert an open F_σ-set between some couple $F \subset G$ where F is closed and G is open.

(b) $\mathcal{K}$ has the perfect Lusin-Menchoff property (w.r.t. ϱ).

Remark. The results of this exercise will be strengthened in Ex. 3.B.21.

3.B.10 Let τ denote either the discrete topology or the density topology on $\mathbb{R}$. Let $\{r_n\}$ be the sequence of all rational numbers. Put

$$g(x) := \begin{cases} -n & \text{for } x = r_n \\ +\infty & \text{for } x \in \mathbb{R} \setminus \mathbb{Q} \end{cases}, \quad f := -\infty \quad \text{on } \mathbb{R}.$$

Show that g is τ-lower semicontinuous, f upper semicontinuous on $\mathbb{R}$, $f < g$ and there is no τ-continuous upper semicontinuous function h for which $f < h < g$ (compare with (v) of Theorem 3.11 and with Ex. 3.E.9).

Hint. Let $h \in B_1(X)$, $f < h < g$. Seek a contradiction with the completeness of $\mathbb{R}$ and the fact $[h = -\infty] = \bigcap_{n=1}^{\infty} [h \leqq -n]$.

3.B.11 Let τ and g be as in the previous exercise. Show that the function $s := g \wedge \mathbb{Q}$ is a real τ-continuous and upper semicontinuous function on the τ-closed set $\mathbb{Q}$ which has not a <u>real</u> τ-continuous and upper semicontinuous extension to all of $\mathbb{R}$ (cf. with (vi) of Theorem 3.11 and Ex. 3.E.10).

Hint. Similarly as in the preceding exercise, show that s has not any real B_1-extension over $\mathbb{R}$.

3.B.12 A subset A of a (real) Banach space is said to be open in the *line topology* τ if for any $x \in A$ and every straight line L through x there is $r > 0$ such that the closed line segment $L \cap \overline{U(x,r)}$ is contained in A.

(a) Show that the restriction of every τ-continuous function to any staight line is continuous.

(b) Show that the line topology (dim $X \geq 2$) is not regular (and, in particular) has not the Lusin-Menchoff property).

Hint. It does no harm to suppose $X = \mathbb{R}^2$. Let $\{a_n\}$ be a sequence of different points of $\{x \in X: \|x\| = 1\}$. Show that the τ-closed set $\{a_1, \frac{1}{2}a_2, \frac{1}{3}a_3, \ldots\}$ cannot be separated from the origin.

3.B.13 Let M be a subset of a Banach space X. Given $x \in X$, $A \subset X$, put

$$\operatorname{cont}(x,A) := \left\{\lim \frac{x_n - x}{\|x_n - x\|} : x_n \in A,\ x_n \to x\right\}.$$

Clearly, $\operatorname{cont}(x,A) \subset S := \{x \in X: \|x\| = 1\}$. Define

$$t_M = \{G \subset X: \operatorname{cont}(x, X\setminus G) \cap M = \emptyset \text{ for every } x \in G\}.$$

(a) Show that t_M forms a (fine) topology on X (t_M will be labelled as the *M-contingent topology*).

(b) If $X = \mathbb{R}^n$, then t_S is the Euclidean topology.

(c) If $X = \mathbb{R}^2$ and $M = \{(1,0), (-1,0), (0,1), (0,-1)\}$, then t_M is strictly in-between the Euclidean topology and the crosswise topology. In particular, in this case any t_M-continuous function is separately continuous.

(d) Show that $\operatorname{der}_{t_M}(A) = \{x \in X: \operatorname{cont}(x,A) \cap M \neq \emptyset\}$.

(e) Prove that t_M has the complete Lusin-Menchoff property.

Hint. Given $A, B \subset X$, $\bar{A} \cap B = \emptyset = A \cap \overline{B}^{t_M}$, put

$$G := \{y \in X: \|y - B\| < \|y - A\|^2\}$$

and show that G is an open set, $G \supset B$, $G \cap A = \emptyset$ and also $A \cap \operatorname{der}_{t_M}(G) = \emptyset$.

REMARK. Some applications of the notion of M-contingent topology appear in Ex. 9.D.3 and 9.D.4. Cf. also Ex. 7.D.5.

3.B.14 Let d be the density topology on $\mathbb{R}$ and d_s the strong density topology on $\mathbb{R}^2$ (cf. Section 9.D.3). If

$$\tau_1 = \{G \subset \mathbb{R}^2: \text{for every } x = (x_1; x_2) \in G \text{ there is a d-neighborhood } U \text{ of } 0 \text{ such that } (\{x_1\} + U) \times \{x_2\} \subset G,\ \{x_1\} \times (\{x_2\} + U) \subset G\}$$

(i.e., τ_1 is the "d-crosswise topology" on $\mathbb{R}^2$), $\tau_2 = d_s \cap \tau_1$, $\tau_3 = d \times d$,

$\tau_4 = d \times e$, then $\tau_4 \subset \tau_3 \subset \tau_2 \subset \tau_1$.

Let $\{r_n\}$ be an enumeration of $\mathbb{Q}$ ($r_i \neq r_j$ for $i \neq j$). Put

$$F = \bigcup_{n=1}^{\infty} (\{r_n\} \times [\tfrac{1}{n}, +\infty)), \quad S = \mathbb{R} \times \{0\}. \quad \text{Show that:}$$

(a) F is τ_4-closed,

(b) if an open set $G \subset \mathbb{R}^2$ contains F, then $S \cap \bar{G}^{\tau_1} \neq \emptyset$.

Deduce that no topology τ_j ($j = 1,2,3,4$) on $\mathbb{R}^2$ has the Lusin-Menchoff property.

Hint. (b) Show that any set $H_n := \{z \in \mathbb{R}: \{z\} \times [\frac{1}{n}, 1] \subset G\}$ is residual and that $(\bigcap_{n=1}^{\infty} H_n) \times \{0\} \subset \bar{G}^{\tau_1}$.

3.B.15 Show that $e^+ \times e^+$ has not the Lusin-Menchoff property (w.r.t. the Euclidean topology on $\mathbb{R}^2$).

Hint. Let $\{(a_n, b_n)\}$ be the collection of component intervals of $[0,1] \setminus C$ where C is the Cantor set. Put $F = \{(x;y) \in \mathbb{R}^2: x \in C, y = -x\}$, $F_\tau = \{(b_n; -a_n): n \in \mathbb{N}\}$. Show that F is closed, F_τ is $e^+ \times e^+$-closed and $F \cap \bar{G}^{e^+ \times e^-} \neq \emptyset$ whenever $G \supset F_\tau$ is (Euclidean) open.

3.B.16 Let a fine topology τ have the Lusin-Menchoff property w.r.t. a metric space (P, ρ). Decide of when the following assertions are true:

(a) $\tau \times \rho$ has the Lusin-Menchoff property w.r.t. $\rho \times \rho$,

(b) $\tau \times \tau$ has the Lusin-Menchoff property w.r.t. $\rho \times \rho$.

Hint. Look at the examples of the previous exercises.

3.B.17 Consider on $\mathbb{R}^2$ the product topology $\tau = \eta \times e$, where the first factor is the discrete topology and the second factor is the usual Euclidean topology.

(a) Show that the fine (metric) topology τ fails the Lusin-Menchoff property w.r.t. the Euclidean topology on $\mathbb{R}^2$.

Hint. Let F_τ be the graph of the Riemann function on $(0,1)$, $F = \mathbb{R} \times \{0\}$. Then F_τ is τ-closed and $\bar{G}^\tau \cap F \neq \emptyset$ for any open subset G of $\mathbb{R}^2$ containing F_τ (cf. also Ex. 3.B.14).

(b) Show that $(\mathbb{R}^2, \tau)$ is "completely conetrizable" in the following sense: Whenever F_τ is a τ-closed set in $\mathbb{R}^2$ and $z \notin F_\tau$, there is an upper semicontinuous and τ-continuous function $f: \mathbb{R}^2 \to [0,1]$ such that $f(z) = 1$ and $f(F_\tau) = 0$.

3.B.18   The topological modification of a σ-topology. Let π be a σ-topology on a set X. The family of all π-open sets is a base for some topology τ_π which will be called the topological modification of π.

Given $M \subset X$, put

$$M_1 := \bigcap \{F \subset X : M \subset F, \ F \text{ is } \pi\text{-closed}\},$$

$$M_2 := \{x \in X : G \cap M \neq \emptyset \text{ whenever } G \text{ is a } \pi\text{-open set containing } x\}.$$

Show that:

(a) $M_1 = M_2 = \bar{M}^{\tau_\pi}$,

(b) π is a topology if and only if $\bar{M}^{\tau_\pi}$ is π-closed for all $M \subset X$.

3.B.19   Let π be a σ-topology on X. A real function f on X is said to be π-continuous at a point $z \in X$ if for each neighborhood V of $F(z)$ there exists a π-open set G containing z such that $f(G) \subset V$.

(a)   Show that f is π-continuous at z if and only if f is τ_π-continuous at z.

(b)   A function which is π-continuous at each point of X need not be π-continuous on X.

Hint.   Consider the σ-topology of all F_σ-subsets of $\mathbb{R}$.

3.B.20   The Lusin-Menchoff property of a σ-topology. Let π be a fine σ-topology on a topological space (X, ρ).

(a)   Prove that the conditions (i) and (iii)-(viii) of Theorem 3.11 are equivalent. If (i) or another equivalent assertion holds, say that π has the Lusin-Menchoff property.

Hint.   Notice that in the proofs only countable operations are used while the " π-closure operation" is not involved. Cf. also Ex. 3.B.5e.

(b)   Define the σ-topology π as the collection of all subsets of $\mathbb{R}$ which differ from open sets by a countable set.

(b$_1$)   Show that π has not the Lusin-Menchoff property.

Hint. Let C be the Cantor set and $S \subset \mathbb{R} \setminus C$ be a countable dense subset of $\mathbb{R}$. Show that C and S cannot be separated.

(b_2) The topological modification $\tau_{\mathcal{T}}$ of $\mathcal{T}$ equals the discrete topology. In particular, $\tau_{\mathcal{T}}$ has the Lusin-Menchoff property.

(b_3) Given a $\mathcal{T}$-closed set $F_{\mathcal{T}}$ and an open set G, $F_{\mathcal{T}} \subset G \subset \mathbb{R}$, there is always an open set H such that $F_{\mathcal{T}} \subset H \subset \bar{H}^{\tau_{\mathcal{T}}} \subset G$.

(c) Let τ be the topology of Ex. 3.B.17 and let Φ be the family of all non--negative bounded functions on $\mathbb{R}^2$ which are τ-continuous and upper semicontinuous.

(c_1) If $\mathcal{T}$ is the σ-topology formed by the collection $\mathrm{Coz}(\Phi)$ of all Φ-cozero sets, then $\mathcal{T}$ has the (perfect) Lusin-Menchoff property (consult (ii) of Ex. 3.B.8c).

(c_2) In light of Ex. 3.B.17b, the topological modification $\tau_{\mathcal{T}}$ of $\mathcal{T}$ is τ. In particular, $\tau_{\mathcal{T}}$ lacks the Lusin-Menchoff property.

3.B.21 Let $\mathcal{T}$ be a fine σ-topology on a topological space (X,ρ) having the Lusin-Menchoff property. Let ω be the σ-topology of all $\mathcal{T}$-open F_σ-subsets of X.

(a) Show that ω has also the Lusin-Menchoff property.

Hint. Use Urysohn's condition of Ex. 3.B.5e for the σ-topology $\mathcal{T}$. Give also the straightforward proof.

(b) Prove that the σ-topology ω is "normal" (cf. Ex. 3.B.5h).

Hint. In view of Zahorski's property of $\mathcal{T}$ (see Ex. 3.B.5k) show that $\omega = \mathrm{Coz}(\Phi)$ where Φ is the family of all $\mathcal{T}$-continuous G_δ-measurable functions. Now cite Ex. 3.B.5h.

REMARKS AND COMMENTS

In the twenties, two soviet mathematicians succeeded to show that the density topology has a significant property which can be situated between the normality and the complete regularity. Clearly, at that time it was impossible to express this property in terms of the density topology. It is curious that neither Lusin nor Menchoff did publish the proof of this result.

The first proof of the property mentioned above (henceforth called the Lusin-Menchoff property) was given by V.J.Bogomolova (1924). In her work she wrote literatelly: "Les theorémes sur les points d'épaisseur étaient démontrés d'abord par M.N.N. Lusin et M. D.E.Menchoff. N'ayant aucune idée de leur méthode j'ai obtenu quelques jours plus tard une autre démonstration." Certain generalization of the

Lusin-Menchoff property with the modified proof can be found in the paper of I. Maximoff (1940). More recent proofs were given by C. Goffman, C. Neugebauer and T. Nishiura (1961) or R.J. O'Malley (1977). See also [Bru] , Th. 6.4.

Further, the Lusin-Menchoff property was studied also in case of many generalized density topologies (see e.g.M. Chaika (1971), R.S.Troyer and W. Ziemer (1963) or G.M.Petersen and T.O. To (1976). Recently this property was considered also in view of other new topologies on the real axis. R. J. O'Malley (1977), (1979a) proved the Lusin-Menchoff property in the case of r-topology and a.e.-topology in $\mathbb{R}$, T. Nishiura (1981) considered similar but weaker properties in the case of a.e.-topology in $\mathbb{R}^n$. The investigation of the analogy of the Lusin-Menchoff property for various Zahorski's classes can be found in S.J.Agronsky (1982).

The idea of considering the Lusin-Menchoff property of the fine topologies of harmonic spaces originated with J. Lukeš (1977a) and J. Malý (thesis, 1980). In Chapter III we shall see that this property plays a key role in the development of the fine potential theory.

As already said, the Lusin-Menchoff property does not express but the binormality of a given space equipped with two topologies. The notion of a bitopological space as well as the investigation of some of its properties is usually connected with the name of J.C. Kelly (1963). However, earlier the spaces with two topologies were studied in another context. For instance, the notion of "espace bitopologique" was introduced by L. Motchane in (1957).

Let us revert to the work of J. C. Kelly (1963). There we find a proof of Tietze's extension theorem which is based on binormality conditions by means of methods completely different from ours. Because the functions are extended only from the sets which are closed in both the topologies, an equivalent binormality condition is not reached.

The notion of the Lusin-Menchoff property can be also introduced in case of structures which are more general than the topology. As we shall see later on, there are nice applications of this approach, too. In Ex. 3.B.5 we examined this property for the class of σ-topologies. Relevant questions were studied by Gh. Bucur (1969) where the notion of a quasi-topology (= σ-topology in our sense) was introduced and where necessary and sufficient conditions were stated for insertion of a function between "semicontinuous" ones in case of a space equipped with two quasi-topologies satisfying the "binormality condition".

Quasitopological notions have their origin in works of M. Brelot and G. Choquet in the field of potential theory and were axiomatized in B. Fuglede (1971a).Further important contribution to applications of quasi-topology appeared in D. Feyel (1978) and (1981) where Urysohn's theorem is also proved.

In a lecture of D. Preiss on the conference at Dědinky in 1972 the notion of a σ-topology was used for extension theorems concerning approximately continuous functions. The conditions equivalent to normality of σ-topologies were already proved by A.D. Alexandrov (1940).

At the end of this section we introduced the complete Lusin-Menchoff property, which will be useful in the text, too. Theorem 3.19 gives sufficient conditions for a fine topology to have this property.

Notice that some results concerning "perfect binormality" and "complete binormality" can be found in E.P. Lane (1967) and C. W. Patty (1967).

The terminology in the literature with respect to topological and bitopological spaces is a little confused. We used the term "binormal" for bitopological spaces which are sometimes called "pairwise normal" while some other writers use the notation "binormal" for those topological spaces which are countably paracompact and normal (e.g. [Will], § 21).

Related papers: K. Nagami (1954), H.A. Priestly (1971).

3.C The Lusin-Menchoff property of base operators

This section is devoted to the investigation of a new concept of the Lusin-Menchoff property of base operators which turns out to be useful in later sections (see e.g. Remark 6.30).

Let b be a fine base operator on a topological space (X,ρ). We say that b has the *Lusin-Menchoff property* (with respect to ρ) if for each closed set $F \subset X$ and for each $M \subset X$ satisfying $F \cap M = F \cap bM = \emptyset$ there is an open set $G \subset X$ such that

$$M \subset G \quad \text{and} \quad F \cap bG = \emptyset .$$

Of course, G can be chosen in such a way that $M \subset G \subset X \setminus F$.

Equivalently, b has the Lusin-Menchoff property if for each open set $U \subset X$ and for each set $M \subset X$ satisfying $M \cup bM \subset U$ there is an open set G such that

$$M \subset G \subset G \cup bG \subset U .$$

We see that a base b has the Lusin-Menchoff property if and only if the corresponding closure operation c_b $(c_b A = A \cup bA)$ has the Lusin-Menchoff property.

In light of these observations it follows easily that a strong base b has the Lusin-Menchoff property if and only if the corresponding b-topology has the Lusin-Menchoff property.

Further, we say that a base operator b has the *complete Lusin-Menchoff property* if for every subset $Y \subset X$ the base operator b_Y $(b_Y A = Y \cap bA$ for each $A \subset Y)$ has the Lusin-Menchoff property (w.r.t. the topology on Y induced by ρ).

It can be easily proved in a similar way as in Theorem 3.18 that b has the complete Lusin-Menchoff property if and only if the following assertion holds: For each couple A, B of subsets of X for which $\bar{A} \cap B = \emptyset = A \cap bB$ there exists an open set G such that

$$B \subset G \subset X \setminus A \quad \text{and} \quad A \cap bG = \emptyset .$$

If b is strong, we see that b has the complete Lusin-Menchoff property if and only if the b-topology has the complete Lusin-Menchoff property.

Dealing as we are with base operator spaces rather than topological spaces, it is important to notice that the Lusin-Menchoff property is invariant with respect to some operations performed on them. Therefore, the base operator spaces have several pleasant and useful properties as we shall see. More precisely, we are going to show the conditions stating that the least upper bound and the composition of finitely many bases conserve the Lusin-Menchoff property. These theorems will have a number of important applications in sections to come. For example, by means of them we can prove the Lusin-Menchoff property of several fine topologies in a short "algebraic" way.

3.20 THEOREM. Let b_1, b_2 be fine bases on a topological space (X,ρ) and let $b = b_1 \vee b_2$.

(a) If b_1 and b_2 have the Lusin-Menchoff property, then so does the base b.

(b) If b_1 and b_2 have the complete Lusin-Menchoff property, then so does the base b.

Proof. (a) Let $F \subset X$ be closed and $F \cap A = \emptyset = F \cap bA$. Then $F \cap b_1 A = \emptyset = F \cap b_2 A$, and therefore there exist open sets $G_1 \supset A$, $G_2 \supset A$ such that $F \cap bG_1 = \emptyset = F \cap bG_2$. Put $G = G_1 \cap G_2$. Obviously, the open set G contains A and $F \cap bG = \emptyset$.

(b) It follows immediately from (a) since

$$b \wedge Y = (b_1 \wedge Y) \vee (b_2 \wedge Y)$$

for any $Y \subset X$.

3.21 THEOREM. Let b_1 be a fine closure operation and b_2 a fine base on a topological space (X,ρ). If b_1 has the complete Lusin-Menchoff property and b_2 has the (complete) Lusin-Menchoff property, then the base $b = b_2 * b_1$ has the (complete) Lusin-Menchoff property, too.

Proof. Assume that b_2 has the complete Lusin-Menchoff property. Let $\bar{B}\cap A = \emptyset = B\cap bA$. Put $A^* = b_1A\setminus\bar{B}$. Since b_1 is a closure operation, $A\subset A^*$. Since b_2 has the complete Lusin-Menchoff property and since $\bar{B}\cap A^* = \emptyset = B\cap b_2A^*$, there is an open set $H\supset A^*$ such that $\bar{B}\cap H = \emptyset = B\cap b_2H$. Put $D = X\setminus(H\cup b_1A)$. Apparently, $\bar{D}\cap A = \emptyset = D\cap b_1A$. The complete Lusin-Menchoff property of b_1 yields the existence of an open set $G\supset A$ for which $D\cap b_1G = \emptyset$. Thus $b_1G\subset H\cup b_1A$, and it follows that

$$bG\subset b_2H\cup b_2(b_1A)\subset X\setminus B.$$

This proof works also in the case when b_2 is supposed to have only the Lusin-Menchoff property and B is a closed set, i.e. $B = \bar{B}$.

EXERCISES

3.C.1 (a) Let b_1, b_2 be fine closure operations on (X,ρ) having the Lusin-Menchoff property. Show that the closure operation $b_2 * b_1$ has also the Lusin-Menchoff property (cf. Theorem 3.21).

Hint. Let F be closed, $F\cap b_2(b_1A) = \emptyset$. Find a open set $H\supset b_1A$ such that $F\cap b_2H = \emptyset$ and an open set $G\supset A$ such that $b_1G\subset H$. Then $F\cap b_2(b_1G) = \emptyset$.

(b) Show that if "b_1 is a fine closure operation" is replaced in Theorem 3.21 by "b_1 is a fine base operator", then the assertion is no longer true.

Hint. Let e be the Euclidean topology on $\mathbb{R}$. Let $A = \mathbb{Q}$ and $F = \{\sqrt{2}\}$. Put $b_1Y := Y\setminus A$, $b_2Z = \bar{Z}^e$. Show that $F\cap b_2(b_1A) = \emptyset$, but $F\cap b_2(b_1G) \neq \emptyset$ for any open $G\supset A$.

(c) Show that the assumption of Theorem 3.21 concerning b_1 cannot be weakened to the requirement that b_1 is a fine closure operation having the Lusin-Menchoff property only.

Hint. Let τ be a fine topology on (X,ρ) having the Lusin-Menchoff property but not the complete one (e.g. Šverák's topology of Ex. 3.B.1.e). Let $\bar{B}\cap A = \emptyset = B\cap\bar{A}^\tau$ and $B\cap\bar{G}^\tau\neq\emptyset$ for any open set $G\supset A$. Set $b_1Z := \bar{Z}^\tau$, $b_2Y := Y\setminus(\bar{B}\setminus B)$, $F = \bar{B}$. Then $F\cap b_2(b_1A) = \emptyset$, but $F\cap b_2(b_1G)\neq\emptyset$ whenever $G\supset A$ is an open set.

(d) Let b_1, b_2 be as in Theorem 3.21, b_2 having the complete Lusin-Menchoff property, and let $M \subset X$. Set $b := (b_2 * b_1) \wedge M$, $b^* := (b_2 \wedge M) * (b_1 \wedge M)$. Then b and b^* have the Lusin-Menchoff property. Show that, in general $b \neq b^*$.

Hint. Consider the density topology d on $\mathbb{R}$ and put $M = \mathbb{Q}$, $b_1 A = \bar{A}$, $b_2 A = = der_d A$.

(e) Let b_1, b_2 be as in Theorem 3.21, b_2 having the Lusin-Menchoff property. If $b_2 \prec b_1$ and b_1 is strong, then the $b_2 * b_1$ - topology has the Lusin-Menchoff property.

Hint. Show that $b_2 * b_1$ is strong.

3.C.2 Let b be a fine base operator on (X, ρ) having the Lusin-Menchoff property. Show that if $Z \subset X$ is closed, then $b \wedge Z$ has the Lusin-Menchoff property.

3.C.3 Let b be a fine base operator on (X, ρ). If there is a locally finite open cover $\mathcal{G}$ of X such that $b \wedge \bar{G}$ has the Lusin-Menchoff property for each $G \in \mathcal{G}$, then b has the Lusin-Menchoff property.

Hint. Let $F \subset X$ be closed, $F \cap A = \emptyset = F \cap bA$. Given $G \in \mathcal{G}$, there is $U_G \subset \bar{G}$ open in $\bar{G}$ such that $A \cap \bar{G} \subset U_G$ and $F \cap \bar{G} \cap bU_G = \emptyset$. Put $U := \bigcup_{G \in \mathcal{G}} (U_G \cap G)$.

3.C.4 Let b be a fine base operator on a regular, paracompact space (X, ρ). If for each $x \in X$ there is a neighborhood $V_x \in \mathcal{V}(x)$ such that $b \wedge V_x$ has the Lusin-Menchoff property, then b has the Lusin-Menchoff property.

Hint. Find an open set $U_x \in \mathcal{V}(x)$, $\bar{U}_x \subset V_x$ and use Ex. 3.C.2 and Ex. 3.C.3.

3.C.5 Show that the "crosswise" base operator b of Example 1.1.d has not the Lusin-Menchoff property (cf. Ex. 2.A.3a).

Hint. Let $F := \mathbb{R} \times \{0\}$ and let A be the graph of the Riemann function on $[0,1]$. Then $F \cap bA = \emptyset$. On the other hand, $F \cap bG \neq \emptyset$ for any open set G containing A.

REMARKS AND COMMENTS

As we shall see the concept of the Lusin-Menchoff property of base operators will play an important role. Sometimes the base operator generating a fine topology has the Lusin-Menchoff property meanwhile the fine topology itself does not have it. (The superdensity topology is a nice example to demonstrate it.) Further, Theorems 3.20 and 3.21 form the basis for proving the Lusin-Menchoff property in case of various modifications of fine topologies.

Considering all this we conclude that this section is extremely important for further applications.

3.D The M-modification of a fine topology

In our later considerations, namely in connection with extension theorems of Tietze's type in Chapter 8, some modifications of a given fine topology can shed light on certain problems involved. We are mainly interested in the Lusin-Menchoff property of these modifications.

Let τ be a fine topology on a topological space (X,ρ), and let $M\subset X$. Define the system τ_M of all subsets G of X with the following property: G is a τ-neighborhood of every point $x\in G\cap M$, and a ρ-neighborhood of every point $x\in G\setminus M$. Then τ_M is a topology which is finer than ρ and coarser than τ. The topology τ_M will be labelled as the *M-modification* of τ (w.r.t. ρ).Notice that we modify in fact the topology τ outside M.

3.22 THEOREM. If τ has the Lusin-Menchoff property with respect to a completely normal topology ρ, then so does τ_M.

Proof. Let $F\cap K=\emptyset$ where F is supposed to be closed and K τ_M-closed. Since K is obviously τ-closed, in view of the Lusin-Menchoff property of τ there are τ-open set G and open set H such that

$$G\cap H=\emptyset\ ,\quad F\subset G,\quad K\subset H.$$

By the definition of τ_M we have $\bar{K}\cap(G\setminus M)=\emptyset$. Obviously, $\overline{G\setminus M}\cap K=\emptyset$. Using the complete normality of ρ we can see that there are open sets H_1, H_2 such that

$$H_1\cap H_2=\emptyset,\quad G\setminus M\subset H_1,\quad K\subset H_2.$$

Put $V:=G\cup H_1$, $W:=H\cap H_2$. Then V is τ_M-open, W is open, and

$$V\cap W=\emptyset,\quad K\subset W,\quad F\subset V.$$

Next we outline another method of proving that the M-modification τ_M has the Lusin-Menchoff property provided τ has it. Analysing the previous proof we can see that after some rearrangement it can be done with help of Theorems 3.20 and 3.21. This new "algebraic" technique will be frequently employed in subsequent sections. We begin with establishing of the formula for the Kuratowski closure corresponding to an M-topology.

3.23 LEMMA. Let τ_M be the M-modification of a fine topology τ on a topological space (X,ρ). Then for every $A \subset X$ we have

$$\bar{A}^{\tau_M} = \overline{(\bar{A}\setminus M)}^{\tau} \cup \bar{A}^{\tau}.$$

Proof. Put $A^* = \overline{\bar{A}\setminus M}^{\tau} \cup \bar{A}^{\tau}$. First, we are going to prove that A^* is τ_M-closed. Obviously, it is τ-closed. Given now $x \in (X\setminus A^*)\cap(X\setminus M)$, we have $x \in X\setminus\bar{A}$, for if $x \in \bar{A}\cap(X\setminus M)$, we must have $x \in \bar{A}\setminus M \subset A^*$. Since $X\setminus\bar{A}$ is an open ρ-neighborhood of x contained in $X\setminus A^*$, we see that A^* is τ_M-closed. On the other hand, since $\bar{A}\setminus M \subset \bar{A}^{\tau_M}$ and $\bar{A}^{\tau} \subset \bar{A}^{\tau_M}$, we get $A^* \subset \bar{A}^{\tau_M}$. We conclude that $A^* = \bar{A}^{\tau_M}$.

Proof of Theorem 3.22. Put $b_2A = \bar{A}^{\tau}$, $b_1A = (\bar{A}\setminus M)\cup A$. In view of Lemma 3.23 and Theorem 3.21, it suffices to verify that the base b_1 has the complete Lusin-Menchoff property. Since the identity base operator $iA = A$ has this property (cf. Remark 3.17) we reduce the proof in light of Theorem 3.20 to prove this assertion for the base operator $A \mapsto \bar{A}\setminus M$. We state it in the following lemma which is slightly more general.

3.24 LEMMA. Let b be a fine base operator on a topological space (X,ρ) and let $M\subset X$. If b has the complete Lusin-Menchoff property, then so does the base

$$b^M:\quad A \longmapsto (bA)\setminus M.$$

Proof. Assume that $\bar{B}\cap A = \emptyset = B\cap b^M A$. Put $B^* := B\setminus M$. Since $\overline{B^*}\cap A = \emptyset = B^*\cap bA$, there is an open set $G\supset A$ such that $(B\setminus M)\cap bG = \emptyset$. Then, of course, $B\cap b^M G = \emptyset$.

Some applications need even a more general approach to the M-modification. Actually the previous lemma leads to a more general result which we restate now as a theorem.

Let τ be a fine topology on a topological space (X,ρ) and let ϑ be another fine topology on (X,ρ) which is coarser than τ. Fix $M\subset X$. Let τ_M denote the M-modification of τ w.r.t. ϑ, i.e. a set G is τ_M-open if each point of $G\cap M$ has a τ-neighborhood contained in G and each point of $G\setminus M$ has a ϑ-neighborhood in G. As above, it can be checked that τ_M is a fine topology on (X,ϑ) which is coarser than τ. Moreover,

$$\bar{A}^{\tau_M} = \overline{(\bar{A}^{\vartheta}\setminus M)\cup A}^{\tau}.$$

Using Lemma 3.24, Theorems 3.20 and 3.21, we get the next theorem immediately.

3.25 THEOREM. Let τ, ϑ, M, (X,ρ) and τ_M be as above. Assume that ϑ has the complete Lusin-Menchoff property w.r.t. ρ. If τ has the (complete) Lusin-Menchoff property (w.r.t. ρ), then so does τ_M.

We close this part with a remark: A real function f on X is τ_M-*continuous* if and only if it is τ-continuous and, in addition, if it is continuous at all points of $X\setminus M$. It should be also clear that f is τ_M-upper semicontinuous if and only if it is τ-upper semicontinuous and, moreover, upper semicontinuous at all points of $X\setminus M$.

EXERCISES

3.D.1 Let A be a subset of a completely normal topological space (X,ρ). Show that the A-modification of the discrete topology on X is also completely normal and has the Lusin-Menchoff property (w.r.t. ρ).

Hint. Use Theorem 3.25.

3.D.2 Let $\tau_1,\dots,\tau_n$ be topologies on a set X and let $A_1,\dots A_n$ be subsets of X. Let $\tau = (\tau_1,\dots,\tau_n, A_1\dots,A_n)$ be a new topology on X defined as follows: A set $G\subset X$ is τ-open if there are τ_i-open sets G_i, $i=1,\dots,n$, such that $G\cap A_i\subset G_i\subset G$ (i.e. $G\cap A_i\subset \operatorname{int}_{\tau_i}(G)$).

(a) Show that a function f is τ-continuous on X if and only if f is τ_i-continuous at all points of A_i.

(b) If, moreover, (X,ρ) is a topological space, $\rho\subset\tau_1\subset\dots\subset\tau_n$, $A_1\subset A_2\subset\dots\subset A_n$ and if the topologies τ_i have the complete Lusin-Menchoff property (w.r.t. ρ), then τ has also the complete Lusin-Menchoff property.

Hint. First, show that $\tau(\tau_1,\tau_2,A_1,X)$ is the $(X\setminus A_1)$-modification of τ_2 w.r.t. τ_1. If d is the discrete topology on X, and if $\tau(\tau_n,d,A_n,X) = t_n$, $\tau(\tau_{n-1},t_n,A_{n-1},X) = t_{n-1},\dots,$ then $t_1 = \tau(\tau_1,\dots,\tau_n,A_1,\dots,A_n)$.

3.D.3 There is a different proof of Theorem 3.22 in case when ρ is a <u>metric</u> topology on X. The alternative proof (which does not use the complete normality of metric spaces) runs as follows: If $F\subset G$ (F ρ-closed and G τ_M-open), there is a τ-open set H such that $F\subset H\subset\bar{H}\subset G$. Given $x\in H\setminus M$, there is $\varepsilon_x>0$ such that $U(x\,\varepsilon_x)\subset G$. Put $K:=H\cup\bigcup_{x\in H\setminus M} U(x,\frac{1}{2}\varepsilon_x)$. Obviously, K is τ_M-open and $F\subset K\subset\bar{K}$. It remains to show that $\bar{K}\subset G$. Let $z\in\overline{\bigcup_{x\in H\setminus M} U(x,\frac{1}{2}\varepsilon_x)}$. There are

$x_n \in H \setminus M$ and $z_n \in U(x_n, \frac{1}{2} \mathcal{E}_{x_n})$ such that $z_n \to z$. If $\lim \mathcal{E}_{x_n} = 0$, then $x_n \to z$ and $z \in \bar{H} \subset G$. In the opposite case, $\limsup \mathcal{E}_{x_n} = c > 0$. Then there is $k \in \mathbb{N}$ such that

$$x_k \in H \setminus M, \quad \rho(z_k, z) < \frac{c}{3} < \frac{1}{2} \mathcal{E}_{x_k}, \quad \rho(x_k, z_k) < \frac{1}{2} \mathcal{E}_{x_k}.$$

Therefore, $z \in U(x_k, \mathcal{E}_{x_k}) \subset G$. Fill in the details of this proof.

3.D.4 Let τ be a fine topology on a topological space (X, ρ) and let $M \subset X$. Show that $\tau_M \wedge M = \tau \wedge M$ provided M is closed. Give an example such that $\tau_M \wedge M \neq \tau \wedge M$.

Hint. Let τ be the density topology on $\mathbb{R}$. If M is a residual subset of $\mathbb{R}$ of measure zero, and if G is a countable dense subset of M, then G is $\tau \wedge M$-open. On the other hand, if $U \supset G$ is τ_M-open in $\mathbb{R}$, then U and $U \cap M$ are residual. Hence, $U \cap M \neq G$.

3.D.5 Let τ be a fine topology on a topological space (X, ρ) and let $A, M \subset X$. Consider the following properties that a function f on X may possess:

(i) f is τ-continuous at all points of $A \cap M$ and ρ-continuous at all points of $A \setminus M$,

(ii) f is τ_M-continuous at all points of A.

Show that (ii)$\Rightarrow$(i). If A is τ-closed, then (i)$\Longleftrightarrow$(ii). Give an example illustrating that, in general, (i)$\nRightarrow$(ii).

Hint.   Use Ex. 3.D.4.

3.D.6 The M-modification of a σ-topology. Let π be a fine σ-topology on a topological space (X, ρ) and let $M \subset X$. Define the M-modification of π as the σ-topology π_M of all π-open subsets G of X for which $(X \setminus M) \cap G \subset \text{int}_\rho G$. Prove the following assertions:

(a)   If π is a topology, then π_M is the (topological) modification of the topology π.

(b)   A function f on X is π_M-upper semicontinuous if and only if it is π-upper semicontinuous and, moreover, f is upper semicontinuous at all points of $X \setminus M$.

(c)   If π has the Lusin-Menchoff property on a completely normal topological space (X,ρ), then so does $\mathcal{T}_M$.

Hint. Imitate the proof of Theorem 3.22.

REMARKS AND COMMENTS

The concept of an M-modification of a given fine topology and, especially, of its Lusin-Menchoff property will be used later in connection with extension theorems. The matter is that the extended τ_M-continuous function will be continuous in the initial topology in all points of the set $X\setminus M$.

3.E B_1 in-between and extension theorems

In this section the key ideas of in-between theorems and of closely related extension theorems are brought sharply into focus. The next lemma provides the first step to our theorems.

The starting point will be a *perfect* (= any open set is F_σ) *topological space* (X,ρ) on which two fine topologies τ_1 and τ_2 are given each of them having the Lusin-Menchoff property (w.r.t. ρ). Underline that there is no relation of pairwise normality between τ_1 and τ_2. Notice also that the perfectness of X implies that each semicontinuous function on X is G_δ-measurable.

Throughout this section we denote by Φ_2^1 the family of all G_δ-measurable functions on X which are simultaneously τ_1-upper semicontinuous and τ_2-lower semicontinuous.

3.26 LEMMA. If τ_1-closed set F_1 and τ_2-closed set F_2 are separated by G_δ-sets, then F_1 is Φ_2^1-separated from F_2.

Proof. Let D_1, D_2 be G_δ-sets, $F_1\subset D_1$, $F_2\subset D_2$, $D_1\cap D_2 = \emptyset$. According to (vii) of Theorem 3.11 there are τ_i-continuous and upper semicontinuous functions f_i on X such that

$$0 \leqq f_i \leqq 1,\quad F_i\subset [f_i = 0]\subset D_i \quad (i = 1,2).$$

Then the function

$$f := f_2.(f_1 + f_2)^{-1}$$

has all properties we need. (Since $(f_1 + f_2)^{-1}$ is τ_2-lower semicontinuous, so is f. On the other hand, $f = 1 - f_1.(f_1 + f_2)^{-1}$, which implies that f is τ_1-upper semicontinuous.)

3.27 ASYMMETRICAL IN-BETWEEN THEOREM. Let t, s be functions on a perfect topological space X, $t \leqq s$, t is supposed to be τ_1-upper semicontinuous and s τ_2-lower semicontinuous. If there is a G_δ-measurable function g such that $t \leqq g \leqq s$ on X, then there is $f \in \Phi_2^1$ such that $t \leqq f \leqq s$ on X.

Proof. No generality is lost with the assumption that $-1 \leqq t \leqq s \leqq 1$. Apparently, real functions from Φ_2^1 form a closed L-cone. Let $a < b$. Since the τ_2-closed set $[s \leqq a]$ and the τ_1-closed set $[t \geqq b]$ are separated by G_δ-sets $[g \leqq a]$ and $[g \geqq b]$, it follows by Lemma 3.26 that $[s \leqq a]$ is Φ_2^1 - separated from $[t \geqq b]$. Hence the assertion follows from Theorem 3.2.

3.28 COROLLARY. Let $M \subset X$ be both τ_1-closed and τ_2-closed. If $f_1 \leqq g \leqq f_2$ on X where f_1 is τ_1-upper semicontinuous on X, f_2 is τ_2-lower semicontinuous on X and g is a function on X such that g/M is both τ_1-upper semicontinuous and τ_2-lower semicontinuous on M, then there is $h \in \Phi_2^1$ such that

$$h = g \text{ on } M \quad \text{and} \quad f_1 \leqq h \leqq f_2 \quad \text{on } X.$$

Proof. Put

$$s(x) = t(x) = g(x) \quad \text{for } x \in M,$$

$$s(x) = f_2(x) \text{ and } t(x) = f_1(x) \qquad \text{for } x \in X \setminus M$$

and apply the Asymmetrical in-between theorem.

Now we derive two important applications of the Asymmetrical in-between theorem. The first one is the following "B_1-extension theorem".

3.29 B_1-EXTENSION THEOREM. Assume that a fine topology τ on a metric space (P,ρ) has the Lusin-Menchoff property (w.r.t. ρ). Let f be a Baire one function on P which is τ-continuous on a τ-closed set $F \subset P$. If s, $-t$ are τ-lower semicontinuous functions on P, $t \leqq f \leqq s$, then there is a τ-continuous B_1-function f^* on P such that

$$f = f^* \text{ on } F \quad \text{and} \quad t \leqq f^* \leqq s \quad \text{on } P.$$

Proof. It follows immediately putting $\tau_1 = \tau_2$ in the preceding corollary. (Notice also that G_δ-measurable functions coincide with B_1-functions in metric spaces.)

REMARKS. (a) In Section 8 of Chapter II, the B_1-extension theorem will play the crucial role in many applications of the Lusin-Menchoff property.

(b) As noticed, B_1-extension theorem leads to many interesting applications. For example, it may be used to prove the existence of extension of functions defined on null sets in $\mathbb{R}$ over the whole space resulting in derivatives (generally unbounded). This is, in fact, the reason why we need the B_1-extension theorem to be furnished with the control functions s and t. There is another approach to in-between and extension theorems based on the extension procedure of Tietze's type. Remark that B_1-extension theorem can be easily derived from Theorem 3.10 or it can be proved directly by Tietze's type proof. Moreover, if we can do it without control "$t \leqq f \leqq s$", the proof can be essentially simplified (cf. Lukeš (1977a) Th. 4).

(c) Remark that the B_1-extension theorem remains "true" if we suppose that P is a perfect space provided we replace B_1-functions by G_δ-measurable functions.

Our second application of the Asymmetrical in-between theorem concerns extension theorem with preservation of continuity and semicontinuity. We use substantially the M-modification of fine topologies (cf. Section 3.D).

3.30 THEOREM. Let τ be a fine topology on a metric space (P,ρ) having the Lusin-Menchoff property. Let $M \subset P$ be τ-closed. Let f_1 and $-f_2$ be upper semicontinuous functions on P and let g be a Baire one function on P such that $g \upharpoonright M$ is τ-continuous on M and $f_1 \leqq g \leqq f_2$ on P. Then there is a Baire one function h on P which is simultaneously τ-continuous such that

(a) $h = g$ on M,

(b) h is continuous on $P \setminus \bar{M}$,

(c) h is lower semicontinuous at those points of M at which $g \upharpoonright M$ is, and upper semicontinuous at those points of M at which $g \upharpoonright M$ is,

(d) $f_1 \leqq h \leqq f_2$ on P.

Proof. Let U (L, respectively) be the set of all points at which $g \upharpoonright M$ is upper (lower, resp.) semicontinuous. Let τ_1 (τ_2, resp.) be the $\bar{M} \setminus U$ ($\bar{M} \setminus L$, resp.) modification of τ. In light of Theorem 3.22, τ_1 and τ_2 have the Lusin-Menchoff property. Put $s = t = g$ on M, $s = f_2$ and $t = f_1$ on $P \setminus M$. Obviously, s is τ_2-lower semicontinuous and t is τ_1-upper semicontinuous (cf. the final remark of Section 3.D). The existence of a function h with all the properties ascribed to it follows now from the Asymmetrical in-between theorem 3.27, which is what we wanted.

3.31 COROLLARY. Let τ be a fine topology on a metric space (P,ρ) having the Lusin-Menchoff property. Let $M \subset P$ be τ-closed and let G be a Baire one function on P. Then there is a Baire one function h on P such that

(a) $h = G$ on M,

(b) h is continuous and finite on $P \setminus \bar{M}$,

(c) h is lower (or upper) semicontinuous at those points of M at which $G \wedge M$ is.

Proof. Let $f_1 \leqq f_2$ be continuous functions on P, finite on $P \setminus \bar{M}$ and $f_1 = -\infty$, $f_2 = +\infty$ on $\bar{M}$. Put $g := \max(f_1, \min(G, f_2))$ and apply Theorem 3.30.

EXERCISES

3.E.1 Let τ be a fine topology on a completely normal topological space (X,ρ) having the Lusin-Menchoff property. Let h be a τ-continuous and upper semicontinuous function on a τ-closed set $A \subset X$. Show that there is an extension h^* of h to all of X such that

(a) h^* is τ-continuous and upper semicontinuous on X,

(b) h^* is continuous on $X \setminus \bar{A}$,

(c) h^* is continuous at those points of A at which h is continuous.

Hint. Define the functions s,t as in the proof of Theorem 3.11, (v)$\Longrightarrow$(vi). Let τ^* be the $(\bar{A} \setminus C)$-modification of τ where $C \subset A$ denotes the set of all continuity points of h. Observe that s is τ_M-lower semicontinuous. Since τ_M has the Lusin-Menchoff property, it suffices to use (v) of Theorem 3.11.

3.E.2 Let h be a function on a subset M of a completely normal topological space X. Show that h can be extended to a function h^* on X which is continuous at all points of $X \setminus \bar{M}$ and at all points $x \in \bar{M} \setminus M$ at which there is a limit $\limsup_{M \ni y \to x} h(y)$

Hint. Extend h by $h(x) = \limsup_{M \ni y \to x} f(y)$ for $x \in \bar{M} \setminus M$ and denote by D the set of all discontinuity points of $h \wedge \bar{M}$. Now extend continuously $h \wedge \bar{M} \setminus D$ to $X \setminus D$.

3.E.3 Consider the density topology d on $\mathbb{R}$. Denote $M = \mathbb{Q} \setminus \{0\}$ and $f = \chi_M \restriction \mathbb{Q}$. Show that:

(a) f is d-continuous on a d-closed set $\mathbb{Q}$ and continuous at all points of M, moreover, f is the restriction of a Baire one function on $\mathbb{R}$,

(b) f has no extension to a d-continuous function on $\mathbb{R}$ which is simultaneously upper semicontinuous at all points of $\mathbb{R} \setminus M$,

(c) f has no extension to a d-continuous function on $\mathbb{R}$ which is continuous at all points of the set $\{x \in \overline{M} \setminus M\colon$ there is a limit $\lim_{M \ni y \to x} f(y)\}$.

REMARK. (a) Compare (c) with the preceding exercise (3.E.2).

(b) The assertion (b) shows that no theorem generalizing both (v) of Theorem 3.11 and the B_1-extension theorem 3.29 (without control functions) can be established.

3.E.4 Suppose that $\varphi, \varphi_n, \psi, \psi_n$ are real functions on a set X, $\varphi_n \nearrow \varphi$, $\psi_n \searrow \psi$, $\varphi_n < \varphi$, $\psi_n > \psi$ and $\psi \leqq \varphi$.

(a) (F. Hausdorff). If

$$\omega_1 = \varphi_1,\ \omega_2 = \max(\omega_1, \psi_1),\ \omega_3 = \min(\omega_2, \varphi_2),\ \omega_4 = \max(\omega_3, \psi_2), \ldots,$$

then $\omega_1 \leqq \omega_3 \leqq \omega_5 \leqq \ldots, \omega_2 \geqq \omega_4 \geqq \omega_6 \geqq \ldots$ and there is a limit $\omega := \lim \omega_n$. Show that $\psi \leqq \omega \leqq \varphi$, and $\psi(x) < \omega(x) < \varphi(x)$ whenever $\psi(x) < \varphi(x)$.

(b) (S. Kempisty). If $\psi < \varphi$, then $\omega = g - f$ where

$$g = (\psi_1 - \varphi_1)^+ + (\psi_2 - \varphi_2)^+ \ldots,$$

$$f = -\varphi_1 + (\psi_1 - \varphi_2)^+ + (\psi_2 - \varphi_3)^+ + \ldots.$$

Hint. Show that the sequence $\{\omega_n\}$ coincides with partial sums of the series

$$\varphi_1 + (\psi_1 - \varphi_2)^+ + (\psi_2 - \varphi_3)^+ + (\psi_2 - \varphi_2)^+ - \ldots$$

If $\psi(x) < \varphi(x)$, only a finite number of members of this series nonvanishes.

3.E.5 Let $\mathcal{T}$ be a fine topology on a topological space (X,ρ) having the Lusin-Menchoff property. Denote by $\Psi^{\rho}_{\mathcal{T}}$ the family of all real function on X which are simultaneously $\mathcal{T}$-lower semicontinuous and ρ-upper semicontinuous. If $g \leqq s$ are bounded functions on X, g is G_δ-measurable and s is $\mathcal{T}$-lower semicontinuous, then there is $u \in (\Psi^{\rho}_{\mathcal{T}})^{\uparrow}$ such that $g \leqq u \leqq s$.

Hint. According to the Abstract in-between theorem 3.2 (cf. Lemma 3.3) it could be clearly sufficient to verify that any $\mathcal{T}$-closed set F can be $(\Psi^{\rho}_{\mathcal{T}})^{\uparrow}$-separated from any F_σ-set A, $A \cap F = \emptyset$. By assuming $A = \bigcup_{n=1}^{\infty} F_n$ where F_n are closed, there are $f_n \in \Psi^{\rho}_{\mathcal{T}}$ such that $f_n = 0$ on F, $f_n = 1$ on F_n and $0 \leqq f_n \leqq 1$. Now, consider the function $\min\,(1, \sum_{n=1}^{\infty} f_n)$.

3.E.6 Let $\mathcal{T}_1$, $\mathcal{T}_2$ be fine topologies on a topological space (X,ρ) each of them having the Lusin-Menchoff property (w.r.t. ρ). Let $t \leqq s$ where t is $\mathcal{T}_1$-upper semicontinuous and s is $\mathcal{T}_2$-lower semicontinuous. If there is a G_δ-measurable function g such that $t \leqq g \leqq s$ on X, then there is a $\mathcal{T}_1$-upper semicontinuous and $\mathcal{T}_2$-lower semicontinuous function f such that

(a) $t \leqq f \leqq s$,

(b) $f(x) > t(x)$ whenever $g(x) > t(x)$ and $f(x) < s(x)$ whenever $g(x) < s(x)$.

(Compare with the Asymmetrical in-between theorem !)

Hint. It does no harm to suppose that t, s are bounded. In light of Ex. 3.E.5, there are $u \in (\Psi^{\rho}_{\mathcal{T}_1})^{\uparrow}$ and $v \in (\Psi^{\mathcal{T}_2}_{\rho})^{\downarrow}$, $t \leqq v \leqq g \leqq u \leqq s$. Let

$$\varphi_n \in \Psi^{\rho}_{\mathcal{T}_1}, \ \varphi_n < u, \quad \varphi_n \nearrow u, \ \psi_n \in \Psi^{\mathcal{T}_2}_{\rho}, \ \psi_n > v, \ \psi_n \searrow v.$$

Now, define ω_n, ω as in Ex. 3.E.4 and put $f = \omega$.

3.E.7 Let $\mathcal{T}_1$, $\mathcal{T}_2$ and (X,ρ) be as in the previous exercise. If $M \subset X$ is both $\mathcal{T}_1$-and $\mathcal{T}_2$-closed, and if g is a G_δ-measurable function on X such that $g \wedge M$ is both $\mathcal{T}_1$-lower semicontinuous and $\mathcal{T}_2$-upper semicontinuous on M, then there is $f \in \Psi^{\mathcal{T}_2}_{\mathcal{T}_1}$ such that $f = g$ on M and f is finite at $x \in X$ provided $g(x)$ is finite.

Hint. Use the preceding exercise.

3.E.8 Let τ be a fine topology on a metric space (P,ρ) having the Lusin-Menchoff property, and let $M \subset X$ be a τ-closed set. Consider the following properties that a real τ-continuous function f on M may possess:

(i) there is a real τ-continuous extension F of f over X,

(ii) there is a real extension $g \in B_1(P)$ of f over X.

Then (ii)$\Longrightarrow$(i); if τ has the G_δ -insertion property, then (i) and (ii) are equivalent.

Hint. Use Ex. 3.E.7, or (in the proof of (ii)$\Longrightarrow$(i)) proceed as follows: Using B_1 - extension theorem 3.29, find a τ-continuous extension $F^* \in B_1(P)$ of f. Now, there is a τ-zero set $M \subset Z \subset [F^* = g]$ and a positive τ-continuous function h such that $Z = [h = +\infty]$. Put $F = \max(-h, \min(F^*,h))$.

3.E.9 Let τ be a fine topology on a topological space (X,ρ) and let $t \leqq s$ on X where t is upper semicontinuous and s is τ-lower semicontinuous. Show that there is a τ-continuous and upper semicontinuous function f on X such that $t \leqq f \leqq s$ and $t(x) < f(x) < s(x)$ at those points where $t(x) < s(x)$ provided either

(a) τ has the perfect Lusin-Menchoff property (cf. Ex. 3.B.9),

or

(b) ρ is a normal topology, τ has the Lusin-Menchoff property and there is a G_δ-measurable function g such that $t \leqq g \leqq s$ and $t(x) < g(x) <$ $< s(x)$ whenever $t(x) < s(x)$.

Hint. As for (a), use Ex. 3.B.8c. In case of (b), use Ex. 3.E.6.

3.E.10 Let τ be a fine topology on a topological space (X,ρ) and let s be a real τ-continuous and upper semicontinuous function on a τ-closed set $A \subset X$. Show that s can be extended to a real τ-continuous and upper semicontinuous function on X if either

(a) τ has the perfect Lusin-Menchoff property (cf. Ex. 3.B.9)

or

(b) ρ is a normal topology, τ has the Lusin-Menchoff property and there is a real G_δ-measurable extension of s to all of X.

Hint. Use the previous exercise.

3.E.11 Let τ be a fine topology on a perfectly normal topological space (X,ρ) having the Lusin-Menchoff property. The following conditions on a real τ-continuous and upper semicontinuous function s on a τ-closed set $A \subset X$ are equivalent:

(i) s can be extended to a real function on X which is τ-continuous and upper semicontinuous,

(ii) there is a real upper semicontinuous extension of s to all of X,

(iii) there is a real function $f \in B_1(X)$ such that $f = s$ on A.

Hint. Use the previous exercise.

3.E.12 Let $\rho = \tau_1 \subset \tau_2 \subset \ldots \subset \tau_n$ be a collection of topologies on a set X each of them having the complete Lusin-Menchoff property w.r.t. ρ (in particular, ρ is completely normal). Let g be a G_δ-measurable function on X, $M \subset X$ and let $f := g \wedge M$. Prove that there is an extension F of f to all of X with the following properties:

(a) if f is τ_i-upper (lower) semicontinuous at $x \in M$ for some i, then F is τ_i-upper (lower) semicontinuous at x,

(b) F is τ_i-continuous on $X \setminus \overline{M}$ (for $i = 1,\ldots,n$),

(c) if $g(z)$ is finite for some $z \in X$, then $F(z)$ is finite.

Hint. Let L_i (U_i, resp.) denote the set of all points of M at which f is τ_i-lower (upper, resp.) semicontinuous. Put

$$\tau_L = \tau(\tau,\ldots,\tau_n,L_1,\ldots L_n),\quad \tau_U = \tau(\tau_1,\ldots,\tau_n,U_1,\ldots,U_n)$$

(cf. Ex. 3.D.2). If

$$t := \begin{cases} f & \text{on } M \\ -\infty & \text{on } X\setminus M, \end{cases} \qquad s := \begin{cases} f & \text{on } M \\ +\infty & \text{on } X\setminus M, \end{cases}$$

then $t \le g \le s$, t is τ_U-upper semicontinuous and s τ_L-lower semicontinuous. Now, use Ex. 3.E.6.

3.E.13 Let τ be a fine topology on a completely normal topological space (X,ρ) having the complete Lusin-Menchoff property. If f denotes a restriction of a G_δ-measurable function g on X to a τ-closed set $F \subset X$, then there is a function f^* on X which extends f and

(a) f^* is τ-continuous except at those points at which f is not τ-continuous,

(b) $f^*(z)$ is finite whenever $g(z)$ is finite.

Hint. This is an immediate corollary of the preceding exercise.

3.E.14 Let (X,ρ) be a completely normal topological space. If $A \subset X$ and if $t \leq s$ are functions on X, t upper semicontinuous and s lower semicontinuous at all points of A, then there is a function h on X, $t \leq h \leq s$ and h is continuous at all points of A.

Hint. Use Ex. 3.D.1.

3.E.15 Let $t \leq s$ be functions on a completely normal topological space (X,ρ). If there is a G_δ-measurable function g, $t \leq g \leq s$, then there is a function f, $t \leq f \leq s$, such that

(a) f is lower semicontinuous at those points at which s is lower semicontinuous,

(b) f is upper semicontinuous at those points at which t is upper semicontinuous,

(c) $f(x) < s(x)$ whenever $g(x) < s(x)$ and $f(x) > t(x)$ whenever $g(x) > t(x)$.

Hint. Use the same idea as in Ex. 3.E.12.

3.E.16 Let Φ be a family of functions on a set X. A function f is said to be Φ*-bounded* if there are $t,s \in \Phi$ such that $t \leq f \leq s$. If Φ consists of bounded functions only and contains all constants, then f is Φ-bounded if and only if f is bounded.

Given a closed L-cone Φ of real functions on X, the following assertions are equivalent:

(i) there are σ-topologies τ_1 and τ_2 such that (X, τ_1, τ_2) is perfectly binormal (cf. Ex. 3.B.8) and $\Phi = \Phi^*$, where

$$\Phi^* := \{f: f \text{ is } \Phi\text{-bounded, } f \text{ is } \tau_1\text{-upper semicontinuous and } \tau_2\text{-lower semicontinuous}\},$$

(ii) there are σ-topologies τ_1 and τ_2 such that $\Phi = \Phi^*$,

(iii) $\Phi = \Phi^\uparrow \cap \Phi^\downarrow$,

(iv) given $s \in \Phi^\uparrow$, $t \in \Phi^\downarrow$, $t \leq s$, there is $g \in \Phi$ such that $t \leq g \leq s$, and $t < g < s$ on the set $[t < s]$,

(v) given $s \in \Phi^{\uparrow}$, $t \in \Phi^{\downarrow}$, $t \leqq s$, there is $g \in \Phi$ such that $t \leqq g \leqq s$

(for similar results cf. D. Feyel (1978), (1981)).

Hint. The implications (i) $\Rightarrow$ (ii) $\Rightarrow$ (iii) and (iv) $\Rightarrow$ (v) are obvious. For (iii) $\Rightarrow$ (iv) see Ex. 3.E.4a. Assuming (v), put $\tau_1 = \mathrm{Coz}(\Phi^{+})$, $\tau_2 = \mathrm{Coz}(\Phi^{-})$.

3.E.17 Let τ be a fine topology on a metric space (P, ρ).

(a) Show that the following two conditions are equivalent:

(i) τ-continuous functions separate closed sets and τ-closed sets,

(ii) if $t \lneqq s$ where t is upper semicontinuous, s is τ-lower semicontinuous, and if there is a B_1-function g, $t \lneqq g \lneqq s$, then there is a τ-continuous function f, $t \lneqq f \lneqq s$.

Hint. For (i) $\Rightarrow$ (ii), modify the proofs 3.27, 3.26 and (vi) $\Rightarrow$ (vii) of Theorem 3.11.

(b) The following conditions are also equivalent:

(i*) every closed set can be separated from any τ-closed set by a τ-continuous function from the Baire class one,

(ii*) if $t \lneqq s$ are as in (ii), then there is a τ-continuous B_1-function f, $t \lneqq f \lneqq s$ provided there is a B_1-function g, $t \lneqq g \lneqq s$,

(c) Show that, in generàl, (i) does not imply (i*).

(d) If τ has the Lusin-Menchoff property, then (i*) holds. Show that the converse is not true.

Hint. Consider the topology τ on $[0,1]$ defined as follows: A set $A \subset [0,1]$ is τ-open if $A \subset (0,1]$ or if there is $\delta > 0$ such that $[0,\delta] \setminus \{\frac{1}{n}: n \in \mathbb{N}\} \subset A$.

3.E.18 In-between and extension theorems for σ-topologies. Assuming that τ_1, τ_2 and τ are σ-topologies only, show that Theorems 3.26-3.30 remain valid.

Hint. For the proof of Theorem 3.30, use Ex. 3.D.6c.

3.E.19 (For the special case see Z. Grande (1979); a similar idea is used, e.g., in G. Lederer (1960).) Let A be a subset of a set X. Let Φ_A be a vector space of real functions on A and let Φ be a closed vector space of real functions on X. Assume that any function $f \in \Phi_A$ has an extension $F \in \Phi$ such that if the function f carries A to $[-k,k]$, then F has the same property. Then any function on A which is a uniform limit of functions from Φ_A can be extended to all of X resulting in a function from Φ.

Hint. Suppose $\{g_n\} \subset \Phi_A$ is uniformly convergent to g. Find a sequence $\{k_n\}$ such that $|h_n| \leq 1/n^2$ where $h_n := g_{k_{n+1}} - g_{k_n}$. Let $H_n \in \Phi$ be an extension of h_n such that $|H_n| \leq 1/n^2$ and let $G_{k_1} \in \Phi$ be an extension of g_{k_1}. Put

$$G = G_{k_1} + \sum_{i=1}^{\infty} H_i .$$

3.E.20 Let A be a subset of a metric space (P,ρ) and let τ be a topology on P. If any real ρ-upper semicontinuous function on A can be extended to a real ρ-upper semicontinuous and τ-continuous function on P, then any real B_1-function on A can be extended to a τ-continuous B_1-function on P.

Hint. Use Corollary 3.35 and the previous exercise where Φ_A is the family of all differences u_1-u_2 where u_i are real ρ-upper semicontinuous functions on A and Φ is the family of all τ-continuous B_1-functions on P.

3.E.21 Let τ be a fine topology on a metric space P having the Lusin-Menchoff property.

(a) Let M be a τ-isolated G_δ-subset of P. If f is a real Baire one function on M, then f can be extended to a real τ-continuous Baire one function on P.

Hint. Use Ex. 3.B.7f and either Ex. 3.E.8 or Ex. 3.E.20, Corollary 3.35 and Ex. 3.E.10.

(b) If $M \subset P$ is a τ-closed G_δ-set and $f \in B_1(M)$ is real and τ-continuous on M, then f can be extended to a real τ-continuous Baire one function over P.

Hint. Use again Ex. 3.B.7f and either Ex. 3.E.8 or Ex. 3.E.19, Corollary 3.36 and Ex. 3.E.10.

3.E.22 Let $\mathbb{Q} = A_1 \cup A_2$ where A_1, A_2 are disjoint and dense in $\mathbb{Q}$. If $f = \chi_{A_2}$, then f cannot be uniformly approximated by differences $s_1 - s_2$ of bounded upper semicontinuous functions on $\mathbb{Q}$. (Cf. Section 3.G.)

Hint. Use the fact that there is no Baire one extension of f to $\mathbb{R}$.

3.E.23 Let C be a G_δ-subset of a metric space P. Then any function of $B_1(C)$ can be extended to a Baire one function over P with "preservation of semicontinuity" such that the resulting function is real and continuous on $P \setminus \bar{C}$.

Hint. Apply Corollary 3.31 to the discrete topology on P (cf. Remark 3.17 and Corollary 3.8.b.)

REMARKS AND COMMENTS

The basis of this section consists in asymmetrical in-between theorem 3.27 to which we came on deriving the extension theorems where semicontinuity of extended functions is required to be preserved (Theorems 3.30 and 3.31). As we have noticed already, the extension to unbounded derivatives needs, in addition to approximate continuity, a "control" of extended functions by initial estimates.

The asymmetrical in-between theorem is in all a trivial consequence of our abstract in-between theorem.

Another important consequence of Theorem 3.27 consists in the B_1-extension theorem. As already mentioned, for the proof thereof the usual Tietze's method could be used, too (cf. J. Lukeš (1977 a)).

Related papers. J. Ceder and T.L.Pearson (1981), V.Kelar (1983) - both include an extension type theorem in the class of DB_1-functions; P. Vetro (1983).

3.F The equality $B_1(\tau) = B_2(\rho)$

In (1971a), D. Preiss proved that each function in Baire class two is a limit of a sequence of approximately continuous functions. A slight generalization of this result is given in the next theorem.

3.32 THEOREM. Let τ be a fine topology on a metric space (P,ρ) having the Lusin-Menchoff property. The following properties are equivalent:

(i) any Baire two function on P is the limit of a sequence of τ-continuous functions,

(ii) any ρ-F_σ - subset of P is a τ-G_δ - set.

Proof. (i)$\Rightarrow$(ii): Let F be an F_δ - set. Then χ_F is in Baire class two, and hence, by our assumptions, it is the limit of a sequence of τ-continuous functions, i.e. χ_F is a Baire one function in the topology τ. It follows (cf. Theorem 2.12) that χ_F is τ-G_δ-measurable. Thus, F is a τ-G_δ-set.

(ii)$\Rightarrow$(i): Let $a \in \mathbb{R}$. We need to prove that the sets $[f \geq a]$, $[f \leq a]$ are countable intersections of τ-cozero sets (cf. Ex. 3.A.1). By our assumptions, these sets are $F_{\sigma\delta}$. Thus it is sufficient to prove that any set of type F_σ is a countable intersection of τ-cozero sets. Assume that F is an F_σ-set. By the assumptions,

$F = \bigcap_{n=1}^{\infty} G_n$ with G_n being τ-open. Fix $n \in \mathbb{N}$. There are closed sets F^j, $F = \bigcup_{j=1}^{\infty} F^j$. For any $j \in \mathbb{N}$ we can find a τ-cozero set C_n^j such that $F^j \subset C_n^j \subset G_n$ (use Theorem 3.11.vii). Then $F = \bigcap_{n=1}^{\infty} \bigcup_{j=1}^{\infty} C_n^j$, and $\bigcup_{j=1}^{\infty} C_n^j$ is a τ-cozero set.

3.33 REMARK. If any τ-continuous function on P is in Baire class one (cf. Corollary 2.14) and if the assumptions of Theorem 3.32 are satisfied, then the set of all Baire two functions coincides with the set of all limits of sequences of τ-continuous functions.

EXERCISES

3.F.1 Let e^+ be the Sorgenfrey topology on $\mathbb{R}$.

(a) Using Theorem 3.32, show that $B_1(e^+) \neq B_2(e)$.

(b) Show that any function of $B_1(e^+)$ is a honorary Baire two function.

Hint. Any e^+-continuous function is continuous except a countable set.

(c) Using Ex. 4.A.2.e, show that any function of $B_1(e^+)$ is continuous at all points of a residual subset of $\mathbb{R}$. In particular, the Dirichlet function does not belong to $B_1(e^+)$.

REMARKS AND COMMENTS

A. Bruckner, J. Ceder and R. Keston (1968) proved that any function in the second

Baire class is a limit of a sequence of Darboux Baire one functions. Later on, D. Preiss showed (1969) that this sequence can consist even of derivatives, and in (1971a) he strengthened this result as follows: A Baire two function is a limit of a sequence of bounded approximately continuous functions. This result was discovered independently by M. Laczkovich and G. Petruska (1973) and extended to the case of fine topologies having the Lusin-Menchoff property by J. Lukeš (1977a).

Related papers: Z. Grande and M. Topolewska (1982).

3.G Approximation theorem

As an application of our abstract in-between theorem 3.2 we have shown in Section 3.A that any bounded approximately continuous function is a uniform limit of a sequence of differences of lower semicontinuous non-negative approximately continuous functions. In this section we establish a general approximation theorem omitting the condition of boundedness and giving a completely different proof. The next theorem provides the key to our applications.

3.34 APPROXIMATION THEOREM. Let $\mathcal{S}$ be a closed convex cone of bounded non-negative functions on a set X which contains constant functions and which is closed with respect to multiplication. Let $\gamma \in \mathcal{S}$, $0 < \gamma \leq 1$, be fixed. If f is a real function on X and f/γ is $Z(\mathcal{S})$-measurable, then there are positive functions $s, t \in \mathcal{S}$ such that

$$\left|\frac{1}{s} - \frac{1}{t} - f\right| \leq \gamma .$$

Proof. Find $g_k \in \mathcal{S}$ such that

$$[g_k > 0] = [(k-1)\gamma < f < (k+1)\gamma]$$

and

$$0 \leq g_k \leq 2^{-|k|-1}(|k|+1)^{-1}$$

for every integer k. Put

$$s = \frac{\sum_k g_k}{1 + \gamma \sum_{k<0} k g_k}, \quad t = \frac{\sum_k g_k}{1 - \gamma \sum_{k>0} k g_k} .$$

Using the fact that $(1-h)^{-1} = \sum_{n=0}^{\infty} h^n \in \mathcal{S}$ for each $h \in \mathcal{S}$,

$0 \leqq h \leqq 1/2$, it is clear from our assumptions that $s, t \in \mathcal{S}$. Further

$$\left| \frac{1}{s} - \frac{1}{t} - f \right| \leqq \gamma .$$

Indeed, pick $x \in X$ and find an integer p for which $f(x) \in [p\,\gamma(x), (p+1)\gamma(x))$. If $k \notin \{p, p+1\}$, we have $g_k(x) = 0$. Hence

$$\frac{1}{s(x)} - \frac{1}{t(x)} = \gamma(x) \frac{p g_p(x) + (p+1) g_{p+1}(x)}{g_p(x) + g_{p+1}(x)} .$$

We get

$$p\,\gamma(x) \leqq \frac{1}{s(x)} - \frac{1}{t(x)} \leqq (p+1)\,\gamma(x) ,$$

which completes the proof.

This theorem admits several corollaries three of which follow below and some of which appear as exercises.

3.35 COROLLARY. Each real G_δ-measurable function on a topological space is a uniform limit of a sequence each term of which is the difference of two positive lower semicontinuous functions.

Proof. Denote by $\mathcal{S}$ the family of all bounded non-negative upper semicontinuous functions on X. Let $\varepsilon > 0$, and put $\gamma = \varepsilon$. Obviously, $\mathcal{S}$ has all properties ascribed to it in Theorem 3.34. Let F be a G_δ-measurable function on X and let $I \subset \mathbb{R}$ be an open interval. Then $F^{-1}(I)$ is an F_σ-set, and therefore it is an $\mathcal{S}$-cozero set. (Indeed, it suffices to put $g = \sum_{n=1}^{\infty} 2^{-n} \chi_{F_n}$ provided $F^{-1}(I) = \bigcup_{n=1}^{\infty} F_n$ with F_n's closed.) According to Theorem 3.34 there are positive $f_i \in \mathcal{S}$ such that

$$\left| \frac{1}{f_1} - \frac{1}{f_2} - F \right| < \varepsilon .$$

3.36 COROLLARY. Let τ be a fine topology on a metric space (P, ρ) having the Lusin-Menchoff property. Then any real Baire one and τ-continuous function on P can be uniformly approximated by the differences of two positive τ-continuous lower semicontinuous functions.

Proof. We merely outline the idea and invite the reader to fill the details. Take $\mathcal{S}$ to be the family of all non-negative τ-continuous upper semicontinuous functions on P. Using Zahorski's property of τ and following the same line of proof as above, we establish the assertion.

Actually Theorem 3.34 gives also the following improvement of the last corollary:

3.37 PROPOSITION. Let τ be a fine topology on a metric space (P,ρ) having the Lusin-Menchoff property. Then to each couple $g < h$ of real Baire one τ-continuous functions there is a function f which is a difference of two real non-negative lower semicontinuous τ-continuous functions such that $g < f < h$ on P.

Proof. Let $\mathcal{S}$ be as in the proof of Corollary 3.36. By Corolarry 3.36 there are $s_o, t_o \in \mathcal{S}$ such that

$$\left| \frac{2}{h-g} - \frac{1}{s_o} + \frac{1}{t_o} \right| \leq 1.$$

In particular, we have $2(h-g)^{-1} < 1 + 1/s_o$. Hence there are $s,t \in \mathcal{S}$ such that

$$\left| \frac{h+g}{2} - \frac{1}{s} + \frac{1}{t} \right| \leq \frac{s_o}{s_o+1} < \frac{h-g}{2}$$

(Theorem 3.34 with $\gamma = s_o(s_o+1)^{-1} = 1 - (s_o+1)^{-1} \in \mathcal{S}$).
Thus, we get $g < s^{-1}-t^{-1} < h$.

REMARKS. (1) Inserting a desired difference between "f and $f-\varepsilon$" we see that Proposition 3.37 is in fact a generalization of Corollary 3.36.

(2) If τ is the discrete topology, the assertion of the preceding proposition is nothing else than Kempisty's theorem of (1921). Cf. also Ex.3.G.1. Note that the class of all differences of non-negative lower semicontinuous functions (cf. W. Sierpiński (1921a) or S. Mazurkiewicz (1921)) is strictly smaller than the Baire one class. In particular, we cannot replace strict inequality $g < h$ by $g \leq h$ in Proposition 3.37.

EXERCISES

In Ex. 3.G.1 - 3.G.5, Φ denotes a vector lattice of real functions on a set X containing the constant functions. Further, $B_1(\Phi)$ is the family of all pointwise limits of sequences from Φ.

3.G.1 If g,h are in $B_1(\Phi)$, $g < h$, then there are non-negative real functions $s,t \in \Phi^{\uparrow}$ such that $g < s-t < h$ on X.

Hint. Since for a function f in $B_1(\Phi)$, $f = \liminf f_n = \limsup f_n$ where $f_n \to f$ and $f_n \in \Phi$, there are real functions h_n, g_n such that

$$h_n \in \Phi^{\downarrow}, \quad h_n \nearrow h, \quad g_n \in \Phi^{\uparrow}, \quad g_n \searrow g, \quad h_1 \leqq 0.$$

Replacing g_n by $g_n + n^{-1}$, h_n by $h_n - n^{-1}$, it is no restriction to assume that $h_n < h$ and $g_n > g$. Put

$$s = \sum_{i=1}^{\infty} \max (g_i - h_i, 0), \quad t = \sum_{i=1}^{\infty} \max (g_i - h_{i+1}, 0) - h_1.$$

REMARKS. This method is the *Hausdorff-Kempisty method* of Ex. 3.E.4. Notice also that $B_1(\Phi) = \Phi^{\uparrow\downarrow} \cap \Phi^{\downarrow\uparrow}$ even in the case when Φ is supposed to be a lattice only - see W. Sierpiński (1932) (cf. R.D.Mauldin (1974) and see also Lemma 3.5).

3.G.2 Show that $B_1(\Phi)$ is closed under uniform convergence.

Hints. (a) Use Lemma 3.3 and the preceding remark of Ex. 3.G.1.

(b) (F. Hausdorff (1919), cf. [Haus], p. 237.) Let $f_n \in B_1(\Phi)$, $f_n \rightrightarrows f$. Find a subsequence $\{f_{n_k}\}$, such that $|f - f_{n_k}| < 2^{-k}$. Further find $g_j^k \in \Phi$, $|g_j^k| \leqq 2^{-k+1}$ such that

$$\lim_{j\to\infty} g_j^k = f_{n_{k+1}} - f_{n_k}.$$

Put $G_k = \sum_{i=1}^{k} g_k^i$ and show that

$$G_k \longrightarrow \sum_{i=1}^{\infty} (f_{n_{i+1}} - f_{n_i}) = f - f_{n_1} .$$

REMARK. This has been also proved by R. D. Mauldin (1974) or A. Csaszár and M. Laczkovich (1979) (for Φ being a translation lattice).

3.G.3 (F. Hausdorff (1919), Ch. Tucker (1968), R.D.Mauldin (1974)). Show that a real function f is in $B_1(\Phi)$ if and only if f is a uniform limit of a sequence $\{s_n - t_n\}$ where s_n, $t_n \in \Phi^{\uparrow}$ are real non-negative functions.

Hint. The necessity follows from Ex. 3.G.1 putting $g = f - n^{-1}$, $h = f$. To establish the sufficiency use Ex. 3.G.2.

3.G.4 Show that the following two assertions are equivalent:

(i) $f \in B_1(\Phi)$,

(ii) f is $Coz_{\sigma\delta}(\Phi)$-measurable.

Hint. The proof of (i) $\Rightarrow$ (ii) is standard. The implication (ii)$\Rightarrow$(i) can be proved by Abstract in-between theorem similarly as Proposition 3.7 using Ex. 3.G.2.

3.G.5 Proceeding in the following steps, show without use Ex. 3.G.4 that any $Coz_{\alpha\delta}(\Phi)$-measurable function is a uniform limit of a sequence $\{s_n - t_n\}$, where $s_n, t_n \in \Phi^{\uparrow}$ are non-negative:

(a) Let $\tilde{\Phi}$ be the familiy of all uniform limits of sequences from Φ. Show that $\tilde{\Phi}$ is a closed vector lattice and

$$B_1(\tilde{\Phi}) = B_1(\Phi),\ (\tilde{\Phi})^{\uparrow} = \Phi^{\uparrow},\ (\tilde{\Phi})^{\downarrow} = \Phi^{\downarrow},\ Z(\tilde{\Phi}) = Z_{\delta}(\Phi),\ Coz(\tilde{\Phi}) = Coz\ (\Phi).$$

Hint. Cf. Lemma 3.3.

(b) The family $\tilde{\Phi}_b$ of all bounded functions from $\tilde{\Phi}$ is closed under multiplications.

Hint. Since $hg = \frac{1}{2}(|h+g|^2 - |h|^2 - |g|^2)$, it is sufficient to show that $f^2 \in \tilde{\Phi}_b$ whenever $f \in \tilde{\Phi}_b$, $f \geq 0$. But it follows easily e.g. from the Abstract in-between theorem (see Ex. 3.A.7).

(c) Put

$$\mathcal{S} = \{f \in \Phi^{\downarrow} : f \text{ is bounded and non-negative}\}$$

and show that $\mathcal{S}$ is a closed L-cone which is closed under multiplication.

(d) If $f \in \mathcal{S}$ is positive, then $1/f \in \tilde{\Phi}^{\uparrow}$.

Hint. Let $f_n \in \tilde{\Phi}_b$, $f_n \searrow f$. Put $g_n = (\max(f_n, n^{-1}))^{-1}$. Then $g_n \nearrow 1/f$ and $g_n \in \tilde{\Phi}$ by Ex. 3.A.7 applied to $f = -\max(f_n, n^{-1})$ and $g(t) = -t^{-1}$.

(e) Show that $Coz_{\delta}(\tilde{\Phi}) \subset Z(\mathcal{S})$.

Hint. Imitate the proof of Lemma 3.6, or put $f = \sum_{n=1}^{\infty} 2^{-n} \lim_{k\to\infty} (1-f_n)^k$ where $f_n \in \tilde{\Phi}, 0 \leq f_n \leq 1$ and $\bigcap_{n=1}^{\infty} [f_n > 0] \in Coz_{\delta}(\tilde{\Phi})$.

(f) Use Theorem 3.34 for $\mathcal{S}$.

Remark. For an alternative proof of this assertion, within another setting of the abstract approximation theorem, see Ex. 3.G.11.

3.G.6 Let τ have the complete Lusin-Menchoff property with respect to a completely normal topological space (X,ρ) and let $A \subset B \subset X$. Assume further that $f < g$ are real Baire one function on X, f is ρ-continuous at all points of A and g τ-continuous at all points of B. Show that there are non-negative lower semicontinuous functions s,t such that

(a) $f < s - t < g$,

(b) s,t are ρ-continuous at all points of A and τ-continuous at all points of B.

Hint. Let τ_1 be the $X\setminus A$-modification of τ (w.r.t. ρ) and τ_2 the $X\setminus B$-modification of the discrete topology with respect to τ_1. Show that τ_2-continuous functions are those which are ρ-continuous at all points of A and τ-continuous at all points of B. Use Theorem 3.25 and Theorem 3.34.

3.G.7 Reduction theorem. (W. Sierpiński (1934), K. Kuratowski (1936) - cf. [Kur], § 30.VII, Th. 1). Let X be a perfect topological space. If $X = \bigcup_{n=1}^{\infty} A_n$ where A_n are F_σ-subsets of X, then there is a sequence $\{B_n\}$ of pairwise disjoint F_σ-subsets of X such that $X = \bigcup_{n=1}^{\infty} B_n$ and $B_n \subset A_n$.

Hint. Suppose $A_n = \bigcup_{k=1}^{\infty} F_{n,k}$, $F_{n,k}$ closed. Let $\{Z_m\}$ be a sequence formed by the family of all sets $\{F_{n,k}\}$ and put $Z_m^* = Z_m \setminus \bigcup_{k<m} Z_k$. Denote $N_i = \{m \in \mathbb{N};\, Z_m^* \subset A_i\}$, $N_i^* = N_i \setminus \bigcup_{j<i} N_j$ and put

$$B_i = \bigcup_{m \in N_i^*} Z_m^*.$$

REMARK. Cf. with Ex. 3.B.7.

3.G.8 Show that any real G_σ-measurable function f on a perfect topological space X is a uniform limit of a sequence of functions in $CG(X)$ (cf. also Ex. 2.D.14).

Hint. Pick $\varepsilon > 0$ and $k \in Z$. Denote $C_k = [k\varepsilon < f < (k+2)\varepsilon]$. In light of the reduction theorem of Ex. 3.G.7 find F_σ-sets $D_k \subset C_k$ such that $\bigcup_{k \in Z} D_k = X$

and $D_i \cap D_j = \emptyset$ for $i \neq j$. Show that the function $g := \sum_{k \in Z} k\chi_{D_k}$ belongs to $CG(X)$ and $|f-g| < \varepsilon$.

3.G.9 Conserving our earlier notations, let $\mathcal{S}$ be a closed convex cone of real functions on a set X such that

(*) $\quad -\log(-w) \in \mathcal{S}$ for every negative function $w \in \mathcal{S}$.

Assume that f is a real function on X such that

(**) for any reals $a < b < c < d$ there is a bounded function $w \in \mathcal{S}$, $w \leqq 0$ for which $[b \leqq f \leqq c] \subset [w < 0] \subset [a < f < d]$.

Then for every $\varepsilon > 0$ there are positive functions $u, v \in \mathcal{S}$ such that

$$|u - v - f| < \varepsilon .$$

Hint. It would be clearly sufficient to assume that $\varepsilon = 4$. For every $k \in Z$ find $w_k \in \mathcal{S}$ such that

$$[k-1 \leqq f \leqq k] \subset [w_k < 0] \subset [k-2 \leqq f \leqq k+1]$$

and

$$0 \leqq -w_k \leqq \frac{1}{3} e^{-k^2}.$$

Put $\tilde{v} = \sum_{k \in Z} e^k w_k$, $\tilde{u} = \sum_{k \in Z} w_k$ and show that $\tilde{u}, \tilde{v} \in \mathcal{S}$, $0 < -\tilde{u} < 1$, $0 < -\tilde{v} < 1$.

Then, as

$$e^{f(x)-2} \leqq \frac{\tilde{v}}{\tilde{u}} \leqq e^{f(x)+2}$$

for any $x \in X$, the functions

$$u := -\log(-\tilde{u}) \quad \text{and} \quad v := -\log(-\tilde{v})$$

satisfy the assertion.

3.G.10 Give alternative proofs of Corollary 3.35 and Corollary 3.36 based on the abstract theorem of Ex. 3.G.9.

3.G.11 Let Φ denote again a vector lattice of real functions on a set X containing the constant functions. Using Ex. 3.G.9, prove that any $\mathrm{Coz}_{\sigma\delta}(\Phi)$-measurable function is a uniform limit of a sequence $\{s_n - t_n\}$, where $s_n, t_n \in \Phi^{\uparrow}$ are nonnegative (cf. Ex. 3.G.5).

Hint. Apply Ex. 3.G.9 for $\mathcal{S} = \Phi^{\uparrow}$. By Lemma 3.3, $\mathcal{S}$ is closed. Further, for $w < 0$, $w \in \mathcal{S}$ use the equality

$$- \log(-w(x)) = \sup_{r \in \mathbb{Q} \cap (0,\infty)} \left(- \log r + \frac{w(x) + r}{r}\right).$$

To verify that any $\mathrm{Coz}_{\sigma\delta}(\Phi)$-measurable function f satisfies (**) of Ex. 3.G.9 use Ex. 3.G.5e.

3.G.12 We say a family $\mathcal{S}$ of real functions on a topological space X is closed under locally finite sums if $\sum_{\alpha \in A} f_\alpha \in \mathcal{S}$ whenever $\{f_\alpha\} \subset \mathcal{S}$ and for any $x \in X$ there exists a neighbourhood U of x such that all but finitely many f_α vanish on U.

Let $\mathcal{S}$ be a convex cone of real functions on a topological space $(X,\mathcal{T})$ which is closed under $\mathcal{T}$-locally finite sums. If $\mathcal{S}$ satisfies (*) of Ex. 3.G.9 and if f is a real $\mathcal{T}$-continuous function on X for which (**) of Ex. 3.G.9 holds, then for every $\varepsilon > 0$ there are positive functions $u, v \in \mathcal{S}$ such that

$$|u - v - f| < \varepsilon .$$

Hint. Construct u,v as in Ex. 3.G.9.

3.G.13 Let $\mathcal{S}$ be a vector space of real functions on a topological space $(X,\mathcal{T})$ which is closed under $\mathcal{T}$-locally finite sums (see Ex. 3.G.12). Let f be a $\mathcal{T}$-continuous real function such that for any reals $a < b < c < d$ there is a bounded function $w \in \mathcal{S}$, $w \geqq 0$ for which

$$[b \leqq f \leqq c] \subset [w > 0] \subset [a < f < d].$$

Then for every $\varepsilon > 0$ there are $u, v \in \mathcal{S}$ such that $v > 0$ and

$$\left|\frac{u}{v} - f\right| < \varepsilon .$$

Hint. We may again suppose $\varepsilon = 4$. For every $k \in \mathbb{Z}$ find $g_k \in \mathcal{S}$ such that

$$[k-1 \leqq f \leqq k] \subset [g_k > 0] \subset [k-2 < f < k+1]$$

and $0 \leqq g_k \leqq 2^{-|k|}$. Now put $u = \sum_{k \in Z} k\, g_k$ and $v = \sum_{k \in Z} g_k$.

REMARKS AND COMMENTS

W. Sierpiński (1921a) has shown that the class $\hat{B}_1$ of all differences of lower semicontinuous functions coincides with the class of all sums of absolutely convergent series of continuous functions and, moreover, he constructed a Baire one function which is not in $\hat{B}_1$. In a subsequent paper S. Mazurkiewicz (1921) simplified this construction and proved (by means of transfinite induction) that any bounded Baire one function can be uniformly approximated by functions of $\hat{B}_1$. This succession proceeds by another paper of W. Sierpiński (1921b) who proved this theorem without use of transfinite induction. Finally S. Kempisty (1921) dropped the condition of boundedness and proved that the uniform closure of $\hat{B}_1$ equals B_1.

In the same paper, S. Kempisty uses a technique of F. Hausdorff (1919) (for the proof of Hahn's theorem)and proves furthermore that a function of $\hat{B}_1$ can be always inserted between a couple $f < g$ of Baire one functions. This result immediately yields the theorem on uniform approximation of functions of B_1 by functions from $\hat{B}_1$. As an application of his separation theorem, W. Sierpiński in (1924) again proved that $\hat{B}_1$ is dense in B_1 (cf. Ex. 3.G.8).

In [Haus], F. Hausdorff generalized Kempisty's theorem to an abstract concept of complete ordinary systems of functions (cf. also F. Hausdorff (1919)). Ch. Tucker (1968) presented a more general theorem on uniform approximation with different proof. A similar result is presented also by R.D.Mauldin (1974).

Now, a natural question has arisen whether any approximately continuous function can be uniformly approximated by the differences of lower semicontinuous approximately continuous functions. Notice that it does not follow from any of the above mentioned general theorems. A positive solution of this problem was presented by S. Vaněček (1985) for the case of bounded functions. His method of proof (see also Proposition 3.9) uses the Lusin-Menchoff property of the density topology and represents an application of a generalized version of the Stone-Weierstrass theorem. Surprisingly enough, the passage from bounded functions to arbitrary ones is not evident so that the problem continued to be open.

Theorem 3.34 presents a further generalization of the mentioned approximation theorems from which follows both Tucker's version and the solution of the problem for approximately continuous functions. The proof of Theorem 3.34 is rather simple and does not use any preliminary means.

4. BAIRE AND BLUMBERG SPACES

4.A Baire-like properties

Recall that a *Baire space* is a topological space in which every nonempty open subset is second category in itself. A space X is Baire if and only if the intersection of each countable family of dense open sets in X is dense. Every open subspace of a Baire space is a Baire space. In contrast, this assertion fails for closed subspaces.

A space X is called a *strong Baire space* (*hereditarily Baire*, respectively) if every nonempty closed subspace (every subspace, respectively) is second category in itself.

It is easy to check that X is a strong Baire space if and only if every nonempty closed subspace of X is a Baire space. In particular, any strong Baire space is a Baire space. Since a space is second category if and only if the intersection of any sequence of dense open subsets is nonempty, a space X is strong Baire if and only if the intersection of any sequence of dense open subsets of every nonempty closed subspace of X is nonempty.

Complete metric spaces and locally compact Hausdorff spaces are strong Baire.

In Proposition 3.5 we have seen that if a fine topology $\mathcal{T}$ has the Lusin-Menchoff property with respect to a locally compact Hausdorff space (X,ρ), then $(X,\mathcal{T})$ is a Baire space. In this section we shall state more general results.

We say that a topological space $(X,\mathcal{T})$ is *cometrizable* if there is a metric topology ρ on X coarser than $\mathcal{T}$ such that each point of X has a neighborhood base in $(X,\mathcal{T})$ the elements of which are ρ-closed.

The next proposition follows immediately.

4.1 PROPOSITION. Every fine topology having the Lusin-Menchoff property with respect to a metric space (X,ρ) is cometrizable.

We say that a family $\mathcal{F} = \{F_s\}_{s\in S}$ of subsets of a set X has the *finite intersection property* if $\mathcal{F} \neq \emptyset$ and $F_{s_1} \cap F_{s_2} \cap \dots \cap F_{s_n} \neq \emptyset$ for any finite system $s_1, s_2, \dots, s_n \in S$.

A completely regular space X is *Čech-complete* if there is a sequence $\{\mathcal{U}_i\}_{i\in\mathbb{N}}$ of open coverings of X such that any family of closed sets $\{F_a: a\in A\}$ which has the finite intersection property, and which has the property that for each $i\in\mathbb{N}$ there exists an F_a contained in some $U\in\mathcal{U}_i$, has a nonempty intersection.

Any Čech-complete space is a strong Baire space. Every locally compact Hausdorff space and every complete metric space is Čech-complete. A metric space is Čech-complete if and only if it is topologically complete (i.e. if there is a complete metric on it).

For a more detailed study of Čech completeness we refer the reader to [Eng], § 3.9.

4.2 THEOREM. Let τ be a fine topology on a Čech-complete topological space (X,ρ) such that each point of X has a neighborhood base in (X,τ) the elements of which are ρ-closed. Then any τ-G_δ-set is τ-second category (in itself).

Proof. Let $\emptyset \neq A = \bigcap_{n=1}^{\infty} G_n$ where G_n are τ-open sets. Assume that $A = \bigcup_{n=1}^{\infty} A_n$ for a sequence $\{A_n\}$ of nowhere dense subsets of $(A, \tau \wedge A)$. Let $\{\mathcal{U}_i\}_{i\in\mathbb{N}}$ be a sequence of open covers of (X,ρ) from the definition of Čech completeness. Since A_1 is nowhere dense in $(A,\tau\wedge A)$, there is $z_1 \in A \setminus \overline{A_1}^{\tau}$. There is $U_1 \in \mathcal{U}_1$ containing z_1. Hence we can exhibit a ρ-closed τ-neighborhood E_1 of z_1 such that $E_1 \subset U_1 \cap G_1$ and $E_1 \cap A_1 = \emptyset$.

We choose further $z_2 \in (A \cap \operatorname{int}_\tau E_1)\setminus \overline{A_2}^{\tau}$, $U_2 \in \mathcal{U}_2$ with $z_2 \in U_2$, and a ρ--closed τ-neighborhood E_2 of z_2 such that $E_2 \cap A_2 = \emptyset$ and $E_2 \subset U_2 \cap G_2 \cap E_1$.

Inductively we find sequences $\{E_k\}$ and $\{U_k\}$ such that:

E_k are nonempty ρ-closed sets, $U_k \in \mathcal{U}_k$, $E_k \cap A_k = \emptyset$,

$E_k \subset U_k \cap G_k$ and $E_1 \supset E_2 \supset E_3 \supset \dots$.

The Čech completeness of (X,ρ) implies the existence of $a \in \bigcap_{k=1}^{\infty} E_k$.

However, any point of $\bigcap_{k=1}^{\infty} E_k$ belongs to $A \setminus \bigcup_{n=1}^{\infty} A_n$, which is a contradiction.

4.3 COROLLARY. If a fine topology τ has the Lusin-Menchoff property with respect to a Čech-complete topological space (X,ρ), then (X,τ) is a Baire space.

4.4 COROLLARY. If the conditions of Theorem 4.2 are satisfied and if, moreover, for a set $A \subset X$ there is a τ-G_δ-set M_A such that

$$M_A \subset A \subset \bar{M}_A^{\tau},$$

then A is a τ-second category set in itself.

Proof. It should be clear that whenever $D \subset H \subset \bar{D}^{\tau}$ and H is τ-first category in itself, then D is τ-first category in itself.

4.5 COROLLARY. Assume that a fine topology τ is cometrizable with respect to a metric topologically complete space (P,ρ) (in particular, let τ have the Lusin-Menchoff property with respect to (P,ρ)). If for each τ-perfect set $A \subset X$ there is a τ-G_δ-set $M \subset A$ which is τ-dense in A, then (X,τ) is a strong Baire space.

We conclude this part by introducing a useful bitopological notion of *weak Baire spaces* which combines the fine topology with the initial one.

Given a fine topology τ on a topological space (X,ρ), we say that (X,τ) is a *weak Baire space* (with respect to ρ) if the intersection of each countable family of τ-dense and ρ-open sets in X is τ-dense.

If (X,τ) is a Baire space, then it is weak Baire as well.

4.6 THEOREM. Let (X,ρ) be a strong Baire space. If a fine topology τ on X has the G_δ-insertion property, and if (X,τ) is a regular space, then (X,τ) is a weak Baire space (with respect to ρ).

Proof. Suppose $D = \bigcap_{n=1}^{\infty} G_n$, where G_n are τ-dense and ρ-open sets in X. It is to be shown that $\bar{D}^{\tau} = X$. To do this, assume there is $x \in X \setminus \bar{D}^{\tau}$ in order to arrive at a contradiction. The τ-regularity of X yields the existence of a τ-open set U for which $x \in U \subset \bar{U}^{\tau} \subset X \setminus \bar{D}^{\tau}$. Now we apply the G_δ-insertion property. There is a G_δ-set A such that $U \subset A \subset \bar{U}^{\tau}$. Since each set G_n is τ-dense in X, $G_n \cap U$ is τ-dense in U. Hence $G_n \cap U$ is ρ-dense in U. It follows that $G_n \cap U$ is ρ-dense in $\bar{U}$. The space (X,ρ) is strong Baire, therefore the set $A \cap \bigcap_{n=1}^{\infty} G_n = A \cap D$ is ρ-dense in $\bar{U}$. This contradicts the fact that $A \cap D = \emptyset$.

REMARK. Later on we shall use Theorem 4.6 in order to prove that certain fine topology spaces are not regular.

EXERCISES

4.A.1 Following A.R. Todd (1981), topologies τ and ρ on a set X are *S-related* if for each set $A \subset X$, $\mathrm{int}_\rho A \neq \emptyset$ if and only if $\mathrm{int}_\tau A \neq \emptyset$.

Show that τ and ρ are S-related if and only if a set $A \subset X$ is ρ-dense exactly when A is τ-dense.

4.A.2 Let τ be a fine topology on a topological space (X,ρ) S-related to ρ. Prove the following assertions:

(a) A is ρ-nowhere dense if and only if it is τ-nowhere dense.

(b) $\bar{A}^\rho \setminus \bar{A}^\tau$ and $\mathrm{int}_\tau A \setminus \mathrm{int}_\rho A$ are both ρ and τ-nowhere dense for every $A \subset X$.

(c) (X,ρ) is a Baire space if and only if (X,τ) is a Baire space.

(d) A has the ρ-Baire property if and only if it has the τ-Baire property.

(e) The set of all points where a function f on X is τ-continuous and ρ-discontinuous is a ρ-first category set. (In particular, any real τ-Baire one function on X or, more generally, any τ-G_δ-measurable function on X is ρ-continuous at all points of X except a ρ-first category set.)

Hint. If f is τ-continuous and ρ-discontinuous at x, then

$$x \in (\mathrm{int}_\tau [f > r] \setminus \mathrm{int}_\rho [f > r]) \cup (\mathrm{int}_\tau [f < r] \setminus \mathrm{int}_\rho [f < r])$$

for a suitable $r \in \mathbb{Q}$. Now, use Ex. 4.A.2.b and Ex. 2.D.11.

4.A.3 A fine topology τ on a topological space (X,ρ) has the *Slobodnik property* if the intersection of each countable family of τ-open ρ-dense sets in X is ρ-dense.

(a) Prove that τ has the Slobodnik property if and only if $\mathrm{int}_\rho(\bigcup_{n=1}^{\infty} F_n) = \emptyset$ whenever F_n are τ-closed and $\mathrm{int}_\rho F_n = \emptyset$.

(b) If τ has the Slobodnik property, then (X,ρ) is a Baire space.

4.A.4 Let τ be a fine topology on a topological space (X,ρ) S-related to ρ. Prove that the following assertions are equivalent:

(i) (X,ρ) is a Baire space,

(ii) (X,τ) is a Baire space,

(iii) τ has the Slobodnik property.

4.A.5 Let τ be a fine topology on a topological space (X,ρ) having the Slobodnik property. Then any τ-Baire one function f on X (or, more generally, any τ-G_δ measurable function) is ρ-continuous at all points of X except a ρ-first category set.

Hint. Set

$$A_r := [f > r] \setminus \operatorname{int}_\rho [f > r], \quad A^r := [f < r] \setminus \operatorname{int}_\rho [f < r].$$

Then the set D of all points of ρ-discontinuity is an F_σ-set and $D = \bigcup_{r \in \mathbb{Q}} (A_r \cup A^r)$. Now, use Ex. 4.A.3.a to show that $\operatorname{int}_\rho D = \emptyset$.

4.A.6 A fine topology τ on a topological space (X,ρ) has *property M* if the intersection of each countable family of τ-open ρ-dense sets in X is τ-dense. If τ has the property M, then:

(a) τ has the Slobodnik property,

(b) (X,τ) is a Baire space,

(c) if U is nonempty τ-open, then $\operatorname{int} \bar{U}^\tau \neq \emptyset$,

(d) if τ is regular, then ρ and τ are S-related.

Hint. To prove (c), assume $\operatorname{int}_\rho \bar{U}^\tau = \emptyset$. Then $X \setminus \bar{U}^\tau$ is τ-open and ρ-dense, hence it is τ-dense.

4.A.7 Let τ be the crosswise topology on $\mathbb{R}^2$ and let $U \subset \mathbb{R}^2$ be τ-open and dense. Show that there is a residual set $S \subset \mathbb{R}$ such that the set $\{y \in \mathbb{R}: (x,y) \in U\}$ is dense and open for every $x \in S$.

Hint. Let $\{I_n\}$ be a countable base of $\mathbb{R}$ formed by open intervals. Set $G_n := \bigcup_{y \in I_n} \{x: (x,y) \in U\}$ and show that G_n are dense open subsets of $\mathbb{R}$. Put

$$S = \bigcap_{n=1}^{\infty} G_n .$$

4.A.8 Prove that the crosswise topology τ on $\mathbb{R}^2$ has the following properties:

(a) τ has the property M (and, consequently, τ has the Slobodnik property and $(\mathbb{R}^2, \tau)$ is a Baire space),

(b) τ and the Euclidean topology are not S-related,

(c) τ is not regular

(cf. also Ex. 2.A.3).

Hint. (a) Use Ex. 4.A.7.

(b) It suffices to construct by induction a countable set C which is dense in $\mathbb{R}^2$ and has the property that it intersects each straight line parallel with an axis in at most one-point set. Such a set C is even τ-nowhere dense.

(c) Use Ex. 4.A.6.d.

4.A.9 (S. G. Slobodnik (1976)). If a function g is a limit of a sequence of separately continuous functions on $\mathbb{R}^2$, then g is continuous on $\mathbb{R}^2$ except a first category set.

Hint. (a) By Ex. 4.A.8, the crosswise topology τ has the Slobodnik property. Since a function is separately continuous exactly when it is τ-continuous, it suffices to use Ex. 4.A.5.

(b) Let σ be the weak topology on $\mathbb{R}^2$ induced by the family of all separately continuous functions (cf. also Ex. 2.B.3). Since τ has property M, then, obviously, σ has property M, and since σ is regular (cf. Theorem 2.3), the Euclidean topology and σ are S-related (Ex. 4.A.6.d). Now, use Ex. 4.A.2.

4.A.10 Let i be the ideal topology on $\mathbb{R}^n$ determined by the Euclidean topology and by the σ-ideal of all Lebesgue null sets. Show that $(\mathbb{R}^n, i)$ is not a Baire space and i has not the Slobodnik property.

Hint. Use the existence of a dense G_δ-set of measure zero.

4.A.11 If every first category subset of a topological space X is closed, then X is a hereditarily Baire space (cf. proof of Theorem 6.9.e).

4.A.12 If the conditions of Theorem 4.2 are satisfied and if, moreover, for each set $A \subset X$ there is a τ-G_δ-set M_A such that $\mathrm{der}_\tau A \subset M_A \subset \bar{A}^\tau$, then (X,τ) is a strong Baire space.

4.A.13 (a) A G_δ-subset of a (strong) Baire space is a (strong) Baire space.

(b) If every point of a topological space X has a neighborhood which is a (strong) Baire space, then X is a (strong) Baire space.

4.A.14 The following two properties of a metric space P are equivalent:

(i) X is of second category,

(ii) there are no disjoint G_δ-subsets of P both dense in P.

Hint. (i)$\Rightarrow$(ii) obviously (any G_δ dense subset of P is residual!).

(ii)$\Rightarrow$(i). Suppose that $X = \bigcup_{n=1}^{\infty} A_n$ where each A_n is closed and nowhere dense. Given $n \in \mathbb{N}$, let $\{G_n^\alpha\}_{\alpha \in I_n}$ be a locally finite open cover of P by sets with $\operatorname{diam} G_n^\alpha < \frac{1}{n}$. Zorn's lemma now provides systems of elements $\{y_n^\alpha\}_{\alpha \in I_n}$ and $\{z_n^\alpha\}_{\alpha \in I_n}$ such that

$$y_n^\alpha \in G_n^\alpha \setminus \Big(\bigcup_{j=1}^{n} A_j \cup \bigcup_{j=1}^{n-1} \bigcup_{\alpha \in I_j} \{z_j^\alpha\}\Big), \qquad z_n^\alpha \in G_n^\alpha \setminus \Big(\bigcup_{j=1}^{n} \bigcup_{\alpha \in I_j} \{y_j^\alpha\}\Big).$$

Put $Y_n = \bigcup_{\alpha \in I_n} \{y_n^\alpha\}$, $Y = \bigcup_{n=1}^{\infty} Y_n$. Show that Y_n are closed and Y is a dense F_σ-set. Since $\bigcup_{n=1}^{\infty} \bigcup_{\alpha \in I_n} \{z_n^\alpha\} \subset P \setminus Y$, the set $P \setminus Y$ is a dense G_δ-set. On the other hand,

$$P \setminus Y = \bigcup_{n=1}^{\infty} A_n \setminus \Big(\bigcup_{n=1}^{\infty} \bigcup_{\alpha \in I_n} \{y_n^\alpha\}\Big) = \bigcup_{n=1}^{\infty} \Big(A_n \setminus \bigcup_{j=1}^{n-1} \bigcup_{\alpha \in I_j} \{y_j^\alpha\}\Big).$$

Hence $P \setminus Y$ is also an F_σ-set.

REMARKS AND COMMENTS

First and second category spaces were defined by R. Baire (1899). The "Baire category theorem" was first proved independently by W. Osgood (1897) for the real line and by R. Baire (1899) for the spaces $\mathbb{R}^n$.

It was F. Hausdorff (1914) who showed that this theorem holds for all complete metric spaces. The same observation for the class of locally compact spaces is due essentially to R.L.Moore (1924). Nowadays, the spaces where the Baire category theorem holds are called Baire spaces. In addition to complete metric spaces or locally compact spaces, a number of other classes studied in this text are further important examples of Baire spaces: The spaces determined by means of density topologies or fine topologies in harmonic spaces are of this type.

Strong Baire spaces which are occasionally called totally non-meagre were termed by F. Hausdorff as F_{II} spaces and were defined in the category of metric spaces as those for which every nonempty closed subset is second category in itself. It was proved by W. Hurewicz (1928) that a space is an F_{II} space if and only if every nonempty perfect set is uncountable.

Čech complete spaces were defined in Čech's paper (1937). He proved also the Baire category theorem for this class of spaces.

The importance of the concept of cometrizability was brought out only recently, especially while studying various set-theoretic questions (cf. F.D. Tall (1978)).

In the case of the initial topology being locally compact, the Lusin-Menchoff property implies the fine topology to be Baire (J. Lukeš (1977a)).We generalize this theorem for the case when the initial topology is Čech-complete.

The concept of weak Baire spaces is introduced to enable us to formulate a sufficient condition for the given fine topology not to be regular.

4.B Blumberg spaces

In (1922), H. Blumberg showed that for any real function f defined on the real line $\mathbb{R}$ there is a dense subset D of $\mathbb{R}$ such that the restriction of f to D is continuous. We say that a topological space X is a *Blumberg space* if for any real function f on X there is a dense subset D of X such that $f \restriction D$ is continuous. The result of J. C. Bradford and C. Goffman (1960) shows that a metric space P is Blumberg if and only if P is a Baire space. While any topological Blumberg space is Baire, the converse is not true in general.

Using the following elementary theorem, we can produce examples of non-Blumberg spaces. (We shall see that not only the density topology on $\mathbb{R}$ is non-Blumberg. Even more general abstract density topologies as well as fine topologies of harmonic spaces serve as examples of (Baire) non-Blumberg spaces).

4.7 THEOREM. Let (X,τ) be a topological space such that each dense subset of X is of cardinality $2^{\aleph_0}$. If there is a collection $\mathcal{M} \subset \exp X$ of cardinality $\leq 2^{\aleph_0}$ such that τ has the $\mathcal{M}$-insertion property, then (X,τ) is not a Blumberg space.

Proof. Assign to each $B \subset X$ a set $M(B) \in \mathcal{M}$ for which

$$\operatorname{int} \bar{B} \subset M(B) \subset \bar{B}.$$

'or any dense subset A of X, and for any real function f on X we put

$$f_A(y) := \sup\{a:\ y \in M(\{x \in A:\ f(x) \geq a\})\}.$$

ince for $a \in \mathbb{R}$ we have

$$\{y \in X:\ f_A(y) \geq a\} = \bigcap_{\substack{r<a \\ r \in \mathbb{Q}}} M(\{y \in A:\ f(y) \geq r\}),$$

t follows that any function f_A is measurable with respect to a certain system of ets which is of cardinality $\leq 2^{\aleph_0}$. By this observation one reaches the conclusion hat the collection

$$\Phi := \{f_A:\ A \text{ is dense in } X,\ f \text{ is a function on } X\}$$

s of cardinality $\leq 2^{\aleph_0}$. Let Λ (resp. λ) be the first ordinal number of cardi- ality $2^{\aleph_0}$ (resp. $|\Phi|$). Suppose now that $\{x_\alpha\}_{\alpha<\Lambda}$ (resp. $\{\varphi_\gamma\}_{\gamma<\lambda}$) is a trans- inite sequence of all points of X (resp. of all functions from Φ). We can con- truct a function f on X such that

$$f(x_\alpha) \neq \varphi_\gamma(x_\alpha) \quad \text{for each } \gamma<\alpha \ (\gamma<\lambda, \alpha<\Lambda).$$

hen for each $g \in \Phi$, the cardinality of $\{x \in X:\ f(x) = g(x)\}$ is less than $|\Phi| \leq$ $\aleph_0$. Hence, it follows easily that there is no dense subset A of X for which f/A is continuous. If it existed, so $f_A \in \Phi$, and this would lead to a contradic- ion since $f = f_A$ on A and the cardinality of A is $2^{\aleph_0}$.

XERCISES

.B.1 (J. Lukeš and L. Zajíček (1976)). Let τ be a fine topology on a separab- e metric space (P,ρ). If every τ-dense subset of P has cardinality $2^{\aleph_0}$ and -derivative of each set is Borel, then (P,τ) is not a Blumberg space.

int. Use Theorem 4.7.

.B.2 Let $\mathcal{L}$ be an ideal on a separable metric space (P,ρ) such that $P \notin \mathcal{L}$ and aving the property that $\mathcal{L}$ contains each set of cardinality less than $2^{\aleph_0}$. If i s the $\mathcal{L}$-ideal topology, then (P,i) is not a Blumberg space.

int. Use the preceding exercise.

4.B.3 Let (P,ρ) be a separable metric space, card $P = 2^{\aleph_0}$. Show that there exists a function f on P such that for no c-dense subset D of P, $f \restriction D$ is continuous.

Hint. Use Ex. 4.B.2.

4.B.4 Let τ be a fine topology on a topological space (X,ρ) S-related to ρ (cf. Ex. 4.A.1). If (X,ρ) is a Blumberg space, then so is (X,τ).

REMARKS AND COMMENTS

The first examples of non-Blumberg Baire space are due to Jr.H.E. White (1974) (assuming the continuum hypothesis, the density topology on the real line serves as an example) and (1975b) (e.g. any Baire space of cardinality, weight and density character $2^{\aleph_0}$ satisfying the countable chain condition (CCC), in which families first category sets and nowhere dense sets coincide). Further examples were produced by R. Levy (1973) (any η_1-set of cardinality $2^{\aleph_0}$) and (1974). W.A.R. Weiss (1975) constructed even an example of a compact non-Blumberg space.

In J. Lukeš and L. Zajíček (1976) general sufficient conditions are produced which enable us to recognize that the given topological space is not Blumberg. The method was inspired by the fact that the system of all derived sets in the density topology or in the fine topology in potential theory has a small cardinality. Another generalization of this idea is given in Theorem 4.7 which was discovered by consistent use of the $\mathfrak{M}$-insertion property.

Related papers. Jr. H.E. White (1975 a), W.A.R. Weiss (1977).

5. CONNECTIVITY PROPERTIES OF FINE TOPOLOGIES

The basic purpose in this section is to give some conditions for a set A to be connected in a fine topology τ on a topological space (X,ρ). Of course, connectedness of A is a necessary condition for τ-connectedness of A. The important paper of A. Denjoy (1915), p. 181, contains implicitely the assertion that the real line is connected in the density topology. To proceed in giving more general results, we need the following basic topological lemma.

5.1 TOPOLOGICAL LEMMA. Let τ be a fine topology on a strong Baire, connected space (X,ρ). Assume that any subset of X which is simultaneously τ-closed and τ-open is of type G_δ, and the τ-closure of any open set G is dense in the boundary of G. Then X is τ-connected.

Proof. Assume $X = A \cup B$, where A, B are disjoint τ-closed sets, $A \neq \emptyset \neq B$.

It follows that both A and B are G_δ - sets. Denote $M := \bar{A} \cap \bar{B}$ $(= \partial A = \partial B)$. Since X is connected, $M \neq \emptyset$. By the strong Baire property, one of the sets A,B (say A) cannot be dense in M. Thus there is an open set $V \subset X$ such that $\emptyset \neq V \cap M \subset B$. Using the equality $X \setminus M = \text{int } A \cup \text{int } B$, we obtain that

$$V \cap A = V \cap \text{int } A.$$

Hence

$$V \cap M = V \cap \partial A = V \cap \partial(\text{int } A).$$

Since $\overline{\text{int } A}^{\tau}$ is dense in $\partial(\text{int } A)$, we get

$$\emptyset \neq \overline{\text{int } A}^{\tau} \cap \partial(\text{int } A) \cap V \subset \bar{A}^{\tau} \cap M \cap V \subset A \cap B = \emptyset,$$

which is a contradiciton.

5.2 COROLLARY. Let (X,ρ) be a locally connected, strong Baire topological space, and let τ be a fine topology on X having the G_δ-insertion property. Then any open connected set is τ-connected if and only if the τ-closure of any open set G is dense in the boundary of G.

Proof. Assume that the τ-closure of every open set G is dense in the boundary of G, and let $H \subset X$ be an open connected set. Then the subspace H fulfils the assumptions of our topological lemma. In fact, if $A \subset H$ is both τ-closed and τ-open in (H,τ), and G is a G_δ-set such that $\text{int}_\tau A \subset G \subset \bar{A}^{\tau}$, then $A = H \cap G$, and therefore A is a G_δ-subset of H. The remaining two assumptions are clearly satisfied.

Suppose now that there exists an open set G such that the τ-closure of G is not dense in the boundary of G. Then there is $x \in \partial G$ and an open connected neighborhood V of X such that $V \cap \bar{G}^{\tau} \cap \partial G = \emptyset$. Since τ is finer than ρ, $\emptyset \neq V \cap G \neq V$ is both τ-open and τ-closed in V, and thus V is not τ-connected.

REMARK. Notice that we do not use the local connectedness of X in the proof of the "if-part" of the previous corollary.

5.3 COROLLARY. Assume that a fine topology τ has the G_δ-insertion property on a metric space (P,ρ). If

(a) every closed and bounded subset of P is compact,

(b) $B(x,r) := \{y \in P: \rho(y,x) \leq r\} \subset \overline{U(x,r)}^{\tau}$ for each $x \in P$ and $r > 0$, then every open and connected subset of P is also τ-connected.

Proof. Notice that P is strong Baire by (a).

According to the previous corollary and following remark, it suffices to show that the τ-closure of every open set G is dense in the boundary of G. Pick $x \in \partial G$ and $\varepsilon > 0$. There is $y \in G$ and $r > 0$ such that

$$\rho(x,y) < \frac{1}{2}\varepsilon, \quad U(y,r) \subset G, \text{ and } \quad U(y,r') \cap (P \setminus G) \neq \emptyset \quad \text{whenever} \quad r' > r.$$

Thus there is a sequence $\{x_n\} \subset P \setminus G$ such that $\rho(y,x_n) \searrow r$. Since $B(y,r+1)$ is compact, the sequence $\{x_n\}$ admits a subsequence converging to some point $z \in (P \setminus G) \cap B(y,r)$. In light of (b) we have $z \in \overline{U(y,r)}^{\tau}$, and therefore $z \in \partial G \cap \bar{G}^{\tau}$. Since $\rho(x,z) < \varepsilon$, we see that $\bar{G}^{\tau}$ is dense in ∂G.

Let τ be a fine topology on a topological space (X,ρ). As it was remarked above, any τ-connected subset of X is connected. The converse assertion is far from being true. The above topological lemma provides the key to our main results 5.2 and 5.3 giving sufficient conditions under which τ-connectedness of any open and connected subset takes place. In the end of this section we consider some useful properties guaranteeing that more general connected sets were τ-connected. The question is, whether

(1) every τ-open connected set is τ-connected,

or

(2) $U \setminus F$ is τ-connected whenever U is open and connected and F belongs to a certain class of (small) sets.

5.4 PROPOSITION. Consider the following properties that a fine topology τ on (X,ρ) may possess:

(i) every τ-open connected set is τ-connected,

(ii) if V and W are disjoint τ-open connected sets, then $V \cap \bar{W} = \emptyset$,

(iii) if V is τ-open τ-connected and W is τ-open connected, $V \cap W = \emptyset$, then $V \cap \bar{W} = \emptyset$,

Then (i) $\Rightarrow$ (ii) $\Rightarrow$ (iii); if ρ and τ are locally connected, then all four properties are equivalent.

Proof. (i)$\Rightarrow$(ii): The set $U := V \cup W$ is τ-open and τ-disconnected. The sets V and $W \cup (V \cap \bar{W}) \subset \bar{W}$ are connected. By assuming $V \cap \bar{W} \neq \emptyset$ we establish that $U = V \cup (W \cup (V \cap \bar{W}))$ is connected.

The implications (ii) $\Rightarrow$ (iii) $\Rightarrow$(iv.b) hold trivially.

(iii)$\Rightarrow$(iv.a): Let G be a τ-component of a connected open set U and assume $G \neq U$. Since τ is locally connected, G is τ-open, and in light of (iii), $U \setminus \bar{G} = U \setminus G \neq \emptyset$. Let D be a component of $X \setminus \bar{G}$ for which $D \cap U \neq \emptyset$. Local connectedness of ρ yields that D is open, $D \cap G = \emptyset$. Furthermore, D is closed in $X \setminus \bar{G}$, and since (iii) implies that $G \cap \bar{D} = \emptyset$ we get that D is closed in $(X \setminus \bar{G}) \cup \cup G$. It follows that $\emptyset \neq D \cap U \neq U$ is both open and closed in $U \subset (X \setminus \bar{G}) \cup G$, which contradicts the fact that U is connected.

(iv)$\Rightarrow$(i): Assuming (iv.b) and following the same line as in the proof of (iii)$\Rightarrow$(iv.a), we establish (i). In the course of the proof we use (iv.a) to deduce that the set D in consideration is τ-connected.

5.5 PROPOSITION. Let τ be a fine topology on a strong Baire topological space (X,ρ). Suppose that:

(a) τ has the G_δ-insertion property,

(b) $\bar{G}^\tau$ is residual in ∂G whenever $G \subset X$ is open,

(c) (X,τ) has no isolated points.

Let $S \subset X$ be a τ-isolated F_σ-set for which there is a base $\mathcal{L}$ of the topology ρ such that $V \setminus S$ is connected for each $V \in \mathcal{L}$. Then $X \setminus S$ is τ-connected provided it is connected.

Proof. Since S is τ-isolated, we have in light of (c) that $\text{int}_\tau S = \text{int } S = \emptyset$. We prove that $Y := X \setminus S$ satisfies the conditions of the Topological lemma 5.1.

1) (Y,ρ) is a strong Baire space because Y is a G_δ - subset of X and Y is connected by the assumptions.

2) If $A \subset Y$ is both τ-closed and τ-open in Y, then A is a G_δ-set. Indeed, A is τ-open in X and there is a G_δ-set $G \subset X$ such that $A \subset G \subset \bar{A}^\tau$. Hence $A = G \cap Y$ is also a G_δ-set.

3) It remains to show that the τ-closure of any open subset H of Y is dense in the boundary B of H (in Y). Let G be a maximal open subset of X for which $G \cap Y = H$. Let B_X denote the boundary of G in X. Since $\text{int } S = \emptyset$,

we get that $B = B_X \cap Y = B_X \setminus S$. We prove that S is a first category set in B_X. If this is not the case, the fact that S is an F_σ-set yields the existence of $V \in \mathcal{L}$ such that $\emptyset \neq V \cap B_X \subset S$. Then $(V \setminus S) \cap G \neq \emptyset$ is both open and closed in $V \setminus S$. Hence $V \setminus S \subset G$. It follows now that

$$(G \cup V) \cap Y = (G \cup V) \setminus S = G \setminus S = H.$$

This contradicts the maximality of G since $G \subsetneq G \cup V$.

Since S is τ-isolated, we have $\mathrm{der}_\tau H = \mathrm{der}_\tau G$. From (b) it follows that $\mathrm{der}_\tau H$ is residual in B_X. Finally, the fact that S is a first category set in B_X leads to the conclusion that $\mathrm{der}_\tau H$ is residual in $B_X \setminus S = B$.

EXERCISES

5.1 Let $\mathcal{I}$ be an ideal on a connected topological space (X,ρ) satisfying (*). If $\mathcal{I}$ does not contain any nonempty open set, then X is connected in the $\mathcal{I}$-ideal topology.

5.2 Let τ be the crosswise topology on $\mathbb{R}^2$. Show that:

(a) every open connected set is τ-connected,

(b) τ is locally connected,

(c) there is a τ-open connected set which is not τ-connected.

Hint. (a) Give a direct proof, or use Corollary 5.2 (cf. also Example 2.15).

(b) Show that each τ-component of each τ-open set is τ-open.

(c) Put

$$V = \{(0,0)\} \cup \{(x,y): \ y \neq x \ \text{ and } \ y \neq 2x\}.$$

REMARKS AND COMMENTS

We have already mentioned the fact that as early as in Denjoy's paper (1915) an assertion about connectedness of $\mathbb{R}$ in the density topology implicitely appears. The generalization of this result for regions in $\mathbb{R}^n$ was set out implicitely by J. Ridder (1929). C. Goffman and D. Waterman (1961) explicitely formulate this assertion and present an alternative proof. This result was generalized by R.S. Troyer and W.P. Ziemer (1963) for the case of some density topologies formed by means of the cube differentiation basis in $\mathbb{R}^n$ and of Radon measure. Similar questions are examined also by W. Eames (1971) and others.

The question of connectedness of the fine topology of harmonic spaces was raised by Ch. Berg and was settled positively in important papers of B.Fuglede (1970), (1971) for the general theory of harmonic spaces with the presence of the domination axiom D.

A general method for determining the connectedness of fine topologies is presented by J. Lukeš and L. Zajíček in (1977 c). The relatively simple topological lemma can be applied both to the case of the density topology (which leads to an alternative proof of its connectedness) and to the fine topology on harmonic spaces. Moreover, the authors prove that it is the connectedness, which entirely characterizes the ellipticity of a harmonic space and thus leads to a generalization of Fuglede's result.

The topological lemma 5.1 of this section is even more general than the one presented in J. Lukeš and L. Zajíček (1977 c). Instead of residuality only a weaker assumption is needed which, in its consequences, leads to the proof of connectedness for a wider class of fine topologies. In particular, the connectedness of the density topology considered by R.S.Troyer and W.P. Ziemer in (1963) can be proved as we shall see later in Section 6.C.

It is known that in the case of the classical fine topology associated with the Laplace equation a finely open subset of $\mathbb{R}^2$ is finely connected if and only if it is connected with respect to the Euclidean topology of $\mathbb{R}^2$ (see T.W.Gamelin and T.J.Lyons (1983)) and this result has no analogy in $\mathbb{R}^n$ for $n > 2$. Proposition 5.4 presents necessary and sufficient conditions guaranteeing the validity of this assertion in general case of fine topologies.

As just shown, not only open and connected sets can be finely connected. For the case of ordinary density topologies an analogy of the above conclusions does not hold but, as remarked for instance by C. Goffman and D. Waterman (1961), the set $\mathbb{R}^n \setminus S$ where S is countable is density connected. The generalization of this observation presented in Theorem 5.5 enables then further applications in Section 6.C.

Related papers: V. Šverák (1982).

6. DENSITY TOPOLOGIES

A. Density topology on the real line

B. Density topologies on metric spaces

C. Properties of density topologies depending on features of differentiation bases

D. Density and superdensity topologies and the Lusin-Menchoff property

E. Abstract density topologies

| F. Lifting topologies

6.A Density topology on the real line

The density topology on the real line is, perhaps, the most instructive example of a fine topology. Its methods bring more clarity to some questions of the theory of real functions. We sketch the definition and the main properties. Further study of more general "density topologies" will appear in the next sections.

A point $z \in \mathbb{R}$ is said to be a *point of density* of a measurable set $M \subset \mathbb{R}$ if

$$\lim_{h \to 0_+} \frac{1}{2h} \lambda(M \cap (z-h, z+h)) = 1.$$

The next theorem is of great significance.

6.1 THEOREM. (Lebesgue density theorem). Almost every point of any measurable set $M \subset \mathbb{R}$ is a point of density of M.

Proof. It follows from the Lebesgue integration theory that the derivative of the indefinite integral of a locally integrable function g equals g almost everywhere. We apply this theorem to the case of g being the characteristic function of the set M.

REMARK. An elementary proof of Lebesgue density theorem can be found in W. Sierpiński (1923), J.C. Oxtoby (1971) and L. Zajíček (1979).

A measurable set $M \subset \mathbb{R}$ will be termed *d-open* if each point of M is a point of density of M.

We will show that the collection of all d-open sets forms a topology d which will be called the *density topology*. Clearly, this topology is finer than the Euclidean one.

6.2 THEOREM (quasi-Lindelöf property of d). Let $\mathfrak{M}$ be a family of d-open sets. Then $\mathfrak{M}$ contains a countable subfamily $\mathfrak{M}_0$ whose union differs from the union of $\mathfrak{M}$ by a Lebesgue null set.

Proof. We may assume that the set $T := \bigcup_{M \in \mathfrak{M}} M$ is contained in some bounded interval I. Let $\mathcal{S}$ be the collection of all countable unions of sets from $\mathfrak{M}$. There is $S \in \mathcal{S}$ such that

$$\lambda(S) = \sup\{\lambda(M):\ M \in \mathcal{S}\}.$$

Every point $x \in T$ is a density point of some set $M \in \mathcal{M}$. Since $\lambda(M \cup S) = \lambda(S)$, we have $\lambda(M \setminus S) = 0$. Hence x is a density point of $M \cap S$ and, of course, a density point of S. Clearly, x cannot be a density point of $I \setminus S$. By Theorem 6.1, almost every point of $I \setminus S$ is a density point of $I \setminus S$, which yields that $T \setminus S$ is a Lebesgue null set.

6.3 COROLLARY. The family d of all d-open sets forms a topology.

Proof. Somewhat difficult is to prove the fact that an arbitrary union of d-open sets is measurable. But it is enough to apply the quasi-Lindelöf property of d. (Another proof can be found in Ex. 6.A.9. Cf. also Remark 6.16.a.)

6.4 PROPOSITION. The d-interior of any measurable set M consists of all density points of M belonging to M..

Proof. Let U be the set of all density points of M belonging to M. By Lebesgue density theorem, $M \setminus U$ is a null set. Hence U is measurable and any point of U is a density point of U. In other words, U is a d-open subset of M. On the other hand, any d-open subset of M is contained in U.

6.5 COROLLARY. A measurable set M is a d-neighborhood of a point $z \in M$ if and only if z is a density point of M.

A function f defined on a neighborhood of a point $z \in \mathbb{R}$ is termed *approximately continuous at* z provided there is a measurable set $M \subset \mathbb{R}$ such that z is a point of density of M and

$$\lim_{M \ni x \to z} f(x) = f(z).$$

If f is approximately continuous at every point, we say simply that f is approximately continuous.

6.6 THEOREM. A function f is approximately continuous at a point z if and only if it is d-continuous at z. In particular, f is approximately continuous on $\mathbb{R}$ if and only if the sets

$$\{x \in \mathbb{R}: f(x) > a\}, \quad \{x \in \mathbb{R}: f(x) < a\}$$

are d-open for any $a \in \mathbb{R}$.

Proof. By the preceding corollary, any function which is approximately continuous at a point z is also d-continuous at z. Assume now f is d-continuous at z. For any $n \in \mathbb{N}$ there is a d-open set U_n containing z such that the inequality $|f(x) - f(z)| < n^{-1}$ holds for every $x \in U_n$. Let $r_n \searrow 0$ be such a sequence that

$$\frac{1}{2h} \lambda(U_n \cap (z-h, z+h)) > 1 - n^{-1}$$

whenever $h \in (0, r_n]$. Let $M_n := U_n \setminus (z - r_{n+1} n^{-1}, z + r_{n+1} n^{-1})$,

$M := \bigcup_{n=1}^{\infty} M_n$. For $h \in [r_{n+1}, r_n]$ we have

$$\frac{1}{2h} \lambda(M \cap (z-h, z+h)) \geq \frac{1}{2h} \lambda(M_n \cap (z-h, z+h)) \geq 1 - \frac{1}{n} - \frac{2r_{n+1}}{2hn} \geq 1 - \frac{2}{n}.$$

We see that z is a density point of M and

$$\lim_{M \ni x \to z} f(x) = f(z).$$

6.7 COROLLARY. Any approximately continuous function is measurable.

Proof. It follows from the preceding theorem since any d-open set is measurable.

REMARK. Of course, much more can be said. The Denjoy-Stepanoff theorem asserts that a function is approximately continuous almost everywhere if and only if it is measurable (cf. Theorem 6.20).

6.8 THEOREM. (a) Any bounded approximately continuous function is a derivative.

(b) Any approximately continuous function is a Darboux function of Baire one class.

Proof. (a) The proof of this part depends on the observation that, whenever f is a bounded approximately continuous function, then f equals the derivative of its indefinite Lebesgue integral (since f is measurable by Corollary 6.7, it suffices to check that $f = F'$ where $F: \ x \longmapsto \int_a^x f$). For the details see e.g. [Bru].

(b) Since any derivative is a Darboux Baire one function, the assertion follows immediately from the part (a) in the case of a bounded approximately continuous function. If f is not bounded, we consider the function arctg f instead of f.

6.9 THEOREM (properties of the density topology). The density topology d on the real line has the following properties:

(a) $(\mathbb{R},d)$ is a completely regular Hausdorff space.

(b) Every nonempty d-open set is uncountable and each countable set is d-closed. In particular,

(b-1) there are no isolated points in the d-topology,

(b-2) the d-compact sets are exactly the finite sets,

(b-3) if $d\text{-}\lim x_n = x$, then the set $\{n \in \mathbb{N}: x_n \neq x\}$ is finite,

(b-4) the space $(\mathbb{R},d)$ is not separable,

(b-5) no point of $\mathbb{R}$ has a countable fundamental system of d-neighborhoods.

(c) The space $(\mathbb{R},d)$ is not normal.

(d) The nowhere dense sets in the density topology coincide with the Lebesgue null sets. In particular, any first category set in the density topology is d-closed and d-nowhere dense.

(e) $(\mathbb{R},d)$ is a hereditarily Baire space.

(f) Assuming the continuum hypothesis, $(\mathbb{R},d)$ is not a Blumberg space.

(g) The density topology has the complete Lusin-Menchoff property.

(h) Any interval of $\mathbb{R}$ is a d-connected set.

Proof. (a) It follows from Corollary 3.2 and from (g) of this theorem.

(b) It is obvious by the definition of the density topology and by Theorem 2.1.

(c) It follows from Theorem 2.1.h and Theorem 6.8.b.

(d) Obviously, any Lebesgue null set is d-closed (by definition) and its d-interior is empty (by Proposition 6.4). Let D be a d-nowhere dense set. Then $\bar{D}^d$ is measurable and also d-nowhere dense. Hence $\bar{D}^d$ cannot contain any density point of $\bar{D}^d$, and thus must be of measure zero by the Lebesgue density theorem (cf. also Theorem 6.18).

(e) It follows from the fact that the d-first category sets are d-closed. Indeed, any topological space X in which every first category set is closed is hereditarily Baire: Assume that $M \subset X$ is first category in itself. Then M is first category in X. So any subset of M is first category in X and thus closed. Hence M must be isolated, and thus empty.

(f) It follows from Theorem 4.7, using the fact that the topology d has the G_δ-insertion property (see Ex. 6.A.10).

(g) See Section 6.D or Ex. 6.A.8.

(h) It is a consequence of the fact that any approximately continuous function has the Darboux property. For another proof, see Ex. 6.A.11.

EXERCISES

Let $x \in \mathbb{R}$ and $M \subset \mathbb{R}$. Define

$$d^{e,sym}(x,M) = \limsup_{h \to 0_+} \frac{1}{2h} \lambda^*(M \cap (x-h,x+h)),$$

$$d_{e,sym}(x,M) = \liminf_{h \to 0_+} \frac{1}{2h} \lambda^*(M \cap (x-h,x+h)),$$

$$d^e(x,M) = \limsup_{I \to x} \frac{1}{\lambda I} \lambda^*(M \cap I).$$

Analogously we define $d_e(x,M)$ and, replacing λ^* by λ_*, also $d^{i,sym}(x,M)$, $d_{i,sym}(x,M)$, $d^i(x,M)$ and $d_i(x,M)$. If M is measurable, we write simply $\bar{d}(x,M)$ instead of $d^e(x,M)$. Analogously we write $\underline{d}(x,M)$, $d^{sym}(x,M)$ and $d_{sym}(x,M)$. Finally, if $\bar{d}(x,M) = \underline{d}(x,M)$, define $d(x,M)$ as this equal value. The number $d(x,M)$ is called the *metric density* of M at x.

6.A.1 (a) Prove that $d_i(x,M) + d^e(x,M) = 1$ and state an analogous assertion for the symmetric density (cf. Lemma 6.10).

(b) Show that $d^{e,sym}(x,M) = 0$ if and only if $d^e(x,M) = 0$ (in this case we say that x is a *dispersion point* of M).

(c) Show that $d_{i,sym}(x,M) = 1$ if and only if $d_i(x,M) = 1$. In this case we say that x is a *density point* of M.

(d) Show that x is a density point of M if and only if x is a dispersion point of $\mathbb{R} \setminus M$.

6.A.2 Show that $\{x \in M:\ d_e(x,M) < 1\}$ is a null set for each $M \subset \mathbb{R}$.

Hint. Use the Lebesgue density theorem for a measurable covering of M (cf. Lemma 6.13.a).

6.A.3 (C. Goffman (1950)). If $\lambda C = 0$, then there is a measurable set M such that $\underline{d}(x,M) = 0$ and $\bar{d}(x,M) = 1$ for each $x \in C$.

Hint. Let $G_1 \supset G_2 \supset \ldots \supset C$ be a sequence of open sets such that

$$1 \geqq \lambda G_n \to 0 \quad \text{and} \quad \frac{\lambda(G_{n+1} \cap I)}{\lambda I} = \frac{1}{n}$$

for every component I of G_n. Put $M = \bigcup_{k=1}^{\infty} (G_{2k-1} \setminus G_{2k})$.

REMARK. The same assertion is true for $d_{sym}(x,M)$ and $d^{sym}(x,M)$.

6.A.4 Let $M \subset \mathbb{R}$.

(a) (L. Zajíček (1974)). The set
$\{x \in \mathbb{R}:\ d^{e,sym}(x,M) < 1 \text{ and } \lambda^*(M \cap (x-h,x+h)) > 0 \text{ for each } h > 0\}$
is a first category F_σ-set. It need not be of measure zero.

Hint. Show that the set

$$\Big\{x \in \mathbb{R}:\ \lambda^*(M \cap (x-h,x+h)) > 0 \text{ for each } h > 0 \text{ and } \frac{\lambda^*((x-h,x+h) \cap M)}{2h} \leqq 1 - \frac{1}{n} \text{ for each } h \in (0, 1/n)\Big\}$$

is closed. In view of Ex. 6.A.2 it is nowhere dense.

(b) The set

$$\{x \in \mathbb{R}:\ 0 < d^{e,sym}(x,M) < 1\}$$

is a first category set of measure zero. It need not be a subset of an F_σ-set of measure zero.

Hint. Use the preceding exercise. If $F \subset [0,1]$ is a nowhere dense closed set for which every portion of F is of positive measure, and if T is the set of all points of F belonging to infinitely many intervals $(a_n - (b_n-a_n),\ b_n + (b_n-a_n))$ (where $\bigcup_{n=1}^{\infty} (a_n,b_n) = [0,1] \setminus F$), show that

$$T \subset \{x \in \mathbb{R}:\ 0 < d^{e,sym}(x,M) < 1\}$$

where

$$M := \bigcup_{n=1}^{\infty} \left(a_n + \frac{1}{4}(b_n - a_n),\ b_n - \frac{1}{4}(b_n - a_n)\right).$$

Since T is a Baire space (it is a G_δ-subset of $\mathbb{R}$), $\bar{T} = F$ and $\lambda F > 0$, the Baire category theorem implies that T cannot be a subset of an F_σ-set of measure zero.

(c) Prove the same assertion as in (b) replacing $d^{e,sym}$ by d^e.

Hint. Put T to be the set of all $x \in F$ for which the sets

$$\{n \in \mathbb{N}:\ x \in (a_n - (b_n - a_n), b_n)\} \quad \text{and} \quad \{n \in \mathbb{N}:\ x \in (a_n, b_n + (b_n - a_n))\}$$

are infinite.

(d) Prove that

$$\{x \in \mathbb{R}:\ 0 < d_e(x,M) \leqq d^e(x,M) < 1\},$$

$$\{x \in \mathbb{R}:\ 0 < d_{e,sym}(x,M) \leqq d^{e,sym}(x,M) < 1\}$$

are F_σ-sets of measure zero.

Hint. Show that the set

$$A_n := \{x \in \mathbb{R}:\ \frac{1}{n} \leqq \frac{\lambda^*(M \cap (x-h,x+h))}{2h} \leqq 1 - \frac{1}{n} \ \text{for each} \ h \in (0,1/n)\}$$

is closed. Suppose that $\lambda A_n > 0$ and seek a contradiction (consider separately the cases $\lambda^*(A_n \cap M) > 0$ and $\lambda(A_n \cap M) = 0$).

<u>REMARK</u>. C. Goffman (1950) has shown that the set of all points at which the metric density exists but equals neither zero nor one is a first category set.

6.A.5 We say that $A, B \subset \mathbb{R}$ are *d-separated* if there are disjoint density open sets G_A and G_B such that $A \subset G_A$, $B \subset G_B$.

(a) Prove that two countable disjoint sets are always d-separated.

(b) There are two disjoint density closed sets one of them being countable which are not d-separated.

Hint. See J. Lukeš and L. Zajíček (1977a) or F.D.Tall (1978).

(c) There are G_δ-sets A,B with disjoint density closures such that $\bar{A}^d$ and $\bar{B}^d$ are not d-separated.

Hint. See J. Malý (1979a) and Ex. 6.A.10 (cf. M. Laczkovich (1975)).

6.A.6 Let H be a non-empty open subset of $\mathbb{R}$. A disjoint countable collection $\{J_n\}$ of open intervals is said to be an *admissible partition* of H if $\{J_n\}$ is locally finite in H and $H = \bigcup_{n=1}^{\infty} \bar{J}_n$. Obviously, the notion of an admissible partition has an evident geometric interpretation.

(a) If $\{G_\alpha\}$ is an open covering of H, then there is an admissible partition of H which refines $\{G_\alpha\}$.

Hint. Use the Lindelöf property of $\mathbb{R}$ and then proceed by induction.

(b) Let G be a nonempty subset of $\mathbb{R}$, let F be a closed set, $\emptyset \neq G \cap F \neq G$ and let $k > 0$. Prove that there is an admissible partition $\{J_n\}$ of $G \setminus F$ such that

$$\lambda(J_n) < k \operatorname{dist}^2(J_n, F \cap G).$$

6.A.7 (a) Let $G \neq \emptyset$ be open and let F be closed. If F and $M \subset G \setminus F$ are of measure zero and $\alpha \in [0,1)$, then there is an open set U such that $M \subset U \subset G \setminus F$ and $d(x,U) = \alpha$ for every $x \in F \setminus G$.

Hint. Find $\{J_n\}$ as in Ex. 6.A.6.b (for k = 1) and choose an open set U for which $M \subset U \subset G$ and

$$\frac{\lambda(U \cap J_n)}{\lambda J_n} = \alpha \quad \text{if} \quad \alpha > 0$$

or

$$\frac{\lambda(U \cap J_n)}{\lambda J_n} = \frac{1}{n} \quad \text{if } \alpha = 0.$$

(b) (N.F.G. Martin (1960)). Let Z be an F_σ-set of measure zero and let $\gamma \in (0,1)$. Show that there is a measurable set S such that $d(x,S) = \gamma$ for each $x \in Z$.

Hint. Let $Z = \bigcup_{n=0}^{\infty} Z_n$ where Z_n is closed and $\emptyset = Z_o \subset Z_1 \subset Z_2 \subset \ldots$. According to (a) there is an open set $U_1 \subset \mathbb{R} \setminus Z_1$ such that

$$Z \setminus Z_1 \subset U_1 \quad \text{and} \quad d(x,U_1) = 1 - \gamma \text{ for } x \in Z_1.$$

By (a) again find an open set G_1 such that

$$Z \setminus Z_1 \subset G_1 \subset U_1 \quad \text{and} \quad d(x,G_1) = 0 \text{ for } x \in Z_1.$$

Analogously, there are open sets U_2, G_2 such that

$$Z \setminus Z_2 \subset G_2 \subset U_2 \subset \mathbb{R} \setminus Z_2 ,$$

$$d(x,U_2) = 1 - \gamma \text{ and } d(x,G_2) = 0 \text{ for } x \in Z_2 \setminus Z_1.$$

Proceeding by induction we conclude that the set S defined by

$$S := \bigcup_{k=1}^{\infty} (U_k \setminus G_k)$$

has density γ at every point of Z.

6.A.8 (Lusin-Menchoff property). Let $\emptyset \neq F \subset \mathbb{R}$ be a closed set, $M \subset \mathbb{R} \setminus F$ measurable and let $c > 0$. Then there is an open set G such that $M \subset G \subset \mathbb{R} \setminus F$ and

$$\lambda((G \setminus M) \cap I) < c(\lambda I)^2$$

for each interval I whenever $I \cap F \neq \emptyset$ (cf. 6.34.F).

Hint. Use Ex. 6.A.6.b for $k = c/2$. Further, construct G such that

$$\lambda(J_n \cap (G \setminus M)) < \tfrac{1}{2} c \operatorname{dist}^2(J_n,F) \quad \text{for each } n \in \mathbb{N}.$$

6.A.9 (Alternative definitions of the density topology.) Put

$$\tau := \{M \subset \mathbb{R}: \ d_i(x,M) = 1 \text{ for every } x \in M\}.$$

(a) Show that τ forms a topology.

(b) Show that τ consists only of measurable sets and that τ = d.

Hint. Use the fact that the union of d-open sets is a measurable set (cf. Corollary 6.3).

(c) Conversely, taking advantage of the knowledge that τ-open sets are measurable prove Corollary 6.3.

(d) Prove that τ-open sets are measurable using the Vitali covering theorem.

Hint. Assume that M is a bounded τ-open set and pick $\varepsilon > 0$. For each $x \in M$ find $r_x > 0$ and a measurable set $M_x \subset M$ such that

$$\lambda(M_x \cap (x-h, x+h)) \geq 2(1-\varepsilon)h$$

for each $h < r_x$. In light of the Vitali theorem there is a countable set $S \subset M$ and for each $z \in S$ there is $h_z \in (0, r_z)$ such that

$$\lambda(M \setminus \bigcup_{z \in S} [z-h_z, z+h_z]) = 0$$

where the intervals in consideration are pairwise disjoint.

Put

$$M_\varepsilon := \bigcup_{z \in S} [z-h_z, z+h_z] \cap M_z$$

and show that $\lambda^*(M \setminus M_\varepsilon) < \varepsilon \operatorname{diam} M$. Deduce that M is measurable.

6.A.10 Show that the density topology has the G_δ-insertion property. (Hence, according to Corollary 2.14, any approximately continuous function is in Baire class one; cf. Theorem 6.8.b.)

Hint. Let $G \subset F$ where G is d-open and F is d-closed. Put

$$B := \{x \in \mathbb{R}: \text{ for every } n \in \mathbb{N} \text{ there is } m > n \text{ such that } \lambda(G \cap (x-\tfrac{1}{m}, x+\tfrac{1}{m})) > \tfrac{1}{m}\}.$$

Show that B is a G_δ-set and $G \subset B \subset F$.

6.A.11 Show that $\mathbb{R}$ is connected in the density topology (cf. Theorem 6.9.h).

Hint. Let $M \subset \mathbb{R}$ be both density open and density closed. Show, using the definition of the density topology only, that $c_M(x) = g'(x)$ where

$$g(x) = \int_0^x c_M(t)\,dt = \lambda(M \cap (0,x)).$$

6.A.12 Let $bA = \mathrm{der}_d A$ for $A \subset \mathbb{R}$.

(a) Show that bA need not be a G_δ-set.

(b) Provide also an example of a set B such that bB is not an F_σ-set.

(c) Prove that bA is always a $G_{\delta\sigma}$-set (cf. Ex. 6.C.7).

(d) The main goal now is to show that bA need not be even an $F_{\sigma\delta}$-set. This can be established by proving the following two propositions whose details are left to the reader.

Proposition A. Let P be a compact subset of $\mathbb{R}$ of measure zero, let G be open and let $a \in (0,1)$. If $A \subset \mathbb{R}$ is a measurable set such that $\bar{A} \cap G \cap P = \emptyset$

and
$$\frac{1}{2h}\lambda(A \cap (x-h,x+h)) < a$$

for each $x \in P \cap G$ and $h > 0$, then for every $b \in (a,1)$ and $\delta > 0$ there is an open set $H \subset G$ such that

(a) $\overline{A \cup H} \cap G \cap P = \emptyset$,

(b) $\frac{1}{2h}\lambda((A \cup H) \cap (x-h,x+h)) < b$ for each $x \in P \cap G$ and $h > 0$,

(c) given $z \in P \cap G$, there is $0 < h < \delta$ such that

$$\frac{1}{2h}\lambda((A \cup H) \cap (z-h,z+h)) > \frac{a}{8}.$$

Hint. Since the proposition is obvious in case $P \cap G = \emptyset$, restrict attention to the case $P \cap G \neq \emptyset$. Applying Principal lemma 6.32 (for $F = \bar{A}$, $T = P \cap G$, $\omega(x) = \frac{1}{2}(b-a)x$) there is an open set $U \supset P \cap G$ such that

$$(\square) \qquad \frac{\lambda(U \cap I)}{\lambda I} < b-a \quad \text{whenever } I \cap \bar{A} \neq \emptyset.$$

As P is compact and $\mathrm{int}\, P = \emptyset$, there is a countable system of pairwise disjoint open intervals $\{J_k\}_{k=1}^{\infty}$ whose lengths are less than δ, which is locally finite in G and satisfies $P \cap G \subset \bigcup_{k=1}^{\infty} J_k \subset U \cap G$.

Let $\{C_k^i\}_{i=1}^{i_k}$ be a finite collection of components of $J_k \setminus P$ for which

$$\lambda(\bigcup_{i=1}^{i_k} C_k^i) > \frac{1}{2} \lambda J_k .$$

Let I_k^i be the open interval concentric with C_k^i such that $\lambda I_k^i = \frac{a}{2} \lambda C_k^i$. Put

$$H := \bigcup_{k=1}^{\infty} \bigcup_{i=1}^{i_k} I_k^i .$$

As $\{J_k\}_{k=1}^{\infty}$ is locally finite in G, we have (a), as needed. Concerning (b), pick $x \in P \cap G$ and $h > 0$. Use ($\square$), provided $(x-h,x+h) \cap A \neq \emptyset$. In the opposite case show $\frac{1}{2h} \lambda(H \cap (x-h,x+h)) < a$. To prove (c), given $z \in P \cap G$, find $j \in \mathbb{N}$ for which $z \in J_j$ and put $h = \lambda J_j$.

Proposition B. Let P be a compact set of measure zero and let $S \subset P$ be a $G_{\delta\sigma}$-set. Then there is an open set $T \subset \mathbb{R}$ such that $S = P \cap bT$.

Hint. Write $S = \bigcup_{n=1}^{\infty} \bigcap_{k=1}^{\infty} G_k^n$ where $G_1^n \supset G_2^n \supset \dots$ are open. It can be supposed that

$$(\diamond) \qquad \lambda(G_k^n \cap (x-h,x+h)) \leqq 2^{-n-k} h^2$$

for each $x \in P \setminus G_k^n$ and $h > 0$. This follows again from Principal lemma 6.32. Indeed, it implies that given $\varepsilon > 0$ and a open set G in P, there is an open set $H \subset \mathbb{R}$ such that $G = H \cap P$ and

$$\lambda(H \cap (x-h,x+h)) < \varepsilon h^2$$

whenever $x \in P \setminus G$ and $h > 0$ (cf. also 6.34.F).

Now apply Proposition A to obtain for any $n \in \mathbb{N}$ (by induction) a sequence $\{A_k^n\}_{k=1}^{\infty}$ of open subsets of G_k^n fulfilling the following conditions:

(a) $\overline{A_k^n} \cap G_k^n \cap P = \emptyset$,

(b) $\frac{1}{2h} \lambda((\bigcup_{i=1}^{k} A_i^n) \cap (x-h,x+h)) < 2^{-n} (2 - \frac{1}{k})$ for each $x \in P \setminus G_k^n$,

(c) given $z \in P \cap G_k^n$, there is $h \in (0,\frac{1}{k})$ such that

$$\frac{1}{2h}\,\lambda\Big(\Big(\bigcup_{i=1}^{k} A_i^n\Big) \cap (z-h,z+h)\Big) > \frac{1}{8}\cdot 2^{-n}.$$

Now, put $T := \bigcup_{k,n=1}^{\infty} A_k^n$. Obviously, $S \subset bT$. What we must show is that $(P\setminus S)\cap bT = \emptyset$. To this end let $x \in P\setminus S$ be given. For every $n \in \mathbb{N}$, let $k_n := \min\{k \in \mathbb{N}: x \notin G_k^n\}$. In light of $(\lozenge)$, $x \notin b\big(\bigcup_{n=1}^{\infty} \bigcup_{k=k_n}^{\infty} A_k^n\big)$. On the other hand, given $m \in \mathbb{N}$, we have by (a) $x \notin \overline{\bigcup_{n\leq m,\ k\leq k_n} A_k^n}$, and by (b)

$$\frac{1}{2h}\,\lambda\Big(\Big(\bigcup_{n>m,\ k<k_n} A_k^n\Big) \cap (x-h,x+h)\Big) < 2.\sum_{n>m} 2^{-n} = 2^{-m+1}.$$

It follows that $x \notin b\big(\bigcup_{n=1}^{\infty} \bigcup_{k=k_n}^{\infty} A_k^n\big)$.

6.A.13 (Cf. F.D.Tall (1976), Th. 2.9.) Show that the density topology on $\mathbb{R}$ has the Knaster property (K) of Ex. 1.B.2.

Hint. Let $\{G_\alpha\}_{\alpha\in A}$ be an uncountable collection of density open subsets of $\mathbb{R}$ (or, more generally, of measurable sets of positive measure). Let $\{I_n\}_{n\in\mathbb{N}}$ be the sequence of all open intervals with rational endpoints. Given $\alpha \in A$, there is $n(\alpha) \in \mathbb{N}$ such that $\lambda(G_\alpha \cap I_{n(\alpha)}) > \frac{1}{2}\lambda I_{n(\alpha)}$. There must be an index $k \in \mathbb{N}$ such that the set $A_k := \{\alpha \in A: n(\alpha) = k\}$ is uncountable. Show that $\{G_\alpha\}_{\alpha\in A_k}$ is an uncountable subcollection and $G_\alpha \cap G_\beta \neq \emptyset$ for $\alpha,\beta \in A_k$.

6.A.14 (D-continuous functions, J. Malý (1979b).) Let $J \subset \mathbb{R}$ be an interval. A mapping $f: J \to \mathbb{R}^n$ satisfies the *Lusin (N) condition* if f maps sets of Lebesgue measure zero onto sets of measure zero, and f is said to be *D-continuous* if it is continuous with respect to the (ordinary) density topology on the domain and the range.

(a) Show that any monotone D-continuous function is continuous and satisfies the Lusin (N) condition.

REMARK. Note that any continuous monotone function having the Lusin (N) condition is absolutely continuous (the Banach-Zarecki theorem, see [Saks], Chap. VII, § 6).

(b) Suppose that a mapping $z: [0,1] \to [0,1]^2$ satisfies the following two conditions:

(b_1) z is $\frac{1}{2}$-Hölder continuous, i.e. there is $c > 0$ such that $|z(x) - z(y)| \leqq c\sqrt{|x-y|}$ for every couple $x, y \in [0,1]$,

(b_2) z is a measure preserving map, i.e. $\lambda_2(A) = \lambda_1(z^{-1}(A))$ for any Lebesgue measurable subset $A \subset [0,1]^2$.

Show that z is D-continuous.

(c) Let π_1 be the first-coordinate projection of $\mathbb{R}^2$ onto $\mathbb{R}$ (i.e. $\pi_1(x,y) = x$), and let z be any mapping satisfying (b_1) and (b_2). If $\Phi = \pi_1 * z$, show that

(c_1) Φ is D-continuous,

(c_2) Φ does not satisfy the Lusin (N) condition.

(d) Show that the classical Peano curve or the Hilbert curve satisfy the conditions (b_1) and (b_2). Hence, D-continuous functions need not satisfy the Lusin (N) condition.

REMARK. Some further properties of D-continuous functions are studied in K.Ostaszewski (1981), (1983a) and (1983b).

REMARKS AND COMMENTS

Although the concept of approximately continuous functions was intensively studied in the first half of our century since its introduction by A. Denjoy (1915), the density topology has been set up much later. Its establishing in $\mathbb{R}^n$ is being attributed mostly to C. Goffman and D. Waterman (1961) and its properties are investigated in C. Goffman, C. Neugebauer and T. Nishiura (1961). Factually, as early as nearly ten years before the papers of O. Haupt and Ch. Pauc (1952) , (1954) were published where even a general "Denjoy's topology" was being defined by a complete measure and by Possel's differentiation basis.

There are several ways how to define the density topology on $\mathbb{R}$. One of them, more suitable for generalization (cf. 6.B., 6.E), does not a priori assume the measurability of d-open sets, which is therefore to be proved (cf. Ex. 6.A.9, see also Corollary 6.15). By using another method, as we do in this section, it is necessary to prove that d-open sets form a topology. The most difficult step of this proof consists in proving the measurability of an arbitrary union of d-open sets. Our proof makes use of the quasi-Lindelöf property of the density topology (Section 1.B; cf. J.L. Doob (1966)).

Although all results of this part follow from more general theorems contained in 6.B and 6.D, we present here - for reader's convenience - a self-contained and as elementary as possible exposition of the density topology including some proofs.

The proof of the important Lusin-Menchoff property which guarantees the complete regularity of the density topology is outlined in exercises. We present an "elementary" one-dimensional proof.

Theorem 6.9 summarizes the most important properties of the density topology on the real line (a relevant paper is F.G. Tall (1976)). Its further properties are included in exercises. The result presented in Ex. 6.A.12 may be new.

6.B Density topologies on metric spaces

There are many generalizations of the density topology on the real line. In this section we introduce a class of density topologies on metric spaces which are established in a similar way as the density topology on the real line.

There is even a more general approach, namely that of Section 6.E based on the notion of a filter differentiation basis. The main results of this section (except Lemma 6.10b) are still valid in this more general situation. The ultimate reason for our restrictions lies in the fact that in the next two sections we are able to find sufficient conditions guaranteeing reasonable properties of density topologies as fine topologies with respect to the original topology of a given metric space. Thus in order to be able to find some criteria, it is convenient to start with some special differentiation bases on metric spaces.

Before defining the main notions, we need some preliminaries from measure theory. Let m be a complete measure defined on a σ-algebra $\mathfrak{M}$ and suppose that m is σ-finite. We say that a set M is *measurable* if $M \in \mathfrak{M}$. We define the corresponding *inner* and *outer* measures by

$$m_*(A) := \sup\{m(M):\ M \in \mathfrak{M},\ M \subset A\},$$
$$m^*(A) := \inf\{m(M):\ M \in \mathfrak{M},\ M \supset A\}$$

for any set $A \subset P$. Then

$$m_*(M \cap A) + m^*(M \setminus A) = m(M)$$

whenever A is arbitrary and M is measurable.

It is known that for any set $A \subset P$ there are measurable sets J_A, H_A satisfying

$$J_A \subset A \subset H_A, \quad m_*(A \cap M) = m(J_A \cap M), \quad m^*(A \cap M) = m(H_A \cap M)$$

for any measurable set M. The sets J_A and H_A are called a *measurable kernel* and a *measurable cover* of A, respectively. They are not uniquely determined, a measurable set Q is a measurable cover of A, if and only if $Q \supset A$ and $m_*(Q \setminus A) = 0$. A set Q is a measurable cover of A, if and only if $P \setminus Q$ is a measurable kernel of $P \setminus A$.

In this section we consider a complete measure m defined on a σ-algebra $\mathfrak{M}$ containing all Borel sets of a metric space (P,ρ), which is regular with respect to Borel sets and finite on bounded elements of $\mathfrak{M}$. It is easy to see that m is σ-finite and that a measurable cover can be chosen to be a G_δ-subset of P.

We say that a *differentiation basis* $\mathcal{B}$ in P is given if for any $x \in P$ there is a collection $\mathcal{B}_x \subset \mathfrak{M}$.

For every set $A \subset P$ and each $x \in P$ we define the *upper outer density* $d^e(x,A)$ and the *lower inner density* $d_i(x,A)$ of A at x by the formulas

$$d^e(x,A) := \lim_{r \to 0_+} \left(\sup \left\{\frac{m^*(A \cap M)}{m(M)} : M \in \mathcal{B}_x,\ m(M) > 0,\ M \subset U(x,r)\right\}\right),$$

$$d_i(x,A) := \lim_{r \to 0_+} \left(\inf \left\{\frac{m_*(A \cap M)}{m(M)} : M \in \mathcal{B}_x,\ m(M) > 0,\ M \subset U(x,r)\right\}\right)$$

whenever the sets in considerations are nonempty. Otherwise $d^e(x,A)$ and $d_i(x,A)$ are undefined.

In what follows, we assume that a differention basis $\mathcal{B} = \{\mathcal{B}_x\}_{x \in P}$ is given.

For any $A \subset P$, the set

$$b(A) = \{x \in P: \ d^e(x,A) > 0\}$$

is said to be the *base of* A. The set $P \setminus b(P)$ consists of all points in which $d^e(x,P)$ is not defined. In other words, $x \in P \setminus b(P)$ if and only if there is $r > 0$ such that all sets from $\mathcal{B}_x$ contained in $U(x,r)$ are of m-measure zero. Since the set function $A \mapsto d^e(x,A)$ is monotone increasing for each $x \in b(P)$ and finitely subadditive, it is easy to check that b is a base operator in the sense of Section 1.A. The corresponding topology t (a set F is t-closed if $b(A) \subset A$) is called the *(metric) density topology determined by differentiation basis* $\mathcal{B}$. The following lemma guarantees that t is finer than ρ (in other words, $b(A) \subset \bar{A}$ for any A).

6.10 LEMMA. (a) For each set $A \subset P$ and for each $x \in b(P)$

(a 1) $d_i(x,A) + d^e(x,P \setminus A) = 1,$

(a 2) $d_i(x,A) = d_i(x,J_A)$, $d^e(x,A) = d^e(x,H_A)$

whenever H_A and J_A are a measurable cover and kernel of A, respectively.

(b) For each open set $G \subset P$ and every $x \in b(P) \cap G$,

$d_i(x,G) = 1$, $d^e(x,P \setminus G) = 0$.

Proof. (a) The assertion (a1) follows from the equality

$$m_*(A \cap M) + m^*((P \setminus A) \cap M) = m(M)$$

which is true for any measurable set M, (a2) is obvious as well as (b).

A point $z \in b(P)$ for which $d_i(z,A) = 1$ is termed the *density point* of A. Thus the set of all density points of A equals $b(P) \setminus b(P \setminus A)$. Note that a set G is t-open if and only if every point of $G \cap bP$ is a density point of G.

We say that the *density theorem holds for* $\mathcal{B}$ if for every measurable set M almost every point of M is a density point of M. Of course, if the density theorem holds for $\mathcal{B}$, then $m(P \setminus b(P)) = 0$.

REMARK. Notice that most of our proofs of this section do not depend on the definitions of $d^e(x,A)$, $d_i(x,A)$ and bA but they are based on the following two properties only:

(1) b is compatible with m (cf. Section 6.D),

(2) the validity of the density theorem.

As already stated all assertions remain true at least in a more general context of density topologies of Section 6.E.

6.11 EXAMPLES. A list of important examples of differentiation bases follows. Notice that in the most examples one or both of the following conditions hold:

(α_1) $\mathcal{B}_x$ consists of (bounded) sets containing x,

(α_2) for each $r > 0$, $\mathcal{B}_x$ contains a subset of $U(x,r)$ of positive measure.

(B_1) For the *symmetric basis* on (P,ρ), $\mathcal{B}_x$ consists of the system of all open balls with center at x, more precisely, $\mathcal{B}_x = \{U(x,r) : r > 0\}$ for any $x \in P$. If m is a Radon measure on $\mathbb{R}^n$ (or, more generally, on a finite dimensional Banach space), then the base $\mathcal{B}$ has the Vitali property (see Ex. 6.E.4) and, consequently, the density theorem holds for the symmetric basis (Besicovitch (1945), Morse (1947)). Let us remarks that in this case $b(\mathbb{R}^n)$ is nothing else than the support of m. In general metric spaces the density theorem for the symmetric basis may be false (Ex. 6.B.8). Moreover, there is a metric space Lipschitz homeomorphic with $\mathbb{R}$ in which the density theorem does not hold for a Radon measure (cf. D. Preiss (1983)).

(B_2) For the *cube basis* in $\mathbb{R}^n$, $\mathcal{B}_x$ is formed as the collection of all closed cubes in $\mathbb{R}^n$ with center at x whose edges are parallel with the coordinate axes. If m is a Radon measure on $\mathbb{R}^n$, then again $b(\mathbb{R}^n)$ equals $\operatorname{supp}(m)$. In this case the Vitali theorem and the density theorem for the cube basis hold (Morse (1947)). The density topology determined by the cube basis and a Radon measure m with $\operatorname{supp}(m) = \mathbb{R}^n$ was examined in an interesting paper by R. S. Troyer and W. P. Ziemer (1963). Of course, any density topology like that is determined also by a symmetric basis on $\mathbb{R}^n$ which is supposed to be equipped with the maximum metric.

(B_3) We say that $\mathcal{B} = \{\mathcal{B}_x\}_{x\in\mathbb{R}^n}$ is a *basis of convex sets of Morse type* in $\mathbb{R}^n$ if $\mathcal{B}_x$ consists of convex subsets of $\mathbb{R}^n$ and if there is $c = c(x) > 1$ such that for each $B \in \mathcal{B}_x$ we have $U(x,r) \subset B \subset U(x,cr)$ with some $r > 0$. We assume also that $\mathcal{B}_x$ contains sets with arbitrarily small diameters. If m is a Radon measure on $\mathbb{R}^n$, then the density theorem holds for such bases (Morse (1947)). Obviously, symmetric or cube bases are of Morse type.

(B_4) If $\mathcal{B}_x$ consists of all closed intervals (i. e. the sets of the form $[a_1,b_1] \times \dots \times [a_n,b_n]$) in $\mathbb{R}^n$ containing x, then $\mathcal{B} = \{\mathcal{B}_x\}_{x\in\mathbb{R}^n}$ is called the *interval basis*. If $n = 1$ and m is a Radon measure on $\mathbb{R}$, then the density theorem holds for the interval basis (Iseki (1960); let us remark that the interval basis is not of Morse type). This is not true for an arbitrary Radon measure in the case of $n > 1$. But the density theorem holds for the interval basis in $\mathbb{R}^n$ with respect to the Lebesgue measure (see [Saks] where this theorem is labelled as a "density theorem"). Nowadays, it is often called the *strong density theorem* and the functions continuous in the density topology determined by the interval basis and Lebesgue measure are called *strongly approximately continuous*.

(B_5) For the *rectangle basis* in $\mathbb{R}^n$, $\mathcal{B}_x$ is formed for every x by the family of all rectangles (not necessarily intervals) containing x. Even if m is the Lebesgue measure, the density theorem does not hold for the rectangle basis (cf. [Guz], V, §2).

(B_6) Let A be a closed angle in $\mathbb{R}^2$ with vertex at $(0,0)$. For every $x \in \mathbb{R}^2$ denote by A_x the translation of A with its vertex at x. Let $\mathcal{B}_x$ be formed by all sets of the form $A_x \cap U(x,r)$, $r > 0$. The system $\mathcal{B} = \{\mathcal{B}_x\}_{x \in \mathbb{R}^2}$ is termed the *angle basis*. Since the density theorem holds for the symmetric basis, it holds also for the angle basis in case of the Lebesgue measure.

(B_7) The system $\mathcal{B} = \{[x,x+r] : r > 0\}_{x \in \mathbb{R}}$ is termed the *right-density basis*. For each Lebesgue-Stieltjes measure, the density theorem holds for $\mathcal{B}$.

(B_8) The *ideal basis* in any metric space (P,ρ) is defined as follows: $\mathcal{B}_x$ is formed by the family of all Borel sets. A point z is a density point of an m-measurable set M if and only if every open neighborhood of z is of positive measure and if the set $U(z,r) \setminus M$ is of m-measure zero for some $r > 0$. It is easy to see that, in general, the density theorem does not hold for the ideal basis.

(B_9) For each $x \in \mathbb{R}$ we put

$$\mathcal{B}_x := \{M \subset \mathbb{R}:\ M \text{ is measurable and there is } h > 0 \text{ such that } M \subset (x-h,x+h),\ \lambda(M) \geq h^2\}.$$

The collection $\mathcal{B} = \{\mathcal{B}_x\}_{x \in \mathbb{R}}$ is termed the *superdensity basis* on $\mathbb{R}$. Let us consider the Lebesgue measure on $\mathbb{R}$. It is easy to check that

$$d^e(x,A) > 0 \quad \text{if and only if} \quad \limsup_{h \to 0_+} h^{-2} \cdot \lambda^*(A \cap (x-h,x+h)) > 0 .$$

The corresponding so called *superdensity topology* will be studied in details in Section 6.D. Let us remark that the density theorem does not hold for the superdensity basis.

(B_{10}) A basis $\{\mathcal{B}_x\}_{x \in \mathbb{R}^n}$ is called a basis of *comparable convex sets* in $\mathbb{R}^n$ if the collections $\mathcal{B}_x$ have the following properties:

(a) $\mathcal{B}_x$ consists of compact convex sets in $\mathbb{R}^n$ with the non-empty interior containing x,

(b) $\mathcal{B}_x$ contains sets with arbitrarily small diameters,

(c) given $A \in \mathcal{B}_x$, $B \in \mathcal{B}_y$, there is a translation A^* of A such that either

$A^* \subset B$ or $B \subset A^*$.

For this basis and for the Lebesgue measure, the Vitali theorem and, consequently, the density theorem holds (Morse (1947), [Guz], I, §3).

(B_{11}) A basis of *comparable intervals* in $\mathbb{R}^n$ is defined for each $x \in \mathbb{R}^n$ as a collection $\mathcal{B}_x$ of closed intervals with center at x containing sets with arbitrarily small diameters and having the following property: Given $I \in \mathcal{B}_x$, $J \in \mathcal{B}_y$, then either $I + (y-x) \subset J$ or $J \subset I + (y-x)$. The Vitali theorem and, consequently, the density theorem holds in this case for any Radon measure on $\mathbb{R}^n$ (Guzman (1970), cf. [Guz], I, §1).

(B_{12}) A basis $\{\mathcal{B}_x\}_{x \in P}$ is called an *incomplete symmetric basis* on a metric space (P, ρ) if $\mathcal{B}_x$ contains only open balls with center at x. Moreover, we assume that in $\mathcal{B}_x$ there are balls with arbitrarily small diameters. Any incomplete symmetric basis in $\mathbb{R}^n$ is of Morse type.

(B_{13}) *Net structures* ([Saks], p. 153, A.M. Bruckner (1971), p. 34). A *net* in a metric space P is a finite or countable disjoint family of Borel sets which cover P. The sets constituting a net are called *cells* (*meshes* in Saks' terminology). A sequence $\{\mathcal{M}_n\}$ of nets is termed *regular* if

(a) each cell of $\mathcal{M}_{n+1}$ is contained in some cell of $\mathcal{M}_n$,

(b) $\lim_{n \to \infty} \sup \{\text{diam } A\colon\ A \in \mathcal{M}_n\} = 0$.

It is easy to see that there is a regular net structure in P if and only if this space is separable.

Let $\{\mathcal{M}_n\}$ be a regular net structure. A *net basis* $\{\mathcal{B}_x\}_{x \in P}$ is determined as follows: $\mathcal{B}_x$ consists of all cells containing x (obviously, for each n exactly one cell of $\mathcal{M}_n$ contains x). If m is a measure as at the beginning of this section, the density theorem holds for the net basis ([Saks], Th. 15.7, cf. A.M. Bruckner (1971), p. 35).

Notice that the density topology on $\mathbb{R}$ is determined by the Lebesgue measure and by any of the differentiation bases (B_1), (B_2) or (B_4). In case of the Lebesgue measure on $\mathbb{R}^n$, the density topologies determined by differentiation bases (B_1) and (B_2) coincide and they are commonly termed *ordinary density topologies*.

In examples (B_1) and (B_2) we have just seen that the base $b(\mathbb{R}^n)$ is equal to the support of the considered measure. Further, it is easy to check that all points of $P \setminus bP$ are always isolated in the density topology. Hence, we are mostly interested in the case of $bP = P$. Since, moreover, the assertions or the proofs of some of the next theorems can be profitably simplified under the assumption $bP = P$, we add this requirement. We do not feel that a full generality justified the effort involved. In exercises it is to be shown adequate illustrations of general assumptions.

6.12 PROPOSITION. For a differentiation basis $\mathcal{B}$, the following assertions are equivalent:

(i) the density theorem holds for $\mathcal{B}$,

(ii) $m(P \setminus bP) = 0$ and $M \cap b(P \setminus M)$ is a null set for each measurable set M,

(iii) $m(P \setminus bP) = 0$ and $bM \setminus M$ is a null set for each measurable set M.

Proof. Since bP is the set of all density points of P and $\{x \in P: d_i(x,M) \neq 1\} = (P \setminus bP) \cup \{x \in P: d_i(x,M) < 1\} = (P \setminus bP) \cup \{x \in P: d^e(x,P \setminus M) > 0\} = (P \setminus bP) \cup b(P \setminus M)$, the equivalence (i)$\Leftrightarrow$(ii) is straightforward. The equivalence (ii)$\Leftrightarrow$(iii) can be proved by the passage to complements.

6.13 LEMMA. If the density theorem holds for $\mathcal{B}$, then

(a) $A \setminus b(A)$ is a null set for each $A \subset P$,

(b) if $A \subset P$ and H_A is a measurable cover of A, then $b(A) \triangle H_A$ is a null set,

(c) b is an idempotent base operator.

Proof. (a) We have by Lemma 6.10.a2

$$A \setminus b(A) \subset \{x \in A: d^e(x,A) = 0\} \subset \{x \in A: d^e(x,A) < 1\} \subset$$
$$\subset \{x \in H_A: d^e(x,H_A) < 1\} \subset \{x \in H_A: d_i(x,H_A) < 1\}$$

and the density theorem yields the assertion.

(b) Again, by Lemma 6.10.a2, $b(A) = b(H_A)$. Hence

$$b(A) \triangle H_A = b(H_A) \triangle H_A = (b(H_A) \setminus H_A) \cup (H_A \setminus b(H_A)),$$

and we use (a) together with the density theorem.

(c) By (b), $b(b(A)) = b(H_A) = b(A)$.

6.14 PROPOSITION. If the density theorem holds for $\mathcal{L}$, then for each set $A \subset P$

(a) $\bar{A}^t = A \cup b(A)$ is a measurable cover of A,

(b) $\mathrm{int}_t(A) = \{x \in A: d_i(x,A) = 1\} \cup (A \setminus bP)$ is a measurable kernel of A.

Proof. (a) By Theorem 1.2 and Lemma 6.13.c we know that $\bar{A}^t = A \cup b(A)$. If H_A is a measurable cover of A, then

$$H_A \,\Delta\, (A \cup b(A)) \subset H_A \,\Delta\, b(A)$$

and therefore by Lemma 6.13.b, $H_A \,\Delta\, (A \cup b(A))$ is a null set. Consequently, $A \cup b(A)$ is a measurable cover of A.

(b) Obviously, $\mathrm{int}_t A = P \setminus \overline{P \setminus A}^t$. Since $\overline{P \setminus A}^t$ is a measurable cover of $P \setminus A$, $\mathrm{int}_t A$ is a measurable kernel of A. Moreover,

$$\mathrm{int}_t A = P \setminus ((P \setminus A) \cup b(P \setminus A)) = A \cap (P \setminus b(P \setminus A)) =$$

$$= A \cap ((P \setminus bP) \cup \{x \in P: d^e(x, P \setminus A) = 0\}) = (A \setminus bP) \cup \{x \in A: d_i(x,A) = 1\}.$$

6.15 COROLLARY. If the density theorem holds for $\mathcal{L}$, then

$$b(A), \quad \mathrm{int}_t A, \quad \bar{A}^t$$

are measurable sets for each $A \subset P$. Consequently, any t-Borel set is measurable and it is easily seen that a set G is t-open if and only if G is measurable and $d_i(x,G) = 1$ for every $x \in G \cap bP$.

Proof. It is obvious by the preceding proposition and by Lemma 6.13.b.

6.16 REMARK. We have just proved the measurability of t-open sets provided the density theorem holds for $\mathcal{L}$. We would like to notice at this point that even weaker assumptions lead both to the measurability of t-open sets and to some other assertions presented until now as consequences of the density theorem (cf., e.g., Ex.6.B.5).

Suppose that $bP = P$. At the presence of the density theorem, a set U is a t-neighborhood of a point $x \in U$ if and only if x is a *point of density* of U. But this is the case if and only if $d^e(x,P\setminus U) = 0$, in other words, if the complement of U is *thin* at x. Recall that a set A is said to be thin at a point x if $x \notin b(A)$, i.e. in our case if $d^e(x,A) = 0$. In this section we use a frequently used terminology. We say that a point x is a *dispersion point* of a set A instead of "the set A is thin at x". Thus the base $b(A)$ is exactly the set of all non-dispersion points of A.

6.17 THEOREM. If the density theorem holds for $\mathcal{L}$, and if $x \in B$ whenever $B \in \mathcal{L}_x$, then

$$b(M) = \mathrm{der}_t M \cup \{x \in M:\ m(\{x\}) > 0\}$$

for every $M \subset P$. If moreover $m(\{x\}) = 0$ for each $x \in P$, then

$$b(M) = \mathrm{der}_t M, \quad \mathrm{der}_t(\mathrm{der}_t M) = \mathrm{der}_t M.$$

Proof. It follows easily from Theorem 1.2 since obviously $x \in b(\{x\})$ if and only if $m(\{x\}) > 0$.

6.18 THEOREM. Let $bP = P$ and assume that the density theorem holds for $\mathcal{L}$. Consider the following assertions:

(i) A is a t-nowhere dense set,

(ii) A is a null and t-first category set,

(iii) A is a t-first category set,

(iv) A is a null set,

(v) A is a t-isolated set,

(vi) A is a t-discrete set.

Then (i) $\Leftrightarrow$ (ii) $\Leftrightarrow$ (iii) $\Leftrightarrow$ (iv) $\Rightarrow$ (v) $\Rightarrow$ (vi). If $m(\{x\}) = 0$ for each $x \in A$, then all the assertions are equivalent.

Proof. (i) $\Rightarrow$ (ii): If A is t-nowhere dense, then $\bar{A}^t$ is t-nowhere dense. Since $\bar{A}^t$ is measurable,

$$\bar{A}^t = \bar{A}^t \setminus \mathrm{int}\,\bar{A}^t = \{x \in \bar{A}^t:\ d_i(x,\bar{A}^t) < 1\},$$

so $\bar{A}^t$ must be of measure zero in view of Proposition 6.12.

(iii) ⇒ (iv): If A is t-first category, then it is a countable union of t-nowhere dense (and hence null) sets. Thus A is a null set.

(iv) ⇒ (v) et (i) : For each set A of measure zero we have $\mathrm{der}_t A \subset b(A) = \emptyset$. Of course, $\mathrm{int}_t A = \emptyset$ by Proposition 6.14.b.

(vi) ⇒ (iv): Assume $m(\{x\}) = 0$ for each $x \in A$. Since $\mathrm{der}_t A \cap A = \emptyset$ and $b(A) = \mathrm{der}_t A$, $A \setminus b(A) = A$ is of measure zero by Lemma 6.13.a.

Recall that in a topological space, a *set* is said to have the *Baire property* if it belongs to the smallest σ-algebra containing all open sets as well as the first category sets. A real *function* on a topological space has the *Baire property* if it is measurable with respect to the σ-algebra of all sets with the Baire property. It is known that a real function f on a topological space X has the Baire property if and only if there exists a first category set M such that the restriction of f to $X \setminus M$ is continuous (cf. [Kur], Theorem 8.1).

6.19 THEOREM. If bP = P and if the density theorem holds for $\mathcal{L}$, then the following families of sets coincide:

(i) sets of type G_δ in the t-topology,

(ii) t-Borel sets,

(iii) sets with the Baire property in the t-topology,

(iv) measurable sets.

Proof. Since any t-nowhere dense set is t-closed, t-Borel sets coincide with sets with the Baire property in t. Thus, it is sufficient to prove (iv) ⊂ (i) only. But any measurable set M is of the form $H_M \setminus (H_M \setminus M)$, where the measurable cover H_M is of type G_δ and $H_M \setminus M$ is of measure zero. Since any null set is t-closed, M is of type G_δ in t.

6.20 THEOREM (Denjoy-Stepanoff). If bP = P and the density theorem holds for $\mathcal{L}$, then the following families of real functions coincide:

(i) t-Borel functions,

(ii) functions with the Baire property in t,

(iii) functions t-continuous almost everywhere,

(iv) measurable functions.

Proof. A real function f is measurable if and only if it has the Baire property in t, and this is the case if and only if f is t-Borel (Theorem 6.19, (ii) = (iii) = (iv)). Further, f has the Baire property in t if and only if there is a t-first category set M such that the restriction of f to $X \setminus M$ is t-continuous on $X \setminus M$. But this is exactly the case when f is t-continuous at all points of the t-open set $X \setminus M$.

EXERCISES

6.B.1 (a) Assume that the density theorem holds for a differentiation basis $\mathcal{L}$ in P. Set

$$\mathcal{L}_x^* := \{B \cap bP : B \in \mathcal{L}_x\}$$

for $x \in bP$. Let t* be the density topology on bP determined by $\mathcal{L}^*$. Show that (P,t) is the topological sum of (bP,t*) and $P \setminus bP$ equipped with the discrete topology.

(b) Omitting the requirement " bP = P " in Theorem 6.18, the equivalence (i) $\Leftrightarrow$ (ii) $\Leftrightarrow$ (iii) still holds, and (i) $\Leftrightarrow$ (iv*) where

(iv*) A is a null set and $A \subset bP$.

If $m(\{x\}) = 0$ for each $x \in A$, then (iv) $\Leftrightarrow$ (v) $\Leftrightarrow$ (vi).

Hint. Use (a).

(c) Show that Theorem 6.19 and the Denjoy-Stepanoff theorem 6.20 are still valid even omitting the assumption bP = P.

Hint. Use (a) again.

6.B.2 Show that the following assertions are equivalent:

(i) the density theorem holds for $\mathcal{L}$,

(ii) for any set A, almost every point of A is an outer density point of A,

(iii) $m(\overline{M}^t \setminus \mathrm{int}_t M) = 0$ for each measurable set M and $m(P \setminus bP) = 0$.

(A point z is an *outer density point* of A if

$$d_e(z,A) := \lim_{r\to 0_+} \left(\inf\left\{\frac{m^*(A \cap M)}{m(M)} : M \in \mathcal{L}_z,\ m(M) > 0,\ M \subset U(z,r)\right\}\right) = 1.)$$

6.B.3 If the density theorem holds for $\mathcal{L}$, then:

(a) $m^*(\{x \in A:\ d^e(x,A) < 1\}) = 0$ for each $A \subset P$,

(b) $m_*(\{x \in A:\ d_i(x,A) < 1\}) = 0$ for each $A \subset P$,

(c) the following assertions are equivalent:

(i) M is a measurable set,

(ii) $\{x \in M:\ d_i(x,M) < 1\}$ is a null set,

(iii) $\{x \in P \setminus M:\ d^e(x,M) > 0\}$ is a null set,

(iv) $bP \setminus bM$ equals the set of all outer density points of $P \setminus M$.

6.B.4 If the density theorem holds for $\mathcal{L}$, and if for a set $A \subset P$, $d^i(x,A) > 0$ for each $x \in A \cap bP$, then A is measurable.

(Here, $$d^i(x,A) := \lim_{r\to 0_+} \left(\sup\left\{\frac{m_*(A \cap M)}{m(M)} : M \in \mathcal{L}_x,\ m(M) > 0,\ M \subset U(x,r)\right\}\right).)$$

Hint. (a) Define a new differentiation basis $\mathcal{L}^*$ such that A is open in the density topology determined by $\mathcal{L}^*$ and for which the density theorem holds. Then use Corollary 6.15.

(b) Let J_A be a measurable kernel of A. By the assumption, $A \subset bJ_A$. Since $bJ_A \sim J_A$ by 6.13.b, we have $A \sim J_A$.

6.B.5 Let $\mathcal{L}$ be a differentiation basis such that $m(M \setminus bM) = 0$ for each measurable set $M \subset P$. Show that any t-Borel set is measurable.

Hint. Let F be t-closed. Then

$$m^*(H_F \setminus F) \leq m^*(H_F \setminus bF) = m(H_F \setminus bH_F) = 0.$$

REMARK. We say that the *theorem of nondispersion* holds for $\mathcal{L}$ if $m(M \setminus bM) = 0$ for each measurable set M. Of course, the theorem of nondispersion is a consequence of the density theorem (Lemma 6.13.a).

6.B.6 Let t be a metric density topology on P, $bP = P$.

(a) If $\emptyset \neq G$ is t-open, then $m_*(G) > 0$. Consequently, (P,t) satisfies (CCC).

(b) If f is a mapping from P into a metric space Y which is t-continuous at all points of a set $M \subset X$, then $f(M)$ is a separable subspace of Y. (C. Goffman and D. Waterman (1961) for ordinary density topology on $\mathbb{R}^n$.)

Hint. Use (a) and Ex. 1.B.1.c.

6.B.7 Let ν be a Radon measure on $\mathbb{R}^n$, supp $\nu = \mathbb{R}^n$. If $\mathcal{B}$ is a differentiation basis, denote $t(\nu,\mathcal{B})$ the density topology determined by $\mathcal{B}$ and ν.

(a) Let $\mathcal{S}$ be the symmetric basis in $\mathbb{R}^n$ and let $\bar{\mathcal{S}}_x = \{\bar{S}: S \in \mathcal{S}_x\}$. Show that $t(\nu,\mathcal{S}) = t(\nu, \bar{\mathcal{S}})$.

(b) Let $\mathcal{K}$ be the cube basis in $\mathbb{R}^n$ and let $\mathcal{K}_x^{\circ} = \{\text{int } K: K \in \mathcal{K}_x\}$. Show that $t(\nu,\mathcal{K}) = t(\nu, \mathcal{K}^{\circ})$.

(c) If λ_n denotes the Lebesgue measure, then $t(\lambda_n, \mathcal{S}) = t(\lambda_n, \mathcal{K})$.

(d) Find a Radon measure for which $t(\nu,\mathcal{S}) \neq t(\nu,\mathcal{K})$.

6.B.8 Let $P = [0,1] \times [0,1]$, $I = \{0\} \times [0,1]$. Let m be the completion of the Borel measure $\lambda_2 \wedge P + \lambda_1 \wedge I$ where λ_2 denotes two-dimensional Lebesgue measure and λ_1 is one-dimensional (Hausdorff) measure. Given a nondecreasing function $f: (0,1] \to (0,1]$, $\lim_{t\to 0_+} f(t) = 0$, set

$\mathcal{F}_f := \{G: [0,1] \to P$: $G = (x_G, y_G)$ where x_G, y_G are piecewise C^1-smooth functions, $x_G > 0$ on $(0,1)$ and $y_G'(t) = f(x_G(t)).x_G'(t)$ for all $t \in (0,1)$ except a finite set$\}$.

Define

$$\rho_f(u,v) = \inf\{\text{var}_0^1 x_G: G \in \mathcal{F}_f,\ G(0) = u,\ G(1) = v\}$$

($\text{var}_0^1 g$ denotes the variation of a function g on $[0,1]$).

(a) Show that ρ_f is a metric on P equivalent to the Euclidean metric.

Let $\mathcal{S}_f$ be the symmetric basis on (P, ρ_f). Denote by t_f the density topology determined by $\mathcal{S}_f$ and m, and let b_f be the corresponding base operator.

(b) Choose $f(x) = e^{-1/x}$ and prove that in this case:

(b_1) $b_f(I) = \emptyset$, $b_f(P) = P$,

(b_2) there is a t_f-closed subset of P which is not m-measurable,

(b_3) the density theorem does not hold for $\mathcal{L}_f$.

Hint. Pick $x \in I$ and show that

$$m(U(x,r)) \geq \lambda_2(U(x,r)) \geq \frac{1}{2}\left(\frac{r}{3}\right)^2 . e^{-3/(2r)},$$

$$m(U(x,r) \cap I) = \lambda_1(U(x,r) \cap I) \leq 2r\, e^{-2/r}$$

for all $r > 0$ small enough. A moment's reflection shows that (b_2) is a consequence of (b_1).

(c) Define $f(x) = 2^{-2^n}$ for $x \in (4^{-n}, 4^{-n+1}]$ and prove:

(c_1) the density theorem does not hold for $\mathcal{L}_f$,

(c_2) the theorem of nondispersion holds for $\mathcal{L}_f$,

(c_3) each t_f-closed subset of P is m-measurable.

Hint. (c_1) Pick $x \in I$ and $n \in \mathbb{N}$. Show that

$$m(U(x,\ 2.4^{-n})) \geq \lambda_2(U(x,\ 2.4^{-n})) \geq \frac{1}{2}.2^{-2^n}.4^{-2n},$$

$$m(U(x, 2.4^{-n}) \cap I) = \lambda_1(U(x,\ 2.4^{-n}) \cap I) \leq 2^2.4^{-n}.2^{-2^{n+1}}.$$

(c_2) Given $x \in I$ and $n \in \mathbb{N}$, we have

$$\lambda_2(U(x,\ 4^{-n})) \leq 4^{-2n}.2^{-2^{n+1}}.2,$$

$$\lambda_1(U(x,\ 4^{-n}) \cap I) \geq 2.\frac{1}{4}.4^{-n}.2^{-2^{n+1}}.$$

From this deduce that $m((A \cap I) \setminus bA) = 0$ for every m-measurable set $A \subset P$ (do not remember that for $\lambda_1 \wedge I$ the density theorem holds). Further show that $m((A \setminus I) \setminus bA) = 0$.

(c_3) See Ex. 6.B.5.

REMARKS AND COMMENTS

In the literature we find a number of papers dealing with density topologies determined by various differentiation bases. Already the first papers of O. Haupt and Ch. Pauc (1952) and (1954) contained a considerably abstract approach. Among many consequent papers let us mention at least C. Goffman, C. Neugebauer and T. Nishiura (1961), R.S.Troyer and W.P.Ziemer (1963), N.F.G.Martin (1964), W. Eames (1971), M. Chaika (1971), G.M.Petersen and T.O. To (1976).

Very comprehensive is the literature treating the differentiation bases (cf. A.M. Bruckner (1971) or M. de Guzmán (1975) and (1981)). The deep results contained therein concern the validity of covering theorems, density theorem and differentiation of integrals with respect to various differentiation bases. In fact, we are mostly interested in the validity of density theorems only. An interesting and useful criterion for the validity of the density theorem for symmetric basis can be found in D. Preiss (1981).

As early as in (1915) A. Denjoy proved that any Lebesgue measurable function is approximately continuous almost everywhere. The converse was proved by W. Stepanoff (1924) (cf. also W. Sierpiński (1922) and E. Kamke (1927). Our "topological" proof is contained in J. Lukeš (1978).

Related papers: B.K.Lahiri (1977), S. Chakrabarti and B.K. Lahiri (1984), Z.Grande and M. Topolewska (1982).

6.C Properties of density topologies depending on features of differentiation bases

Motivated by the importance of fine topology methods indicated in earlier sections, we devote now the bulk of these sections to some further considerations regarding these topics. In the preceding investigations we have found some criteria guaranteeing certain properties of fine topologies we are interested in. Now we use it to derive several sufficient conditions in the language of differentiation bases. We discuss namely conditions which insure

(a) fine limits property,

(b) G_δ-insertion property,

(c) fine connectivity,

(d) quasi-Lindelöf property,

(e) Lusin-Menchoff property.

With regard to the importance of the Lusin-Menchoff property, we postpone its investigation to the next section.

Fine limits. At first, we are concerned with the question whether a fine limit is equal to a limit with respect to some fine neighborhood, i.e. we want to find conditions leading to (tFL). (The relative question to it is whether t-continuous functions coincide with "approximately continuous" ones.) According to general Proposition 2.10 it occurs if the following condition (bFL) is satisfied:

(bFL) If $\{A_n\}$ is a sequence of sets and x is a dispersion point of each A_n, then there is a sequence $\{U_n\}$ of neighborhoods of x such that x is again a dispersion point of the set $\bigcup_{n=1}^{\infty} (A_n \cap U_n)$.

6.21 THEOREM. Assume that b is a base determined by an upper outer density and let $x \in P$. Suppose that for every $\delta > 0$ there is $r > 0$ such that whenever $B \in \mathcal{L}_x$, $m(B) < r$ then there exists $B^* \in \mathcal{L}_x$ for which

$$B \cap U(x,r) \subset B^* \subset B \cap U(x,\delta).$$

Then the condition (bFL) is satisfied.

Proof. Suppose $\{A_n\}$ is a sequence of sets and $d^e(x,A_n) = 0$ for each $n \in \mathbb{N}$. According to Ex. 2.C.5 we may assume $x \notin A_n$ for $n \in \mathbb{N}$. Thus, for every $n \in \mathbb{N}$ there is $\delta_n > 0$ such that

$$\frac{m^*(A_n \cap B)}{m(B)} < 2^{-n}$$

whenever $B \in \mathcal{L}_x$, $B \subset U(x, \delta_n)$ and $m(B) > 0$. Let $r_n > 0$ correspond to δ_n as required by assumptions. Find $\lambda_n \in (0,r_n)$ such that

$$m(U(x,\lambda_n) \setminus \{x\}) < 2^{-n} r_n .$$

Putting $U_n := U(x, \lambda_n)$ we intend to prove that $\bigcup_{n=1}^{\infty} (A_n \cap U_n)$ is thin at x.

Given $B \in \mathcal{L}_x$, $m(B) > 0$ we have

(*) $m^*(B \cap A_n \cap U_n) \leq 2^{-n}\, m(B)$.

Indeed, in case of $m(B) \geq r_n$, (*) follows from

$$m^*(B \cap A_n \cap U_n) \leq m(U_n \setminus \{x\}) < 2^{-n} r_n.$$

If $m(B) < r_n$, then there is $B^* \in \mathcal{L}_x$ such that $B \cap U(x,r_n) \subset B^* \subset B \cap U(x, \delta_n)$. Hence

$$\frac{m^*(B \cap A_n \cap U_n)}{m(B)} \leq \frac{m^*(B^* \cap A_n)}{m(B)} \leq 2^{-n}.$$

We get for each $k \in \mathbb{N}$

$$d^e(x, \bigcup_{n=1}^{\infty} (A_n \cap U_n)) \leq \sum_{n=1}^{k} d^e(x, A_n \cap U_n) + d^e(x, \bigcup_{n=k+1}^{\infty} (A_n \cap U_n)) =$$

$$= d^e(x, \bigcup_{n=k+1}^{\infty} (A_n \cap U_n)) \leq \sum_{n=k+1}^{\infty} 2^{-n} = 2^{-k},$$

which yields the conclusion.

6.22 COROLLARY. The condition (tFL) holds provided one of the following conditions is satisfied:

(P_1) For every $\delta > 0$,

$$\inf\{m(B):\ B \in \mathcal{L}_x,\ m(B) > 0,\ B \setminus U(x,\delta) \neq \emptyset\} > 0.$$

(P_2) There is a local base $\{U_n\}$ at x such that $B \cap U_n \in \mathcal{L}_x$ provided $B \in \mathcal{L}_x$.

(P_3) For every $\delta > 0$ there is $r > 0$ with the property: whenever $B \in \mathcal{L}_x$, there is $B^* \in \mathcal{L}_x$ such that

$$B \cap U(x,r) \subset B^* \subset B \cap U(x,\delta).$$

REMARK. Using the preceding criteria and the fact that (bFL) is satisfied if $x \notin bP$, we get immediately that the condition (tFL) is satisfied for most of density topologies determined by our concrete differentiation bases and by an arbitrary measure satisfying our assumptions. More precisely, in case of:

B_1, B_2 .... (P_1) or (P_2) can be used,

B_3 .... (P_1) only is suitable,

B_4 .... (P_2) is satisfied,

B_5 .... (P_3) only comes in handy,

B_6, B_7, B_8 .... (P_2) is satisfied,

B_9 with the Lebesgue measure .. (P_1) only is suitable,

B_{11}, B_{12}, B_{13} .... (P_1) only is suitable.

REMARK. If the assumptions of Theorem 6.21 are satisfied, then a function f is t-continuous at x if and only if x is t-isolated or there is a t-neighborhood U of x such that $f(x) = \lim_{U \ni y \to x} f(y)$.

For some negative results see Ex. 6.C.2 and 6.C.4.

G_δ-insertion and Baire one functions. The question whether any t-continuos function is of the Baire class one in P can be answered by the next theorem which yields a sufficient condition general enough.

6.23 THEOREM. Assume that the following condition (Ba1) is satisfied:

(Ba 1) Given $x \in P$, $x_n \to x$, $B \in \mathcal{B}_x$, there are $B_n \in \mathcal{B}_{x_n}$ such that

$$\operatorname{diam}_{x_n} B_n \to \operatorname{diam}_x B \quad \text{and} \quad m(B \,\Delta\, B_n) \to 0$$

where $\operatorname{diam}_x A := \inf\{r > 0 : A \subset U(x,r)\}$.

If $bP = P$ then the t-topology has the G_δ-insertion property, and consequently each t-continuous function of P is a Baire one function.

Proof. To prove the assertion, we need to find a set A^* of the type G_δ for which $\operatorname{int}_t(A) \subset A^* \subset \bar{A}^t$ whenever $A \subset P$. Putting

$$A^* := \{x \in P: \text{ for every } n \in N \text{ there is } B \in \mathcal{L}_x \text{ such that}$$

$$\operatorname{diam}_x B < 1/n,\ m^*(A \cap B)/m(B) > 1/2\},$$

we have that A^* is inserted between $\operatorname{int}_t(A)$ and $\bar{A}^t$. Clearly, $A^* = \bigcap_{n=1}^{\infty} A_n$ where

$$A_n := \{x \in P: \text{ there is } B \in \mathcal{L}_x \text{ with } \operatorname{diam}_x B < 1/n \text{ and}$$
$$m^*(A \cap B)/m(B) > 1/2\}.$$

To complete the proof we have to show that every A_n is open. If this is not the case, there are $x \in A_n$ and $x_k \in P \setminus A_n$ such that $x_k \to x$. By the condition (Bal) there are $B_k \in \mathcal{L}_{x_k}$ such that

$$\operatorname{diam}_{x_k} B_k \to \operatorname{diam}_x B, \quad m(B \,\Delta\, B_k) \to 0$$

where $B \in \mathcal{L}_x$ is a set from the definition of A_n. Since

$$|m^*(A \cap B_k) - m^*(A \cap B)| \le m^*((A \cap B_k) \,\Delta\, (A \cap B)) \le$$
$$\le m\,(B \Delta B_k),$$

we have

$$\lim \frac{m^*(A \cap B_k)}{m(B_k)} = \frac{m^*(A \cap B)}{m(B)}$$

which contradicts the fact that $x_k \in P \setminus A_n$ and diam $B_k < 1/n$ for k large enough.

6.24 EXAMPLES. If $bP = P$, then it is not too hard to see that the condition (Bal) is satisfied in cases of differentiation bases

$$B_1,\ B_2,\ B_4,\ B_5,\ B_8,\ B_9.$$

E.g., for B_1, given $x_n \to x$ and $B \in \mathcal{L}_x$, it suffices to put

$$B_n = U(x_n,\ \max\,(r - \rho(x_n, x),\ r/2)),$$

where r is the radius of B. In cases of remaining bases the proofs are similar (cf. Ex. 6.C.9).

Note that for the right-density basis B_7 the condition (Bal) holds for the case of the Lebesgue measure but counterexamples exist showing that this is not true for an arbitrary measure. In fact, there are measures for which not all B_7-approximately continuous functions are in the Baire class one (see Ex. 6.C.11).

Something like that can occur in the case of incomplete symmetrical basis B_{12}. There is even a basis B_{12} on $\mathbb{R}$ such that (Bal) does not hold for the Lebesgue measure and there are B_{12}-approximately continuous functions which are not Baire one (see Ex. 6.C.12).

Connectivity properties. Using Corollary 5.3 we can derive a general theorem on density-connectivity of domains in P. Observe that, by definition, a point x is said to be a *dispersion point* for a set E if $d^e(x,E) = 0$.

6.25 PROPOSITION. Let (P,ρ) be a strong Baire metric space in which every closed ball is a compact set. If a metric density topology d on P has the G_δ-insertion property, and if for each $x \in P$ and for each $r > 0$ no point of $B(x,r)$ is a dispersion point for $U(x,r)$, then every open and connected set of P is d-connected.

Proof. Since

$$B(x,r) \subset \{z \in P: d^e(z, U(z,r)) > 0\} \subset \overline{U(x,r)}^{\,d},$$

Corollary 5.3 yields the result.

It is of interest to apply the preceding results to the study of connectivity propertis of some differentiation bases. Certain results admit several corollaries some of which appear in the problems for this section.

For example, the result of R.S. Troyer and W.P. Ziemer (1963) (connectedness in case of density topology corresponding to a cube differentiation basis B_2) is an immediate consequence of our theorems (see Ex. 6.C.14).

As pointed out by C. Goffman and D. Waterman in (1961), there are more density-connected sets under the ordinary density topology in $\mathbb{R}^n$ than these which are open and connected. In Ex. 6.C.20 we sketch the proof of the following more general result: Let $F \subset \mathbb{R}^n$ be an F_σ and Lebesgue null set which does not separate any open ball of $\mathbb{R}^n$. Then $\mathbb{R}^n \setminus F$ is connected in the ordinary density topology.

Surprisingly, the density topology d in $\mathbb{R}^n$ determined by Lebesgue measure and by any incomplete symmetrical basis B_{12} is connected (i.e. every open connected subset of $\mathbb{R}^n$ is also connected in this topology). The proof of this assertion cannot be done by means of Proposition 6.25 since, in general, the condition (Bal) is violated and thus we cannot verify in light of (Bal) that any both d-closed and d-open set is a G_δ-set. Nevertheless, in Ex. 6.C.16 we sketch a different proof of this assertion. It is similar to Ridder's proof in (1929). He proved there (of course, in other language) that the density topology in $\mathbb{R}^n$ is connected. Hence, we obtain especially an alternative proof of connectedness of the density topology in $\mathbb{R}^n$ which seems to be simpler than that of C. Goffman and D. Waterman (1961).

Finally, we are going to exhibit further important properties of a (metric) density topology. First, however, let us agree that $\mathcal{Z}_\mu$ stands for the σ-ideal of μ-null sets.

6.26 THEOREM. If the density theorem holds for $\mathcal{L}$, then P equipped with the density topology t:

(a) is hereditarily Baire,

(b) has the $\mathcal{Z}_\mu$-quasi-Lindelöf property.

Proof. (a) In light of Theorem 6.18 any t-first category set is t-closed. Following the same line of proof as in Theorem 6.9.e we establish the assertion.

(b) Since (P,t) is a Baire space in which the σ-ideal $\mathcal{Z}_\mu$ equals that one of t-first category sets, P has the $\mathcal{Z}_\mu$-quasi-Lindelöf property if and only if P satisfies the countable chain condition (CCC) (see Ex. 1.B.1). The proof that (CCC) holds in (P,t) is an easy consequence of the fact that μ is σ-finite and that any nonempty open set is of positive μ-measure.

EXERCISES

6.C.1 Show that the condition (P_2) of Corollary 6.22 implies (P_3).

6.C.2 Let $\mathcal{L}$ be a differentiation basis in $\mathbb{R}^2$ such that

$$\mathcal{L}_{(0;0)} = \{B \subset \mathbb{R}^2 :\ B \text{ is } \lambda_2\text{-measurable and there is } a > 0 \text{ such that } (a;0) \in B \text{ and } B \subset \{(x;y) : 0 \le x \le a,\ 0 \le y \le ax\}\}.$$

Then neither the condition (bFL) nor the condition (tFL) holds for the corresponding base operator and the metric density topology determined by $\mathcal{L}$ and λ_2.

Hint. Consider the sequence $A_n = \{(x;y) : 0 \le x \le 1/n,\ x/n \le y \le n^{-2}\}$.

6.C.3 Let $\mathcal{L}$ be the differentiation basis in $\mathbb{R}^n$ for which $\mathcal{L}_x$ consists of all open balls containing x. If m is a Radon measure on $\mathbb{R}^n$ and if $z \in \operatorname{int}(\operatorname{supp} m)$, then (tFL) holds at z for the corresponding metric density topology.

Hint. Verify (P_1) of Corollary 6.22.

6.C.4 Let $\mathcal{L}$ be the basis in $\mathbb{R}^3$ as in the previous exercises. Then there is a Radon measure m on $\mathbb{R}^3$ and a point $z \notin \operatorname{supp} m$ such that neither (bFL) nor (tFL) holds at z.

Hint. First, prove the following "geometric" assertion: There exist closed balls $K_1, K_2, K_3, \ldots$ with radii $r_i \searrow 0$ and circular arcs $k_1, k_2, \ldots$ starting from the origin, such that each k_i is a part of a principal circle of K_i, $k_i \cap K_j = \{0\}$ for $i \neq j$, and k_i is not tangent to K_j for $i \neq j$.

Now construct new curves k_i^* starting again from the origin, where $k_i^* \cap K_j = \{0\}$ for all i,j, and k_i and k_i^* have the contact of 3rd order at the origin. It does no harm to suppose that the sum of all lengths of the curves k_i and k_i^* is finite.

On k_i find a Borel set $T_i \subset k_i$ such that $0 < H_1(T_i \cap U(0,r)) < r^2$ for any $r > 0$ (here H_1 denotes the one-dimensional Hausdorff measure).

Take for m the sum of one-dimensional Hausdorff measures on all T_i and k_i^*.

Neither (bFL) nor (tFL) is valid, it suffices to put $A_i = k_i$.

The proof that every A_i is thin: Consider a small ball containing the origin which is intersected by k_i. Then k_i and k_i^* "leave" this small ball almost at the same place. We see that even $(\mathbb{R}^3 \setminus A_i) \cup \{0\}$ is a density neighborhood of 0.

Now suppose that $\bigcup_{i=1}^{\infty} A_i \cap U(0, \delta_i)$ is thin for some $\{\delta_i\}$. Let $\delta > 0$ be given. Take i such that $\delta > r_i$ and consider a ball which is very close to K_i and cuts out on k_i the arc $k_i \cap U(0, \delta_i)$.

6.C.5 Show that the conclusion of Theorem 6.23 remains valid if the hypothesis "$\operatorname{diam}_{x_n} B_n \to \operatorname{diam}_x B$" is weakened to "$\limsup \operatorname{diam}_{x_n} B_n \leq \operatorname{diam}_x B$".

6.C.6 If the condition (Bal) holds for $\mathcal{L}$, then the corresponding t-topology satisfies the essential radius condition (cf. Ex. 2.D.16).

Hint. Given $x \in X$ and a t-open t-neighborhood U of x, there is $p(x,U) > 0$ such that $m(U \cap B) \geq \frac{3}{4} m(B)$ whenever $B \in \mathcal{L}_x$, and $B \subset U(x, 2p(x,U))$. Proceeding by contradiction, show that there is $r(x,U) > 0$ such that $U \cap V \neq \emptyset$ provided $\rho(x,y) < r(x,U)$, V is a t-open t-neighborhood of y and $p(y,V) \geq p(x,U)$.

6.C.7 If the condition (Bal) is satisfied, then bA is a $G_{\delta\sigma}$-set for each $A \subset X$.

Hint. Show that

$$\{x \in P\colon \text{for each } n \in \mathbb{N} \text{ there is } B \in \mathcal{L}_x \text{ such that}$$
$$\operatorname{diam} B < 1/n \text{ and } m^*(A \cap B)/m(B) > 1/k\}$$

is a G_δ-set.

6.C.8 Let P be a separable metric space, card $P = 2^{\aleph_0}$, $bP = P$ and $m\{x\} = 0$ for each $x \in P$. If the condition (Bal) is satisfied, then the t-topology is non-Blumberg.

Hint. Use Ex. 6.C.7 and Ex. 4.B.1.

6.C.9 If $\mathcal{L}_x$ consists of $\overline{U(x,r)}$ for each $x \in P$, then (Bal) is satisfied.

Hint. Given $x_n \to x$, $B = \overline{U(x,r)} \in \mathcal{L}_x$, put $B_n = \overline{U(x_n, r + \rho(x_n,x))}$.

6.C.10 If $\mathcal{L}$ satisfies the condition (Bal) and if $f\colon P \to Y$ (where (Y,σ) is a metric space) is t-continuous at all points of a set $M \subset P$, then $f \wedge M$ is G_δ-measurable (for each closed set F in Y, $(f \wedge M)^{-1}(F)$ is a G_δ-set in M).

Hint. Show that $(f \wedge M)^{-1}(U(y,r))$ is an F_σ-set for every $y \in Y$ and $r > 0$. (Use Ex. 2.D.12 for $g * f$ where $g(z) = \sigma(z,x)$ and Ex. 6.B.6.)

6.C.11 Let $\mathcal{L}$ be the right-density basis B_7 on $\mathbb{R}$. There exists a Radon measure μ with supp $\mu = \mathbb{R}$ such that t-continuous functions need not be in Baire class one.

Hint. (By D. Preiss). Let $P \neq \emptyset$ be a perfect nowhere dense subset of $\mathbb{R}$ such that the component intervals (a_n,b_n) that are complementary to P are of length < 1 ($\mathbb{R} \setminus P = \bigcup_{n=1}^{\infty} (a_n,b_n)$). Put

$$\mu = \lambda + \sum_{n=1}^{\infty} n^{-2}(b_n - a_n)\, \varepsilon_{b_n}$$

(ε_z denotes the Dirac measure at z). Define

$$f(x) = \begin{cases} 1 & \text{if } x = b_n \text{ for some } n, \\ 0 & \text{otherwise.} \end{cases}$$

Now, $f \notin B_1(\mathbb{R})$ (use Theoreme 2.12, (Ba) or (D-P)) although f is t-continuous.

6.C.12 There is an incomplete symmetric basis $\mathcal{L}$ on $\mathbb{R}$ such that $b\,\mathbb{R} = \mathbb{R}$, and such that there exist t-continuous functions (t is formed with respect to $\mathcal{L}$ and the Lebesgue measure) that are not Borel functions.

Hint. (By M. Chlebík) (a) Show that the following assertions are equivalent for a function f on $\mathbb{R}$:

(i) f is t-continuous for some metric density topology on $\mathbb{R}$ determined by the Lebesgue measure and by an incomplete symmetric basis such that $b\,\mathbb{R} = \mathbb{R}$,

(ii) $d^{i,sym}(a, f^{-1}(f(a)-\varepsilon, f(a)+\varepsilon)) = 1$ for each $a \in \mathbb{R}$ and $\varepsilon > 0$.

(b) Construct a continuous function f on $(0,1)$ such that

$$\bar{d}_+(0, f^{-1}(0)) = \bar{d}_+(0, f^{-1}(1)) = 1.$$

($\bar{d}_+$ is defined in an obvious manner, cf. Lemma 2.16 and Ex. 6.A.1.).

(c) Fix a function f satisfying properties of (b), and let $h_k \searrow 0$ be a sequence such that

$$\lambda((0,h_k) \cap f^{-1}(0)) > (1 - 1/k)h_k \quad \text{whenever } k \text{ is odd,}$$
$$\lambda((0,h_k) \cap f^{-1}(1)) > (1 - 1/k)h_k \quad \text{whenever } k \text{ is even.}$$

Let Φ_k^+, Φ_k^- be a collection of intervals having the following properties:

(1) $\Phi_1^+ = \{(2n, 2n+1) : n \in \mathbb{Z}\}$, $\Phi_1^- = \{[2n+1, 2n+2] : k \in \mathbb{Z}\}$,

(2) the intervals of Φ_k^- are formed by components of $\bigcup_{j=1}^{k} [\,\cup\, \{I :\ I \in \Phi_j^+\}]$,

(3) given $k \in \mathbb{N}$ and $I = [a,b] \in \Phi_k^-$, choose h_m such that $h_m < \frac{1}{2}(b-a)$ and m is even or odd provided k is of this type; let $J_I = (c,d)$ be the interval for which

$$\frac{1}{2}(d+c) = \frac{1}{2}(b+a) \quad \text{and} \quad k(b-d) = h_m .$$

Now, $\Phi^+_{k+1} = \{J_I : I \in \Phi^-_k\}$. If $(p,q) \in \bigcup_{i=1}^{\infty} \Phi^-_i$, define

$$g(x) = \begin{cases} f(x-p) & \text{whenever} \quad x \in (p, \frac{p+q}{2}), \\ f(q-x) & \text{whenever} \quad x \in [\frac{p+q}{2}, q) \end{cases}$$

and extend g on $\mathbb{R}$ by values 0 or 1 (arbitrarily). Using (a), prove that g is t-continuous on $\mathbb{R}$ for some incomplete symmetric basis.

(d) Using cardinality arguments and (c), prove the assertion.

6.C.13 Let t be a metric density topology on $\mathbb{R}^n$ (determined by some differentiation basis by a Radon measure) having the G_δ-insertion property. Let $G \subset b(\mathbb{R}^n)$ be an open connected set. Suppose that there is a bounded open set $T \subset \mathbb{R}^n$ with the following property: If $T^* \subset G$ is a result of a translation or of a homothety mapping applied on T, then no point of ∂T^* is a dispersion point of T^*. Prove that G is t-connected.

Hint. Pick $x \in \partial G$ and $\delta > 0$. Find a closed ball $B(y,r) \subset G$ not containing x such that $y \in U(x, \frac{\delta}{2})$ and a set T^* homothetic with T such that $y \in T^* \subset U(y,r)$. Find maximal $\lambda > 0$ for which $T^* + \lambda_0(x-y) \subset G$ for each $0 \leq \lambda_0 \leq \lambda$. Then $M := \partial(T^* + \lambda(x-y)) \cap \partial G \neq \emptyset$. Let $z \in M$. By assumption, any point of M is not a dispersion point of $T^* + \lambda(x-y)$. Hence $z \in \bar{G}^t$. The proof will be finished by proving $\rho(z,x) \leq \delta$ (see Corollary 5.2).

6.C.14 (R.J.Troyer and W.P.Ziemer (1963), Th. 5.2). Let m denote a Radon measure over $\mathbb{R}^n$ and let t be a metric density topology on $\mathbb{R}^n$ determined by m and by the cube differentiation basis B_2. If $S \subset \mathbb{R}^n$ is an open connected set having the property that each point $x \in S$ is not a dispersion point of any of 2^n open "octants" with vertex in x (op.cit., p.493), then S is t-open.

Hint. Use Ex. 6.C.13 or Proposition 6.25 applied to the basis B_1 in $\mathbb{R}^n$ equipped with the maximum metric.

6.C.15 Show that Ex. 6.C.14 remains valid in the following cases:

(a) $\mathcal{L}$ equals the symmetric basis B_1 in $\mathbb{R}^n$,

(b) $\mathcal{L}$ is a translation invariant incomplete symmetric basis B_{12} in $\mathbb{R}^n$ and $m(\partial B) = 0$ for each $B \in \mathcal{L}_x \in \mathcal{L}$,

(c) $\mathcal{L}$ is a translation invariant "incomplete cube basis" in $\mathbb{R}^n$ and, again, $m(\partial B) = 0$ for all $B \in \mathcal{L}_x$ (in particular, this condition is satisfied if m vanishes on all hyperplanes).

Hint. Show that (Bal) is satisfied and use Ex. 6.C.13.

6.C.16 Show that the density topology d in $\mathbb{R}^n$ determined by the Lebesgue measure λ_n and by an incomplete symmetric basis B_{12} has the property that any open connected set is also d-connected.

Hint. Assume, in order to obtain a contradiction, that S is an open connected subset of $\mathbb{R}^n$ which is not d-connected.

(a) There is a measurable set $\emptyset \neq M \neq S$ such that

$$d^{sym}(x,M) = 1 \quad \text{for all} \quad x \in M \quad \text{and}$$

$$d^{sym}(x,S \setminus M) = 1 \text{ for all } x \in S \setminus M,$$

(where d^{sym} denotes the upper density with respect to the basis B_1).

(b) Using the density theorem and the arcwise connectivity of S, there is a closed cube $K_1 \subset S$ with edges of the length $r > 0$ such that $\lambda_n(K_1 \cap M) = \frac{1}{2}\lambda_n(K_1)$.

(c) Construct a sequence of cubes $K_1 \supset K_2 \supset K_3$ ... such that the length of edges of K_i is $r2^{1-i}$ and $\lambda_n(K_i \cap M) = \frac{1}{2}\lambda_n K_i)$. (Divide, e.g., K_1 into 2^n cubes with the length of the edges equal $\frac{r}{2}$ and show that for one of them, say K^*, $\lambda_n(K^* \cap M) \geq \frac{1}{2}\lambda_n(K^*)$ and for another ohne, say K_*, $\lambda_n(K_* \cap M) \leq \frac{1}{2}\lambda_n(K_*)$.)

(d) Prove that $\bigcap_{n=1}^{\infty} K_n = \{z\} \in S$ and that

$$d_{sym}(z,M) > 0, \; d_{sym}(z, S \setminus M) > 0$$

which contradicts (a).

6.C.17 Let f be a function on $\mathbb{R}^n$ such that

$$d^{e,sym}(a, f^{-1}((f(a)-\varepsilon, f(a)+\varepsilon))) = 1$$

for any $a \in \mathbb{R}^n$ and $\varepsilon > 0$. Show that

(a) f has a connected graph,

(b) in case $n = 1$, f is a Darboux function.

Hint. Use Ex. 6.C.16 (cf. Hint of Ex. 6.C.12).

REMARK. In view of Ex. 6.C.12, f can be chosen to be a non-Borel function on $\mathbb{R}^n$.

6.C.18 Assume that the density theorem holds for a differentiation basis $\mathcal{L}$ on (P, ρ) and that $x \in B$ whenever $B \in \mathcal{L}_x$. If for any $x \in bP$ and $r > 0$,

$$\alpha_r := \inf\{m(B) : B \in \mathcal{L}_x, \text{ diam } B \geq r\} > 0,$$

then any point of P has a t-neighborhood basis formed by ρ-closed sets. In particular, the density topology t is cometrizable (and, of course, regular).

Hint. Assume that G is a t-open set containing a point $x \in bP$. Given $n \in \mathbb{N}$, find a closed set $F_n \subset U(x, \frac{1}{n}) \cap G$ such that $m((G \cap U(x, \frac{1}{n})) \setminus F_n) \leq \frac{1}{n} \alpha_{n+1}$. Show that $\{x\} \cup \bigcup_{n=1}^{\infty} F_n$ is a ρ-closed t-neighborhood of x contained in G.

6.C.19 Let d be the ordinary density topology on $\mathbb{R}^n$ and let G be an open subset of $\mathbb{R}^n$. Prove that $\bar{G}^d$ is residual in ∂G (cf. J. Lukeš and L. Zajíček (1977c)).

Hint. Set $G^* := \{ x \in \mathbb{R}^n$: for any $k \in \mathbb{N}$ there is $m > k$

such that $\lambda(G \cap U(x, \frac{1}{m})) > \frac{1}{3} \lambda(U(x, \frac{1}{m}))\}$.

It is not difficult to see that G^* is a G_δ-set (cf. Theorem 6.23). Of course, $G \subset G^* \subset \bar{G}^d$. To prove that G^* is dense in ∂G proceed as in Corollary 5.3 or Ex. 6.C.13 (choose T as an open ball).

6.C.20 Let d be the ordinary density topology on $\mathbb{R}^n$. Let $S \subset \mathbb{R}^n$ be an F_σ-set of Lebesgue measure zero such that $U \setminus S$ is connected whenever U is an open ball in $\mathbb{R}^n$. Let G be an open connected subset of $\mathbb{R}^n$. Show that $G \setminus S$ is d-connected.

Hint. The assertion follows from Proposition 5.5 if we use the previous exercise and observe that $G\setminus S$ is connected.

6.C.21 Give an alternative proof of Ex. 6.C.20 using the idea similar to that of Ex. 6.C.16.

Hint. Let $S = \bigcup_{n=1}^{\infty} F_n$ where F_n are closed. If $\emptyset \neq M \neq G\setminus S$ is both d-open and d-closed subset of $G\setminus S$, construct a sequence of closed cubes $G \supset K_1 \supset K_2 \supset \dots$ such that $K_i \cap F_i = \emptyset$, $\lambda_n(K_i \cap M) = \frac{1}{2}\lambda_n(K_i)$ and $\operatorname{diam} K_i \to 0$.

6.C.22 Let d be the ordinary density topology on $\mathbb{R}^n$ and let S be an F_σ-subset of $\mathbb{R}^n$ of $(n-1)$-dimensional Hausdorff measure zero. If G is an open connected subset of $\mathbb{R}^n$, then $G\setminus S$ is d-connected.

Hint. Use Ex. 6.C.20 (cf. Ex. 6.C.21).

REMARK. It is interesting to remark that by M. Chlebík (1984) it is to be shown the condition that S is an F_σ-set can be dropped. The proof of this assertion is rather complicated.

REMARKS AND COMMENTS

Properties of the ordinary density topology on $\mathbb{R}^n$ were investigated in detail in C. Goffman, C. Neugebauer and T. Nishiura (1961). The density topology determined by cube differentiation basis on $\mathbb{R}^n$ and by a Radon measure of support $\mathbb{R}^n$ was examined by R.S. Troyer and W.P. Ziemer (1963). The purpose of this section was to generalize their results using the methods of proofs based on general theorems of Chap. I.

6.D Density and superdensity topologies and the Lusin-Menchoff property

In this section we derive simple sufficient conditions on differentiation bases which insure the Lusin-Menchoff property of the corresponding density topology. Guided by interesting subsequent examples 6.27 and 6.31, we consider their analogue in a more general setting. The reason for doing so is that we want to include also the case of the superdensity topology (cf. also the differentiation basis B_9). Thus we will be concerned in this section with the investigation of the Lusin-Menchoff property of these generalized density topologies and with the Lusin-Menchoff property of corresponding base operators.

We start with an interesting example.

6.27 EXAMPLE (the superdensity topology on $\mathbb{R}$).

(a) We say that a set $G \subset \mathbb{R}$ is s-open if for every $x \in G$

$$\lim_{h \to 0_+} h^{-2} \cdot \lambda^*((x-h,x+h) \setminus G) = 0.$$

It is not difficult to check that the collection of all s-open sets is actually a topology which will be labelled as the *superdensity topology*. Let us notice that the superdensity topology is coarser than the density topology, and therefore any s-open set is Lebesgue measurable. Of course, the superdensity topology is finer than the Euclidean topology.

(b) It is not difficult to prove that the superdensity topology has the G_δ-*insertion property* (cf. 6.23 and 6.24).

(c) If we put

$$bA := \{x \in \mathbb{R}: \limsup_{h \to 0_+} h^{-2} \cdot \lambda^*((x-h,x+h) \cap A) > 0\},$$

we can verify that b is a base operator. A set $F \subset \mathbb{R}$ is s-closed if and only if it is b-closed in the sense that $bF \subset F$. Obviously,

$$bA = b(H_A)$$

whenever H_A is a measurable cover of A.

(d) For each closed interval I and for each $\varepsilon > 0$ there is an open set $M \subset I$ such that $bM = I$, $\lambda(M) \leq \varepsilon$.

Proof. It is no restriction to assume that $I = [0,1]$, $\varepsilon < 1$. Put

$$M := \bigcup_{n=1}^{\infty} \bigcup_{k=1}^{2^n} M_n^k,$$

where

$$M_n^k = \left(\frac{k-1}{2^n}, \frac{k-1}{2^n} + \frac{\varepsilon}{4^n}\right).$$

Then M is an open set in $(0,1)$ and

$$\lambda(M) \leq \sum_{n=1}^{\infty} \sum_{k=1}^{2^n} \frac{\varepsilon}{4^n} = \varepsilon.$$

Let $x \in [0,1]$, $h = 2^{-n}$. Then there is $k \in \mathbb{N}$ such that $M_n^k \subset [x-h,x+h]$.

hus

$$h^{-2}.\ \lambda(M \cap (x-h,x+h)) \geq h^{-2}.\ \lambda(M_n^k) = \varepsilon.4^{-n}.(2^n)^2 = \varepsilon.$$

ence, by definition, $x \in bM$. Obviously, $bM \subset [0,1]$.

(e) For each $\varepsilon > 0$ there is an open set G such that $bG = \mathbb{R}$, $\lambda(G) < \varepsilon$.

roof. We put $G = \bigcup_{n=1}^{\infty} M_n$ where $M_n \subset [-n,n]$ are open sets,

$$b(M_n) = [-n,n], \quad \lambda(M_n) < \frac{\varepsilon}{2^n}.$$

f course, $\lambda(G) < \varepsilon$ and for every $k \in \mathbb{N}$ we have

$$bG \supset b(\bigcup_{n=1}^{k} M_n) = \bigcup_{n=1}^{k} b(M_n) = \bigcup_{n=1}^{k} [-n,n] = [-k,k] .$$

hus $bG = \mathbb{R}$.

(f) There is an open set $A \subset [0,1]$ such that $bA = (0,1]$.

roof. By (c) there are open sets $A_n \subset [\frac{1}{n+1}, \frac{1}{n}]$ with

$$b(A_n) = [\frac{1}{n+1}, \frac{1}{n}], \quad \lambda(A_n) \leq (n+1)^{-3}2^{-n}.$$

utting $A = \bigcup_{n=1}^{\infty} A_n$, we can easily see that A is an open set satisfying $(0,1] \subset bA \subset [0,1]$. We are to show that $0 \notin bA$. Let $h \in (0,1)$. Then there is $n \in \mathbb{N}$ for which $h \in (\frac{1}{n+1}, \frac{1}{n}]$ and

$$\lambda((-h,h) \cap A) \leq \lambda(\bigcup_{k=n}^{\infty} A_k) \leq \sum_{k=n}^{\infty} (k+1)^{-3}2^{-k}$$

$$\leq (n+1)^{-3} \sum_{k=n}^{\infty} 2^{-k} \leq (n+1)^{-3} \leq h^3,$$

thus $0 \notin bA$ by definition.

(g) The operator b *is not a strong base operator*. Indeed, let A be as in (f). Then $bA = (0,1]$ and $bbA = [0,1]$.

(h) For the superdensity topology the *density theorem does not hold*. If A is the set constructed in the proof of (f) and $M := \mathbb{R}\setminus A$, then the set $M \cap b(\mathbb{R}\setminus M) = (0,1] \setminus A$ is not of measure zero.

(i) The s-closure of the set A from (f) is not equal to $A \cup bA$: $A \cup bA = (0,1] \neq [0,1] = \bar{A}^s$.

(j) $(\mathbb{R},s)$ is not a weak Baire space. Consequently, $(\mathbb{R},s)$ *is not a Baire space.*

Proof. By (e), there is an increasing sequence $\{G_n\}$ of open sets satisfying $b(G_n) = \mathbb{R}$, $\lambda(G_n) \to 0$. Put

$$F = \bigcap_{n=1}^{\infty} G_n, \quad D = \mathbb{R}\setminus F.$$

Since F is a null set, D is a nonempty s-open set. We have $\mathbb{R} = b(G_n) \subset G_n \cup b(G_n) \subset \overline{G_n}^s$, so G_n are s-dense in $\mathbb{R}$. On the other hand, $F \cap D = \emptyset$, and thus $(\mathbb{R}, s)$ is not a weak Baire space.

(k) The topological space $(\mathbb{R},s)$ *is not regular*. This follows immediately from Theorem 4.6 by (b) and (j).

(l) $(\mathbb{R},s)$ *has not the Lusin-Menchoff property*. This follows obviously from Proposition 3.16 (because $(\mathbb{R},s)$ is not a Baire space) or from Corollary 3.13 (because s is not regular). But even more can be said: The s-closed set F constructed in (j) cannot be separated from any nonempty closed set in $\mathbb{R}$.

6.28 REMARK. The superdensity topology yields in a natural way the following generalization of the concept of density in a metric space:

Let $\mathcal{L} = \{\mathcal{L}_x\}$ be a differentiation basis in a metric space P equipped with a measure m as in Section 6.B and let σ be a nonnegative set function defined on $\bigcup_{x\in P} \mathcal{L}_x$. Let D be the set of all $x \in X$ such that every $U(x,r)$ contains some set $B \in \mathcal{L}_x$ with $\sigma(B) > 0$. For each set $A \subset P$ and $x \in D$ define

$$D^{\sigma}(x,A) := \lim_{r\to 0_+} \left(\sup\left\{\frac{m^*(A\cap B)}{\sigma(B)}:\ B \in \mathcal{L}_x,\ \sigma(B) > 0,\ B \subset U(x,r)\right\}\right).$$

It is not difficult to check that the operator

$$b:\ A \longmapsto \{x \in D:\ D^{\sigma}(x,a) > 0\}$$

is a base operator. The corresponding b-topology is called the (σ)-superdensity topology (determined by $\mathcal{L}$). Particularly, if

$$\mathcal{L}_x := \{(x-h,x+h) : h > 0\}, \quad \sigma((x-h,\ x+h)) := h^2,$$

then the (σ)-superdensity topology is that of Example 6.27 and it coincides with the density topology for a suitable differentiation basis (cf. the superdensity basis

B_9 of Example 6.11). Actually, the following more general result can be obtained.

6.29 PROPOSITION. Let $\mathcal{L} = \{\mathcal{L}_x\}$ be the symmetric differentiation basis on P, m as above and let σ be a nonnegative monotone real function on $\bigcup_{x\in P} \mathcal{L}_x$.

Assume that $D = P$ and

(a) m has the intermediate value property,

(b) for each $x \in P$ there is $r > 0$ such that $\sigma(B) \leq m(B)$ whenever $B \in \mathcal{L}_x$, $B \subset U(x,r)$.

Then there is a differentiation basis $\mathcal{U} = \{\mathcal{U}_x\}_{x\in P}$ such that

$$\min (1, D^{\sigma}(z,A)) = d^e(z,A)$$

for each $A \subset P$ and each $z \in P$ ($D^{\sigma}(z,A)$ is formed with respect to $\mathcal{L}$, $d^e(z,A)$ is taken with respect to $\mathcal{U}$). Hence the (σ)-superdensity topology and the density topology (determined by $\mathcal{U}$) coincide.

Proof. Put

$$\mathcal{U}_x := \{M \in \mathcal{M}: \text{there is } B \in \mathcal{L}_x \text{ such that } M \subset B,\ m(M) \geq \sigma(B)\}.$$

Let $z \in P$, $A \subset P$. Passing to a measurable cover we may suppose $A \in \mathcal{M}$. Let $B \in \mathcal{L}_z$, $\sigma(B) \leq m(B)$ and $\frac{m(A \cap B)}{\sigma(B)} = c$. If $c \geq 1$, we put $M = A \cap B$. In this case $m(M) \geq \sigma(B)$. Otherwise $m(A \cap B) \leq \sigma(B) \leq m(B)$ and there is a set $M \in \mathcal{M}$ such that $A \cap B \subset M \subset B$, $m(M) = \sigma(B)$. In each case

$$\frac{m(A \cap M)}{m(M)} \geq \min \left(1, \frac{m(A \cap B)}{\sigma(B)}\right),$$

and, therefore, $\min (1, D^{\sigma}(z,A)) \leq d^e(z,A)$. For the converse inequality consider $z \in D$, $A \in \mathcal{M}$ and $M \in \mathcal{U}_z$ such that $M \subset U(z,r)$. Using the definition of $\mathcal{U}_z$ and monotonicity of σ we get $B \in \mathcal{L}_z$ such that $M \subset B \subset U(z,r)$ and $m(M) \geq \sigma(B) > 0$. Consequently, $\frac{m(A \cap M)}{m(M)} \leq \frac{m(A \cap B)}{\sigma(B)}$.

6.30 REMARK. In the example of the superdensity topology we have seen that b is not a strong base operator and the topology determined by b has not the Lusin-Menchoff property. In the next Chapter 7 we assign to any fine base operator b on a metric space its new topological modifications. In addition to the b-topology we introduce the notions of an r-topology and of an a.e.-topology. It can happen that either the a.e.-topology or the r-topology has the Lusin-Menchoff property while the b-topology has not. Remark that it happens exactly in the case of the superdensity topology.

Therefore, it seems quite natural to consider the notion of the Lusin-Menchoff property in the more general case of base operators in a way which could be useful for the investigation of the Lusin-Menchoff property of the mentioned modifications. Dealing as we are with the Lusin-Menchoff property of base operators rather than of fine topologies, it is important to know that the base operator arising from the superdensity topology has this Lusin-Menchoff property. After all this was the reason for the extension of the Lusin-Menchoff property over base operators.

Before formulating our main theorem in a general form we add another example justifying our considerations. This example goes back to W. Sierpiński (1927) and A.S. Besicovitch (1928) who examined the questions of linear density of planar sets.

6.31 EXAMPLE.

(a) Let Γ be a Caratheodory (= Hausdorff) outer "linear" measure on $\mathbb{R}^2$. We define for each $x \in \mathbb{R}^2$ and for each $A \subset \mathbb{R}^2$ the upper and lower *outer linear densities* by

$$D^*(x,A) := \limsup_{r \to 0_+} \frac{1}{2r} \Gamma(A \cap U(x,r)),$$

$$D_*(x,A) := \liminf_{r \to 0_+} \frac{1}{2r} \Gamma(A \cap U(x,r)).$$

(b) A.S. Besicovitch (1928) proved the following "density theorems": Let $A \subset \mathbb{R}^2$ be a Γ-measurable set of finite Γ-measure. Then:

<u>Theorem A:</u> For Γ-a.e. point $x \in A$,
$$\frac{1}{2} \leq D^*(x,A) \leq 1, \quad 0 \leq D_*(x,A) \leq 1.$$

<u>Theorem B:</u> For Γ-a.e. point $x \in \mathbb{R}^2 \setminus A$, $D^*(x,A) = 0$.

(c) Let $X \subset \mathbb{R}^2$ be a Γ-measurable set of finite Γ-measure. Let us consider the following "linear" base operator:

$$b^*A := \{x \in X: \; D^*(x,A) > 0\}.$$

Let $A \subset X$ be fixed, let H_A be a Γ-measurable cover of A. Of course, $b^*A =$

$=b^*(H_A)$. By Besicovitch's Theorem B, $b^*A \setminus H$ is of Γ-measure zero. Hence $\Gamma(b^*A \Delta H_A) = 0$. Let d^* be the "*linear density topology*" determined by the base b^*. As above, the following theorems can be deduced:

(i) b^* is an idempotent base,

(ii) $A^{d^*} = A \cup b^*A$,

(iii) d^*-Borel sets are Γ-measurable

(compare with Lemma 6.13, Proposition 6.14 and Corollary 6.15).

(d) Notice that d^* is not given as a "density topology" in the sense of Section 6.B (for "one-dimensional" density topologies of this type on the plane see Ex.6.D.7). However, b^*A is of the type $\{x: D^{\sigma}(x,A) > 0\}$ with $\sigma(U(x,r)) = 2r$. The natural question arises whether the linear density topology d^* has the Lusin-Menchoff property.

Now, we are ready to deduce the general form of the Lusin-Menchoff theorem which contains as a special case the theorem concerning the Lusin-Menchoff property of density topologies. We start with the following lemma which provides the crucial step in the proof of the Lusin-Menchoff property.

In what follows, the symbol Ω will be reserved for the collection of all real, positive increasing functions on $(0,+\infty)$.

6.32 PRINCIPAL LEMMA. Let $F \subset P$ be closed, $T \subset P$ be measurable, $F \cap T = \emptyset$, and let $\omega \in \Omega$. Then there is an open set U such that

$$T \subset U \subset P \setminus F \quad \text{and} \quad m(U(x,r) \cap (U \setminus T)) \leqq \omega(r)$$

whenever $x \in F$ and $r > 0$.

Proof. Let ρ be a metric on P, $n \in \mathbb{N}$. Denote

$$R_n := \{x \in P: \quad \rho(x,F) > \tfrac{1}{n}\}.$$

For each $n \in \mathbb{N}$ there is an open set $U_n \subset R_n$ such that

$$T \cap R_n \subset U_n \quad \text{and} \quad m(U_n \setminus T) < \varepsilon_n,$$

where $\{\varepsilon_n\}$ is a sequence of positive numbers satisfying

$$\sum_{j=k}^{\infty} \varepsilon_j < \omega(\tfrac{1}{k})$$

for each $k \in \mathbb{N}$. We set

$$U := \bigcup_{n=1}^{\infty} U_n.$$

Obviously,

$$T \subset U \subset \bigcup_{n=1}^{\infty} R_n = P \setminus F.$$

Let $x \in F$ and $r > 0$. There is the smallest $n \in \mathbb{N}$ for which $1 \leq nr$. Then

$$m(U(x,r) \cap (U \setminus T)) \leq \sum_{k=n}^{\infty} m(U_k \setminus T) \leq \sum_{k=n}^{\infty} \varepsilon_k < \omega(\tfrac{1}{n}) \leq \omega(r).$$

Now, we are able to pass to our general form of the Lusin-Menchoff property for the base operators. Let b be a fine base operator on a topological space (X,ρ). Recall that b has the complete Lusin-Menchoff property if and only if the following holds: For each couple A,B of subsets of X for which $\bar{A} \cap B = \emptyset = A \cap bB$ there exists an open set G such that $B \subset G \subset X \setminus \bar{A}$ and $A \cap bG = \emptyset$.

Finally, we introduce the basic definition of compatibility of a base operator with a measure. Let $(P,\mathfrak{M},m)$ be the same as in Section 6.B. We say that a base operator b is *compatible* with m if for every set $A \subset P$ and for each measurable cover H_A of A the equality

$$bA = b(H_A)$$

holds. Obviously, if b is compatible with m, then $bM = \emptyset$ for any null set $M \subset P$.

6.33 THEOREM (the complete Lusin-Menchoff property of b). Let b be compatible with m. Suppose that there exists a function $\omega \in \Omega$ such that for every $A \subset P$ and each $x \in P$ the following implication holds:

if $m^*(A \cap U(x,r)) \leq \omega(r)$ for each $r > 0$, then $x \notin bA$.

Then b has the complete Lusin-Menchoff property.

Proof. Suppose $A,B \subset P$, $\bar{A} \cap B = \emptyset = A \cap bB$. Let H_B be a measurable cover of B. We may assume $\bar{A} \cap H_B = \emptyset$ (otherwise, $H_B \setminus \bar{A}$ is again a measurable cover of B). By Principal lemma 6.32 there is an open set U such that $B \subset H_B \subset U \subset P \setminus \bar{A}$ and $m(U(x,r) \cap (U \setminus H_B)) \leqq \omega(r)$ whenever $x \in \bar{A}$ and $r > 0$. Hence $\bar{A} \cap b(U \setminus H_B) = \emptyset$, and therefore

$$A \cap bU = A \cap (b(U \setminus H_B) \cup b(H_B)) = A \cap b(H_B) = A \cap bB = \emptyset.$$

6.34 APPLICATIONS OF 6.32 AND 6.33.

(A) Superdensity topology. Recall that

$$bA := \{x \in \mathbb{R}: \limsup_{r \to 0_+} r^{-2} . \lambda^*((x-r,x+r) \cap A) > 0\}$$

is a base operator for the superdensity topology on $\mathbb{R}$. Taking $\omega(r) = r^3$ in Theorem 6.33 and making use of the fact that b is compatible with λ (see 6.27.c), we can see that the base operator b has the complete Lusin-Menchoff property. This property turns out to be important later on. Namely, using the Lusin-Menchoff property of the *superdensity base operator* we are able to deduce that the *a.e.-modification of the superdensity topology* has the Lusin-Menchoff property.

(B) Density (and superdensity) topologies on metric spaces. Consider now a (σ)-superdensity topology on a metric space P as in Remark 6.28 and let $\mathcal{L}$, D, b have also the same meaning.

Assume that for each $r > 0$

$$(*) \qquad \inf\{\sigma(B): B \in \mathcal{L}_x, \sigma(B) > 0, x \in D, B \setminus U(x,r) \neq \emptyset\} > 0.$$

Then the base b has the complete Lusin-Menchoff property. To prove it, put

$$\omega(r) := r.\min(1, \inf\{\sigma(B): B \in \mathcal{L}_x, \sigma(B) > 0, x \in D, B \setminus U(x,\tfrac{r}{2}) \neq \emptyset\}).$$

Then $\omega \in \Omega$. Let $A \subset P$ and $x \in D$ be such that

$$m^*(A \cap U(x,r)) \leqq \omega(r) \quad \text{for every } r > 0.$$

Consider $B \in \mathcal{L}_x$, $\sigma(B) > 0$. If $B = \{x\}$ or $B = \emptyset$, then $\dfrac{m^*(A \cap B)}{\sigma(B)} =$
$= m^*(A \cap B) = 0$. In the opposite case suppose that $B \subset U(x,r)$ and $B \setminus U(x,\tfrac{r}{2}) \neq \emptyset$. Then

$$\frac{m^*(A \cap B)}{\sigma(B)} \leq \frac{m^*(A \cap U(x,r))}{\sigma(B)} \leq \frac{r.\omega(r)}{\omega(r)} = r.$$

Thus we have $D^\sigma(x,A) = 0$. Since b is compatible with m, the assertion follows from Theorem 6.33.

More generally, assume that $D = \bigcup_{n=1}^{\infty} A_n$ and that for each $n \in \mathbb{N}$ and each $r > 0$,

$$i(n,r) := \inf \{ \sigma(B) : B \in \mathcal{b}_x, \ \sigma(B) > 0, \ x \in A_n, \ B \setminus U(x,r) \neq \emptyset \} > 0.$$

Again, it is simple to verify that b has the complete Lusin-Menchoff property. Indeed, it suffices to put

$$\omega(r) := r.\min (1, \min (i(n,r) : n \in \mathbb{N}, \ nr < 1)).$$

Using this condition it can be proved that a density topology has the Lusin-Menchoff property if it is determined by one of the following differentiation bases:

(a) B_1 and B_{12} under conditions that P is σ-compact and the density theorem holds,

(b) B_2, B_3, B_{11} and, if supt $m = \mathbb{R}$, also B_7,

(c) B_4 for $n = 1$ provided supt $m = \mathbb{R}$,

(d) B_6 in the case when the density theorem holds and supt $m = \mathbb{R}^2$ (in particular, in the case of Lebesgue measure).

We merely outline the main idea of the proofs in cases of (a) and B_3 and invite the reader to fill in the details in remaining cases.

$\underline{B_1 \text{ and } B_{12}}$: Let $D = \text{supt } m = \bigcup_{n=1}^{\infty} A_n$ where A_n are compact. We claim that $i(n,r) > 0$ for each $n \in \mathbb{N}$ and $r > 0$. For contrast, consider $n \in \mathbb{N}$ and $r > 0$ for which $i(n,r) = 0$. Then there are $x_k \in A_n$ and $r_k \geq r$ such that $m(U(x_k,r_k)) \to 0$. No generality is lost with the assumption that $x_k \to x$. But then we easily establish a contradiction because $m(U(x,\frac{r}{2})) > 0$ and $U(x,\frac{r}{2}) \subset U(x_k,r_k)$ for k large enough.

$\underline{B_3}$: Let $B_{k,m}$ denote the set of all points $x \in$ supt m for which $c_x \leq k$ and $|x| \leq m$. Now it only needs to be observed that the family $B_{k,m}$ can be arranged into a sequence $\{A_n\}$.

(C) Let $X \subset \mathbb{R}^2$ be a set of finite linear measure. Applying (B) to Example 6.31.c of the linear density topology we see that the induced d^*-topology on X has the complete Lusin-Menchoff property.

(D) The Lusin-Menchoff property of finer topologies. Let b_1, b_2 be base operators compatible with m. Assume that

(i) there is $\omega \in \Omega$ such that for each $A \subset P$, $x \in P$ we have $x \notin b_1 A$ whenever $m^*(A \cap U(x,r)) \leq \omega(r)$ for every $r > 0$,

(ii) $b_2 M \subset b_1 M$ for every $M \subset P$.

Then b_2 has the complete Lusin-Menchoff property. (It is almost a trivial consequence of Theorem 6.33 since $x \notin b_1 A$ implies $x \notin b_2 A$.) This property will be also applied later on to the case of fine topologies.

(E) We use our Principal lemma 6.32 combining it with (B). It enables us to state explicitely the following assertion concerning the ordinary density topology on $\mathbb{R}^n$: Let $M \subset \mathbb{R}^n$ be a measurable set, $F \subset M$ be closed. Then there is a closed set F^*, $F \subset F^* \subset M$ such that for every $x \in F$ we have $d^e(x, M \setminus F^*) = 0$ (cf. Ex. 6.D.4).

(F) Finally, we deduce the following ("stronger") lemma of Lusin and Menchoff: Let $E \subset \mathbb{R}$ be a measurable set, $F \subset E$ be closed. Then for every $c > 0$ there is a closed set H, $F \subset H \subset E$ such that

$$\lambda((x-h, x+h) \cap (E \setminus H)) \leq c\, h^2$$

whenever $x \in F$ and $h > 0$. (We may replace $c\,h^2$ by an arbitrary function $\omega(h)$ from Ω and the interval $(x-h, x+h)$ by interval $(x, x+h)$ where h is, possibly, negative.)

Proof. This is practically Principal lemma 6.32 applied to the case of the Lebesgue measure on $\mathbb{R}$.

EXERCISES

6.D.1 (a) Let d_s be the strong density topology on $\mathbb{R}^2$ (d_s is determined by the basis B_4 and by the Lebesgue measure). Let G be a d_s-open set and $(x,y) \notin \overline{G}^{d_s}$. If $G_x = \{z \in \mathbb{R}: (x,z) \in G\}$ and d denotes the density topology on $\mathbb{R}$, then $y \notin \bar{G}_x^d$.

Hint. Assume $d^{e,sym}(y,G_x) = \varepsilon > 0$. Given $\delta > 0$, there is $r \in (0,\delta)$ such that $\lambda_1^*(G_x \cap (y-r,y+r)) > \varepsilon r$. Pick $t \in G_x \cap (y-r,y+r)$. There is $h_t > 0$ such that $\lambda_2(G \cap I) > \frac{1}{2}\lambda_2(I)$ whenever I is an interval with center (x,y) and diam $I < h_t$. The family

$$\mathcal{V} = \{(t-h,t+h) \subset (y-r,y+r) : t \in G_x \cap (y-r,y+r),\ 0 < h < \tfrac{1}{2}h_t\}$$

is a Vitali covering of G_x $(y-r,y+r)$. Hence, there is a disjoint countable system $\{(y_n-h_n, y_n+h_n)\} \subset \mathcal{V}$ such that $\lambda_1(\bigcup_{n=1}^{\infty} (y_n-h_n,y_n+h_n)) > \varepsilon r$. Deduce that

$$\lambda_2([(x-p,x+p) \times (y-r,y+r)] \cap G) \geq \sum_{n=1}^{\infty} \lambda_2([(x-p,x+p)\times(y_n-h_n,y_n+h_n)] \cap G)$$
$$> \frac{1}{2}\varepsilon pr$$

for p small enough, but this contradicts the assumption $(x,y) \notin \bar{G}^{d_s}$.

(b) Formulate and prove analogous assertions in $\mathbb{R}^n$.

6.D.2 Show that the strong density topology d_s on $\mathbb{R}^n$ is Hausdorff but it is not regular. In particular, d_s has not the Lusin-Menchoff property.

Hint. Use Ex. 6.D.1 and show that, say for $n = 2$, the point $(0,0)$ cannot be separated from the d_s-closed set $(\{0\} \times \mathbb{R}) \setminus \{(0,0)\}$.

6.D.3 Let d_s be again the strong density topology on $\mathbb{R}^2$ and let τ^* be the topology determined by the family of all d_s-open sets G for which all sections of G (i.e. the sets $\{x: (x,y) \in G\}$, $y \in \mathbb{R}$ and $\{y: (x,y) \in G\}$, $x \in \mathbb{R}$) are open in the (one-dimensional) density topology. If τ_g is the weak topology induced by the family of all d_s-continuous functions, then $\tau_g \subset \tau^* \subsetneq d_s$.

In particular, if f is strongly approximately continuous on $\mathbb{R}^2$, then the functions $f(.,y)$ and $f(x,.)$ are approximately continuous on $\mathbb{R}$.

Hint. Use Ex. 6.D.1.

<u>REMARK</u>. It is an interesting question of whether it is true that $\tau_g = \tau^*$ (cf. C.Goffman, C. Neugebauer and T. Nishiura (1961)).

6.D.4 Prove directly the assertion (E) of 6.34.

Hint. Let $R_k = \{x \in \mathbb{R}^n : \frac{1}{k+1} < \rho(x,F) \leq \frac{1}{k}\} \cap M$ and find a closed set $F_k \subset R_k$ such that $m(R_k \setminus F_k) < 2^{-k}$. Then put $F^* = \bigcup_{k=1}^{\infty} F_k \cup F$.

6.D.5 Let $\mathcal{L} = \{\mathcal{L}_x\}_{x \in \mathbb{R}^2}$ be a differentiation basis in $\mathbb{R}^2$ where each $\mathcal{L}_x$ is formed by a family of some rectangles containing x. Consider on $\mathbb{R}^2$ the Lebesgue measure and assume that $b\,\mathbb{R}^2 = \mathbb{R}^2$.

If I is a rectangle in $\mathbb{R}^2$, denote by $s(I)$ and $r(I)$ the shorter and longer sides respectively. We say that $\mathcal{L}$ is *uniformly regular* if there is a positive and increasing real-function ω on $(0,+\infty)$ such that for every $z \in \mathbb{R}^2$

(*) there is $r_0 > 0$ such that $\frac{s(I)}{r(I)} \geq \omega(r(I))$

whenever $I \in \mathcal{L}_z$ and $r(I) \geq r_0$.

Show that if $\mathcal{L}$ is uniformly regular then the corresponding base operator has the complete Lusin-Menchoff property.

Hint. Using Theorem 6.33, proceed as in 6.34.B.

REMARK. Ex. 6.D.5 is a slight generalization of Th. 2.1 of A.P.Baisnab and G.M. Petersen (1971). By modifying their example, we give the next exercise.

6.D.6 Let $\mathcal{L}$ be a differentiation basis in $\mathbb{R}^2$ as in the preceding exercise. Assume that there is a point $z = (z_1;z_2)$ such that $\mathcal{L}_z$ consists of intervals only and that there is no function ω satisfying (*) of Ex. 6.D.5. Show that the corresponding base operator b has not the Lusin-Menchoff property.

Hint. Set $M := (\{z_1\} \times \mathbb{R}) \cup (\mathbb{R} \times \{z_2\}) \setminus \{z\}$. Clearly, $z \notin bM$. If $G \supset M$ is an open set, construct a function ω (depending on G) in order to show that $d^e(z,G) = 1$.

6.D.7 (Notation as in Example 6.31.) Let $P \subset \mathbb{R}^2$ be a Γ-measurable set of finite Γ-measure. Let Γ_P be the restriction of Γ to the family $\mathcal{M}$ of all Γ-measurable subsets of P. Then $(P,\mathcal{M},\Gamma_P)$ satisfies the assumptions of Section 6.B. Let d be the (metric) density topology on P determined by the symmetric differentiation basis B_1. Prove the following assertions:

(a) d has the complete Lusin-Menchoff property,

(b) there is a set $N \subset P$ of Γ_p-measure zero such that d^* on $P \setminus N$ is finer than d on $P \setminus N$,

(c) there is a set $P \subset \mathbb{R}^2$ such that $d^* \wedge P \setminus N \neq d \wedge P \setminus N$ for each $N \subset P$ with $\Gamma_p(N) = 0$.

Hint. To prove (a), use 6.34.B and the fact that d is the restriction of a density topology on $\mathbb{R}^2$ determined by some Radon measure and the symmetric differentiation basis B_1. To establish (b), set $N := \{x \in P: D^*(x,P) = +\infty\}$ and use the Besicovitch Theorem A. The key observation for the proof of (c) is the fact that there is $P \subset \mathbb{R}^2$ such that $D_*(x,P) = 0$ for each $x \in P$.

6.D.8 Let $\mu_1,\ldots,\mu_k$ be mutually equivalent Radon measures on $\mathbb{R}^n$, $\text{supt}\,\mu_i = \mathbb{R}^n$. Let b_i and τ_i be the base operator and the density topology, respectively, determined by the measure μ_i and by the differentiation basis $\mathcal{L}$ where $\mathcal{L}$ is either the symmetric basis B_1 or, in the $n = 1$ case, the interval basis B_4. Finally, let $b := \sup(b_1,\ldots,b_k)$ and $\tau := \inf(\tau_1,\ldots,\tau_k)$. Prove that:

(a) τ equals the b-topology,

(b) b is idempotent,

(c) b has the complete Lusin-Menchoff property,

(d) τ has the complete Lusin-Menchoff property.

Hint. In order to show (a) use Ex. 1.A.4c. The proof of (b) is based on Lemma 6.13.c, and for the proof of (c) use 6.34.B and Theorem 3.20. Now, for (d) it remains to refer to Section 3.C and (b).

REMARKS AND COMMENTS.

In view of the importance of the Lusin-Menchoff property of density topologies we devote this special section to investigation thereof.
We start with an instructive example of a superdensity topology (determined by the superdensity differentiation basis) which itself does not possess the Lusin-Menchoff property. This topology, however, is determined by the superdensity base operator having the Lusin-Menchoff property. This fact shows itself as important for further considerations and leads to a proof of the Lusin-Menchoff property for

certain modifications of the superdensity topology.

The main idea of the proof of the Lusin-Menchoff property consists in Principal lemma 6.32 yielding in its application very simple proofs of this property of many density topologies. Here, the discovery of (*) in 6.34.B as an appropriate criterion of the Lusin-Menchoff property was crucial. As to the history of the Lusin-Menchoff property and the proofs thereof we refer to our comments to Section 3.B.

6.E Abstract density topologies

In Section 6.B we considered density topologies determined by differentiation bases for which a special "differentiation process" was described. It consists in letting $r \to 0$ in the inclusion $B \subset U(x,r)$ (B represents an element of $\mathcal{L}_x$). In this section we apply a more general approach based on the notion of a filter. We begin again taking a measurable space (X,Σ,μ) where μ is supposed to be σ-finite and complete. We denote $\Sigma^b := \{M \in \Sigma : 0 < \mu M < +\infty\}$, $\mathcal{N} := \{M \in \Sigma : \mu M = 0\}$.

REMARK (G). Throughout this section we need not to take advantage of the σ-finiteness of the measure μ. All proofs go well even we impose the following properties on (X,Σ,μ) only :

(a) μ is a monotone and subadditive real function on a σ-algebra Σ, $\mu\emptyset = 0$.

(b) $\mathcal{N} := \{N \in \Sigma : \mu N = 0\}$ is a σ-ideal and any set $A \in \Sigma \setminus \mathcal{N}$ is a (possibly uncountable) union of sets from $\Sigma^b := \{M \in \Sigma : 0 < \mu M < +\infty\}$

(c) For every $A \subset X$ there is a "measurable cover" H_A of A (i.e. H_a is a set from Σ containing A such that $H_A \setminus B \in \mathcal{N}$ whenever $B \in \Sigma$, $A \subset B$).

Note that $\mu^*A = \mu H_A$ where, as usual, $\mu^*A := \inf\{\mu M : M \in \Sigma,\ A \subset M\}$.

We say that $\mathcal{F} = \{\mathcal{F}_x\}_{x \in D(\mathcal{F})}$ is a *filter differentiation basis* on X with *domain* $D(\mathcal{F}) \subset X$ if for each $x \in D(\mathcal{F})$ the family $\mathcal{F}_x$ is a non-trivial (i.e. $\emptyset \notin \mathcal{F}_x$) nonempty filter defined on the set Σ^b.

For each set $A \subset X$ and each $x \in D(\mathcal{F})$ define

$$\overline{D}_{\mathcal{F}}(x,A) := \limsup_{\mathcal{F}_x} \frac{\mu^*(A \cap F)}{\mu F} ,$$

$$\underline{D}_{\mathcal{F}}(x,A) := 1 - \overline{D}_{\mathcal{F}}(x, X \setminus A) .$$

(Concerning the notions of upper and lower limits with respect to a filter see e.g. [Bour] , IV.5.6.)

REMARK. If the function μ is a measure (cf. the above Remark (G)), then

$$\underline{D}_{\mathcal{F}}(x,A) = \liminf_{\mathcal{F}_x} \frac{\mu_*(A \cap F)}{\mu F} .$$

For each $A \subset X$, the set

$$bA := \{x \in D(\mathcal{F}) : \bar{D}_{\mathcal{F}}(x,A) > 0\}$$

is termed the *base* of A. As in Section 6.B, it is easy to verify that $b_{\mathcal{F}}: A \mapsto bA$ is a base operator in the sense of Section 1.A. The corresponding $b_{\mathcal{F}}$-topology is denoted by $d_{\mathcal{F}}$ and is labelled as a *topology determined by a filter differentiation basis*. It is almost obvious that a set G is $d_{\mathcal{F}}$-open if and only if $\underline{D}_{\mathcal{F}}(x,G) = 1$ whenever $x \in G \cap D(\mathcal{F})$.

Let us recall (Section 6.C) that a base operator b is *compatible* with μ if for every set $A \subset X$, $bA = b(H_A)$ holds whenever H_A is a measurable cover of A.

Now, the following proposition characterizes the bases compatible with μ.

6.35 PROPOSITION. The base operator $b_{\mathcal{F}}$ is compatible with μ. Conversely, let b be a base operator compatible with μ. Then there exists a filter differentiation basis $\mathcal{F}$ such that $b = b_{\mathcal{F}}$.

Proof. Obviously, $b_{\mathcal{F}}$ is compatible with μ. Let now b be compatible with μ. Put $D(\mathcal{F}) = bX$ and let for every $x \in D(\mathcal{F})$ the filter $\mathcal{F}_x$ on Σ^b be determined by the filter base

$$\{\{F \in \Sigma^b : F \subset U\} : U \in \Sigma ,\ x \in CbCU\}.$$

Obviously, $\mathcal{F}_x$ is a non-trivial filter on Σ^b. Indeed, if $x \in D(\mathcal{F})$, $U \in \Sigma$ and $x \in CbCU$, then $x \in bX \setminus bCU \subset bU$, and hence $\mu U > 0$. It follows that $\emptyset \notin \mathcal{F}_x$. To prove $b = b_{\mathcal{F}}$, fix $A \subset X$ and assume $x \notin bA$. If $x \notin D(\mathcal{F})$, then $x \notin b_{\mathcal{F}}A$ by definition. If $x \in D(\mathcal{F}) \setminus bA$, then also $x \notin b(H_A)$ (H_A is a measurable cover of A) and

$$\sup\left\{\frac{\mu^*(A \cap F)}{\mu F} : F \in \Sigma^b,\ F \subset CH_A\right\} = 0.$$

It follows that $x \notin b_{\mathcal{F}}A$. Suppose now $x \in bA$. Let H_A be a measurable cover of A) and $U \in \Sigma$, $x \in CbCU$. Putting $B = U \cap H_A$ we get

$$x \in bA \setminus bCU = b(H_A) \setminus bCU = (b(H_A \cap U) \cup b(H_A \setminus U)) \setminus bCU \subset bB.$$

Since $bM = \emptyset$ for every $M \in \Sigma$ with $\mu M = 0$, we have $\mu B > 0$. Find $E \in \Sigma^b$, $E \subset B$. Then $E \subset U$ and we obtain

$$\frac{\mu^*(A \cap E)}{\mu E} = \frac{\mu(H_A \cap E)}{\mu E} = \frac{\mu E}{\mu E} = 1.$$

Hence $\bar{D}_{\mathfrak{F}}(x,A) = 1$, which proves that $bA = b_{\mathfrak{F}}A$.

6.36 EXAMPLES.

(F_1) Let $\mathcal{L} = \{\mathcal{L}_x\}$ be a differentiation basis on a metric space (P,ρ). Define the corresponding filter basis $\mathfrak{F}$ in the following way. First of all, the domain of $\mathfrak{F}$ is defined as

$$D(\mathfrak{F}) := \{x \in P: \text{ for every } r > 0 \text{ there is } B \in \mathcal{L}_x \text{ such that } \mu B > 0 \text{ and } B \subset U(x,r)\}$$

(i.e. $D(\mathfrak{F})$ equals bP in the terminology of Section 6.B). Further, let for each $x \in D(\mathfrak{F})$, $\mathfrak{F}_x$ be the filter with the base

$$\{\{B \in \mathcal{L}_x: \mu B > 0,\ B \subset U(x,r)\} : r > 0\}.$$

It is easy to see that the upper outer density of a set M at x with respect to $\mathcal{L}$ or to $\mathfrak{F}$ coincide (i.e. $d^e(x,M) = \bar{D}_{\mathfrak{F}}(x,M)$), and therefore the corresponding base operators and topologies coincide as well.

(F_2) Let $(X,\mathfrak{M},\nu)$ be a measure space. Assume that for each $x \in X$ the collection $\mathcal{A}_x \subset \mathfrak{M}$ is given. Define the filter differentiation basis as follows: The domain of $\mathfrak{F}$ is the set

$$D(\mathfrak{F}) := \{x \in X: \text{ for every } r > 0 \text{ there is } M \in \mathcal{A}_x \text{ such that } 0 < \nu(M) < r\}.$$

For $x \in D(\mathfrak{F})$ let $\mathfrak{F}_x$ be the filter determined by the base

$$\{\{M \in \mathcal{A}_x: 0 < \nu(M) < r\} : r > 0\}.$$

(F_3) Let $(X,\mathfrak{M},\nu)$ and $\mathcal{A}_x$ be the same as in the previous example (F_2). Let

ρ be a topology on X. Define

$$D(\mathfrak{F}) := \{x \in X: \text{ for every } \rho\text{-neighborhood } U \text{ of } x \text{ there is } M \in \mathcal{A}_x \text{ such that } 0 < \nu(M) < +\infty,\ M \subset U\}$$

and for $x \in D(\mathfrak{F})$ let the filter $\mathfrak{F}_x$ be determined by the base

$$\{\{M \in \mathcal{A}_x:\ 0 < \nu(M) < +\infty,\ M \subset U\}:\ U \text{ is a } \rho\text{-neighborhood of } x\}.$$

(F_4) *De Possel differentiation basis* in abstract spaces: The notion of this basis is introduced usually by means of contracting nets or sequences rather than the filters. But in fact it is equivalent to the notion of filter differentiation basis as we shall see at once.

Let (X, Σ, μ) be as above, and let $D \subset X$. We say that a *net differentiation basis* $\mathcal{S} = \{S_x\}_{x \in D}$ with the domain D is given if

(a) for each $x \in D$, S_x is a nonempty system of nets consisting of the sets of Σ^b,

(b) if $I \in S_x$ and J is a cofinal subnet of I, then $J \in S_x$.

If $I = \{I_\alpha\} \in S_x$, we say that *I contracts to* x (in abbreviation $I_\alpha \Rightarrow x$).

For $x \in D$ and $M \subset X$ define

$$\bar{D}_{\mathcal{S}}(x,M) := \sup\{\limsup \frac{\mu^*(M \cap I_\alpha)}{\mu(I_\alpha)}\}$$

(where the expression in brackets denotes the upper limit for any net $\{I_\alpha\}$ contracting to x, and the supremum is taken over all such nets). Now, we define in an obvious way the base operator and the density topology corresponding to a net differentiation basis.

We show that the notions of filter differentiation basis and net differentiation basis are equivalent. Let us say that $\mathfrak{F}$ and $\mathcal{S}$ are *associated* if $\mathfrak{F}$ and $\mathcal{S}$ have the same domains, and $F \in \mathfrak{F}_x$ if and only if for each $\{I_\alpha\} \in S_x$ there is a tail of $\{I_\alpha\}$ which is a subset of F. If $\mathfrak{F}$ and $\mathcal{S}$ are associated, then $\bar{D}_{\mathfrak{F}} = \bar{D}_{\mathcal{S}}$, and therefore the corresponding density topologies coincide. Then, obviously, we also have the same notions of Vitali covering and Vitali property (Ex. 6.E.3 and 6.E.4).

Given a net basis $\mathcal{S}$, there is a unique filter basis $\mathfrak{F}$ associated with $\mathcal{S}$. Conversely, a filter basis can be associated with a number of net bases but there is at least one, namely the maximal associated net basis.

REMARK. Analogously we define the notion of a differentiation basis determined by sequences. In many cases this approach suffices to describe the notion of a density topology (cf. the following example (F_5)).

(F_5) *Morse differentiation basis.* Given $x \in R^n$ and $\lambda \in (0,1)$, we say that a closed set $M \subset R^n$ belongs to the family $\mathcal{M}(x,\lambda)$ if there is $r > 0$ such that

$$B(x,\lambda r) \subset M \subset B(x,r)$$

and, moreover, if the convex hull of $B(x,\lambda r)$ and $\{y\}$ is a subset of M for every $y \in M$.

We say that $A_k \Rightarrow x$ (resp. $I_\alpha \Rightarrow x$) if there is $\lambda > 0$ such that $A_k \in \mathcal{M}(x,\lambda)$ (resp. $I_\alpha \in \mathcal{M}(x,\lambda)$) and diam $A_k \to 0$ (resp. diam $I_\alpha \to 0$)). In these cases the density theorem holds for every Radon measure on R^n.

We have seen that a density topology, even in the metric case, can be deprived of many important properties. For example, there are differentiation bases for which $d_{\mathcal{F}}$-open sets need not be measurable (Ex.6.B.8b). If the density theorem holds for $\mathcal{F}$ (i.e. if for each measurable set $M \in \Sigma$ we have $\underline{D}_{\mathcal{F}}(x,M) = 1$ for almost every $x \in M$), the situation is much better. On the other hand, some interesting properties can be derived even under weaker assumptions (cf. Ex. 6.B.5 and 6.E.5). In the sequel, we characterize those topologies determined by a filter differentiation basis for which both $D(\mathcal{F}) = X$ and the density theorem hold. These topologies are nothing else than the topologies determined by a lower density studied in the lifting theory.

Let (X, Σ, μ) be as above. We denote $A \sim B$ if $\mu(A \triangle B) = 0$. A mapping $S: \Sigma \to \Sigma$ is said to be a *lower density* if

(S_1) $S(A) \sim A$,

(S_2) $A \sim B \Rightarrow S(A) = S(B)$,

(S_3) $S(\emptyset) = \emptyset$, $S(X) = X$,

(S_4) $S(A \cap B) = S(A) \cap S(B)$.

Obviously, any lower density satisfies $S(A) \subset S(B)$ whenever $A \subset B$. Let S be a lower density. Denote

$$\tau_S := \{A \in \Sigma: \quad A \subset S(A)\} .$$

6.37 PROPOSITION. The family τ_S forms a topology on X and

$$\tau_S = \{S(A) \setminus N:\ A \in \Sigma,\ \mu N = 0\}.$$

Proof. Assume that $\{A_i:\ i \in I\} \subset \tau_S$ and let $A = \bigcup_{i \in I} A_i$. If M denotes a measurable kernel of A, there are $N_i \sim \emptyset$ such that $A_i \subset M \cup N_i$. Hence $S(A_i) \subset S(M)$, so that

$$M \subset A \subset \bigcup_{i \in I} S(A_i) \subset S(M).$$

By (S_1), $M \sim S(M)$, and therefore A is measurable. Since $A \subset S(M) \subset S(A)$, $A \in \tau_S$. If $A, B \in \tau_S$, then by (S_4)

$$A \cap B \subset S(A) \cap S(B) = S(A \cap B),$$

thus $A \cap B \in \tau_S$. If $B \in \tau_S$, then $B \subset S(B)$ and $B = S(B) \setminus (S(B) \setminus B)$. By (S_1) we have $\mu(S(B) \setminus B) = 0$. Conversely, if $B = S(A) \setminus N$, where $A \in \Sigma, \mu N = 0$, then by (S_1) and (S_2),

$$B = S(A) \setminus N \subset S(A) = S(S(A)) = S(B),$$

and thus $B \in \tau_S$.

Let τ be a topology on X. We say that τ is *determined by a lower density* (or *by a filter differentiation basis*) if there is a lower density S (or a filter differentiation basis $\mathfrak{F}$) such that $\tau = \tau_S$ (or $\tau = d_{\mathfrak{F}}$).

Further, a filter differentiation basis $\mathfrak{F}$ is said to satisfy the *density theorem* if for every set $M \in \Sigma$ almost every point of M is a density point of M. Here, a point $x \in X$ is a *density point* of a set A if $x \in D(\mathfrak{F})$ and $\underline{D}_{\mathfrak{F}}(x,A) = 1$.

Before stating our main theorem, we need the following lemma.

6.38 LEMMA. Let b be a base operator compatible with μ such that $bX = X$ and $bA \sim A$ for any $A \in \Sigma$. Then

(a) b is idempotent,

(b) $\bar{A}^b = A \cup bA$, $\operatorname{int}_b A = A \cap CbCA$ for each $A \subset X$,

(c) each b-Borel set is measurable.

Proof. (a) Obviously, if b is compatible with μ and $A \sim B$, then $bA = bB$. Hence

$$bbA = bbH_A = bH_A = bA$$

for every $A \subset X$ (here, H_A denotes a measurable cover of A).

(b) We have $\bar{A}^b = A \cup bA$ by general Theorem 1.2. It follows that $\mathrm{int}_b A = C\overline{CA}^b = C(CA \cup bCA) = A \cap CbCA$.

(c) Let F be a b-closed set, i.e. $bF \subset F$. Then

$$H_F \setminus F \sim bH_F \setminus F = bF \setminus F = \emptyset,$$

and therefore F is measurable.

6.39 THEOREM. Let τ be a topology on X. Then the following assertions are equivalent:

(i) τ is determined by a lower density,

(ii) there is a base operator b compatible with μ such that $bX = X$, $bA \sim A$ for each $A \in \Sigma$, and τ is exactly the b-topology,

(iii) the topology τ has the following three properties:

(a) every μ-null set is τ-closed,

(b) every nonempty τ-open set contains a measurable subset of positive μ-measure,

(c) $A \sim \mathrm{int}_\tau A$ $(\sim \bar{A}^\tau)$ for every $A \in \Sigma$,

(iv) the topology τ has the following two properties:

(a) a set is of μ-measure zero if and only if it is τ-closed and τ-nowhere dense,

(b) a set is measurable if and only if it has the τ-Baire property,

(v) τ is determined by a filter differentiation basis $\mathfrak{F}$ for which the density theorem holds and which satisfies $D(\mathfrak{F}) = X$. (In addition, for each $\mathfrak{F}_x$ the base can be chosen such that all its elements contain x.)

Proof. (i)$\Leftrightarrow$(ii): Let S be a lower density on Σ. Define the so-called *upper density* $U: \Sigma \to \Sigma$ by

$$U(M) := CS(CM).$$

Then it is easy to see that

(U_1) $U(A) \sim A$,

(U_2) $A \sim B \Rightarrow U(A) = U(B)$,

(U_3) $U(\emptyset) = \emptyset$, $U(X) = X$,

(U_4) $U(A \cup B) = U(A) \cup U(B)$.

Conversely, it can be simply verified that any upper density (i.e. a mapping satisfying (U_1)-(U_5)) can be constructed from a lower density. Put $bA := U(H_A)$ (H_M denotes a measurable cover of B). By (U_2), the definition of bA is correct. Since $H_A \cup H_B$ is a measurable cover of $A \cup B$, we have

$$b(A \cup B) = U(H_A \cup H_B) = U(H_A) \cup U(H_B) = bA \cup bB.$$

Obviously, $b\emptyset = \emptyset$ and $bX = X$. Further, b is compatible with μ and $bA \sim A$ for each $A \in \Sigma$.

Let b be a base operator satisfying (ii). If U denotes the restriction of b to Σ, then U has the properties (U_1)-(U_5) and, of course, $bA = bH_A = U(H_A)$.

We have just proved that there is a one-to-one correspondence between lower densities, upper densities and base operators satisfying the conditions of (ii).

If b corresponds to S, then

$$(*) \qquad A \cap bCA = A \cap CS(CCA) = A \setminus S(A)$$

holds for each set $A \in \Sigma$, and thus any τ_S-open set is also b-open (recall that A is τ_S-open if $A \subset S(A)$ and b-open if $A \cap bCA = \emptyset$). Suppose now that A is b-open, i.e. $A \cap bCA = \emptyset$. Then $A \in \Sigma$ by Lemma 6.38.c and $(*)$ yields that A is τ_S-open.

(ii) $\Rightarrow$ (iii): If $A \sim \emptyset$, then A is a measurable cover of $\emptyset$ and hence $bA = \emptyset \subset A$, i.e. A is τ-closed. Let now $V \neq \emptyset$ be τ-open. Then V is measurable in accordance with Lemma 6.38.c and, since

$$\emptyset \neq V \subset CbCV = bX \setminus bCV \subset bV,$$

V is a set of positive μ-measure. Finally, by Lemma 6.38.b,

$$A \setminus \operatorname{int}_\tau A = A \setminus CbCA, \quad \overline{A}^\tau \setminus A = bA \setminus A,$$

and, of course, $bA \setminus A \sim \emptyset \sim A \setminus CbCA$.

(iii) $\Rightarrow$ (iv) : At first we prove

(b*) every nonempty τ-open set is measurable and of positive μ-measure.

Indeed, let $\emptyset \neq G$ be τ-open and let H be a measurable cover of CG. By (iii.b), $A \cap \operatorname{int} H = \emptyset$. Therefore, $CH \subset A \subset C(\operatorname{int} H)$. In light of (iii.c), G is measurable. If $A \sim \emptyset$, then by (iii.a) A is τ-closed and by (b*) A is τ-nowhere dense. If F is τ-closed and τ-nowhere dense, then $F \sim \operatorname{int}_\tau F = \emptyset$. Hence, we proved (iv.a). If a set M is measurable, then $M \sim \operatorname{int}_\tau M$ by (iii.c), and since each null set is τ-nowhere dense, it follows that M has the τ-Baire property. Conversely, let A be a set having the τ-Baire property, i.e. $A = (G \setminus T_1) \cup T_2$ where G is τ-open and T_1, T_2 are τ-first category. We need only to show that each τ-nowhere dense set Y is measurable. To prove that we observe that $\overline{Y}^\tau$ is τ-nowhere dense if Y is. Therefore Y is a null set, so Y is measurable.

(iv) $\Rightarrow$ (v): Suppose that a topology τ has all the properties enlisted in (iv). For any $x \in X$, let $\mathcal{F}_x$ be a filter on Σ^b having the base

$$\left\{\{M \in \Sigma^b : x \in M \subset G\} : G \text{ is a } \tau\text{-open neighborhood of } x\right\}.$$

Since every nonempty τ-open set is of positive μ-measure, we can see that $\mathcal{F}_x$ is non-trivial and, of course, $\mathcal{F}_x \neq \emptyset$.

We shall demonstrate that

(α) $\tau = d_{\mathcal{F}}$,

(β) the density theorem holds for $\mathcal{F}$.

(α) First let G be τ-open, $x\in G$. Then for each $M\in\Sigma^b$ for which $x\in M\subset G$ we have $\mu(M\cap G)=\mu M$, so that $\underline{D}_{\mathcal{F}}(x,G)=1$. Therefore G is $d_{\mathcal{F}}$-open. Conversely, assume that H is $d_{\mathcal{F}}$-open and choose $x\in H$. It is sufficient to find a τ-open neighborhood U of x for which $\mu(U\setminus H)=0$. Indeed, then $U\cap H=U\setminus(U\setminus H)$ is a τ-open neighborhood of x contained in H ($U\setminus H$ is τ-closed by (iv.a)). Since $\underline{D}_{\mathcal{F}}(x,H)=1$, there is a τ-open neighborhood G of x such that $M\in\Sigma^b$, $x\in M\subset G$ implies $\mu_*(M\cap H)>\frac{1}{2}\mu M>0$. If $\mu(G\setminus H)=0$, we are through. Suppose that $\mu^*(G\setminus H)>0$. Let B be a measurable cover of $(G\ H)\cup\{x\}$. Then $x\in B\cap G\in\Sigma$ and $\mu(B\cap G)>0$. So, there is $M\in\Sigma^b$ such that $x\in M\subset B\cap G$. Having $\mu_*(M\cap H)>0$, we can find $A\in\Sigma^b$, $A\subset M\cap H$. Since A has the τ-Baire property, it follows from (iv.a) that there is a τ-open set U, $U\sim A$. Then simultaneously $A\cap U\sim A$ and $A\cap U\in\Sigma^b$. Of course, $x\in U$ (otherwise, B and $B\setminus(A\cap U)$ are measurable covers of $(G\setminus H)\cup\{x\}$ with $\mu(A\cap U)>0$) and $\mu(U\setminus H)=0$.

(β) Let now A be measurable. Proceeding as above, we get that there is a τ-open set U such that $A\sim U$. Then for each $x\in A\cap U$ we have $\underline{D}_{\mathcal{F}}(x,A)=\underline{D}_{\mathcal{F}}(x,U)=1$, and thus almost every point of A is a density point of A.

(v) $\Longrightarrow$ (ii): Let b be a base operator determined by $\mathcal{F}$. We know that b is compatible with μ (Proposition 6.35), and $bX=X$ since $\mathcal{F}_x\neq\emptyset$ for each $x\in X$. Let $A\in\Sigma$ and let $A_1\subset A$ (resp. $A_2\subset CA$) be the set of all density points of A (resp. of CA). Then $A_1\subset bA\subset CA_2$, and $\mu(CA_2\setminus A_1)=0$ implies $A\sim bA$.

Each topology satisfying the conditions of Theorem 6.39 is termed an *abstract density topology* (on (X,Σ,μ)).

In particular, we draw an attention to the conditions (iii) and (iv) which yield an intrinsic characterization of an abstract density topology.

6.40 REMARKS. (A) Notice that an abstract density topology need not be T_1, e.g. if Σ does not separate the points of X. But even in the case when Σ separates the points of X, this topology need not be T_1 (cf. Ex. 6.E.10).

(B) Let τ be an abstract density topology. Then the base operator b from Theorem 6.39.ii (and also the lower density of (i)) which generates τ is uniquely determined. Indeed, $bA=\overline{\mathrm{int}_\tau A}^\tau$ for each measurable set $A\in\Sigma$ (by 6.38, $bA=b(\mathrm{int}_\tau A)\subset\overline{\mathrm{int}_\tau A}^\tau$; on the other hand, $\overline{\mathrm{int}_\tau A}^\tau\subset\overline{bA}^\tau\subset bA$ since b is idempotent).

(C) In a similar way as in Section 6.B we are able to derive some interesting properties of abstract density topologies (e.g. the quasi-Lindelöf property, the Denjoy-Stepanoff theorem, X equipped with an abstract density topology is a hereditarily Baire space, etc.). The reader is referred to Ex. 6.E.16-18.

The following proposition will be used later.

6.41 PROPOSITION. Let τ be an abstract density topology on X. Assume τ^* is a topology on X finer than τ such that each nonempty τ^*-open set contains a measurable set of positive μ-measure. Then τ^* is an abstract density topology.

Proof. It follows immediately from Theorem 6.39.iii.

REMARK. In this section, we studied the concept of an abstract density topology. We have also seen that in the proofs the central role was played by the existence of a measurable cover while the role of the measure was reduced. Actually it is possible to generalize the notion of the abstract density topology in the following way: Let Σ be a σ-algebra of "measurable sets" on a set X and let $\mathcal{N} \subset \Sigma$ be a σ-ideal of "null sets". Suppose that for any $A \subset X$ there exists a "measurable cover" H_A, i.e. a set $H_A \supset A$, $H_A \in \Sigma$, such that $H_A \setminus P \in \mathcal{N}$ whenever $A \subset P \in \Sigma$. Then we can in an obvious way define the following notions: Equivalence of sets, lower density on $(X, \Sigma, \mathcal{N})$, topology determined by a lower density, base operators compatible with $(X, \Sigma, \mathcal{N})$. If we translate the preceding propositions and their proofs to this new context where no measure μ on Σ is defined (we write "a set lies in $\Sigma \setminus \mathcal{N}$" instead of "a set is a measurable set of positive μ-measure", etc), we obtain that 6.37, 6.38 and 6.41 hold. Also the conditions (i) — (iv) of Theorem 6.39 are equivalent. For the proof of (iv) $\Longrightarrow$ (ii) see Ex.6.E.12.

Another possibility of proofs consists in the following idea: Define the set function μ on Σ by $\mu A = 1$ for $A \notin \mathcal{N}$ and $\mu A = 0$ for $A \in \mathcal{N}$. Then μ has all properties of Remark (G) and therefore all assertions of this section are true.

A topology which is determined on X by some lower density on $(X, \Sigma, \mathcal{N})$ will be called also an *abstract density topology on* $(X, \Sigma, \mathcal{N})$.

An interesting particular case is that of $(X, \Sigma, \mathcal{N})$ where Σ is the family of all sets with the Baire property on X and $\mathcal{N}$ is the σ-ideal of all first category subsets of X (cf.Ex.6.E.30). The corresponding abstract density topologies will be labelled as *category density topologies* on X. A notable example of a category density topology on $\mathbb{R}$ was investigated by W.Wilczyński et al. (cf. Ex.6.E.28).

EXERCISES

6.E.1 Let $\mathcal{S}$ be a net Morse differentiation basis (F_5). If $\mathcal{F}$ is a (unique) filter basis associated with $\mathcal{S}$ and if $\hat{\mathcal{S}}$ is the maximal net basis associated with $\mathcal{F}$, then $\mathcal{S} \subset \hat{\mathcal{S}}$ and $\mathcal{S} \neq \hat{\mathcal{S}}$.

6.E.2 (a) For the basis (F_1) of Example 6.36, the maximal associated net basis is defined as follows:

$$I_\alpha \Rightarrow x \text{ if } I_\alpha \in \mathcal{L}_x,\ \mu I_\alpha > 0 \text{ and diam } I_\alpha \longrightarrow 0.$$

(b) In case of (F_2) we have

$$I_\alpha \Rightarrow x \text{ if and only if } I_\alpha \in \mathcal{A}_x,\ \nu I_\alpha > 0 \text{ and } \nu I_\alpha \rightarrow 0.$$

6.E.3 Let $\mathcal{F}$ be a filter differentiation basis on $(X, \mathfrak{M}, \nu)$ and let $A \subset X$. We say that a system $\mathcal{V}$ of sets is an $\mathcal{F}$-*Vitali covering* of A if for each $x \in A \cap D(\mathcal{F})$ and each $F \in \mathcal{F}_x$ there is $I \in \mathcal{V} \cap F$.

Analogously, a system $\mathcal{W}$ is an $\mathcal{S}$-*Vitali covering* of A (where $\mathcal{S}$ is supposed to be a net differentiation basis) provided for each $x \in A \cap D(\mathcal{S})$ there is a net $I_\alpha \Rightarrow x$ such that $I_\alpha \in \mathcal{W}$.

Show that if $\mathcal{F}$ and $\mathcal{S}$ are associated, then $\mathcal{L}$ is an $\mathcal{F}$-Vitali covering if and only if it is an $\mathcal{S}$-Vitali covering.

REMARK. We do not assume that a Vitali covering of A is even a covering of A !

6.E.4 Let $\mathcal{T}$ be a filter or a net differentiation basis. We say that $\mathcal{T}$ has the *strong Vitali property* if the following (VCT) condition holds:

(VCT) If $\mathcal{V}$ is a $\mathcal{T}$-Vitali covering of a set $A \subset X$ and if $\varepsilon > 0$, then there is a sequence $I_1, I_2, \ldots$, chosen from $\mathcal{V}$, such that

(a) $\nu(A \setminus \bigcup_k I_k) = 0$,

(b) $I_m \cap I_n = \emptyset$ if $m \neq n$,

(c) $\nu(\bigcup_k I_k \setminus H_A) < \varepsilon$, where H_A is a measurable cover for A.

Further we say that $\mathcal{T}$ has *the density property* if the following (LDT) condition holds:

(LDT) If $A \subset X$, then for almost every $x \in A$, x is a point of density for A. (That is $\underline{D}_{\mathcal{T}}(x,A) = 1$.)

Prove that $\mathcal{T}$ has the density property provided it has the strong Vitali property.

Hint. Obviously, $D(\mathcal{J}) \sim X$. By assuming that (LTD) does not hold, there is $n \in \mathbb{N}$ and $A \subset X$ such that $\nu^*(A) > 0$ and $\underline{D}_{\mathcal{J}}(x,A) < 1-\frac{1}{n}$ for every $x \in A$. Now, find a $\mathcal{J}$-Vitali covering $\mathcal{V}$ of A such that $\nu^*(A \cap I) < (1-\frac{1}{n})\ \nu(I)$ for all $I \in \mathcal{V}$ and seek a contradiction.

6.E.5 Let $\mathcal{F}$ be a filter differentiation basis on (X,Σ,μ). Show that all statements of 6.12-6.16 and 6.18-6.20 and of Ex. 6.B.1-6.B.5 (with natural definitions of d_e, d^e, d_i, d^i) hold for $b = b_{\mathcal{F}}$ and $d = d_{\mathcal{F}}$.

6.E.6 Let $\mathcal{F}$ be the filter differentiation basis on (X,Σ,μ) constructed in the course of the proof (iv)$\Rightarrow$(v) of Theorem 6.39. Prove that $\mathcal{F}$ has the strong Vitali property.

Hint. Let $\mathcal{V}$ be an $\mathcal{F}$-Vitali covering of a set A. By Zorn's lemma there is a maximal disjoint sequence $I_1, I_2, \ldots$ from $\mathcal{V}$ such that $I_i \subset H_A$ for every i. Suppose that $\mu^*(A \setminus \bigcup_k I_k) > 0$ to derive a contradiction with the maximality of $\{I_i\}$.

6.E.7 Construct an abstract density topology on some (X,Σ,μ) having the property that $\mu(U) = +\infty$ for each neighborhood U of a certain point $z \in X$.

Hint. Let τ be the family of all open sets U in the ordinary density topology on $\mathbb{R}$ for which $\lambda(\mathbb{R}\setminus U) = 0$ whenever $0 \in U$. Using 6.39.iii, show that τ is an abstract density topology on $\mathbb{R}$. Put $z = 0$.

6.E.8 Let τ^* denote the restriction of the topology τ of Ex. 6.E.7 to $[0,1]$. Show that:

(a) τ^* is an abstract density topology on $[0,1]$ (with the Lebesgue measure λ),

(b) τ^* is determined by a sequence differentiation basis such that $K_n \Rightarrow x$ implies $\lambda K_n \to 0$,

(c) any τ^*-neighborhood of 0 has Lebesgue measure one.

In particular, the example of τ^* yields a negative answer to a query of N.F.G.Martin (1964), p.11.

6.E.9 Let τ be an abstract density topology on (X,Σ,μ) (μ finite). For every $x \in X$ let $\mathcal{J}_x$ be the filter having the base

$$\{\{M \in \Sigma^b : M \subset U,\ M \text{ is a } \tau\text{-neighborhood of } x\} : U \text{ is a } \tau\text{-neighborhood of } x\}$$

(cf. proof of (iv)$\Rightarrow$(v) of Theorem 6.39 and Ex. 6.E.6). Show that:

(a) $\mathcal{T}$ has the strong Vitali property,

(b) $d_{\mathcal{T}}$ is an abstract density topology finer than τ,

(c) it can happen that $d_{\mathcal{T}} \neq \tau$.

Hint. One proves (i) exactly like assertion of Ex. 6.E.6, (ii) follows directly from (i) and Ex. 6.E.4. For (iii), consider $X = (0,2)$ with the Lebesgue measure λ and the ordinary density topology d on X. Let τ be the collection of all d-open sets U such that $\lambda((0,1)\setminus U) = 0$ whenever $1 \in U$. Using 6.39.iii show that τ is an abstract density topology. Now the set $(0,1]$ is $d_{\mathcal{T}}$-open but it is not τ-open.

6.E.10 Show that an abstract density topology need not be T_1 even in case when Σ separates the points.

Hint. Put $X = \{0,1\}$, $\Sigma = \exp X$, μ = the Dirac measure at 0 and $\tau = \{X, \emptyset, \{0\}\}$.

6.E.11 Using Proposition 6.35, prove directly the implication (ii)$\Rightarrow$(v) of Theorem 6.39.

Hint. Let $\mathcal{F} = \{\mathcal{F}_x\}$ be the filter differentiation basis corresponding to b by Proposition 6.35 and let $\mathcal{T}_x$ be a filter having the base $\{F \in \mathcal{F}_x : x \in F\}$. The implication will follow, provided we can prove that $x \in b_{\mathcal{F}} A$ if and only if $x \in b_{\mathcal{T}} A$. This is trivial if $\mu(\{x\}) = 0$. If $\mu^*(\{x\}) > 0$, show that $x \notin bCH_{\{x\}}$ (where $H_{\{x\}}$ is a measurable cover of $\{x\}$) and that $x \in B$ whenever $B \subset H_{\{x\}}$, $B \in \Sigma^b$.

6.E.12 Let Σ be a σ-algebra on a set X and let $\mathcal{N} \subset \Sigma$ be a σ-ideal. Suppose that every $A \subset X$ has a "measurable cover" H_A (cf. the last remark of this section). Let τ be a topology on X such that:

(a) $A \in \mathcal{N}$ if and only if A is τ-closed and nowhere dense,

(b) $A \in \Sigma$ if and only if A has the τ-Baire property.

For $A \subset X$, put

$$bA = \mathrm{der}_{\tau} A \cup \{x \in A : \{x\} \notin \mathcal{N}\} \ .$$

Prove the following assertions:

(a) If $\mathcal{I} \subset \mathcal{N}$ is an ideal containing all finite sets from $\mathcal{N}$, then b is the ideal base operator determined by τ and $\mathcal{I}$.

(b) b is a base operator compatible with $(X, \Sigma, \mathcal{N})$ (i.e. $bA = bH_A$), $bX = X$, $bA \sim A$ for each $A \in \Sigma$ and τ is exactly the b-topology. Moreover, b is uniquely determined by these conditions.

(c) $bA = \overline{\mathrm{int}_{\tau} A}^{\tau} = \bigcap_{B \sim A} \bar{B}^{\tau}$.

Hint. The assertions (a) and (b) follow easily if we observe that any nonempty τ-open set lies in $\Sigma \setminus \mathcal{N}$, and that $A \setminus \mathrm{int}_{\tau} A \in \mathcal{N}$ for every $A \in \Sigma$. Concerning the uniqueness of b and (c), cf. Remark 6.40.B.

6.E.13 Let τ be an abstract density topology on $(X, \Sigma, \mathcal{N})$. Then τ is determined by a unique lower density L on Σ. In fact,

$$L(A) = \{x \in X : x \in \mathrm{int}_{\tau}(A \cup \{x\})\} \ .$$

Hint. You can use the preceding exercise.

6.E.14 Let τ_1, τ_2 be abstract density topologies on $(X, \Sigma, \mathcal{N})$ and let L_1, L_2 and b_1, b_2 be corresponding lower densities and base operators (cf. Ex. 6.E.12 and 6.E.13). If τ_2 is finer than τ_1, then $b_2 A \subset b_1 A$ $(A \subset X)$ and $L_1(S) \subset L_2(S)$ $(S \in \Sigma)$.

6.E.15 Let τ be an abstract density topology on $(X, \Sigma, \mathcal{N})$. For each $x \in X$, let $\mathcal{L}_x \subset \Sigma \setminus \mathcal{N}$ be a filter basis (of a filter $\mathcal{F}_x$) containing all τ-open neighborhoods of x. If b is the base operator corresponding to $\{\mathcal{F}_x\}_{x \in X}$ (cf. Ex. 1.A.2), then:

(a) b-topology is an abstract fine topology finer than τ,

(b) $\{F \in \mathcal{F}_x : x \in F\}$ is the system of all b-neighborhoods of x,

(c) b is the base operator corresponding to the b-topology by Ex. 6.E.12 provided $X \setminus \{x\} \in \mathcal{L}_x$ for each $x \in X$ such that $\{x\} \in \mathcal{N}$.

Hint. For the proof of (a) use Theorem 6.39.iii.

6.E.16 (Denjoy-Stepanoff). Let τ be an abstract density topology on $(X,\Sigma,\mathcal{N})$. Show that:

(a) $A \in \mathcal{N}$ if and only if A is a τ-first category set,

(b) a real function f on X is Σ-measurable if and only if f has the τ-Baire property, which is the case if and only if f is τ-continuous at all points except a set from $\mathcal{N}$.

Hint. Proceed as in the proof of Theorem 6.20.

6.E.17 (Quasi-Lindelöf property.) Let τ be an abstract density topology on $(X, \Sigma, \mathcal{N})$. If any disjoint family of sets of $\Sigma \setminus \mathcal{N}$ is countable, then τ has the $\mathcal{N}$-quasi-Lindelöf property. In particular, any abstract density topology on a σ-finite complete measure space (X, Σ, μ) has the $\mathcal{N}$-quasi-Lindelöf property for $\mathcal{N} = \{A \in \Sigma : \mu A = 0\}$.

Hint. Proceed as in the proof of Theorem 6.26.

6.E.18 Any abstract density topology on $(X,\Sigma,\mathcal{N})$ is hereditarily Baire.

Hint. Cf. Theorem 6.26.

6.E.19 Let τ be a topology on X. Then the following assertions are equivalent:

(i) every τ-first category set is τ-closed,

(ii) τ is an abstract density topology on $(X,\Sigma,\mathcal{N})$ for some Σ and $\mathcal{N}$.

Hint. For (ii)$\Longrightarrow$(i), see Ex. 6.E.16. If (i) holds, then τ is a category density topology on (X,τ).

6.E.20 Let $\mathcal{N}$ be the σ-ideal of all τ-first category sets of a Baire space (X,τ). If i is the ideal topology determined by τ and $\mathcal{N}$, then i is an abstract category density topology on (X,τ). In fact, it is the coarsest category density topology on X finer than τ.

6.E.21 If i is the ideal topology on $\mathbb{R}$ determined by the Euclidean topology and by the σ-ideal of Lebesgue null sets, then i is not an abstract density topology on $(\mathbb{R},\mathcal{M},\lambda)$ ($\mathcal{M}$ is the family of all Lebesgue measurable sets).

6.E.22 Following J.C. Oxtoby (1961), a measure μ on a topological space is a *category measure* if the μ-null sets and the first category sets agree, and if the family of all μ-measurable sets coincides with the family of all sets having the Baire property.

(a) If τ is an abstract density topology on (X,Σ,μ), then μ is a τ-category measure.

(b) There is a topology τ on $\mathbb{R}$ such that the Lebesgue measure λ is a τ-category measure but τ is not an abstract density topology on $(\mathbb{R},\mathcal{M},\lambda)$.

Hint. Consider τ as the $\mathbb{R}\setminus\{0\}$-modification of the ordinary density topology d on $\mathbb{R}$ w.r.t. the Euclidean topology, i.e. a set G is τ-open if it is d-open and $0 \in \operatorname{int} G$ whenever $0 \in G$.

6.E.23 Show that in the condition (iv) of Theorem 6.39 it is not possible to replace (a) by

(a*) a set is of μ-measure zero if and only if it is τ-nowhere dense,

nor to omit the condition (b).

Hint. Consider the topology τ of Ex. 6.E.22.b. Concerning the condition (b), let $N \subset [0,1]$ be a Lebesgue nonmeasurable set for which $\lambda_* N = 0$, $\lambda^* N = 1$. Define a set $A \subset [0,1]$ to be τ-closed if it has one of the following forms: $[0,1]$, S_1, $N \cup S_2$, $([0,1]\setminus N) \cup S_3$ where S_i are Lebesgue null sets. Show that τ satisfies the condition (a) of 6.39.iv, $[0,1/2]$ lacks the τ-Baire property, and the set N has the τ-Baire property.

6.E.24 Put

(a) $X = \mathbb{N} \cup \{\infty\}$, $\Sigma = \exp X$, $\mu = \varepsilon_\infty$, $\tau = \{[n,\infty) \cap \mathbb{N}:\ n \in \mathbb{N}\}$,

(b) $X = \mathbb{Q}$, $\Sigma = \exp X$, $\mu \equiv 0$, $\tau =$ the Euclidean topology on $\mathbb{Q}$.

Show that in both cases μ is a τ-category measure but τ is not an abstract density topology on (X,Σ,μ). In (a), (X,τ) is a Baire space but in (b) it is not.

6.E.25 (a) If τ is a compact abstract density topology on $(X,\Sigma,\mathcal{N})$, then any set of $\mathcal{N}$ is finite. If, moreover, any singleton belongs to Σ, then X is countable.

(b) If $X = \{0\} \cup \{1/n: n \in \mathbb{N}\}$, $\Sigma = \exp X$ and $\mu A = \text{card}\,(A \setminus \{0\})$, then the Euclidean topology on X is a compact Hausdorff abstract density topology on an infinite set.

(c) If $\Sigma = \{\mathbb{R},\emptyset\}$, $\mathcal{N} = \{0\}$, then the indiscrete topology on $\mathbb{R}$ is a compact (non-Hausdorff) abstract density topology on an uncountable space.

6.E.26 (S. Wroński, (1978)). Let (X,Σ,μ) be again a complete σ-finite measure space and $\mu(\{x\}) = 0$ for any $x \in X$. Show that a topology τ on X is an abstract density topology if and only if τ is T_1 and the τ-derivative der_τ fulfils the conditions:

(a) $A \cup \text{der}_\tau A \in \Sigma$ for every $A \subset X$,

(b) $A \sim \text{der}_\tau A$ for each $A \in \Sigma$,

(c) $\text{der}_\tau A = \emptyset$ provided $\mu A = 0$,

(d) $A, B \in \Sigma$, $A \sim B$ implies $\text{der}_\tau A = \text{der}_\tau B$,

(e) $\text{der}_\tau X = X$.

Hint. Show that $S(A) := X \setminus \text{der}_\tau(X \setminus A)$ is a lower density which determines τ.

6.E.27 Let τ be an abstract density topology on $(X,\Sigma,\mathcal{N})$, let b be the corresponding base operator and let $\emptyset \neq Y \subset X$. Denote

$$\Sigma_Y = \{A \cap Y:\ A \in \Sigma\},\ \mathcal{N}_Y = \Sigma_Y \cap \mathcal{N},\ \tau_Y = \tau / Y.$$

Show that for $(Y,\Sigma_Y,\mathcal{N}_Y)$ the existence of measurable covers is guaranteed and that the following conditions are equivalent:

(i) $Y \subset bY$,

(ii) $\mathcal{N}$ does not contain any nonempty τ_Y-open set,

(iii) τ_Y is an abstract density topology on $(Y,\Sigma_Y,\mathcal{N}_Y)$.

Show also that (i) - (iii) are satisfied if Y is τ-open or, more generally, if Y has not τ-isolated points.

6.E.28 Let $A \subset \mathbb{R}$ be a set having the Baire property. We say that 0 is a *W-density point* of A if every subsequence $\{f_{n_k}\}_k$ of the sequence $f_n := \chi_{nA \cap [-1,1]}$ contains a subsequence $\{f_{n_{k_m}}\}_m$ which converges to 1 for all points of $[-1,1]$ except a first category set. Further, $x \in \mathbb{R}$ is a W-density point of A if 0 is a W-density point of $A-x$.

Let $W(A)$ denote the set of all W-density points of A. Show that W is a categorial lower density (defined on the family of sets with the Baire property).

Let τ be the categorial density topology on $\mathbb{R}$ determined by W, and let i be the categorial ideal topology on $\mathbb{R}$ (cf. Example 1.5.d). Prove that:

(a) τ is strictly finer than i,

(b) any τ-open set is of the form $G \cup T$ where G is i-open, T is nowhere dense and $T \subset \bar{G}$.

Hint. For (a), consider a sequence $k_n \nearrow \infty$ such that $k_n/k_{n+1} \to 0$ and show that

$$H := \{0\} \cup \bigcup_{n=1}^{\infty} \{x \in \mathbb{R}: \frac{1}{k_{n+1}} < |x| < \frac{1}{k_n} - \frac{1}{k_{n+1}}\}$$

is τ-open.

6.E.29 Let a be the a.e.-modification of the ordinary density topology on $\mathbb{R}$. Show that the topology τ given by the collection of all sets of the form $H \setminus N$ where H is a-open and N is a first category set is a categorial density topology and it satisfies also (a) and (b) of Ex. 6.E.28.

6.E.30 (cf. H.I. Miller (1981) for $X = \mathbb{R}$). Let X be a topological space. A subset H_A of X is called a *Baire cover* of a set $A \subset X$ if $A \subset H_A$, H_A has the Baire property, and if any subset of $H_A \setminus A$ that has the Baire property is of first category. Show that for any set $A \subset X$ there is always a Baire cover of A. Of course, H_A is a "measurable cover" of A in the sense of the final Remark to Section 6.E.

Hint. Put $H_A = A \cup \{x \in X: A$ is of the second category at $x\}$. In other words, $H_A = \bar{A}^i$ where i is the "ideal base operator" of Example 1.5d.

6.E.31 Let τ be an abstract density topology on $(X, \Sigma, \mathcal{N})$. Show that the following conditions on $A \subset X$ are equivalent:

(i) $A \in \Sigma$,

(ii) A is a τ-Borel set,

(iii) A has the τ-Baire property,

(iv) $A = G \cup F$ where G is τ-open and F is τ-closed

(cf. Theorem 6.19). Further show that, without additional hypothesises, none of these conditions implies that A is a τ-G_δ-set.

Hint. Look at the proof of Theorem 6.39, and use the example of Ex. 6.E.7.

6.E.32 (a) Let τ be a (metric) density topology as in Section 6.B such that the density theorem holds and bP = P. Then τ is an abstract density topology on $(P, \mathfrak{M}, m)$.

(b) Let τ_i, $i = 1,\dots,n$ be abstract density topologies on $(X, \Sigma, \mathcal{N})$. Then $\tau := \inf(\tau_1,\dots,\tau_n)$ is an abstract density topology on $(X, \Sigma, \mathcal{N})$ as well.

(c) Show that the topology τ of Ex. 6.D.8 is an abstract density topology on $\mathbb{R}^n$ (w.r.t. every μ_i).

Hint. For (b) use the equivalence (i)$\Leftrightarrow$(iii) of Theorem 6.39 in the more general context of the Remark following Proposition 6.41. For (c) use (a) and (b).

REMARKS AND COMMENTS

In this section we study the abstract density topologies determined by filter differentiation bases. The most important result consists in Theorem 6.39 which characterizes these topologies satisfying furthermore the density theorem as the topologies determined by a lower density, or those determined by a base operator possessing some further properties. Moreover, Theorem 6.39 gives an intrinsic characterization of these topologies.

Abstract density topologies have been examined by a number of authors, e.g. by A. Ionescu Tulcea and C. Ionescu Tulcea (1969) or by J.C. Oxtoby (1971) as topologies determined by a lower density and, in fact, originally by O.Haupt and Ch. Pauc (1952), (1954), too, as topologies determined by Possel's differentiation bases and satisfying, in addition, the density theorem.

An even more general approach which includes apart from density topologies also some fine topologies of potential theory is used in B. Fuglede (1971a).

In the remarks of this section we indicated further generalizations of abstract density topologies to cover in the new abstract theory even "category density topologies" originated in papers by W. Wilczyński.

Related papers: W. Wilczyński (1982), (1984), W. Wilczyński and V. Aversa (1984), B. Aniszczyk and R. Frankiewicz (1985).

6.F Lifting topologies

We say that a lower density S is a *lifting* if it satisfies

(S_5) $S(A \cup B) = S(A) \cup S(B)$ for all $A, B \in \Sigma$.

Hence, a lifting is simultaneously a lower and an upper density. The corresponding topology τ_S is termed the *lifting topology*. We intend to prove a theorem characterizing lifting topologies in the class of all abstract density topologies on X. For this purpose we need the following "calculation" lemma.

6.42 LEMMA. Let the topology τ be determined by a lower density S. Then for each $A \in \Sigma$,

$$A \cap S(A) = \mathrm{int}_\tau A \subset A \cup S(A) \subset \bar{A}^\tau = A \cup CS(CA),$$

and all these sets are (mutually) equivalent. If, moreover, S is a lifting, then

$$\bar{A}^\tau = A \cup S(A).$$

Proof. Let b be the base operator given as $bA := CS(CA)$. According to the proof of Theorem 6.39, (i)$\Longleftrightarrow$(ii), the b-topology is exactly τ. By Lemma 6.38.b we get

$$\mathrm{int}_\tau A = A \cap CbCA = A \cap S(A),$$

$$\bar{A}^\tau = A \cup bA = A \cup CS(CA).$$

Since $A \cap CA = \emptyset$, we take in consideration the properties of lower density and obtain $S(A) \cap S(CA) = S(\emptyset) = \emptyset$, so that $S(A) \subset CS(CA)$. By (S_1) all sets in question are equivalent. If, in addition, S is a lifting, then $A \cup CA = X$ implies $S(A) \cup S(CA) = S(X) = X$, and this decomposition of X is disjoint. Hence,

$$\bar{A}^\tau = A \cup CS(CA) = A \cup S(A).$$

6.43 THEOREM. Let τ be a topology on X. Then the following four assertions are equivalent:

(i) τ is a lifting topology,

(ii) there is a base operator b compatible with μ such that $bA \sim A$, $bCA = CbA$ for each $A \in \Sigma$ and τ is exactly the b-topology,

(iii) τ is an abstract density topology and the τ-closure of any τ-open set is τ-open,

(iv) τ is an abstract density topology and there is no abstract density topology strictly finer than τ.

Proof. (i)$\Longrightarrow$(iv): Let L be a lifting on Σ which determines τ. Assume that t is an abstract density topology finer than τ determined by a lower density S. Let $A \subset X$ be t-closed. Then A is measurable and Lemma 6.42 yields $A \sim \bar{A}^{\tau}$. Denoting $B := C\bar{A}^{\tau}$, and using once more Lemma 6.42 we have

$$CA \subset S(CA) = S(B) \subset \bar{B}^{t} \subset \bar{B}^{\tau}.$$

Hence

$$C(A \cup B) = CA \setminus B \subset \bar{B}^{\tau} \setminus B \subset L(B),$$

$$C(A \cup B) = CB \setminus A = \bar{A}^{\tau} \setminus A \subset L(A),$$

and thus

$$C(A \cup B) \subset L(A) \cap L(B) = L(A \cap B) = L(\emptyset) = \emptyset.$$

We have shown that $\bar{A}^{\tau} = A$, from which $t = \tau$.

(iv)$\Longrightarrow$(iii): Suppose that we have a τ-open set G for which $\bar{G}^{\tau}$ is not τ-open. Define the new topology t on X: A set H is t-open if $H = (\bar{G}^{\tau} \cap U_1) \cup U_2$, where $U_1, U_2 \in \tau$. Obviously, t is a topology which is strictly finer than τ ($\bar{G}^{\tau} \in t$!) By Proposition 6.41 we need only to show the fact that each non-empty set $H \in t$ is of positive measure. Obviously, any set $H := (\bar{G}^{\tau} \cap U_1) \cup U_2$ is measurable. If $U_2 \neq \emptyset$, then $\mu(H) \geq \mu(U_2) > 0$. If $U_1 \cap \bar{G}^{\tau} \neq \emptyset$, then also $U_1 \cap G \neq \emptyset$ and $\mu(H) \geq \mu(U_1 \cap G) > 0$ (let us notice that in each topology determined by a lower density, each non-empty open set is of positive measure in view of Theorem 6.39.iii.)

(iii)$\Longrightarrow$(ii): In accordance with Theorem 6.39 it suffices to show that $bCA = CbA$ for any $A \in \Sigma$ whenever b determines τ and satisfies 6.39.ii. Suppose, on the contrary, the existence of a measurable set A with $bCA \cap bA \neq \emptyset$. It follows from Remark 6.40.b that $\overline{\mathrm{int}_\tau A}^\tau = bA$. Since $A \sim bA$, we have $CA \sim CbA$. It follows that

$$\emptyset \neq bCA \cap bA = bCbA \cap bA,$$

which is inconsistent with the assumption that $\overline{\mathrm{int}_\tau A}^\tau = bA$ is τ-open, and thereby completing the proof.

(ii)$\Longrightarrow$(i): Since $bX = bC\emptyset = Cb\emptyset = X$, the base b corresponds uniquely to a lower density L. We now wish to prove (S_5). To do this, it is necessary and sufficient to show that $b(A \cap B) = bA \cap bB$ for each $A, B \in \Sigma$. But

$$b(A \cap B) = bC(CA \cup CB) = Cb(CA \cup CB) = C(bCA \cup bCB) = bA \cap bB.$$

REMARK. Let τ be a lifting topology determined by a lifting L. If G is a τ-open set, by Lemma 6.42

$$\bar{G}^\tau = G \cup L(G) \subset L(G) \subset L(\bar{G}^\tau)$$

holds, so that $\bar{G}^\tau$ is also τ-open. In such a way we have just proved the implication (i)$\Longrightarrow$(iii) of Theorem 6.43 directly. The property (iii) expresses nothing else than the fact that lifting topologies are precisely *extremally disconnected* abstract density topologies.

It is an interesting well-known fact that each abstract density topology can be enlarged to a lifting topology. We propose now an alternative "topological" proof of this fact.

6.44 THEOREM. Let τ_0 be an abstract density topology on X. Then there is a lifting topology on X which is finer than τ_0.

Proof. Let $\mathcal{T}$ be a linearly ordered set of abstract density topologies on X finer than τ_0. Let t be a collection of all sets $G := \bigcup_{\tau \in \mathcal{T}} G_\tau$, where G_τ are τ-open. Obviously, t is a topology finer than τ_0. Let I_G be a measurable kernel of a set $G = \bigcup_{\tau \in \mathcal{T}} G_\tau$. Then for every $\tau \in \mathcal{T}$ we have

$$G_\tau \setminus \bar{I}_G^{\tau_0} \sim G_\tau \setminus I_G \sim \emptyset$$

(Theorem 6.39.iii) so that $G_\tau \setminus \bar{I}_G^{\tau_0}$ is a τ-open set of measure zero. Therefore $G_\tau \subset \bar{I}_G^{\tau_0}$ and we have

$$I_G \subset G \subset \bar{I}_G^{\tau_0} \sim I_G.$$

We see that any non-empty t-open set is measurable and of positive measure, so t is an abstract density topology by Proposition 6.41. By Zorn lemma there is a maximal topology in the ordering under consideration. Using Theorem 6.43.iv, we conclude that this topology is a lifting topology we are looking for.

We can now employ the existence of a lifting topology just accomplished to provide one of the important results in lifting theory.

6.45 THEOREM. Let U be an upper (lower, resp.) density on Σ. Then there is a lifting L such that $L(A) \subset U(A)$ ($L(A) \supset U(A)$, resp.) for each $A \in \Sigma$.

Proof. Let τ_0 be an abstract density topology determined by U. In light of Theorem 6.44 we conclude that there exists a lifting topology τ which is finer than τ_0. The lifting L corresponding to τ has the desirable properties (see Ex.6.E.14).

REMARK. The abstract density topologies as well as the lifting ones are usually examined on spaces equipped with an additional metric or topological structure. We shall throw light upon this topic in Section 7.C.

EXERCISES

6.F.1 Following the final Remark of Section 6.E, define the notion of a *lifting* on $(X, \Sigma, \mathcal{N})$ and prove that all assertions of Section 6.F remain true also in this more general setting.

6.F.2 Show that there exists a *categorial lifting* on any Baire space (X, τ).

Hint. By Ex. 6.E.20, there is a categorial density topology on (X, τ). Now, use Theorem 6.44 and the preceding exercise.

6.F.3 Keep the notations of Ex. 6.E.15. If, for each x, $\mathcal{B}_x$ is a maximal filter basis consisting of sets from $\Sigma \setminus \mathcal{N}$, then τ^* is a lifting topology and b equals the base operator of Theorem 6.43.

6.F.4 Let τ be a lifting topology on $(X, \Sigma, \mathcal{N})$.

(a) The filter basis $\{B \in \Sigma : B$ is a τ-neighborhood of $x\}$ is maximal among all filter bases of sets from $\Sigma \setminus \mathcal{N}$ containing x.

(b) Let b be the base operator of Theorem 6.43. If $\{\mathcal{F}_x\}_{x \in X}$ is the corresponding family of filters (cf.Ex. 1.A.2), then $\mathcal{B}_x := \mathcal{F}_x \cap \Sigma$ is a maximal filter basis among all bases consisting of sets from $\Sigma \setminus \mathcal{N}$.

6.F.5 Let τ be an abstract density topology on $(X, \Sigma, \mathcal{N})$. Prove that the following assertions are equivalent (cf. Theorem 6.43):

(i) τ is a lifting topology,

(v) any Σ-measurable function has a τ-limit at each non-τ-isolated point,

(vi) for any Σ-measurable function f there exists a τ-continuous function g such that $[f \neq g] \in \mathcal{N}$,

(vii) for any $A \in \Sigma$ there is a both τ-closed and τ-open set H such that $H \sim A$.

Hint. (i)$\Longrightarrow$(v). Let b be the corresponding base operator (Theorem 6.43). Let x be a non-τ-isolated point at which the τ-limit of f does not exist. By compactness of $\overline{\mathbb{R}}$, the cluster set $\mathrm{Cl}_\tau(f,x)$ contains at least two different points y_1, y_2. If $y_1 < c < y_2$, then

$$x \in \mathrm{der}_\tau [f \geq c] \subset b[f \geq c],$$

and simultaneously

$$x \in \mathrm{der}_\tau [f < c] \subset b[f < c],$$

which is a contradiction.

(v)$\Longrightarrow$(vi). Put

$$g(x) := \begin{cases} f(x) & \text{if } x \text{ is } \tau\text{-isolated}, \\ \lim\limits_{t \to x} f(t) & \text{in the opposite case.} \end{cases}$$

By Ex. 6.E.16, $[f \neq g] \in \mathcal{N}$ and therefore $[f \neq g]$ is τ-isolated. Hence, g is τ-continuous.

(vi)$\Longrightarrow$(vii). Apply (vi) for $f = \chi_A$ and find g. Show that $g(X) \subset \{0,1\}$ and put $H := [g = 1]$.

(vii)$\Longrightarrow$(i). Let G be τ-open. Find a both τ-open and τ-closed set H for which $H \sim G$. Show that $\overline{G}^\tau = H$ and use Theorem 6.43.

6.F.6 Let τ be a lifting topology on $(X,\Sigma,\mathcal{N})$. Since τ is extremally disconnected (cf. Remark following Theorem 6.43), τ is *zero-dimensional* provided it is regular. Show that the topology τ of Ex. 6.E.10 is a lifting topology which is not zero-dimensional.

6.F.7 Let τ be a (metric) density topology determined by a differentiation basis on a metric space (P,ρ) as in Section 6.B. Prove that τ is a lifting topology if and only if $bP = P$ and $m(\{x\}) > 0$ for every $x \in P$.

Hint. Suppose the existence of $x \in bP$ with $m(\{x\}) = 0$. By induction, construct $\alpha_k \searrow 0$ such that for the set

$$A := \bigcup_{n=1}^{\infty} \{y \in P: \alpha_{2n} \leq \rho(y,x) < \alpha_{2n+1}\}$$

we have $x \in bA \cap bCA$.

6.F.8 Let X be a commutative group and let τ be an abstract density topology on $(X,\Sigma,\mathcal{N})$. If Σ, $\mathcal{N}$ and τ are translation invariant, then there exists a lifting topology τ^* finer than τ which is also translation invariant.

Hint. Let $z \in X$ and let $\mathcal{B}_z$ be a maximal filter basis consisting of sets of $\Sigma \setminus \mathcal{N}$ containing all τ-open τ-neighborhoods of z. Put $\mathcal{B}_x := \{B + (x-z): B \in \mathcal{B}_z\}$ and use Ex. 6.F.3.

6.F.9 Let $\mathcal{N}$ be the σ-ideal of all Lebesgue null sets on $\mathbb{R}^n$ and let Σ be the family of all Lebesgue measurable sets.

(a) Show that there is a translation invariant lifting topology on $(\mathbb{R}^n,\Sigma,\mathcal{N})$ (cf. Scheinberg's topologies in Section 7.C).

Hint. Use the preceding exercise.

(b) There is no lifting topology on $\mathbb{R}^n$ which is invariant with respect to any isometry on $\mathbb{R}^n$.

Hint. Reflect upon that $(0,1]$ and $[1,2)$ are isometric.

6.F.10 Show that there is a translation invariant categorial lifting on any commutative topological group.

6.F.11 Let τ be a lifting topology on $(X,\Sigma,\mathcal{N})$. If $\emptyset \neq Y \subset X$ satisfies the condition (i) of Ex. 6.E.27, then $\tau_Y := \tau \wedge Y$ is a lifting topology on $(Y, \Sigma_Y, \mathcal{N}_Y)$.

6.F.12 If L is a lifting on $(X, \Sigma, \mathcal{N})$ and τ is the corresponding lifting topology, then for any $A \in \Sigma$, $L(A)$ is both τ-open and τ-closed.

Hint. Use Proposition 6.37 and the equality $L(CA) = CL(A)$.

REMARKS AND COMMENTS

Theorem 6.43 characterizes the abstract density topologies which are determined by a lifting. Indeed, it concerns just those ones which do not admit any further refinement, or which are extremally disconnected. As an immediate consequence of our considerations we obtain the assertion that every abstract density topology can be refined to a lifting topology, or also the known assertion (Proposition 6.45) on the existence of a lifting. We indicate the possibility of a topological approach to some results in the lifting theory. Instead of filters and ultrafilters it is possible to employ topologies and maximal topologies as well. (Cf. also the proof of the existence of a strong lifting in separable spaces in Ex. 7.C.6.)

Related paper : P.A. Fillmore (1966).

7. FURTHER EXAMPLES OF FINE TOPOLOGIES

7.A The a.e.-modification of a fine topology

In (1977), R.J.O'Malley introduced a new topology on the real line which he labelled as the a.e.-topology. This one is the weak topology induced on $\mathbb{R}$ by the collection of all approximately continuous functions which are continuous almost everywhere. The a.e.-open subsets of $\mathbb{R}$ are these ones which are of the form $G \cup M$, where G is open in the Euclidean topology and each point of M is a density point of G.

In this section we examine the *a-modification* of an arbitrary *base operator* on a topological space and also the *a.e.-modification* of a fine *topology*. Of course, starting with the density topology on the real line we should like to obtain the R.J.O' Malley's a.e.-topology as a particular case.

Let τ be a fine topology on a topological space (X,ρ). It is easy to see that the collection of all τ-open sets of the form $U \cup \{x\}$ where U is a ρ-open subset of X and $x \in X$ together with the empty set determines a base for a new topology which will be termed the a.e.-modification of τ, or shortly, the *a.e.-topology*.

By definition, the a.e.-topology is coarser than τ but it is still finer than the ρ-topology.

7.1 PROPOSITION. Let τ be a fine topology on (X,ρ), $M \subset X$. Then the following assertions are equivalent:

(i) M is a.e.-open,

(ii) for each $x \in M$, the set $\{x\} \cup \mathrm{Int}_\rho M$ is τ-open,

(iii) M is τ-open and there are disjoint sets G, R, such that G is ρ-open, R is τ-discrete and $M = G \cup R$,

(iv) M is τ-open and the set $M \setminus \mathrm{Int}_\rho M$ is τ-discrete.

Proof. (i)$\Rightarrow$(ii): Let $x \in M \setminus \mathrm{Int}\, M$. Then there is a ρ-open set $U \subset M$ such that $\{x\} \cup U$ is τ-open. Hence the set

$$\{x\} \cup \mathrm{Int}_\rho M = (\{x\} \cup U) \cup \mathrm{Int}_\rho M$$

is τ-open.

(ii)$\Rightarrow$(iv): Of course, $M = \bigcup_{x \in M} (\{x\} \cup \mathrm{Int}_\rho M)$ is τ-open. We put

$$G = \mathrm{Int}_\rho M, \ R = M \setminus G.$$

Suppose R is not τ-discrete, i.e. there is $x \in R \cap \mathrm{der}_\tau R$. Then in view of $(R \setminus \{x\}) \cap (\{x\} \cup G) = \emptyset$, the set $\{x\} \cup G$ cannot be τ-open.

(iv)$\Rightarrow$(iii): Obvious.

(iii)$\Rightarrow$(i): Let $M = G \cup R$ be a τ-open set where G is ρ-open, R is τ-discrete and $G \cap R = \emptyset$. Obviously, $M = \bigcup_{x \in M} (\{x\} \cup G)$ and it is sufficient to check that $\{x\} \cup G$ is τ-open whenever $x \in M$. But this is obvious because $M \setminus \overline{R \setminus \{x\}}^{\tau}$ is a τ-neighborhood of x contained in $\{x\} \cup G$.

In Section 1.A we assigned to any base operator b its new modification $\bar{b}$ (recall that $\bar{b}A: = \overline{bA}^b = b\bar{A}^b$). The operator $\bar{b}$ turns out to be a strong base and it represents the smallest strong enlargement of b which determines the same topology as b. Now, we shall slightly modify the definition of $\bar{b}$ to obtain the desired a-modification.

Let b be a fine base operator on a topological space (X,ρ). Define

$$aA = b\bar{A} \quad (= b\bar{A}^{\rho}).$$

Obviously, a is again a base operator on X and it is termed the *a-modification* of the *base* b. Remark that the a-topology is coarser than the b-topology but it is still a fine topology on (X,ρ).

7.2 PROPOSITION. The base operator a is strong and $\bar{A}^a = A \cup b\bar{A}$ for each $A \subset X$.

Proof. We have

$$aaA = b\,\overline{b\bar{A}} \subset b\bar{A} = aA.$$

An application of Theorem 1.2 completes the proof.

7.3 PROPOSITION. The following four assertions concerning a set $M \subset X$ are equivalent:

(i) M is a-open,

(ii) there is a ρ-open set $G \subset M$ such that $M \cap bCG = \emptyset$,

(iii) $M = G \cup H$, where G is ρ-open and $G \cap H = \emptyset = H \cap bCG$,

(iv) $M \cap b(\mathrm{Cint}_{\rho}M) = \emptyset$.

Proof. By definition, a set M is a-open if $aCM \subset CM$. But this is the case if and only if $b\overline{CM}^{\rho} \subset CM$, which is true if and only if $\emptyset = M \cap b\overline{CM}^{\rho} = M \cap b(\mathrm{Cint}_{\rho}M)$. Hence, (i) $\Longleftrightarrow$ (iv).

(ii) $\Longrightarrow$ (iii): Put $H = M \setminus G$. Then $G \cap H = \emptyset$ and

$$H \cap bCG \subset M \cap bCG = \emptyset.$$

(iii)$\Longrightarrow$(iv): We have

$$M \cap b(C\,\mathrm{int}_\rho M) \subset M \cap bCG = (G \cap bCG) \cup (H \cap bCG) = \emptyset .$$

(iv)$\Longrightarrow$(ii): Putting $G = \mathrm{int}_\rho M$, the assertion is trivial.

Let us consider a fine base operator b on (X,ρ). If a is the a-modification of b, the question arises whether or not the a-topology equals the a.e.-modification of the b-topology. In other words, given the diagram

$$\begin{array}{ccccc} b & \longmapsto & a \ (= \text{a-modification of } b) & & \\ \downarrow & & & \searrow & \\ \text{b-topology } \tau & \longmapsto & \text{a.e.-modification of } \tau & \overset{?}{=} & \text{a-topology} \end{array}$$

the problem is whether or not this diagram commutes. The following example illustrates that, in general, this is not the case, Let τ be a fine topology on (X,ρ), $bA = \bar{A}^{\tau}$. Then $aA = \bar{A}^{\rho}$, so that the a-topology equals ρ. On the other hand, the a.e.-modification of τ is, in general, different from ρ (e.g. the a.e.-modification of the discrete topology on the real line). The following theorem gives an answer to our problem.

7.4 THEOREM. Let τ be a fine topology determined by a base operator b on X. Then

(a) the a.e.-modification of τ is finer than the a-topology,

(b) if $b\{x\} = \emptyset$ for each $x \in X$, then both these topologies coincide.

Proof. (a) Let M be a-open, $x \in M$. By Proposition 7.3.iv, $x \notin b(C(\mathrm{int}_\rho M))$, hence $\{x\} \cup \mathrm{int}_\rho M$ must be τ-open and by Proposition 7.1 we see that M is a.e.-open.

(b) Let M be a.e.-open, $x \in M$. By Proposition 7.1, $x \notin bC(\{x\} \cup \mathrm{int}_\rho M)$. Since $x \notin b\{x\}$ and $C(\mathrm{int}_\rho M) \subset C(\{x\} \cup \mathrm{int}_\rho M) \cup \{x\}$, we have $x \notin b(C(\mathrm{int}_\rho M)$. Hence $M \cap b(\mathrm{int}_\rho M) = \emptyset$ and applying Proposition 7.3 we are through.

REMARK. Let τ be a T_1 fine topology on X. Then, of course, τ is determined by means of a base operator $A \mapsto \mathrm{der}_\tau A$ which is strong. From Theorem 7.4.b it follows that the a.e.-modification of τ coincides with the "a-modification" of this special base operator.

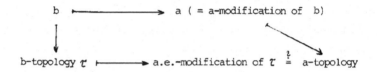

Our principal aim is the Lusin-Menchoff property of the a-topology. We recall briefly some facts. We said that a *base operator* b has the *Lusin-Menchoff property* (with respect to ρ) if for each open set U and for each set A such that $A \cup bA \subset U$ there is an open set G satisfying $A \subset G \subset G \cup bG \subset U$. We also know that a strong base operator b has the Lusin-Menchoff property if and only if the b-topology has it so. In particular, the a-modification of b (which is strong independently of strongness of b) has the Lusin-Menchoff property if and only if the a-topology has it as well.

The methods of the proofs of the Lusin-Menchoff property are based on "algebraic" properties of base operators stated in Theorem 3.21.

7.5 THEOREM. (The Lusin-Menchoff property of an a-topology.) Let a be the a-modification of a fine base operator b having the (complete) Lusin-Menchoff property with respect to a completely normal topological space (X,ρ). Then the corresponding a-topology has also the (complete) Lusin-Menchoff property.

Proof. As mentioned, it suffices to prove the assertions for the base operator a. To do this we apply Theorem 3.21 where we put

$$b_1A = \bar{A}\ ,\ b_2 = b$$

(cf. also the second proof of Theorem 3.22).

Analogously as in section 3.D we can take a slightly deeper look on the previous theorem. We give its improvement (cf. Theorem 3.25).

7.6 THEOREM. Assume that a fine topology ϑ has the complete Lusin-Menchoff property with respect to a topological space (X,ρ). If b is a fine base operator on (X,ϑ) which has the (complete) Lusin-Menchoff property w.r.t. ρ, then the a-modification of b (w.r.t. ϑ !) has the (complete) Lusin-Menchoff property (w.r.t. ρ).

Proof. Use again Theorem 3.21 putting $b_1A = \bar{A}^{\vartheta}$, $b_2 = b$.

REMARK. Since a is obviously strong the a-topology has also the (complete) Lusin-Menchoff property.

Now, Theorem 7.5 yields immediately the following important property of the a.e.-topology. Of course, appealing to Theorem 7.6, a more general result could be stated.

7.7 COROLLARY. (The Lusin-Menchoff property of an a.e.-topology). If a fine topology τ has the Lusin-Menchoff property with respect to a completely normal topological space (X,ρ), so does the a.e.-modification of τ.

As for the complete Lusin-Menchoff property, the similar assertion is valid.

Proof. Use Theorem 7.5 for $b = \mathrm{der}_{\tau}$.

It is easy to derive Corollary 7.7 from Theorem 3.22. We sketch now the main idea: Let F be ρ-closed, F_a a.e.-closed, $F \cap F_a = \emptyset$. By Proposition 7.1, $F_a = P \setminus D$ where P is ρ-closed and D is τ-discrete, $D \subset P$. Let us consider the P-modification of τ. By definition, F_a is τ_P-closed. Using now Theorem 3.22, there is a ρ-open set $G \supset F_a$ and a τ_P-open set $H \supset F$ such that $G \cap H = \emptyset$. It remains to show that H is a.e.-open. We use again Proposition 7.1. Obviously, H is τ-open and $H \setminus P$ is ρ-open. Since $H = (H \setminus P) \cup (H \cap P)$ and $H \cap P \subset D$ is a τ-discrete set, the proof is complete.

This proof suggests that there is a close relation between the a.e.-modification and the M-modification of a fine topology. For further illustration we refer to Ex. 7.A.13.

Let now τ be a fine topology on a topological space (X, ρ), and let τ_a be its a.e.-modification. Given a real function f on X and $r, s \in \mathbb{R}$, we denote

$$A_{r,s} = \{x \in X: r < f(x) < s\}.$$

7.8 PROPOSITION. Let f be a real τ_a-continuous function on X. Then there is a σ-discrete set D (in τ) such that f is ρ-continuous on $X \setminus D$.

Proof. Put

$$D := \bigcup_{\substack{r < s \\ r,s \text{ rational}}} (A_{r,s} \setminus \mathrm{int}_{\rho} A_{r,s}).$$

Using (iv) of Proposition 7.1, D is σ-discrete in τ. Moreover, it is easy to see that f is ρ-continuous at all points of

$$X \setminus D = \bigcap_{\substack{r < s \\ r,s \text{ rational}}} \{x \in X: x \in \mathrm{int}_{\rho} A_{r,s} \text{ whenever } x \in A_{r,s}\}.$$

7.9 PROPOSITION. Let $D \subset X$ be a τ-discrete set. If a τ-continuous real functions f on X is ρ-continuous at every point of $X \setminus D$, then f is τ_a-continuous.

Proof. Let $G \subset \mathbb{R}$ be open. Denoting C_f the set of all ρ-continuity points of f, we have

$$f^{-1}(G) \cap C_f \subset \operatorname{int} f_\rho^{-1}(G).$$

Hence

$$M := f^{-1}(G) \setminus \operatorname{int}_\rho f^{-1}(G) \subset X \setminus C_f \subset D.$$

Therefore M is τ-discrete. Since $f^{-1}(G)$ is τ-open, it is also a.e.-open in light of Proposition 7.1.

7.10 COROLLARY. Assume that τ-discrete subsets of X form a σ-ideal. Then a real τ-continuous function on X is τ_a-continuous if and only if it is ρ-continuous except a τ-discrete set.

REMARKS. (a) Exercises 7.A.1 and 7.A.2 show that discrete and σ-discrete sets cannot be replaced in the assertions of Propositions 7.8 and 7.9.

(b) We conclude this section giving a characterization of an a.e.-topology which generalizes O'Malley's original concept. If a fine topology τ has the Lusin-Menchoff property with respect to a metric space (X,ρ) and if the family of all τ-discrete subsets of X forms a σ-ideal, then the a.e.-modification of τ is the weak topology induced on X by the collection Φ of all τ-continuous functions which are ρ-continuous except a τ-discrete set, and Φ is exactly the family of all a.e. continuous functions.

EXERCISES

7.A.1 (a) Show that the a.e.-modification of the Sorgenfrey topology e^+ (w.r.t. the Euclidean topology on $\mathbb{R}$) is e^+.

(b) Show that there exists an e^+-continuous function f for which the set of all discontinuity points of f is not e^+-discrete. (Compare with Remark (a) following Corollary 7.10).

Hint. Let $\{r_n\}$ be an enumeration of all rationals. Put $f(x) = \sum_{r_n \leq x} 2^{-n}$.

7.A.2 Let $\{S_n\}_{n=1}^{\infty}$ be a partition of the set $\mathbb{Q}$ of all rationals into disjoint dense subsets. Define a set $G \subset \mathbb{R}$ to be τ-open if for every $n \in \mathbb{N}$ and every $x \in G \cap S_n$ there is an Euclidean neighborhood V of x such that $V \cap \bigcup_{i>n} S_i \subset G$. Show that:

(a) τ is a (fine) topology on $\mathbb{R}$,

(b) $\mathbb{R}$ is a σ-isolated set in the topology τ,

(c) the Dirichlet function $\chi_{\mathbb{Q}}$ is τ-continuous although it is not a.e.-continuous at any point $x \in \mathbb{Q}$ (compare with Remark (a) following Corollary 7.10).

7.A.3 Let τ be the restriction of the Sorgenfrey topology e^+ on $\mathbb{Q}$ and let τ_a be the a.e.-modification of τ (w.r.t. the Euclidean topology on $\mathbb{Q}$). Show that:

(a) $\tau_a = \tau$,

(b) there exists a τ_a-continuous function on $\mathbb{Q}$ which is not continuous at any point.

Hint. Consider $f \wedge \mathbb{Q}$ where f is as in Ex. 7.A.1.

7.A.4 Let τ be a fine topology on (X,ρ). If τ has no isolated points, then the a.e.-modification of τ is S-related to ρ (cf.Ex.4.A.1). In particular, in this case,

(a) a set $A \subset X$ is dense (nowhere dense, first category, resp.) if and only if A is dense (nowhere dense, first category, resp.) in the a.e.-topology,

(b) (X,ρ) is a Baire space if and only if X is Baire in the a.e.-topology,

(c) the a.e.-modification of τ is a Blumberg topology provided (X,ρ) is a Blumberg space.

7.A.5 Let τ be a fine topology on (X,ρ) having no isolated points. If f is a Baire one function in the a.e.-modification of τ, then there exists a ρ-first category set P such that f is ρ-continuous at all points of $X \setminus P$. Consequently, if (X,ρ) is a Baire space, then f is continuous at any point of a dense subset of X (compare with Ex. 7.A.3).

Hint. Use Ex. 7.A.4 and Ex. 4.A.2.

7.A.6 Let τ_a be the a.e.-modification of a fine topology τ on (X,ρ) and let $M \subset X$. Show that:

(a) the a.e.-modification of $\tau \wedge M$ need not coincide with $\tau_a \wedge M$ even if M has no τ-isolated points,

(b) if M is τ-open, then the a.e.-modification of $\tau \wedge M$ equals $\tau_a \wedge M$,

(c) if M is ρ-closed and τ-discrete, then the a.e.-modificatoin of $\tau \wedge M$ is the discrete topology and equals $\tau_a \wedge M$.

Hint. For (a), consider the density topology on $\mathbb{R}$. Put $G = \bigcup_{n=1}^{\infty} (2^{-2n-1}, 2^{-2n})$ and construct a measurable set M such that $\lambda(M \cap I) > 0$ for any open interval I and $\lim_{h \to 0_+} \frac{\lambda(M \cap G \cap (-h,h))}{\lambda(M \cap (-h,h))} = 1$. Prove that the set $\{0\} \cup (G \cap M)$ is open in the a.e.-modification of $\tau \wedge M$ but it is not $\tau_a \wedge M$-open.

7.A.7 Let τ_a be the a.e.-modificatoin of a fine topology τ on (X,ρ). Let $M \subset X$ have no τ-isolated points. Then $(M, \tau_a \wedge M)$ is S-related to $(M, \rho \wedge M)$.

7.A.8 Let τ be a ρ-fine abstract density topology (cf. Section 7.C) on a strong Baire topological space (X,ρ). If τ has the G_δ-insertion property, then the a.e.-modification of τ is a strong Baire topology.

Hint. Given an a.e.-closed set $A \subset X$, it needs to show that A is a.e.-second category (in itself). It would be clearly sufficient to assume that A has no a.e.-isolated points. Deduce that in this case A has no τ-isolated points as well. Find a G_δ-set D, $\mathrm{int}_\tau A \subset D \subset A$. Since $\overline{\mathrm{int}_\tau A}^{\tau} = A$, show that D has no τ-isolated points. Taking into account Ex. 7.A.7, show that D is a.e.-second category in itself. The assertion follows now if we observe that D is τ-dense (and, consequently, a.e.-dense) in A.

7.A.9 Let τ be the a.e.-modification of the ordinary density topology d on $\mathbb{R}^n$. Show that:

(a) τ has the complete Lusin-Menchoff property,

(b) a function f is τ-continuous if and only if f is approximately continuous everywhere and continuous almost everywhere and τ_a is the weak topology induced by the family of all approximately continuous functions which are continuous almost everywhere.

(c) $(\mathbb{R}^n, \tau)$ is a strong Baire and Blumberg space,

(d) any τ-Baire one function is continuous at all points of a residual set.

Hint. Use Corollary 7.7, Corollary 7.10 and exercises 7.A.4, 7.A.5, 7.A.8.

7.A.10 Let τ be a fine topology on a topological space (X,ρ). Given $M \subset X$, denote $\tau_{a,M}$ the a.e.-modification of τ w.r.t. τ_M (= the M-modification of τ w.r. t. ρ). If any σ-τ- discrete set is τ-isolated, then

(a) the collection $\mathcal{F}$ of all τ-continuous functions which are continuous at all points of $X \setminus M$ except a τ-isolated set is precisely the family of all $\tau_{a,M}$-continuous functions,

(b) $\tau_{a,M}$ is the weak topology induced by $\mathcal{F}$.

Hint. Using 3.D.5, show that $\mathcal{F}$ is the set of all τ-continuous functions which are τ_M-continuous at all points except a τ-isolated set. Then use Corollary 7.10.

7.A.11 Let a be the a-modification of a fine base operator b on a topological space (X,ρ). If a^* is the a-modification of a, then $a = a^*$.

7.A.12 Let τ_a be the a.e.-modification of a fine topology τ on a topological space (X,ρ). If τ^* is the a.e.-modification of τ_a, then $\tau_a = \tau^*$.

Hint. Show that each τ_a-basis set is τ^*-open.

7.A.13 Let b be a fine base operator on a topological space (X,ρ) and let $M \subset X$. Define the base operator a_M as

$$a_M(A) := b(\bar{A} \setminus M) \cup bA.$$

Show that:

(a) a_M has the Lusin-Menchoff property provided b has,

(b) a_M is strong provided b is strong,

(c) if $M = \emptyset$, then a_M is the a-modification of b,

(d) if τ is a fine topology on (X,ρ), and if

(d_1) $b = cl_{\tau}$, then $a_M = cl_{\tau_M}$ where τ_M is the M-modification of τ,

(d_2) $b = der_{\tau}$, then the a_M-topology is the topology $\tau_{a,M}$ of Ex. 7.A.10.

7.A.14 Let ν_* denote the inner Jordan-Peano content on $\mathbb{R}^n$. A point $x \in \mathbb{R}^n$ is said to be a *JP-density point* of a set $A \subset \mathbb{R}^n$ if

$$\lim_{r \to 0_+} \frac{\nu_*(U(x,r) \cap A)}{\nu_*(U(x,r))} = 1.$$

(a) Show that the collection of all subsets of $\mathbb{R}^n$ having each of their points as a JP-density point forms a topology τ_ν.

(b) Prove that τ_ν is the a.e.-modification of the ordinary density topology on $\mathbb{R}^n$.

REMARKS AND COMMENTS

The first mention of the a.e. topology on an interval is due to R.J.O'Malley (1977). This topology is formed by all density open sets having the property that their Lebesgue measure is the same as the measure of their Euclidean interiors. In the same paper a proof is given that every approximately continuous function which is continuous almost everywhere is continuous with respect to this topology. In the following O'Malley's paper (1979a) the Lusin-Menchoff property of the a.e. topology is proved as well as the fact that the class of all continuous functions in this topology is formed exactly by the family of all approximately continuous functions which are continuous almost everywhere. Thus a.e.-topology is the weak topology induced on $\mathbb{R}$ by this family of functions. The latter property is examined also in $\mathbb{R}^n$ in T. Nishiura's paper (1981), where the a.e.-topology is defined (equivalently with R.J. O'Malley for $n = 1$) as the collection of all subsets U of $\mathbb{R}^n$ which are density open and are of the form $U = G \cup Z$, where G is open and $\lambda(Z) = 0$. T. Nishiura among others proves that the a.e.-topology is completely regular and is not normal; on the other hand, he does not prove the Lusin-Menchoff property.

Our principal result consists in Theorem 7.5 according to which the a-modification of a fine base operator having the Lusin-Menchoff property conserves this property as well. Theorem 7.5 together with the characterization of a.e.-continuous functions (Corollary 7.11) enables us to apply extension theorems of Tietze's type to those functions.

Let us point out here again that the a.e. topology may possess the Lusin-Menchoff property (combination of Theorem 7.5 and Theorem 7.4.b) even if the initial fine topology does not have it. Recall again the case of the superdensity topology.

7.B The r-modification of a fine topology

The r-topology on the real line was defined by R. J. O'Malley in (1977) as the coarset topology making all approximately differentiable functions continuous. O'Malley himself proved that the r-topology (which is strictly coarser than the density topology) has the Lusin-Menchoff property. Besides, in the same paper it was shown that the base for the r-topology is formed by density open sets which are ambivalent. As is often the case, this characterization becomes for us the basis for a definition of the r-topology in a more general context.

Let τ be a fine topology on a topological space (X,ρ). A subset of X is called *ambivalent* if it is both an F_σ - set and a G_δ-set. A set $A \subset X$ is an *r-basis set* if it is τ-open and ambivalent. The *r-modification* of τ (or shortly the *r-topology*) is defined as the topology having for its (topological) base the collection of all r-basis sets, i.e. a set G is r-open if G is a union of r-basis sets. If b is a fine base operator on (X,ρ), then the *r-modification* of b (also called the r-topology) is defined as the r-modification of the b-topology.

We are going to derive theorems concerning the Lusin-Menchoff property of the r-topology. More precisely, we shall prove that the Lusin-Menchoff property of a base operator b on a normal topological space X implies the complete regularity of its r-modification, or even the Lusin-Menchoff property of the r-topology provided the so-called r-basis condition (B) is satisfied. Moreover, we show that the r-topology is the weak topology induced by some subfamilies of b-continuous functions and we present an approximation theorem. Underline that we speak of the Lusin-Menchoff property of a base operator rather than of a fine topology. The example of the superdensity shows that the b-topology can lack the Lusin-Menchoff property even if b has it.

If τ is a fine topology on a normal T_1-space (X,ρ), then the r-topology is *finer than* ρ. Indeed, let $U \subset X$ be open and pick $x \in U$. By normality, there is a sequence $\{V_n\}$ of open sets such that $x \in V_1 \subset \bar{V}_1 \subset V_2 \subset \bar{V}_2 \subset \dots \subset U$. The set $\bigcup_{n=1}^{\infty} V_n$ is an r-basis set containing x and contained in U.

Recall that a function f belongs to $B_1^d(X)$ (cf. Ex. 2.D.14) if there is a sequence $\{f_n\}$ of continuous functions on X such that the set $\{n \in \mathbb{N}: f_n(x) \neq f(x)\}$ is finite for every $x \in X$. We use the fact that any function of $B_1^d(X)$ is ambivalent (cf. Ex. 2.D.15). Hence, any τ-continuous B_1^d-function is r-continuous.

7.11 LEMMA. Let b be a fine base operator having the Lusin-Menchoff property on a normal T_1-space (X,ρ). Given a closed set F and an r-basis set A, $F \subset A \subset X$, there is an upper semicontinuous, r-continuous B_1^d-function f such that

$$f = 0 \text{ on } X\setminus A,\ f = 1 \text{ on } F \text{ and } 0 < f \leq 1 \text{ on } A.$$

Proof. Using the Lusin-Menchoff property of b, there is a sequence $\{F_n\}$ of closed sets such that

$$F_1 = F,\ A = \bigcup_{n=1}^{\infty} F_n \quad \text{and} \quad F_n \subset F_{n+1} \setminus bCF_{n+1}$$

for every $n \in \mathbb{N}$. Further, let $\{G_n\}$ be a decreasing sequence of open sets such that $A = \bigcap_{n=1}^{\infty} G_n$. There are continuous functions f_n on X,

$$0 \leq f_n \leq 1,\ f_n = 0 \text{ on } X\setminus G_n,\ f_n > 0 \text{ on } A \text{ and } f_n = 1 \text{ on } F_n.$$

Put $f := \prod_{n=1}^{\infty} f_n$. Let $c > 0$. If $x \in [f > c]$, then $x \in F_n$ for some $n \in \mathbb{N}$. Find an open neighborhood W of x such that $f_1 f_2 \dots f_{n+1} > c$ on W. Since

$$b[f \leq c] \subset b(CW \cup CF_{n+1}) \subset X \setminus (F_n \cap W),$$

we can easily see that $[f > c] \cap b\,[f \leq c] = \emptyset$. Therefore f is b-lower semicontinuous. Apparently, f has all properties ascribed to it.

7.12 THEOREM. If b is a fine base operator having the Lusin-Menchoff property on a normal T_1-space (X,ρ), then the r-topology has the following properties:

(a) The r-topology is completely regular.

(b) The r-topology is the coarsest topology making all b-continuous B_1^d-functions on X continuous.

(c) If the following r-basis condition (B) is satisfied,

(B) Given a closed set F and r-open set $G \supset F$, there is an r-basis set B such that $F \subset B \subset G$,

then the r-topology has also the Lusin-Menchoff property.

Proof. It follows immediately from Lemma 7.11.

In what follows, let τ be a fine topology on a metric space (P,ρ). A function

f on P is termed τ-Lipschitzian at a point $x \in P$ if

$$\tau\text{-}\limsup_{y \to x} \frac{|f(y)-f(x)|}{\rho(x,y)} < +\infty .$$

The collection of all functions on P which are τ-Lipschitzian at every point of P will be denoted by $\mathcal{L}$.

7.13 THEOREM. Let b be a fine base operator on P having the Lusin-Menchoff property and let τ be the b-topology. Then the following assertions hold:

(a) Every r-basis set A is a cozero set of a nonnegative bounded upper semicontinuous function from $\mathcal{L} \cap B_1^d$.

(b) If, moreover, every element of $\mathcal{L}$ is ambivalent, then the r-topology is the weak topology induced on P by $\mathcal{L}$.

(c) (Approximation theorem). If the r-basis condition (B) of Theorem 7.12.c is satisfied, then $\mathcal{L} \cap \mathcal{C}(P,r) \cap B_1(P)$ is dense in $\mathcal{C}(P,r) \cap B_1(P)$.

Proof. (a) We follow the idea of the proof of Lemma 7.11. Using the Lusin-Menchoff property of b, there is a sequence $\{F_n\}$ of closed sets and a decreasing sequence $\{G_n\}$ of open sets such that

$$A = \bigcup_{n=1}^{\infty} F_n = \bigcap_{n=1}^{\infty} G_n$$

and

$$F_n \subset F_{n+1} \setminus bCF_{n+1}$$

for every $n \in \mathbb{N}$. Put

$$f_n := \frac{\operatorname{dist}(x,CG_n)}{\operatorname{dist}(x,CG_n) + \operatorname{dist}(x,F_n)} .$$

Then f_n are locally Lipschitzian functions, $0 \leq f_n \leq 1$, $f_n = 0$ on CG_n, $f_n > 0$ on A and $f_n = 1$ on F_n. Put

$$f := \prod_{n=1}^{\infty} f_n .$$

According to the proof of Lemma 7.11, f is an upper semicontinuous r-continuous B_1^d-function. Clearly f is bounded and τ-Lipschitzian at every point of CA. Let $x \in F_n$. Denote

$$g_n = \prod_{j=1}^{n+1} f_j .$$

Then g_n is a locally Lipschitzian function and hence there is $K \in \mathbb{R}$ and an open set V containing x such that

$$|g_n(x) - g_n(y)| < K \varrho(x,y)$$

for every $y \in V\setminus\{x\}$. Set

$$M = \{x\} \cup \{y \in P: \quad |f(x)-f(y)| < K \varrho(x,y)\}.$$

Since $f = g_k$ on F_{n+1}, we have

$$x \in \mathrm{CbC}F_{n+1} \cap V \subset \mathrm{CbC}(F_{n+1} \cap V) \subset \mathrm{CbC}M.$$

From the τ-continuity of f it follows

$$M\setminus\{x\} \subset \mathrm{CbC}M.$$

We see that the set M is τ-open, so that f is τ-Lipschitzian at x.

(b) It follows from (a), since every ambivalent function from $\mathcal{L}$ is r-continuous.

(c) At first, we prove the following assertion: Given an r-open set $U \subset P$ and an ambivalent set $A \subset U$, there is an r-basis set B such that $A \subset B \subset U$. Indeed, if

$$A = \bigcup_{n=1}^{\infty} F_n = \bigcap_{n=1}^{\infty} G_n ,$$

where $\{F_n\}$ is an increasing sequence of closed sets and $\{G_n\}$ is a decreasing sequence of open sets, there is an r-basis set B_n such that

$$F_n \subset B_n \subset G_n \cap U.$$

Set $B := \bigcup_{n=1}^{\infty} B_n$. Then B is obviously a τ-open F_σ-set. Since $B = \bigcap_{n=1}^{\infty} (B_1 \cup B_2 \cup \dots \cup B_{n-1} \cup G_n)$, we see that B is a G_δ-set, too.

To conclude the proof, pick $f \in \mathcal{C}(P,r) \cap B_1(P)$ and $\varepsilon > 0$. Taking into account Sierpiński's separation theorem (cf. Ex. 3.B.7.b) and the just proved assertion, there are r-basis sets B_k $(k \in \mathbb{Z})$ such that

$$[(k-1)\varepsilon \leq f \leq k\varepsilon] \subset B_k \subset [(k-2)\varepsilon < f < (k+1)\varepsilon] .$$

Further, by (a), there are nonnegative upper semicontinuous r-continuous functions $g_k \in B_1(P) \cap \mathcal{L}$ such that

$$g_k \leq 2^{-|k|} , \quad B_k = [g_k > 0] .$$

Put

$$g := \Big(\sum_{k \in Z} k g_k \Big) \Big(\sum_{k \in Z} g_k \Big)^{-1} \varepsilon .$$

Then $|g - f| < 3\varepsilon$, g is r-continuous and belongs to $B_1(P)$. For each $k \in \mathbb{Z}$ we have

$$g = \Big(\sum_{j=k-2}^{k+2} j g_j \Big) \Big(\sum_{j=k-2}^{k+2} g_j \Big)^{-1} \varepsilon .$$

on B_k and thus g is τ-Lipschitzian at every point of B_k.

REMARK. A convenient criterion guaranteeing that every function of $\mathcal{L}$ is ambivalent (or, even that $\mathcal{L} \subset B_1^d(P)$) is given in Ex. 2.D.17.

OBSERVATION. Let us consider again a differentiation basis $\mathcal{L}$ on a metric space (P,ρ) and a measure m as in Section 6.B. Denote by b the corresponding base operator, by t the density topology and by r its r-modificatoin.

If the topology t has the $\mathcal{Z}_m$-quasi-Lindelöf property (by Theorem 6.26 this is the case if the density theorem holds) and if $H \subset P$ is r-open, then there is a sequence $\{B_n\}$ of r-basis subsets of H such that $m(H \setminus \bigcup_{n=1}^{\infty} B_n) = 0$. Moreover, given $\varepsilon > 0$ there exists an r-basis set $B \subset H$ such that $m(H \setminus B) < \varepsilon$. Indeed, pick $x \in P$ and set $A_n := \{y \in P\colon n-2 < \rho(x,y) < n\}$. Since $mA_n < +\infty$ and $A_n \cap H$ is r-open, there are r-basis sets $B_n^* \subset A_n \cap H$ such that $m((A_n \cap H) \setminus B_n^*) < \varepsilon 2^{-n}$. Set $B := \bigcup_{n=1}^{\infty} B_n^*$. Then $m(H \setminus B) < \varepsilon$. Since B is locally ambivalent, it is an r-basis set (cf. [Kur], § 30.X, Th. 1).

Now, if t has the $\mathcal{Z}_m$-quasi-Lindelöf property and if there is $\omega \in \Omega$ (cf. Section 6.D) such that whenever $A \subset P$, $x \in P$ and

(b) $m^*(A \cap U(x,r)) \leq \omega(r)$ for each $r > 0$ implies $x \notin bA$,

then the r-basis condition (B) holds for the r-topology.

The idea of the proof is straightforward: Let $F \subset U$, F closed, U r-open. Set

$$V_o := \{x \in P : 1/2 < \rho(x,F)\} ,$$

$$V_n := \{x \in P : \quad 1/(n+2) < \rho(x,F) < 1/n\}$$

for $n \in \mathbb{N}$. Find $\varepsilon_j > 0$ such that $\sum_{j=k-1}^{\infty} \varepsilon_j < \omega(1/k)$ for each $k \in \mathbb{N}$. As stated above, there is an r-basis set $B_n \subset V_n \cap U$ such that $m((V_n \cap U) \setminus B_n) < \varepsilon_n$. Set

$$B := F \cup \bigcup_{n=0}^{\infty} B_n .$$

Pick $x \in F$ and $s > 0$. There is the smallest $n \in \mathbb{N}$ for which $1/n \leq s$. Then

$$m(U(x,r) \cap (U \setminus B)) \leq \sum_{k=n-1}^{\infty} m((V_n \cap U) \setminus B_n) \leq \sum_{k=n-1}^{\infty} \varepsilon_k <$$

$$< \omega(1/n) \leq \omega(s) .$$

Therefore, $x \notin b(U \setminus B)$ and it follows that B is d-open. Further, since $\bigcup_{n=0}^{\infty} B_n$ is locally ambivalent, it is ambivalent (cf. [Kur] , § 30.X, Th. 1). Thus B is an r-basis set, $F \subset B \subset U$.

Now, it may be worth while to note for the interested reader that there is a close connection between the r-topology and the class of all "$\mathcal{T}$-differentiable functions" (which has from this point of view a similar behaviour as $\mathcal{L}$) provided the space P is endowed also by a linear metric structure. It seems that these relations motivated the introduction of the r-topology by R. J. O'Malley (1977). We will be solely concerned in the rest of this section with the class $\mathcal{D}$ of all approximately differentiable (i.e. d -differentiable) functions which is apparently of the greatest interest.

THEOREM. Let r be the r-modification of the ordinary density topology on $\mathbb{R}^n$. Then:

(a) r has the Lusin-Menchoff property,

(b) r is the weak topology induced on $\mathbb{R}^n$ by $\mathcal{D}$,

(c) any r-continuous function can be uniformly approximated by functions from $\mathcal{D}$.

Proof. (a) By the above observation and by 6.34.B, the r-basis condition (B) holds for r. Now, the assertion follows from Theorem 7.12.c.

(b) Any approximately differentiable function is ambivalent (see Ex. 2.D.17 and Ex. 6.C.6 or Ex. 2.D.19). Now, we proceed analogously as in the proof of Theorem 7.13 (a) and (b) with only change in its mechanics concerning the construction of functions f_n: Given an open set G and a closed set $F \subset G$, what we seek is a differentiable function $0 \leq f \leq 1$ such that

$$f = 0 \text{ on } \mathbb{R}^n \setminus G, \quad f > 0 \text{ on } G \text{ and } f = 1 \text{ on } F.$$

To this end let $\{B_k\}$ be a locally finite covering of G by open balls with $\bar{B}_k \subset G$, and let $\{h_k\}$ be a sequence of differentiable functions on $\mathbb{R}^n$ such that $h_k = 0$ on $\mathbb{R}^n \setminus B_k$, $h_k > 0$ on B_k and $h_k < 2^{-k} \operatorname{dist}^2(B_k, \mathbb{R}^n \setminus G)$. Set

$$h = \sum_k h_k .$$

Then h is differentiable, $h = 0$ on $\mathbb{R}^n \setminus G$ and $h > 0$ on G. In similar fashion we can construct a differentiable function g such that $g = 0$ on F and $g > 0$ on $\mathbb{R}^n \setminus F$. Now, it suffices to put

$$f = \frac{h}{h + g} .$$

(c) We know that any approximately continuous function is in Baire class one. Hence, any r-continuous function belongs to $B_1(\mathbb{R}^n)$. Now, replacing "τ-Lipschitzian" by "approximately differentiable" we can follow the proof of Theorem 7.13.c. We can use also the results of Ex. 3.G.12 or Ex. 3.G.13 for $\mathcal{S} = \mathcal{D}$.

EXERCISES

7.B.1 Let (X,ρ) be a strong Baire, locally connected and normal T_1-topological space. If every open connected set is τ-connected, then each nonempty r-open set has nonempty interior.

Hint. Let A be an r-basis set. Then $\bar{A} \setminus A$ and A are disjoint G_δ-subsets of $\bar{A}$. Hence $\bar{A} \setminus A$ is not dense in $\bar{A}$. Conclude that there is a connected open set V such that $\emptyset \neq \bar{A} \cap V = A \cap V$. As V is τ-connected, we get $V \subset A$.

7.B.2 Let r be the r-modification of the ordinary density topology on $\mathbb{R}^n$. Then r and the Euclidean topology are S-related (cf. Ex. 4.A.1). In particular, $(\mathbb{R}^n, r)$ is a Blumberg space.

Hint. Use Proposition 6.25 and the preceding exercise.

7.B.3 (a) Show that the r-modification of the density topology determined by Morse basis B_3 of convex sets in $\mathbb{R}^n$ and by a Radon measure of support $\mathbb{R}^n$ has the Lusin-Menchoff property.

Hint. Show that the condition (b) of the Observation is satisfied (cf. 6.34.B).

(b) Show that the r-modification of the superdensity topology on $\mathbb{R}$ has the Lusin-Menchoff property.

Hint. Use again 6.34.A and the $\mathcal{Z}_\mu$-quasi-Lindelöf property of the superdensity topology.

7.B.4 Let τ denote the r-modification of the ordinary density topology on $\mathbb{R}^n$. Then any τ-Baire one funtion on $\mathbb{R}^n$ is continuous at the points of a residual set. In particular, $B_1(\mathbb{R}^n,\tau) \subsetneq B_2(\mathbb{R}^n)$.

Hint. Use Ex. 7.B.2 and Ex. 4.A.2e.

7.B.5 Let b be a fine base operator on a metric space (P,ρ) having the Lusin-Menchoff property. Show that the r-modification of b is "cometrizable" (w.r.t. ρ).

Hint. Use Lemma 7.11.

7.B.6 The r-σ-topology. Let b be a fine base operator on a topological space (X,ρ). By an r-σ-topology r^* we mean the collection of all countable unions of r-basis sets.

Show that r^* is finer than ρ provided (X,ρ) is a normal T_1-space.

(a) If b has the Lusin-Menchoff property on a normal topological T_1-space (X,ρ), then r^* is a normal σ-topology.

Hint. Let Φ be the collection of all nonnegative r^*-continuous and upper semicontinuous functions on X. Show that Φ is closed under uniform convergence, and that $r^* = \operatorname{Coz}(\Phi)$ in view of Lemma 7.11. Now, use Ex. 3.B.5h.

(b) If b has the Lusin-Menchoff property on a perfectly normal T_1-space (X,ρ), then the σ-topology r^* has also the Lusin-Menchoff property.

Hint. For a closed set F and an r^*-closed set F_1 which does not meet F find

$f_1 \in \Phi$ such that $F_1 = [f_1 = 0]$ and a ρ-continuous function f such that $F = [f = 0]$. Now consider the function $\frac{f}{f + f_1}$.

7.B.7 Let b,t and r have the same meaning as in the Observation. If t has the $\mathcal{Z}_m$-quasi-Lindelöf property and if the condition (b) holds, then the following assertions are true:

(a) Let F be a closed set, G an r-open set and assume that $F \cup G$ is t-open. Then there is an r-basis set B such that

$$F \subset B \subset F \cup G.$$

Hint. Follow the same line of proof as in the Observation.

(b) The following statements are equivalent:

(i) G is an r-open set,

(ii) G is a t-open set and $G = H \cup N$ where N is a null set and H is open in the r-δ-topology,

(iii) G is a t-open set and $G = H \cup D$ where H is open in the r-δ-topology and D is a t-discrete set.

Hint. Use (a) .

(c) If F is an r-closed set, then $\mathrm{der}_t F = \mathrm{der}_r F$.

Hint. Use (a) again.

7.B.8 Let τ be a fine topology on a locally connected, strong Baire metric space. Assume that every open connected set is also τ-connected and that there exists a nonempty τ-open set with empty interior.

(a) Show that τ is strictly finer than its r-modification r.

Hint. Use Ex. 7.B.1.

(b) If, in addition, τ has the Lusin-Menchoff property, then there is a τ-continuous Baire one function which is not a uniform limit of a sequence of τ-continuous CG-functions (cf.Ex. 3.G.8).

Hint. Cite (iv) of Theorem 3.11 and Ex. 7.B.1 to get a τ-continuous B_1-function which is not r-continuous. Now use the fact that any τ-continuous CG-function is r-continuous.

7.B.9 Let τ be a fine topology on a topological space X.

(a) If X is perfect, then the r-modification of τ is finer than the a.e.-modification of τ.

(b) Let X be a separable metric space. If any countable subset of X is τ-isolated, and if there is a closed nowhere dense set $F \subset X$ which is not τ-isolated, then the r-modification of τ is strictly finer than the a.e.-modification of τ.

Hint. Find a countable set $S \subset X \setminus F$ such that der S = F and show that $X \setminus S$ is r-open but it is not a.e.-open.

7.B.10 Let (P,ρ) be a σ-compact complete metric space and let τ be a metric density topology on P determined by the symmetric differentiation basis B_1 and by a measure m for which the density theorem holds. Show that P equipped with the r-modification of τ is a strong Baire space.

Hint. Use Ex. 6.B.1 to reduce the problem to the case bP = P. In light of Example 6.24 conclude that τ has the G_δ-insertion property. Moreover, τ has the Lusin-Menchoff property (see 6.34.B - (a)), and therefore r is "cometrizable" (w.r.t. ρ) - cf. Ex. 7.B.5. Thanks to Corollary 4.5 it is sufficient to prove that any r-perfect set A contains an r-G_δ-subset D which is r-dense in A. By Ex. 7.B.7c, A is τ-perfect, and therefore $\text{int}_\tau A$ is τ-dense in A. Using the G_δ-insertion property of τ, there is a G_δ-set D such that $\text{int}_\tau A \subset D \subset A$. Clearly, D has all desired properties.

7.B.11 Let r be the r-modification of the ordinary density topology on $\mathbb{R}^n$. If $G \subset \mathbb{R}^n$ is an r-open set, then for every $x \in G$, $\text{contg}_x(\text{int } G) = \{x \in \mathbb{R}^n: \|x\| = 1\}$. In particular, if n = 1, then every point of G is a bilateral accumulation point of int G.

Hint. Use Ex. 7.B.1.

7.B.12 Let τ be a fine topology on a normal T_1-topological space (X,ρ), and let r be the r-modification of τ. Show that the r-modification of r (w.r.t. ρ) equals r.

7.B.13 Let b be a fine base operator on a strong Baire metric space (P,ρ) which has the Lusin-Menchoff property. Without use of Lemma 7.11, prove the following assertion: Whenever $F \subset B \subset X$, F is closed and B is an r-basis set, then there is an r-basis set U such that $F \subset U \subset \bar{U} \subset B$.

Hint. Denote by $\mathcal{Y}$ the class of all open sets G satisfying the following condition:

Given a closed set $T \subset B$, there is an r-basis set $B(T,G)$ such that $T \cap G \subset B(T,G) \subset G$ and $\overline{B(T,G)} \subset B$.

Now, show that:

(a) $\mathcal{Y}$ is a hereditary class,

(b) $\bigcup\{G:\ G \in \mathcal{Y}\} = X$,

(c) $X \in \mathcal{Y}$.

For (b), put $H := \bigcup\{G:\ G \in \mathcal{Y}\}$ and suppose that $Z := X \setminus H \neq \emptyset$. Since Z is a Baire space, there are $z \in Z$ and $r > 0$ such that

either (α) $B \cap Z \cap U(z,2r) = \emptyset$

or (β) $Z \cap U(z,2r) \subset B$.

In both cases, by considering (a) and the existence of an open cover $\mathcal{Y}' \subset \mathcal{Y}$ of H which is locally finite in H, it can be shown that $U(z,r) \in \mathcal{Y}$ which is inconsistent with $z \notin H$.

REMARKS AND COMMENTS

The notion of the r-topology has been first examined in R.J.O'Malley's papers (1976b) and (1977). O'Malley's proof of the Lusin-Menchoff property of the r-topology is one-dimensional and makes use of relation to approximately differentiable functions. Our approach is quite abstract; we show under which conditions an r-topology is completely regular and under which it possesses the Lusin-Menchoff property. We specially find out again that O'Malley's r-topology on $\mathbb{R}^n$ has the Lusin-Menchoff property.

Related paper : Z. Grande (1978)

7.C Fine abstract density topologies

As before, let (X,Σ,μ) be a complete measure space with σ-finite measure μ.

Given an abstract density topology $\mathcal{T}$ on X in the sense of Section 6.E, sometimes another (original) topology ϑ is considered on X. We say that $\mathcal{T}$ is a ϑ*-fine* abstract density topology if $\mathcal{T}$ is finer than ϑ.

Starting with ϑ, let us recall that we can define an ideal topology i on X with respect to the σ-ideal $\mathcal{Z}$ of all μ-null sets. Since an abstract density topology has the $\mathcal{Z}$-quasi-Lindelöf property (see Remark 6.40.C), a set G is i-open if and only if $G = U \setminus Z$ where $U \in \vartheta$ and $\mu(Z) = 0$ (Proposition 1.8).

7.14 THEOREM. Let $\mathcal{T}$ be an abstract density topology on X, and let ϑ be a topology on X. Then the following assertions are equivalent:

(i) $\mathcal{T}$ is ϑ-fine,

(ii) there is a lower density S on Σ which determines $\mathcal{T}$ such that $U \subset S(U)$ for each ϑ-open set U,

(iii) $\mathcal{T}$ is finer than the ideal topology i.

Proof. The equivalence of (i) and (ii) follows immediately from definitions. The implication (iii)$\Rightarrow$(i) is trivial; in the proof of (i)$\Rightarrow$(iii) we use the fact that each set of μ-measure zero is $\mathcal{T}$-closed (cf. Theorem 6.39).

REMARK. Of course, any ϑ-fine abstract density topology $\mathcal{T}$ is Hausdorff provided ϑ is. On the other hand, $\mathcal{T}$ need not be regular even in the case that $\mathcal{T}$ is a lifting topology on the real line. Nevertheless, we can find conditions guaranteeing that $\mathcal{T}$ is completely regular (Theorem 7.15) or, moreover, that it has the Lusin-Menchoff property (Theorem 7.16). But first of all we start with the example we have just promised.

EXAMPLE. Let $\mathcal{T}$ be an e-fine (where e equals the Euclidean topology) lifting topology on $\mathbb{R}$. Its existence follows from Theorem 6.43 applied to the classical density on $\mathbb{R}$. Let b be a base operator determining $\mathcal{T}$ (see Remark 2 following Theorem 6.39). We define b_1 as follows: for $x \neq 0$, $x \in b_1A$ if and only if $x \in bA$; $0 \in b_1A$ if the set $\{n \in \mathbb{N}: 1/n \in bA\}$ is infinite. It is easy to check that b_1 is a base operator satisfying conditions of Theorem 6.39ii. Let $\mathcal{T}_1$ be an abstract density topology determined by b_1. Again, by Theorem 6.43 there is a lifting topology $\mathcal{T}_2$ on $\mathbb{R}$ finer than $\mathcal{T}_1$. Obviously, $\mathcal{T}_2$ is e-fine. Since the restrictions of b and b_1 coincide on $Y := \mathbb{R}\setminus\{0\}$, $\mathcal{T}_1/Y = \mathcal{T}/Y$. Clearly, $\mathcal{T}/Y$ is a lifting topology (cf. Theorem 6.42), and because $\mathcal{T}_2/Y$ is also a lifting topology finer than $\mathcal{T}/Y$, Theorem 6.42 yields $\mathcal{T}/Y = \mathcal{T}_2/Y$. The set $F := \{1/n: n \in \mathbb{N}\}$ being of measure zero is $\mathcal{T}_2$-closed. Let $U \subset Y$ be a $\mathcal{T}_2$-open set containing F. By the definition of b_1, $0 \notin b_1(\mathbb{R} \setminus U)$. Hence $0 \notin \mathrm{der}_{\tau_1}(\mathbb{R} \setminus U) \supset \mathrm{der}_{\tau_2}(\mathbb{R} \setminus U)$. Since 0 is not a τ_2-isolated point, it is $0 \in \mathrm{der}_{\tau_2} U$ and we see that 0 and F cannot be separated. Thus, $(\mathbb{R}, \tau_2)$ is not regular.

Now, we prove two important theorems guaranteeing that the fine abstract density topology is under certain conditions at least completely regular.

7.15 THEOREM. Let ϑ be a completely regular abstract density topology on X. If τ is a ϑ-fine lifting topology, then τ is completely regular.

Proof. Let $F \subset X$ be τ-closed, $x \in X \setminus F$. If L is the lifting determining τ, then there is $A \in \Sigma$ and a μ-null set N such that $F = L(A) \cup N$ (cf. Proposition 6.37). Since N is ϑ-closed (cf. Theorem 6.39), we can find a ϑ-continuous functions $F: X \to [0,1]$ such that $f(x) = 1$ and $f = 0$ on N. Further, $L(A)$ is both τ-open and τ-closed (cf. Ex. 6.F.12). Hence, the characteristic function $\chi_{L(CA)} = \chi_{CL(A)}$ is τ-continuous. Therefore, the function $f \cdot \chi_{L(CA)}$ is τ-continuous and separates F and x.

7.16 THEOREM. Let (X, ρ) be a metric space. Assume that

(a) τ_1, τ_2 are abstract density topologies on X,

(b) the base operator b_1 corresponding to τ_1 satisfies the condition (ω) of Section 6.C,

(c) τ_2 is finer than τ_1.

Then τ_2 has the complete Lusin-Menchoff property (w.r.t. ρ).

Proof. Let b_2 correspond to τ_2. Then $b_2 A \subset b_1 A$ for any $A \subset X$ (cf. Ex. 6.E.14) and we can use 6.34.D.

EXAMPLE. (Scheinberg's $\mathcal{U}$-topologies). Let $\mathcal{U}$ be an ultrafilter in the family of all Lebesgue measurable subsets of $\mathbb{R}$ containing the filter of all measurable density neighborhoods of the point 0. Let $\mathcal{U}_x := \{\mathcal{U}+x\}$, $x \in \mathbb{R}$, where $\mathcal{U}+x$ means the translation of $\mathcal{U}$ at x. The topology $\tau_{\mathcal{U}}$ determined by neighborhood basis $\{\mathcal{U}_x\}_x$ is a lifting topology finer than the density topology. Using 6.34.B and Theorem 7.16 we get immediately that $\tau_{\mathcal{U}}$ has the complete Lusin-Menchoff property (Cf. also Ex. 6.E.15 and 6.F.9.)

The question whether density topologies are normal can be mostly answered negatively by the following simple proposition.

7.17 PROPOSITION. Let μ be a complete regular Borel measure on a separable metric space (P,ρ) and suppose that there is a μ-null set of cardinality c. Let b be a base on P compatible with μ, $bP = P$. Then the b-topology is not normal.

Proof. We put $\mathcal{A} = \{bB: B \subset P$ is a Borel set$\}$. Given a b-open set G and a b-closed set F, $G \subset F$, we have

$$G \subset bG = bH_G \subset bF \subset F$$

(H_G is a measurable cover of G which can be chosen to be a Borel set) and Theorem 2.2 yields the result.

EXERCISES

7.C.1 Let τ be a lifting topology on $(\mathbb{R}^n, \mathcal{M}_n, \lambda_n)$ (e.g., let τ be the Scheinberg topology). Prove that not every τ-continuous function is in Baire class one. In particular, τ has not the G_δ-insertion property.

Hint. Use Ex. 6.F.5 and the fact that there is a Lebesgue measurable function which is not equivalent to any Baire one function.

7.C.2 Construct a fine abstract density topology τ on $(\mathbb{R}, \mathcal{M}, \lambda)$ for which there exists a τ-continuous function which is not Borel measurable.

Hint. Use the example of Ex. 6.C.12.

7.C.3 Construct a fine category density topology τ on $(\mathbb{R}, \mathcal{M}, \lambda)$ for which there exists a non-Baire τ-continuous function.

Hint. Let i be as in Example 1.5d on $\mathbb{R}$ and let $C \subset [0,1]$ be the Cantor set. Construct a real non-Baire function f on $\mathbb{R}$ which is continuous on $\mathbb{R} \setminus C$ and which has the property that for each $x \in \mathbb{R}$ there is a filter basis $\mathcal{B}_x$ consisting of second category sets with the Baire property such that $\mathcal{B}_x$ contains all i-open neighborhoods of x as well as all sets of the form $f^{-1}((f(x)-\varepsilon, f(x)+\varepsilon))$, $\varepsilon > 0$. Now use Ex. 6.E.15.

7.C.4 If τ is the category density topology on $\mathbb{R}$ from Ex. 6.E.25, then τ has the G_δ-insertion property.

7.C.5 (Kupka's adjustment theorem; J. Kupka (1983)). Let τ be an abstract density topology on $(X,\Sigma,\mathcal{N})$ and let ρ be a topology on X having the property that any nonempty ρ-open set is in $\Sigma\setminus\mathcal{N}$. Let i be the $\mathcal{N}$-ideal topology corresponding to ρ. If $N\in\mathcal{N}$ and if τ_N is the topology for which G is τ_N-open if and only if $N\cap G \subset \operatorname{int}_i G$ and $G\setminus N \subset \operatorname{int}_\tau G$, then τ_N is an abstract density topology on $(X,\Sigma,\mathcal{N})$.

Hint. Show that (iii) of Theorem 6.39 is satisfied.

7.C.6 (Kupka's proof of the existence of the strong lifting on second countable topological spaces; J. Kupka (1983)). Let (X,ρ) be a second countable topological space and let τ be an abstract density topology on $(X,\Sigma,\mathcal{N})$. Then there is a ρ-fine abstract density topology τ^* on $(X,\Sigma,\mathcal{N})$.

Hint. Let $\{G_n\}$ be a countable base for (X,ρ). Put $N := \bigcup_{n=1}^{\infty} (G_n \setminus \operatorname{int}_\tau G_n)$ and define τ^* as τ_N of Ex. 7.C.5. Prove that any G_n is τ^*-open.

REMARKS AND COMMENTS

In this section we take account of an abstract density topology which is finer than a given topology ϑ on some measure space. In the case of the lifting topology this situation is equivalent to the question, whether this topology arose from a strong lifting.

The example showing that a fine lifting topology on $\mathbb{R}$ needs to be regular was inspired by Theorem 2 of Chap. V,4 from the book of A.Ionescu-Tulcea and C.Ionescu-Tulcea (1969) where enough attention is paid to these problems.

The conditions ensuring the complete regularity of a fine abstract density topology and even its Lusin-Menchoff property, as well as the sufficient conditions for the nonnormality follow mostly from preceding results.

Scheinberg's $\mathcal{U}$-topology was introduced in his paper (1971). F.D.Tall in (1978) asked for the cometrizability of this topology. An answer was given by L. Zajíček in J. Lukeš paper (1977a), where it was proved that the $\mathcal{U}$-topology has even the Lusin-Menchoff property. In this section we present a simplified proof of this assertion using our abstract theorems.

7.D Fine boundary topology

In this section we are concerned with the investigation of a "boundary topology" that will be used further in our study of boundary behaviour of functions. Let F be a closed subset of a metric space (X,ρ) and let $\omega\in\Omega_0$. By Ω_0 we denote the set of all continuous increasing functions ω on $[0,+\infty)$ satisfying $\omega(0) = 0$ which are positive on $(0, +\infty)$. We define the *fine boundary topology* $\tau = \tau(F,\omega)$ on X by saying that a set $G \subset X$ is τ-open if $G \setminus F$ is ρ-open and

$$\lim_{n\to\infty} \frac{\rho(x_n,F)}{\omega(\rho(x_n,x))} = 0$$

whenever $x_n \xrightarrow{\rho} x$, $x_n \notin G$, $x \in G \cap F$. It can be proved without much effort that τ is actually a topology.

REMARKS. (a) Of course, $G \subset X \setminus F$ is τ-open if and only if G is ρ-open. Moreover, τ induces on F the discrete topology.

(b) If $X = \{(x,y) \in \mathbb{R}^2 : y \geq 0\}$, $F = \{(x,y) \in \mathbb{R}^2 : y = 0\}$ and $\omega(t) = t$, then a point $z \in F$ is a τ-interior point of a set G if and only if the following condition is satisfied: Given any (closed) Stolz angle S with vertex at z, there is $r > 0$ such that $S \cap U(z,r) \subset G$. (We postpone the definition of a Stolz angle to Section 9.A.) In this special case, a real function f on X is τ-continuous if and only if f is continuous on $X \setminus F$ and

$$f(z) = \operatorname*{S-lim}_{X\setminus F \ni x \to z} f(x)$$

for each $z \in F$. (Of course, by S-lim we mean a limit with respect to any Stolz angle.) We say that f on F is a *Stolz angle boundary function* of f on $X \setminus F$.

7.18 LEMMA. Let $x \in A$, $A \subset X$. Then $x \notin \bar{A}^{\tau}$ if and only if the following condition (T) holds:

(T) For each $\varepsilon > 0$ there is $r > 0$ such that
$\rho(y,F) < \varepsilon\,\omega(\rho(y,x))$ whenever $y \in U(x,r) \cap A$.

Proof. It is similar to the proof of the Heine theorem concerning the definition of the limit of a function.

7.19 THEOREM. The fine boundary topology $\tau = \tau(F,\omega)$ has the complete Lusin-Menchoff property (with respect to ρ).

Proof. Let $\omega_0 \in \Omega_0$ be a function for which

(i) $$\lim_{t\to 0_+} \frac{\omega_0(t)}{\omega(t)} = 0,$$

(ii) $\omega(\omega_0(t) + t) < 2\,\omega(t)$ whenever $t \in (0,1)$.

Such a function actually exists. Since a function

$$\varphi(t) := \omega^{-1}(2\omega(t)) - t$$

is positive and continuous on $(0,+\infty)$, $\lim_{t\to 0_+} \varphi(t) = 0$, we need only to put

$$\omega_o(t) = \min\left(\frac{1}{2}\inf_{\xi\geq t} \varphi(\xi),\ \omega^2(t)\right) \quad \text{for} \quad t > 0,$$

$$\omega_o(0) = 0.$$

To verify (iii) of Theorem 2.8, let $A,B\subset X$ be biseparated sets (i.e. $\bar{A}\cap B = \emptyset = A\cap \bar{B}^{\tau}$). Put

$$G := \{y\in X: \ \rho(y,B) < \omega_o(\rho(y,A))\}$$

and $U := X\setminus \bar{G}^{\tau}$. Of course, G is an open set and since $\bar{A}\cap B = \emptyset$, we have $B\subset G$. Obviously, the τ-open set U satisfies $U\cap G = \emptyset$. To conclude the proof it suffices to check that $A\subset U$. Let $x\in A$ and choose $\varepsilon>0$. Since $x\notin \bar{B}^{\tau}$, there is $r\in(0,1)$ such that

$$\rho(z,F) \leq \varepsilon\omega(\rho(z,x)) \quad \text{for any} \quad z\in U(x,r+\omega_o(r))\cap B$$

(Lemma 7.18) and

$$\omega_o(t) < \varepsilon\omega(t) \quad \text{whenever} \quad t\in(0,r).$$

Let $y\in U(x,r)\cap G$. Then

$$\rho(y,B) < \omega_o(\rho(y,A)) \leq \omega_o(\rho(y,x)),$$

and therefore there is $z\in B$ such that

$$\rho(y,z) < \omega_o(\rho(y,x)).$$

We have

$$\rho(x,z) \leq \rho(x,y) + \rho(y,z) < \rho(x,y) + \omega_o(\rho(y,x)) < r + \omega_o(r),$$

and using the inequality (ii),

$$\rho(z,F) \leqq \varepsilon\omega(\rho(z,x)) < \varepsilon\omega(\rho(x,y) + \omega_0(\rho(x,y))) < 2\varepsilon\omega(\rho(x,y)).$$

Finally,

$$\rho(y,F) \leqq \rho(y,z) + \rho(z,F) < \omega_0(\rho(x,y)) + 2\varepsilon\omega(\rho(x,y)) <$$

$$< 3\varepsilon\omega(\rho(x,y)).$$

We have shown that $x \notin \bar{G}^{\tau}$ (cf. Lemma 7.18), which establishes our claim.

EXERCISES

7.D.1 Let $\tau = \tau(F,\omega)$ be a fine boundary topology on X. Show that

$$\mathrm{der}_{\tau} A = \{x \in X: \limsup_{A \ni y \to x} \frac{\rho(y,F)}{\omega(\rho(y,x))} > 0\},$$

$$\mathrm{int}_{\tau} A = (\mathrm{int}_{\rho} A \setminus F) \cup \{x \in F \cap A: \text{for every } \varepsilon > 0 \text{ there is } r > 0 \text{ such that } U(x,r) \cap \{y \in X: \rho(y,F) \geqq \varepsilon\omega(\rho(y,x))\} \subset A\}.$$

7.D.2 Let X,F and ω be as in Remark (b) of this section. Prove in details the assertion of this Remark and show that no point of F has a countable τ-local base.

7.D.3 Let again X and F be as in Remark (b). Given $\theta \in (0, \pi/2)$ and $z = (x,0) \in F$, denote by S_z° the (open) standard Stolz angle $\mathrm{int}(S_x(\theta))$ (cf. Section 9.A). Define the topology η on X as follows: A set $G \subset X$ is η-open if for each $z \in G \cap F$ there is $r > 0$ such that $S_z^\circ \cap U(z,r) \subset G$. Verify that this gives a topology and show that neither η nor the F-modification η_F of η (w.r.t. the Euclidean topology) has the Lusin-Menchoff property.

Hint. Assume $\theta = \pi/4$. Let $\{(a_n,b_n)\}$ be the collection of all bounded component intervals of $\mathbb{R} \setminus C$ where C is the Cantor set. Put

$$V := \{(\tfrac{1}{2}(a_n+b_n); \tfrac{1}{2}(b_b-a_n)) \in \mathbb{R}^2 : n \in \mathbb{N}\}.$$

Then V is closed in both η and η_F, $V \cap \{(x;0): x \in C\} = \emptyset$ and $\bar{G}^{\eta} \cap C \neq \emptyset$ whenever G is an open set containing V.

7.D.4 Let $\tau = \tau(F,\omega)$ be a fine boundary topology on X, let f be a function on X and let $z \in X$.

(a) If $z \in X \setminus F$, then f is τ-upper semicontinuous at z if and only if f is upper semicontinuous thereon.

(b) If $z \in F$, then f is τ-upper semicontinuous at z if and only if $\limsup f(x_n) \leq f(z)$ whenever $x_n \to z$, $x_n \neq z$ and $\liminf \dfrac{\rho(x_n,F)}{\omega(\rho(x_n,z))} > 0$.

(c) If $z \in F$, then f is τ-continuous at z if and only if $\lim f(x_n) = f(z)$ whenever $x_n \to z$ and there is $\varepsilon > 0$ such that $\rho(x_n,F) > \varepsilon\, \omega(\rho(x_n,z))$.

7.D.5 Let (P,ρ) be a metric space, $\omega \in \Omega_0$. A set $A \subset P$ is said to be *ω-semitangent* to a set $B \subset P$ at a point $x \in P$ if there are $a_n \in A$, $a_n \to x$ such that $\lim \dfrac{\rho(a_n,B)}{\omega(\rho(a_n,x))} = 0.$

Suppose now that in P a collection $\mathcal{S}_x$ of subsets of P is assigned to each $x \in P$, let $\mathcal{S} = \{\mathcal{S}_x\}$. Call a subset G of P *$(\mathcal{S},\omega)$-open* if for each $x \in G$ and each $F \in \mathcal{S}_x$ the set $P \setminus G$ is not ω-semitangent to F at x. Verify that the collection of all $(\mathcal{S},\omega)$-open sets is a topology for P, called the *$(\mathcal{S},\omega)$-contingent topology*.

(a) Show that the concept of an $(\mathcal{S},\omega)$-contingent topology generalizes the notion of an M-contingent topology of Ex. 3.B.13.

(b) Show also that the notion of an $(\mathcal{S},\omega)$-contingent topology is a successful generalization of a boundary topology $\tau(F,\omega)$.

Hint. Given F and ω, define $\mathcal{S}_x$ as the system of all sets $\{a_1,a_2,\dots\}$ such that $a_n \notin F$ and $\liminf \rho(a_n,F)/\omega(\rho(a_n,x)) > 0$. Show that (F,ω) is an $(\mathcal{S},\omega_0)$-contingent topology for a suitable $\omega_0 \in \Omega_0$ (small enough).

(c) Prove that any $(\mathcal{S},\omega)$-contingent topology τ has the complete Lusin-Menchoff property.

Hint. Proceed as in the proof of Theorem 7.19: Given $\bar{A} \cap B = \emptyset = A \cap \bar{B}^{\tau}$, consider

$$G = \{y \in P: \ \rho(y,B) < \tilde{\omega}(\rho(y,A))\}$$

for a "sufficiently small" function $\tilde{\omega} \in \Omega_0$.

7.D.6 Let X and F be again as in Remark (b) of this section. Define the topology τ on X as follows: The local base of each point in $X\setminus F$ is formed by the usual Euclidean neighborhoods, while the local base at $z \in F$ if formed by the sets $\{z\} \cup A$, where A is an open disc in $X\setminus F$, tangent to F at z. With this topology X is called the *Niemytzki plane* or sometimes the *Moore plane*. (This space was defined - and attributed to V. Niemytzki - by P.S. Alexandroff and H.Hopf in (1935).)

(a) Show that $y_n \xrightarrow{\tau} x$ ($x, y_n \in X$, ρ is the Euclidean metric) if and only if $\lim \rho(y_n, F) / \rho^2(y_n, x) = +\infty$.

(b) Prove that τ is an $(\mathcal{S},\omega)$-contingent topology on X for some $\mathcal{S}$ and some $\omega \in \Omega_0$. In particular, τ has the Lusin-Menchoff property.

Hint. Proceed as in Ex. 7.D.5b.

REMARKS AND COMMENTS

In this Section we introduced the notions of a fine boundary topology and its generalization which we call an $(\mathcal{S},\omega)$-contingent topology. As these topologies have the complete Lusin-Menchoff property, they can be successfully employed in examinations of a boundary behaviour of functions as well as for constructions of some functions. Notice that L.E.Snyder in (1967b) defined the notion of a generalized Stolz cone which is closely related to our notion of a fine boundary topology.

8. EXTENSION THEOREMS

A. Constructions of bounded derivatives

B. Constructions of unbounded derivatives

C. Miscellaneous applications

D. Extensions from F_σ-sets

E. Derivatives of additive interval functions in $\mathbb{R}^n$

It is well known that any continuous function without derivative serves as an example of a nowhere monotone function. But there are also nowhere monotone everywhere differentiable functions. It seems that in the original Köpcke's papers at the end of the nineteenth century, the construction of such functions appeared for the first time. Later, a sequence of articles followed containing a study of derivatives which change often their sign. Nevertheless, constructions of functions with desirable properties have been rather complicated and some of the original papers contained errors.

A function f on an open interval I is said to be of the *Pompeiu type* if f has a bounded derivative, and if the sets on which f' is zero or does not vanish, respectively, are both dense in I. A *Köpcke function* is any function of Pompeiu type such that the sets on which f' is positive or negative, respectively, are dense in I. In the investigations of G. Petruska and M. Laczkovich (1974) certain "topological" methods for construction of functions of Pompeiu or Köpcke types are suggested and rather delicated results are derived.

In this chapter we shall show, among others, how to profit from the Lusin-Menchoff property of the density topology for the constructions of functions with "bad" behaviour.

8.A Constructions of bounded derivatives

Nowadays, Denjoy's theorem stating that a bounded approximately continuous function is a derivative belongs already to the mathematical matters of course (cf. Theorem 6.8.a). If we can construct bounded approximately continuous functions which exhibit certain specified behaviour, we get also some informations about certain behaviour of differentiable functions. Zahorski property of the density topology (Corollary 3.14) offers one of the possible approaches to our constructions, the use of the B_1-extension theorem is another possibility. We state it explicitely now.

8.1 THEOREM. (a) Let F be a density-closed subset of $\mathbb{R}$ (in particular, let F be of Lebesgue measure zero) of type G_δ. Then there is a bounded approximately continuous function $f \geq 0$ such that

$$F = [f = 0] .$$

Moreover, f can be chosen as upper semicontinuous (cf. also Prologue).

(b) Let $M \subset \mathbb{R}$ be a (Lebesgue) null set, and let f be a bounded Baire one function on $\mathbb{R}$. Then there is a bounded approximately continuous function f^* on $\mathbb{R}$ such that $f = f^*$ on M.

REMARKS. (1) Some concrete applications of these theorems are given in exercises.

(2) If M is a null set of type G_δ and $f \in B_1(M)$ is bounded, then f can be extended to a bounded Baire one function on $\mathbb{R}$ (cf. Corollary 3.8). Hence there is a bounded approximately continuous function f^* which extends f over $\mathbb{R}$.

Moreover, making use of Theorem 3.11.vi we get immediately the following result.

8.2 PROPOSITION. Let f be a bounded lower semicontinuous function on a null set

$M \subset \mathbb{R}$. Then there is a bounded lower semicontinuous derivative on $\mathbb{R}$ which extends f.

Using our methods of Section 3.E, we get now the next theorem.

8.3 THEOREM. Let f be a bounded function on a null set $M \subset \mathbb{R}$ being the restriction of a Baire one function on $\mathbb{R}$. Then there is an approximately continuous bounded derivative f^* on $\mathbb{R}$ which extends f such that

(a) f^* is continuous at all points of $\mathbb{R} \setminus \bar{M}$,

(b) f^* is upper (lower) semicontinuous at those points of M at which f is.

EXERCISES

8.A.1 Prove that there exists a Köpcke function on $(0,1)$.

Hint. Let $C = \{c_n : n \in \mathbb{N}\}$ and $D = \{d_n : n \in \mathbb{N}\}$ be disjoint dense subsets of $(0,1)$. Put $f(c_n) = 1/n$, $f(d_n) = -1/n$, $f = 0$ on $(0,1) \setminus (C \cup D)$ and use Theorem 8.1b.

8.A.2 Let F be a nowhere dense closed subset of $\mathbb{R}$ such that $\lambda(F \cap (c,d)) > 0$ whenever $F \cap (c,d) \neq \emptyset$.

(a) (Cf. [Bru], Theorem 6.8.) There exists an increasing differentiable function f on $\mathbb{R}$ which is constant on each interval contiguous to F but not on any open interval intersecting F, and f' is bounded, approximately continuous and upper semicontinuous.

(b) There is a differentiable function g which is again constant precisely on each interval contiguous to F such that g is monotone on no open interval intersecting F and g' is a bounded approximately continuous function.

Hint. Let d be the density topology on $\mathbb{R}$. Prove that there exist countable disjoint subsets C and D of F such that $(C \cup D) \cap \overline{\mathbb{R} \setminus F}^d = \emptyset$. Concerning (a), use Remark 3.12 ($F_\tau = \overline{\mathbb{R} \setminus F}^d$ and $A = \mathbb{R} \setminus (C \cup D)$), for (b) use Corollary 3.31 (put $M = \overline{\mathbb{R} \setminus F}^d \cup C \cup D$, $C = \{c_n\}$, $D = \{d_n\}$, and $G(c_n) = 1/n$, $G(d_n) = -1/n$, $G = 0$ on $\overline{\mathbb{R} \setminus F}^d$).

8.A.3 (V. Kelar (1980), cf. [Bru], Theorem II.9.1 and A. Bruckner (1983)). Let A and B be disjoint countable subsets of $\mathbb{R}$. Z. Zalcwasser (1927) showed the existence of a differentiable function g on $\mathbb{R}$, having a bounded derivative, such that A is exactly the set of strict local maxima of g and B is the set of strict local minima of g. Direct constructions of functions having the desired properties are quite complicated.

Construct an example of Zalcwasser's function using Theorem 8.1a and an observation that there is a Lipschitz function ψ on $\mathbb{R}$ for which A (B, resp.) is the set of strict local maxima (minima, resp.) of ψ. (The reader should find it with some effort to construct ψ as a limit of an inductively defined sequence of piecewise linear functions, cf. V. Kelar (1980).)

Hint. Let C be the set of all nondifferentiability points of ψ. Citing Theorem 8.1a, find an increasing function F on $\mathbb{R}$, having a bounded derivative, such that $F'(y) = 0$ for each $y \in \psi(C)$. Next look at the function $F * \psi$.

REMARKS AND COMMENTS

An interesting application of the Lusin-Menchoff property in the theory of real functions consists in construction of derivatives with peculiar properties. This observation is far from being new. Already C. Goffman (1975) used the complete regularity of the density topology for the construction of Köpcke functions. Another idea of constructions of these functions is due to M. Laczkovich and G. Petruska (1974) and it is based on an extension of Baire one functions from sets of measure zero resulting in bounded approximately continuous functions.

As we have already mentioned, constructions of functions of Pompeiu or Köpcke type were produced by many authors. We mention only the deeply founded study of A. Denjoy (1915), the papers of Z. Zalcwasser (1927), D. Pompeiu (1906), S. Marcus (1963), and Y. Katznelson and K. Stromberg (1974). J. Blažek, E. Borák and J. Malý (1977) presented interesting elementary constructions which did not use the Lusin-Menchoff property. Here it was also indicated how to construct Köpcke functions from Pompeiu functions using some simple facts concerning monotone differentiable transformations on $\mathbb{R}$. Another method how to proceed from Pompeiu functions to Köpcke functions is described in the mentioned paper (1915) by A. Denjoy. C.E. Weil (1976) obtained Köpcke functions from Pompeiu ones by the Baire category method.

The history of differentiable nowhere monotone functions is outlined in A. Bruckner and J. Leonard (1966). See also [Bru], Sections II.6 and XIV.2 (including Laczkovich and Petruska's result) or A. Bruckner (1983).

In this section we improve the above mentioned results taking, in addition, into account possibilities of preservation of continuity and semicontinuity.

Related papers: P. Vetro (1983), Z. Grande and M. Topolewska (1982).

8.B Constructions of unbounded derivatives

It is far from being true that any unbounded approximately continuous function is a derivative. Therefore for constructions of unbounded (or even infinite) derivatives we cannot use directly topological methods issued from the Lusin-Menchoff property. Dealing as we are with general derivatives rather than bounded ones, it is important to know that some constructions of such derivatives have been studied for a number of years. Let us mention only the results of Z. Zahorski (1941) and G. Choquet (1947) (no matter whatever G_δ- set of measure zero is given, there is a derivative having the value $+\infty$ exactly on this set) or that given in [Saks] (given a closed set $F \subset [a,b]$ of measure zero, there is a derivative belonging to $\mathcal{L}^1([a,b])$ continuous outside F and equal $+\infty$ on F). In (1974), G. Petruska and M. Laczkovich proved the theorem that every Baire one function on a G_δ - set of measure zero can be extended to an approximately continuous function or to a derivative. The question whether there is a simultaneous extension resulting in an approximately continuous derivative was settled positively later on using uniform methods by D. Preiss and J. Vilímovský (1980). We propose in the sequel another approach based in fact on the use of topological methods, namely on the B_1 - extension theorem with "control functions" (Theorem 3.29). As a by-product we obtain some interesting theorems most of which, in fact, are not new but the approach to them is quite different of that known in literature.

Suppose that $f \in \mathcal{L}^1_{loc}(a,b)$ and that f is finite at $x \in (a,b)$. Then x is called a *Lebesgue point* for f if $\lim_{h\to o} h^{-1} \int_x^{x+h} |f(t)-f(x)|\,dt = 0$. If x is a Lebesgue point for f, then the derivative of the indefinite integral of f at x equals $f(x)$. It is also known that for f locally bounded at x, a point x is a Lebesgue point for f if and only if f is approximately continuous at x. Moreover, it is almost obvious that x is a Lebesgue point for f if for any $c > f(x)$ and for any $d < f(x)$ we have

$$\lim_{h\to o} h^{-1} \int_x^{x+h} (f(t)-c)^+ dt = \lim_{h\to o} h^{-1} \int_x^{x+h} (f(t)-d)^- dt = 0.$$

A function $f \in \mathcal{L}^1_{loc}(a,b)$ is said to be *upper Lebesgue point* at a point x if either $f(x) = +\infty$ or if there is $c \in R$ such that $\lim_{h\to o} h^{-1} \int_x^{x+h} (f(t)-c)^+ dt = 0$. Analogously define that f is *lower Lebesgue point* at x.

If $f \leqq g$ are locally integrable functions on (a,b), $g(x)$ finite and g is upper Lebesgue bounded at x, then, of course, f is upper Lebesgue bounded at x as well.

<u>REMARKS.</u> (1) If f is lower semicontinuous at x or, locally lower bounded at x, then f is lower Lebesgue bounded at x.

(2) It is far from being true that x is a Lebesgue point for f provided f is simultaneously upper and lower Lebesgue bounded at x. But, adding the condition

of approximate continuity, we have the following remarkable result.

8.8 LEMMA. Let $f \in \mathcal{L}^1_{loc}(a,b)$ be an approximately continuous function at $x \in (a,b)$. If f is simultaneously upper and lower Lebesgue bounded at x, then $f(x)$ equals the derivative of the indefinite integral of f at x. If, in addition, $f(x)$ is finite, then x is even a Lebesgue point for f.

Proof. Assume f to be finite at x. Let $c > f(x) > d$ be chosen such that

$$\lim_{h\to 0} \frac{1}{h} \int_x^{x+h} ((f(t)-c)^+ + (f(t)-d)^-)\, dt = 0.$$

Putting

$$f^*(t) := \max(\min(f(t),\ c),\ d),$$

we can see that f is locally bounded and approximately continuous at x. Hence x is a Lebesgue point for f^*. Since

$$|f(t)-f(x)| = |f^*(t)-f(x)| + (f(t)-c)^+ + (f(t)-d)^-,$$

it easily follows that x is a Lebesgue point for f, too, and we are through. Assume now $f(x) = +\infty$. Let $d < f(x)$ satisfy

$$\lim_{h\to 0} \frac{1}{h} \int_x^{x+h} (f(t)-d)^- dt = 0.$$

Fix $K > d$ and put $f_K(t) := \min(K, f(t))$. Since f_K is approximately continuous at x and $(f(t)-d)^- = (f_K(t)-d)^-$, we have

$$\liminf_{h\to 0} \frac{1}{h} \int_x^{x+h} f(t)\, dt \geq \liminf_{h\to 0} \frac{1}{h} \int_x^{x+h} f_k(t)\, dt = K.$$

It follows that $\liminf_{h\to 0} \frac{1}{h} \int_x^{x+h} f = +\infty$ which, without much effort, establishes the assertion.

REMARK. Let $N \subset (a,b)$ be of measure zero, and let f be a function on $X := (a,b) \setminus N$. We define in a natural way the notion of approximate continuity of f at $x \in X$. It is also clear what we shall mean saying that f is either upper or lower Lebesgue bounded at x. It is easy to check that Lemma 8.8 remains valid if we replace (a,b) by X.

8.9 LEMMA. Let X be as above, let $M \subset X$ be a G_δ-set of measure zero and assume that $f \in B_1(M)$. Then there are functions $f^* \in B_1(X)$, $g,h \in \mathcal{L}^1_{loc}(X)$ with the following properties:

(a) f^*,g,h are finite on $X \setminus M$,

(b) h is lower semicontinuous and upper Lebesgue bounded on X, g is upper semicontinuous and lower Lebesgue bounded on X,

(c) $f^* = f$ on M, $g \leqq f^* \leqq h$ on X.

Moreover, $h(x) < +\infty$ $(g(x) > -\infty)$ whenever $f(x) < +\infty$ $(f(x) > -\infty)$.

Proof. We know that there is a function $\tilde{f} \in B_1(X)$ which extends f and is finite on $X \setminus M$ (see the hint of Ex. 3.B.7f, cf. [Kur]). Putting $A := X \setminus M$, for every $n \in \mathbb{N}$ there are closed sets $F_{n,k}$ in X such that

$$\{x \in X: \tilde{f}(x) < n\} \cup A = \bigcup_{k=1}^{\infty} F_{n,k}$$

and $F_{1,1} \neq \emptyset$. By 6.34.F, for each $m \in \mathbb{N}$ there is a closed set H_m in X such that

$$\bigcup_{n,k \leqq m} F_{n,k} \subset H_m \subset \{x \in X: \tilde{f}(x) < m\} \cup A$$

and such that for each $x \in \bigcup_{n,k \leqq m} F_{n,k}$ and for each $d \neq 0$ we have

$$\lambda((x,x+d) \cap [(\{t \in X: \tilde{f}(t) < m\} \cup A) \setminus H_m]) < 2^{-m}d^2,$$

or equivalently,

$$(*) \qquad \lambda((x,x+d) \cap (X \setminus H_m)) < 2^{-m}d^2.$$

Put

$$h(x) = \inf\{m: x \in H_m\}.$$

Obviously, $1 \leqq h \leqq +\infty$, h is lower semicontinuous on X (H_m are closed in X !) and $h(x) = +\infty$ if and only if $\tilde{f}(x) = +\infty$.
Further $f(x) < h(x) < +\infty$ whenever $f(x) < +\infty$.
Assume that $a,b \in \mathbb{R}$ and use $(*)$ for $d = b-a$.

Since

$$\int_a^b h \leq \lambda(H_1) + \sum_{m=2}^{\infty} m\,\lambda(H_m \setminus H_{m-1}) \leqq (b-a) + 2\sum_{m=2}^{\infty} m2^{1-m}(b-a)^2,$$

we have $h \in \mathcal{L}^1(X)$. To conclude the proof it suffices to check that h is upper Lebesgue bounded on X. Pick $x \in X$ with $h(x) < +\infty$. There is m_o such that $x \in \bigcup_{n,k \leq m} F_{n,k}$ for every $m \geqq m_o$. In light of (*), we obtain

$$\lambda((x,x+d) \cap (X \setminus H_m)) < 2^{-m}d^2$$

for $m \geqq m_o$ and $d \neq 0$. Since $\{t \in X\colon h(t) = m\} \subset X \setminus H_{m-1}$, we get

$$\lambda(Y_m) < 2^{1-m}\, d^2 \qquad (m \geqq m_o,\ d \neq 0)$$

where $Y_m = (x,x+d) \cap \{t \in X\colon h(t) = m\}$. Finally,

$$\lim_{d\to 0} \frac{1}{d} \int_x^{x+d} (h(t)-m_o)^+ dt = \lim_{d\to 0} \frac{1}{d} \sum_{m=m_o+1}^{\infty} \int_Y (m-m_o)\, dt \leq$$

$$\leqq \lim_{d\to 0} \frac{1}{d} \sum_{m=m_o+1}^{\infty} (m-m_o)\, 2^{1-m} d^2 = 0.$$

In a similar way we define a function g with the required properties. Now, put $f^* = \max(g, \min(\tilde{f},h))$.

Now, we are able to prove the main theorem.

8.10 THEOREM. Let f be a Baire one function (finite or infinite) on a G_δ-set $M \subset (a,b)$ of measure zero. Then f can be extended over the whole (a,b) resulting in an approximately continuous derivative $\tilde{f}$. Moreover, the extension $\tilde{f}$ can be found continuous on $(a,b) \setminus \bar{M}$, finite on $(a,b) \setminus M$ and upper (lower) semicontinuous at those points of M at which f is.

Proof. Put $X = (a,b)$ and find functions f^*, g, h as in Lemma 8.9.
Let τ denote the density topology on (a,b).
Since M is τ-closed and f is τ-continuous on M, by the B_1-extension theorem there is a τ-continuous function $\tilde{f}$ extending f over (a,b) such that $g \leqq \tilde{f} \leqq h$. Moreover, $\tilde{f}$ can be chosen to be continuous on $X \setminus \bar{M}$ and upper (lower) semicontinuous at those points at which f is. Obviously, $\tilde{f}$ is finite on $(a,b) \setminus M$, $\tilde{f} \in \mathcal{L}^1_{loc}(a,b)$ and $\tilde{f}$ is upper and lower Lebesgue bounded on (a,b). In view of Lemma 8.8 we see that $\tilde{f}$ is a derivative.

8.11 COROLLARY. Let $F \subset (a,b)$ be a closed set of measure zero. Then every function $f \in B_1(F)$ can be extended to an approximately continuous derivative which is continuous on $(a,b) \setminus F$ and at those points of F at which f is.

REMARK. Similar results are true for extension of semicontinuous functions to semicontinuous derivatives (cf. Ex. 8.B.1 and 8.B.2).

8.12 COROLLARY. Let g be a (finite or infinite) function on (a,b) having the Baire property. Then there is a continuous function G on (a,b) and a residual set $M \subset (a,b)$ such that $G' = g$ on M. Moreover, G can be chosen in such a way that G' exists on the whole (a,b), G' is approximately continuous on (a,b) and G' is continuous at all points of M.

Proof. Since g has the Baire property, there is a residual set B of the type G_δ such that $g \wedge B$ is continuous (cf. the text following Theorem 6.18). Let $H \subset (a,b)$ be a residual set of measure zero and put $M = B \cap H$. By Theorem 8.10 we can extend the function $f = g \wedge M$ over the whole (a,b) resulting in an approximately continuous derivative $\tilde{f}$ being continuous at any point of M. Taking G to be an indefinite integral of $\tilde{f}$, G has all of the desired properties.

REMARK. It is interesting to notice that Corollary 8.12 has its counterpart in terms of measurable functions. The Lusin characterization of measurable functions states that an almost everywhere finite function f on (a,b) is measurable if and only if there is a function F such that $F' = f$ a.e. on (a,b) (cf. [Saks]).

8.13 THEOREM. Let $M \subset (a,b)$ be a set of measure zero and $f \in B_1(M)$. Then there is a continuous function F having a finite derivative F' almost everywhere on (a,b) such that $F' = f$ on M.

Proof. Let $G \supset M$ be a measurable cover of M of type G . Put $X = (a,b) \setminus (G \setminus M)$. Then M is a G_δ - subset of X. Hence there are functions f^*, g, h as in Lemma 8.9. Since the density topology τ has the complete Lusin-Menchoff property we can use the B_1-extension theorem to the space X with the restricted topology $\tau \wedge X$. In such a way obtain an approximately continuous function $\tilde{f}$ on X which extends f such that $g \leqq \tilde{f} \leqq h$ on X. Obviously, $\tilde{f} \in \mathcal{L}^1_{loc}(a,b)$ is upper and lower Lebesgue bounded at each point of X (cf. Remark following Lemma 8.8). As it follows from Lemma 8.8, the function $F(x) := \int_c^x \tilde{f}$, where $c \in (a,b)$, has the desired properties.

8.14 COROLLARY (Eilenberg-Saks, (1935)). Let $\{a_n\}$ be a sequence of different reals and $\{\lambda_n\}$ a sequence of extended reals. Then there exists a continuous function

F on $\mathbb{R}$ such that $F'(a_n) = \lambda_n$.

Proof. We put $M := \{a_1, a_2, \dots\} \subset \mathbb{R}$. Function f defined by $f(a_i) = \lambda_i$ is obviously a Baire one function on M.

EXERCISES

8.B.1 Let M be a subset of $\mathbb{R}$ of measure zero. If f is lower semicontinuous and lower bounded on M, then f can be extended to a lower semicontinuous (and approximately continuous) derivative F on all of $\mathbb{R}$.

Hint. Define

$$f^*(x) := \begin{cases} \liminf\limits_{\delta \to 0_+} \{f(t) : \ t \in U(x,\delta) \cap M\} & \text{if} \quad x \in \bar{M}, \\ c := \inf\limits_{t \in M} f(t) & \text{for} \ x \in \mathbb{R} \setminus \bar{M} \end{cases}$$

and reduce the problem to the case when M is a G_δ-set. There is a lower semicontinuous and upper Lebesgue bounded $\tilde{h} \in \mathcal{L}^1_{loc}(\mathbb{R})$ such that $\tilde{h} \geq f$ on M, $\tilde{h} > c-1$ on $\mathbb{R}$ (the construction is almost the same as that given in Lemma 8.9). Put $h = \min(\tilde{h}, f^*)$, $g(x) = c-1$ for $x \in \mathbb{R} \setminus M$, $g = f$ on M. The function g is approximately upper semicontinuous and $g \leq h$. Hence, citing (v) of Theorem 3.11, there is a lower semicontinuous and approximately continuous function F on $\mathbb{R}$ such that $g \leq F \leq h$. The assertion now follows from Lemma 8.8.

8.B.2 Show that the function F of the precedent exercise can be chosen to be real and continuous on $\mathbb{R} \setminus \bar{M}$ and at all points of M where f is continuous.

Hint. Construct a function as follows: Let τ be the $\bar{M} \setminus C$-modification of the density topology (where $C \subset M$ is the set of all continuity points of f). Finish the proof thank to an observation that the function g of Ex. 8.B.1 is τ-upper semicontinuous.

8.B.3 Show that the condition of lower boundedness of f in Ex. 8.B.1 cannot be replaced by the finiteness of f.

Hint. It is not too hard to find an example illustrating that it can happen $\liminf f = -\infty$ on a set of positive measure. (Indeed, let $D \subset \mathbb{R}$ be a closed nowhere dense set of positive measure. If $\{(a_n, b_n)\}$ is the family of all bounded component intervals of $\mathbb{R} \setminus D$, put $s_n = \frac{1}{2}(a_n + b_n)$, $M = \{s_1, s_2, \dots\}$, $f(s_n) = -n$.) On the other hand, the set $\{x: \varphi'(x) = -\infty\}$ is always of measure zero.

REMARKS AND COMMENTS

In this section we extend (in general unbounded) functions from sets of measure zero resulting in (unbounded) derivatives using the results of Section 3.E, namely the extension theorem with control. Moreover, the produced derivatives are approximately continuous. Theorem 8.10 is due to D. Preiss and J. Vilímovský (1980). We follow, in fact, their method (Lemma 8.9), setting it free of the language of uniform spaces. (In addition, we investigate the possibilities of preservation of continuity and semicontinuity.) The history of this problem is described in the introduction to this section.

One of Lusin's theorems says that a function f is a derivative a.e. of a continuous function if and only if f is measurable and finite a.e. The categorial counterpart of Lusin's theorem was stated as an open problem in N.N. Lusin (1951) and was settled by E.M. Landis in notes contained therein. Our methods also yield this assertion (Corollary 8.12) relatively easily.

It is interesting to remark that the complete Lusin-Menchoff property of the density topology produces also a nontrivial generalization of Eilenberg-Saks' theorem (Theorem 8.13). Notice also that the paper (1974) by G. Petruska and M. Laczkovich contains results leading to another generalization of this Eilenberg-Saks theorem (cf. [Bru] , Section XIV. 2).

8.C Miscellaneous applications

C.1. Extension from closed τ-isolated sets. In (1979), Z. Grande has shown that any finite Baire one function defined on a closed subset F of $\mathbb{R}$ of measure zero can be extended to a finite approximately continuous function over $\mathbb{R}$ which is, in addition, continuous almost everywhere. Actually, our extension theorem 3.30 yields a stronger result - the extended function can be constructed being continuous on $\mathbb{R} \setminus F$ (in fact, Grande's proof works in the same way). Moreover, Theorem 3.30 guarantees preservation of semicontinuity and therefore gives further generalization of Grande's theorem.

It is of interest to apply the results of preceding sections to another direction of generalization of the mentioned result. Our point of view is to consider extended functions not only to be approximately continuous but to be continuous in a stronger sense (e.g. superdensity continuous or "approximately continuous" with respect to another measure). The existence of such extensions is a consequence of the following general Proposition 8.15 the gist of which lies in use of the a-modification of a base operator b with the Lusin-Menchoff property. Notice that in Ex. 8.C.5 it is to be shown that a slightly deeper result is valid: Using more complicated methods of the r-modification of b we show that the ρ-closedness of F can be replaced by the ambivalency of F.

8.15 PROPOSITION. Let $b_1,\dots,b_n$ be fine base operators on a metric space (P,ρ) each of them having the Lusin-Menchoff property and satisfying $b_i\{x\} = \emptyset$ for each $i = 1,\dots,n$ and each $x \in P$. Denote by τ_i the b_i-topology. Let $F \subset P$ be ρ-closed and τ_i-isolated ($i=1,\dots,n$). If t and $-s$ are upper semicontinuous functions on P and if f is a Baire one function on F such that $t \lneqq f \lneqq s$ on F, then there is a Baire one function h on P which is τ_i-continuous for every $i=1,\dots,n$ and

(a) $h = f$ on F,

(b) h is continuous on $P \setminus F$,

(c) h is lower semicontinuous at those points of F at which f is, and upper semicontinuous at those points of F at which f is,

(d) $t \lneqq h \lneqq s$ on P.

Proof. Let $b = \sup(b_1,\dots,b_n)$. In light of Theorem 3.20, b has the Lusin-Menchoff property. Let ϑ be the a-topology corresponding to the a-modification of b (cf. Theorem 7.4.b). According to Theorem 7.5, ϑ has the Lusin-Menchoff property. Obviously, F is ϑ-isolated so that any function on F is ϑ-continuous. The assertion now follows from Theorem 3.30 if we observe that ϑ is coarser than each τ_i.

Now, let us transfer some concepts of Section 6.D to Euclidean spaces $\mathbb{R}^n$. Let σ be a positive increasing function on $(0, +\infty)$, $\lim_{x\to 0+} \sigma(x) = 0$. It is not difficult to check that the family of all sets $G \subset \mathbb{R}^n$ with the property

$$\lim_{h\to 0_+} \frac{\lambda^*(U(x,h) \setminus G)}{\sigma(h)} = 0$$

for every $x \in G$ forms actually a topology on $\mathbb{R}^n$ which is said to be the σ-*superdensity topology*. The functions which are continuous in this topology will be called σ-*approximately continuous*. Let us underline the fact that even in the simple case $n = 1$ and $\sigma(h) = h^2$, the σ-superdensity topology lacks the Lusin-Menchoff property. That is why we prefer to consider the base operator

$$b_\sigma A := \left\{x \in \mathbb{R}^n : \limsup_{h\to 0_+} \frac{\lambda^*(U(x,h) \cap A)}{\sigma(h)} > 0\right\}.$$

Using 6.34.B, we can see that the base operator b_σ has the (complete) Lusin-Menchoff property and, moreover, we can check that the b_σ-topology coincides with the σ-superdensity topology. Now, Proposition 8.15 admits several corollaries two of which follow.

8.16 COROLLARY. Let $F \subset \mathbb{R}^n$ be a closed set of measure zero, and let $f \in B_1(F)$. Then there is a σ-approximately continuous function f^* on $\mathbb{R}^n$ which is continuous on $\mathbb{R}^n \setminus F$ such that $f^* = f$ on F.

REMARK. In Ex. 8.D.2 we shall further generalize the above result as well as the next corollary (cf. also Ex. 8.C.11).

8.17 COROLLARY. Let $\mu_1, \ldots, \mu_k$ be Radon measures on $\mathbb{R}^n$ and let $F \subset \mathbb{R}^n$ be a closed set of μ_i-measure zero (i=1,...,k). Then any Baire one function on F can be extended over $\mathbb{R}^n$ resulting in a function which is continuous on $\mathbb{R}^n \setminus F$ and μ_i-approximately continuous (w.r.t. the symmetric basis) for each $i = 1,\ldots,n$.

REMARK. A few remarks should be made about these corollaries. A particular case of Corollary 8.16 (or, in fact, of Grande's proof) is the following assertion which is (perhaps) somewhat surprising: Let F be a "one-dimensional line" in $\mathbb{R}^2$. Then any Baire one function on F can be extended to an approximately continuous function over $\mathbb{R}^2$ which is continuous outside F. We have seen that the extended function can be constructed also to be h^4-superdensity continuous. Furthermore, in Theorem 9.21 and in Ex. 9.D.5 we derive a more general result: If the line in consideration is not parallel with any coordinate axis, we can extend our function resulting in a function which is also "strongly approximately continuous". Finally, note that we will apply the above results and their generalizations to the problem of extension to derivatives.

C.2. Laczkovich S_2-property. In (1979a), R.J.O'Malley proved that the family of all approximately continuous functions which are continuous almost everywhere has the Laczkovich S_2-property (see M. Laczkovich (1975), cf. also the proof of Proposition 8.18). Since this class of functions is nothing else than the collection of all a.e.-continuous functions, the assertion follows immediately from the following proposition.

8.18 PROPOSITION. If a fine topology τ has the Lusin-Menchoff property with respect to a metric space (X,ρ), then the system $\mathcal{C}$ of all τ-continuous real functions on X has the Laczkovich S_2-property.

Proof. Since τ is completely regular (Corollary 3.13), it follows from Theorem

1.1.2 of M. Laczkovich (1975) that the property S_2 is equivalent to the following assertion:

"For every pair of disjoint τ-closed sets A,B of type G_δ there is a τ-continuous function $h: X \longrightarrow [0,1]$ such that $h = 0$ on A and $h = 1$ on B."

Using Corollary 3.14 we can easily see that $\mathcal{T}$ has the property S_2, even there is a τ-continuous function h in such a way that $A = h^{-1}(0)$, $B = h^{-1}(1)$.

C.3. On Ward's result. In (1933), A.J. Ward proved the following theorem on "approximately angle points" of approximately continuous functions:

THEOREM. Let $E \subset (0,1)$ be a set of measure zero. Then there is an approximately continuous function f satisfying $\underline{f}_{+ap}(x) > \overline{f}_{-ap}(x)$ for each $x \in E$. (Here, of course, $\overline{f}_{+ap}$ denotes the right-hand upper approximate derivative of f.)

It is worth mentioning that, in the proof of his theorem, A.J. Ward uses the technique of the "Lusin-Menchoff insertion of sets" combined with the usual "Urysohn procedure" paying heed to special further properties of inserted sets. In fact, this method was used later many times in subsequent papers. Our aim is to divide the proofs of this type into the topological part (using topological theorems equivalent to the Lusin-Menchoff property or its consequences) and the additional part not requiring the Lusin-Menchoff property. Our method is now illustrated by the following proposition a special case of which is the Ward's theorem.

8.19 PROPOSITION. Let $E \subset (0,1)$ be a set of measure zero. Then there is an approximately continuous function f for which $\underline{f}_{+ap} = +\infty$ and $\overline{f}_{-ap} = -\infty$ on E. Moreover, f can be chosen to be upper semicontinuous.

Proof. Let $F \supset E$ be a G_δ - set of measure zero. The Zahorski property of the density topology (Corollary 3.14) affirms the existence of an approximately continuous function $g \geqq 0$ such that $F = [g = 0]$. For $\alpha > 0$ put

$$\varphi(\alpha) := \lambda([g \leqq \alpha]).$$

Obviously, φ is increasing on (0,1) and $\lim_{\alpha \to 0_+} \varphi(\alpha) = 0$. Let ψ be an increasing homeomorphism of (0,1) onto itself for which

$$\lim_{\alpha \to 0_+} \frac{\varphi(\psi(\alpha))}{\alpha} = 0.$$

Put $f := \psi^{-1} * g$. Clearly, f is a nonnegative approximately continuous function.

Since

$$\lambda([f \leq \alpha]) = \lambda([g \leq \psi(\alpha)]) = \varphi(\psi(\alpha)),$$

we have

$$\lim_{\alpha \to 0_+} \alpha^{-1} \lambda([f \leq \alpha]) = 0.$$

Let $x_0 \in E$ and choose $K > 0$. Then

$$\lim_{\delta \to 0_+} \delta^{-1} \lambda\left(\left\{ x: \frac{f(x) - f(x_0)}{x - x_0} \leq K \right\} \cap (x_0, x_0+\delta) \right) \leq$$

$$\leq \lim_{\delta \to 0_+} (K\delta)^{-1} \lambda([f \leq K.\delta]).K = 0.$$

Hence $\underline{f}_{+ap}(x_0) = +\infty$. Similarly, $\overline{f}_{-ap}(x_0) = -\infty$.

EXERCISES

8.C.1 Let $\mu_1, \ldots, \mu_k$ be mutually equivalent Radon measures on $\mathbb{R}^n$, supt $\mu_i = \mathbb{R}^n$. Let τ_i be (metric) density topologies as in Ex. 6.D.8, and let $\tau = \inf(\tau_1, \ldots, \tau_k)$ again. (If f is a function on $A \subset \mathbb{R}^n$, then, as $\tau = \bigcap_{i=1}^{k} \tau_i$, f is τ-continuous if and only if f is τ_i-continuous for each $i = 1, \ldots, k$.)

Let f be the restriction to a τ-closed subset A of $\mathbb{R}^n$ of a (real) Baire one function. Let f be τ-continuous on A. Show that there exists a (real) extension f^* of f to $\mathbb{R}^n$ such that f^* is τ-continuous, f^* is continuous on $\mathbb{R}^n \setminus \bar{A}$ and f^* "preserves semicontinuity" at the points of A.

Hint. Use Ex. 6.D.8 and Corollary 3.31. The reader should consult Ex. 3.E.12 (for ρ to be the Euclidean topology and τ_2 therein equals τ) in the case of real extension.

8.C.2 Let $h_1, \ldots, h_k: \mathbb{R} \to \mathbb{R}$ be absolutely continuous homeomorphisms such that $h_1^{-1}, \ldots, h_k^{-1}$ are also absolutely continuous. If A is a Lebesgue measure zero set and f is a bounded function on A being the restriction to A of a Baire one function, then there is a bounded (approximately continuous) derivative F which extends f such that $F*h_1,\dots,F*h_k$ are (approximately continuous) derivatives. Moreover, F can be chosen to be continuous on $\mathbb{R} \setminus \bar{A}$ and preserving semicontinuity at points of A.

Hint. Apply Ex. 8.C.1 to λ, $h_1(\lambda),\dots,h_k(\lambda)$ when $n = 1$ and $\mathcal{L}$ is the B_4-interval basis.

8.C.3 Show that the conclusion of the previous exercise remains valid if the hypothesis concerning the boundedness is violated. Further show that F can be taken real provided f is the restriction to A of a real Baire one function.

Hint. There is a G_δ-set M containing A such that M, $h_1^{-1}(M),\dots,h_k^{-1}(M)$ are Lebesgue null sets. Let $\tilde{f}$ be a (real, if it exists) Baire one extension of f to M. Apply Lemma 8.9 to $X = \mathbb{R}$ and to couples $\tilde{f},M$ and $\tilde{f}*h_i$, $h_i^{-1}(M)$ to get lower semicontinuous and upper Lebesgue bounded functions $h,t_1,\dots,t_k$ (it is no restriction to assume that $\min(h,t_1,\dots,t_k) \geqq 1$) and upper semicontinuous and lower Lebesgue bounded functions $g,g_1,\dots,g_k$ ($\max(g,g_1,\dots,g_k) \leqq -1$) as furnished by Lemma 8.9. Now, proceed as in Ex. 8.C.2 via Ex. 6.D.8 and Theorem 3.30 (where $f_2 = \min(h,t_1*h_1^{-1},\dots,t_k*h_k^{-1})$ and $f_1 = \max(g,g_1*h_1^{-1},\dots,g_k*h_k^{-1})$.

8.C.4 Let b be a fine base operator on a perfectly normal T_1-topological space X having the Lusin-Menchoff property. If A is a b-isolated ambivalent subset of X and f is a Baire one function on A, then f can be extended over X resulting in a b-continuous Baire one function.

Hint. This will follow from Ex. 3.B.5j taking into account the fact that the r-σ--topology is normal (Ex. 7.B.6a).

8.C.5 Let b again be a fine base operator on a metric space X having the Lusin-Menchoff property. Let A be a b-isolated ambivalent subset of X, let $f \in B_1(A)$ and let s, $-t$ be lower semicontinuous functions on X. If $t \leqq f \leqq s$ on A, then there is a b-continuous Baire one function F such that

$$F = f \text{ on } A, \quad t \leqq F \leqq s \quad \text{on } X,$$

F is continuous on $X \setminus \bar{A}$ and semicontinuous (lower or upper) at all points of A where f is semicontinuous.

Hint. Use Ex. 7.8.6b and Ex. 3.E.18.

REMARK. If the r-basis condition (B) of Theorem 7.12 holds, you can use directly the r-topology instead of r-σ-topology.

8.C.6 Let μ be a Radon measure on $\mathbb{R}^n$ with support $\mathbb{R}^n$ and let $\sigma \in \Omega$. Set

$$b_{\sigma,\mu} : A \longmapsto \{x \in \mathbb{R}^n : \limsup_{h\to 0} \frac{\mu^*(U(x,h) \cap A)}{\sigma(h)} > 0\},$$

and define the (σ,μ)-superdensity topology as the $b_{\sigma,\mu}$ -topology. Prove the following assertions:

(a) The base operator $b_{\sigma,\mu}$ has the complete Lusin-Menchoff property.

(b) Let μ_i $(i=1,\ldots,k)$ be Radon measures on $\mathbb{R}^n$, $\operatorname{supt}\mu_i = \mathbb{R}^n$ and let $\sigma_i \in \Omega$.

If M is a closed subset of $\mathbb{R}^n$ of μ_i-measure zero for each i, and if f is the restrictions to M of a Baire one function on $\mathbb{R}^n$, then there is a function F on $\mathbb{R}^n$ with the properties:

$F = f$ on M, F is (σ_i,μ_i)-continuous for each $i = 1,\ldots,k$,
F is continuous and finite on $\mathbb{R}^n \setminus \bar{M}$ and
F is semicontinuous (lower or upper) at those points of M where f is.

(c) Drop the requirement that M is closed in (b) and prove a generalization of (b) by imposing the condition that the set M is ambivalent only.

Hint. Assertion (a) follows from Theorem 6.33. Citing Proposition 8.15, we have (b). If M is ambivalent only, then (c) is a consequence of Ex. 8.C.5 (put $b = \sup(b_{\sigma_1,\mu_1},\ldots,b_{\sigma_k,\mu_k})$).

8.C.7 Show that the conclusions of the previous exercise remain valid replacing

$$\limsup_{h\to 0_+} \frac{\mu^*(U(x,h) \cap A)}{\sigma(h)} \quad \text{by} \quad \limsup_{h\to 0_+} \frac{\mu^*(U(x,h) \cap A)}{\sigma(\mu(U(x,h)))} .$$

Hint. Verify the conditions of 6.34.B and use Theorem 3.20.

8.C.8 Let μ_i $(i=1,\ldots,k)$ be Radon measures of support $\mathbb{R}$, and let τ_i be the (metric) density topology determined by μ_i and interval differentiation basis B_4. If A is a closed subset of $\mathbb{R}$ of μ_i-measure zero for each $i = 1,\ldots,k$, and if f is the restriction to A of a Baire one function on $\mathbb{R}$, then there is a function

F on $\mathbb{R}$ such that $F = f$ on A, F is τ_i-continuous for each i, F is continuous and finite on $\mathbb{R}\setminus\bar{A}$ and F preserves the semicontinuity at points of A.

Making use of Ex. 8.C.5, prove the assertion for an ambivalent set A.

REMARK. Cf. with Ex. 8.C.1.

8.C.9 Let $h_1,\ldots,h_k: \mathbb{R}\to\mathbb{R}$ be homeomorphisms. If A is an ambivalent subset of $\mathbb{R}$ such that $\lambda(A) = \lambda(h_1^{-1}(A)) = \ldots = \lambda(h_k^{-1}(A)) = 0$, and if $f \in B_1(\mathbb{R})$ is bounded on A, then there exists a bounded derivative F such that

$F = f$ on A, F is continuous on $\mathbb{R}\setminus\bar{A}$,

$F * h_i$ are derivatives for all $i=1,\ldots,k$, and

F preserves continuity at points of A.

Hint. Proceed as in Ex. 8.C.2 using Ex. 8.C.8.

8.C.10 Show that the assumption "f is bounded on A" in Ex. 8.C.9 can be omitted.

Hint. Cf. Ex. 8.C.3.

8.C.11 Let μ_i $(i=1,\ldots,k)$ be Radon measures of support $\mathbb{R}^2$, and let τ_i be the (metric) density topology determined by μ_i and symmetric differentiation basis B_1 or let τ_i be a (σ_i,μ_i)-superdensity topology (defined as in Ex. 8.C.6 or Ex. 8.C.7) for some $\sigma_i\in\Omega$. If A is an ambivalent set such that $\lambda_2(A) = \mu_1(A) = \ldots = \mu_k(A) = 0$ and $M\cap \operatorname{cont}_x A = \emptyset$ for each $x\in\mathbb{R}^2$ (where $M = \{(1;0), (0;1), (-1;0), (0;-1)\}$), and if f is the restriction to A of a Baire one function on $\mathbb{R}^2$, then f can be extended over $\mathbb{R}^2$ resulting in a function F which is τ_i-continuous for each $i = 1,\ldots,k$, it is strongly approximately continuous, continuous on $\mathbb{R}^2\setminus\bar{A}$ and preserves semicontinuity at points of A.

Hint. Proceed as in Ex. 9.D.5, use Ex. 8.C.5 instead of Proposition 8.15.

REMARKS AND COMMENTS

In this section we offer some applications of the Lusin-Menchoff property and some extension theorems that can be derived from it. In the first part, while extending from closed finely isolated sets, we show the advantage of the use of the a-modification of a base operator which has the Lusin-Menchoff property although superdensity topologies in which we do extensions lack this property. Note that further generalizations can be found in Ex. 8.C.5.

The application of the Lusin-Menchoff property to Laczkovich S_2-property or to the generalization of an older Ward's result is one of many further illustrations of fine topology methods in analysis. In the following self-contained Section 8.E we demonstrate further application of extension theorems to derivatives of additive interval functions.

8.D Extensions from F_σ - sets

In the previous sections we developed extension theorems based largely on the Zahorski property or on the B_1-extension theorem which both depend essentially on the Lusin-Menchoff property of a suitable fine topology. The reason for doing so lies in the fact that we use the Lusin-Menchoff property for separation of sets in our abstract in-between theorem 3.2 and we need not repeat the usual "Urysohn's procedure".

In particular, using these extension methods in conjunction with the methods of a-modifications we derived the result on extension of a Baire one function from a closed subset of $\mathbb{R}$ of measure zero resulting in a function which is continuous in the superdensity topology.

Surprisingly it is possible to give a more general result concerning with the extension from F_σ-subsets of measure zero. However, we are not able to apply neither the Lusin-Menchoff property for the separation of sets nor the "Urysohn's procedure". We use a quite different method which is similar to that given in Blažek, Borák and Malý's paper (1978).

Thus, in this section we indicate a further extension method which can be applied also in some cases of fine topologies deprived of the Lusin-Menchoff property. However, our new theorem does not cover the old results. Before giving the main result it will be convenient to have some lemmas.

8.20 LEMMA. Let m be a measure on a metric space (P,ϱ) as in Section 6.B, and let b be a fine base operator on P compatible with m for which there is a function $\omega \in \Omega$ such that $x \notin bA$ whenever

$$m^*(A \cap U(x,r)) \leqq \omega(r)$$

for each $r > 0$ (cf. Theorem 6.33). Let B be a b-closed F_σ-set and let F be a closed set, $B \cap F = \emptyset$. Then there exists a b-continuous and upper semicontinuous function w such that

$$0 \leqq w \leqq 1, \quad w = 0 \text{ on } B \text{ and } w = 1 \text{ on } F.$$

WARNING: In Theorem 6.33 we showed that the base operator b has the Lusin-Menchoff property. But this fact does not imply that the b-topology itself has this property!

Proof. Choose closed sets A_n in such a way that $B = \bigcup_{n=1}^{\infty} A_n$ and $\rho(A_n, F) \geq \frac{1}{n}$. Pick $n \in \mathbb{N}$ and put $r_n^0 = \frac{1}{n}$. If r_n^k is determined for $0 \leq k < n$, let $r_n^{k+1} \in (0, r_n^k)$ be such that

$$m(\{x \in P: \rho(x, A_n) \leq r_n^{k+1}\} \setminus A_n) < 2^{-n} \omega(r_n^k - r_n^{k+1}).$$

Let $\varphi_n : [0, +\infty) \longrightarrow [0,1]$ be a continuous increasing function, $\varphi_n(r_n^k) = \frac{n-k}{n}$ $(k = 0,1,\dots,n)$. Put

$$w_n(x) := \varphi_n(\rho(x, A_n)), \quad w := \inf w_n.$$

Obviously, w is upper semicontinuous, $0 \leq w \leq 1$, $w = 0$ on B and $w = 1$ on F. To prove the assertion it suffices to show that the set $[w > c]$ is b-open for every $c \in [0,1]$. So, fix $c \in [0,1]$ and $z \in [w > c]$. Let $d \in (c, w(z))$. There is $n_0 \in \mathbb{N}$ such that $\frac{2}{n_0} < w(z) - d$. Let $r > 0$ and $n > n_0$. Find k, $0 \leq k < n-1$ such that

$$d < \frac{n-k-1}{n} < \frac{n-k}{n} < w(z).$$

We prove that

$$m(U(z,r) \cap \{x \in P \setminus B: w_n(x) \leq d\}) \leq 2^{-n} \omega(r).$$

Let us consider two cases: (a) If $r \geq r_n^k - r_n^{k+1}$, then

$$m(U(z,r) \cap \{x \in P \setminus B: w_n(x) \leq d\}) \leq m(\{x \in P \setminus A_n: w_n(x) < \frac{n-k-1}{n}\}) \leq$$

$$\leq m(\{x \in P \setminus A_n: \rho(x, A_n) \leq r_n^{k+1}\}) \leq 2^{-n} \omega(r_n^k - r_n^{k+1}) \leq 2^{-n} \omega(r).$$

(b) If $r < r_n^k - r_n^{k+1}$, then

$$U(z,r) \cap \{x \in P: w_n(x) \leq d\} = \emptyset.$$

Indeed, assuming the existence of $y \in P$ for which $\rho(z,y) \leq r$ and $w_n(y) \leq d$, we have

$$\rho(z, A_n) = \rho(z,y) + \rho(y, A_n) \leq r_n^k - r_n^{k+1} + r_n^{k+1} = r_n^k,$$

hence

$$w(z) \leq w_n(z) \leq \frac{n-k}{n},$$

which is a contradiction.

Since $z \notin bB$, also $z \notin b\{x \in B: \inf_{n \geq n_0} w_n(x) \leq c\}$ and for every $r > 0$ we have

$$m(U(z,r) \cap \{x \in P \setminus B: \inf_{n \geq n_0} w_n(x) \leq c\}) \leq$$

$$\leq \sum_{n=n_0}^{\infty} m(\{x \in U(z,r) \setminus A: w_n(x) \leq d\}) \leq \sum_{n=n_0}^{\infty} 2^{-n} \omega(r) \leq \omega(r),$$

which implies $z \notin b\{x \in P \setminus B: \inf_{n \geq n_0} w_n(x) \leq c\}$. Summarizing, we get $z \notin b\,[w \leq c]$. Thus

$$b\,[w \leq c] \subset [w \leq c]$$

and the proof is complete.

REMARK. It is not difficult to show that w is continuous on $P \setminus \bar{B}$.

8.21 LEMMA. With the same notation as in Lemma 8.20, let $\mathcal{F}$ be the family of all b-continuous Baire one functions on P. If $\mathrm{Sep}(\mathcal{F})$ denotes the collection of all $A \subset P$ with the property that A is $\mathcal{F}$-separated from any closed set disjoint with A, then

(a) $\mathrm{Sep}(\mathcal{F})$ is closed under finite unions,

(b) $A \in \mathrm{Sep}(\mathcal{F})$ is $\mathcal{F}$-separated from $B \in \mathrm{Sep}(\mathcal{F})$ provided A, B are G_δ-separated.

Proof. (a) It follows from the fact that the product of finitely many functions of $\mathcal{F}$ is again in $\mathcal{F}$.

(b) Let D_A, D_B be G_δ-sets, $A \subset D_A$, $B \subset D_B$, $D_A \cap D_B = \emptyset$. If

$$P \setminus D_A = \bigcup_{i=1}^{\infty} F_i^A, \quad P \setminus D_B = \bigcup_{i=1}^{\infty} F_i^B$$

where F_i^A, F_i^B are supposed to be closed, then there are f_i^A, $f_i^B \in \mathcal{F}$ such that

$$0 \leq f_i^A \leq 1, \quad f_i^A = 0 \text{ on } A, \quad f_i^A = 1 \text{ on } F_i^A,$$

$$0 \leq f_i^B \leq 1, \quad f_i^B = 0 \text{ on } B, \quad f_i^B = 1 \text{ on } F_i^B .$$

Put

$$f_A = \sum_{i=1}^{\infty} 2^{-i} f_i^A, \quad f_B = \sum_{i=1}^{\infty} 2^{-i} f_i^B .$$

Then the function

$$\varphi := \frac{f_A}{f_A + f_B}$$

is in $\mathcal{F}$, $0 \leq \varphi \leq 1$, $\varphi = 0$ on A and $\varphi = 1$ on B.

REMARK. In light of Lemma 8.20, the system Sep($\mathcal{F}$) contains all F_σ-sets of measure zero (or, more generally, all b-closed F_σ-sets) as well as all closed subsets of P. If the b-topology has the Lusin-Menchoff property, all b-closed sets are in Sep($\mathcal{F}$) as well.

8.22 THEOREM. Let m, b and (P,ρ) have the same meaning as in Lemma 8.20. Let $M \subset P$ be an F_σ-set of measure zero and let f_1, $-f_2$ be upper semicontinuous functions on P. If g is a Baire one function on P, $f_1 \leq g \leq f_2$, then there is a b-continuous function $h \in B_1(P)$ such that

$$h = g \text{ on } M \text{ and } f_1 \leq h \leq f_2 \text{ on } P.$$

Proof. It may be assumed without loss of generality that f_1 and f_2 are bounded on P. Put

$$t = \begin{cases} g & \text{on } M \\ f_1 & \text{on } P \setminus M \end{cases}, \quad s = \begin{cases} g & \text{on } M \\ f_2 & \text{on } P \setminus M \end{cases}.$$

In light of in-between theorem 3.2, the assertion will follow, provided we can prove that, if $a < b$, then the set

$$[t \geq b] = ([g \geq b] \cap M) \cup [f_1 \geq b]$$

is $\mathcal{F}$-separated from

$$[s \leq a] = ([g \leq a] \cap M) \cup [f_2 \leq a] .$$

Choose $a < c < d < b$. Clearly, it suffices to prove now that the set

$$T := ([g > d] \cap M) \cup [f_1 \geq b]$$

is $\mathcal{F}$-separated from the set

$$S := ([g < c] \cap M) \cup [f_2 \leq a].$$

By the preceding remark, all the sets $[g > d] \cap M$, $[f_1 \geq b]$, $[g < c] \cap M$, $[f_2 \leq a]$ belong to $\mathrm{Sep}(\mathcal{F})$. Hence, according to Lemma 8.21.a, $T, S \in \mathrm{Sep}(\mathcal{F})$. As T and S are separated by G_δ-sets $[g \geq d]$ and $[g \leq c]$, Lemma 8.21.b concludes the proof.

EXERCISES

8.D.1 Show that the function h in Theorem 8.22 can be chosen to be continuous on $P \setminus \bar{M}$.

Hint. Fix $M \subset P$. Show that Lemma 8.21 remains true replacing $\mathcal{F}$ by the family $\mathcal{F}_M$ of all functions from $\mathcal{F}$ which are continuous on $P \setminus \bar{M}$. By the remark following Lemma 8.20, every F_σ-subset of M belongs to $\mathrm{Sep}(\mathcal{F}_M)$. Using this fact, restate now the proof of Theorem 8.22.

Remark. It may be worth while to note for the interested reader that the technique of our methods does not permit to preserve semicontinuity in Theorem 8.22. So, we now introduce exercises obviously related to those of Section 8.C, slightly but not completely stronger, where the requirement that the set in consideration is closed or ambivalent is relaxed to require that it is only an F_σ-set.

8.D.2 Let τ_i be a (σ_i, μ_i)-superdensity topology defined as in Ex. 8.C.6 or Ex. 8.C.7 and let $\tau = \inf(\tau_1, \dots, \tau_k)$. Let M be an F_σ-subset of $\mathbb{R}^n$ of μ_i-measure zero for each i, and let f be the restriction to M of a Baire one function on $\mathbb{R}^n$. Then f can be extended over $\mathbb{R}^n$ resulting in a τ-continuous function which is continuous on $\mathbb{R}^n \setminus \bar{M}$.

Hint. Use Ex. 8.D.1. Verify that $b = \sup(b_{\sigma_1, \mu_1}, \dots, b_{\sigma_k, \mu_k})$ satisfies the condition of Lemma 8.20 for $m = \mu_1 + \dots + \mu_k$.

8.D.3 Let τ_i be the (metric) density topology as in Ex. 8.C.8, $\tau = \inf(\tau_1, \dots, \tau_k)$.

Let M be an F_σ-subset of $\mathbb{R}$. If $f \in B_1(\mathbb{R})$, then there is a $\mathcal{T}$-continuous function F on $\mathbb{R}$ which is continuous on $\mathbb{R} \setminus \bar{M}$ and which extends f.

Hint. Use again Ex. 8.D.1 (compare Ex. 8.C.8).

8.D.4 Let $h_1,\dots,h_k: \mathbb{R} \to \mathbb{R}$ be homeomorphisms. If A is an F_σ-subset of $\mathbb{R}$ such that $\lambda(A) = \lambda(h_1^{-1}(A)) = \dots = \lambda(h_k^{-1}(A)) = 0$, and if f is a bounded function on A being the restriction of a Baire one function on $\mathbb{R}$, then there is a bounded (approximately continuous) derivative F which extends f over $\mathbb{R}$ such that F is continuous on $\mathbb{R} \setminus \bar{A}$ and $F * h_1,\dots,F * h_k$ are (approximately continuous) derivatives.

Hint. Proceed as in Ex. 8.C.2 using Ex. 8.D.3.

REMARK. Using the method of Ex. 8.C.3 show that in this exercise the condition "f is bounded" can be dropped.

REMARKS AND COMMENTS

The purpose of this section is to show the use of another technique which is nontopological and different in principle from the previous ones for extension of functions from F_σ-sets (using control functions).

8.E Derivatives of additive interval functions in $\mathbb{R}^n$.

The question of replacing $\mathbb{R}$ in extension theorems of this chapter by more complicated spaces $\mathbb{R}^n$ is a comprehensive subject. By various authors, a number of results concerning derivatives and Dini derivatives of additive interval functions have been generalized from the theory of functions of one real variable. The interested reader should consult Saks' excellent monograph. Recently, V. Aversa and M. Laczkovich (1982) studied the problem of extension of Baire one functions to derivatives of additive interval functions. Extending (unbounded) functions, the authors used "nontopological" procedures to produce derivatives which need not be approximately continuous.

Contrary to their methods we turn now to techniques of Section 8.B the results of which are easily transferred to the context of "ordinary" derivatives of additive interval functions, and their proofs are almost literally the same as in the case of functions on $\mathbb{R}$. It is even possible to go further, replacing the "ordinary" differentiation by a suitable "$\mathcal{G}$-differentiation". The reason lies in the fact that it is of interest to apply the results of preceding sections (the Lusin-Menchoff property of the a.e.-modification of the $\mathcal{G}$-superdensity topology, extensions from F_σ-sets) to obtain extended functions which would be approximately continuous. If further hypotheses are imposed, these functions are then stronger than "ordinary" derivatives of their indefinite integrals.

Throughout σ will be a fixed function from Ω_o. A sequence $\{I_k\}$ of closed intervals in $\mathbb{R}^n$ is called *σ-regular* at a point x provided there is $c > 0$ and a sequence $\{r_k\}$ of positive numbers tending to 0 such that

$$I_k \subset U(x,r_k) \quad \text{and} \quad \lambda I_k > c\,\sigma(r_k).$$

A function f defined in $\mathbb{R}^n$ is said to be *weakly σ-approximately continuous* at a point $x \in \mathbb{R}^n$ if for each $\varepsilon > 0$

$$\lim_{h\to 0_+} \frac{1}{\sigma(h)} \lambda^*(\{t: |f(t)-f(x)| > \varepsilon\} \cap U(x,h)) = 0.$$

8.23 REMARKS. (a) In the case where $\sigma(t) = t^n$ we get the usual notion of "ordinary" approximate continuity, i.e. the continuity w.r.t. the ordinary density topology.

(b) If f is σ-approximately continuous at x (cf. Section 8.C.1), then f is also weakly σ-approximately continuous at x. In Ex. 8.E.1a it is shown that, without additional hypotheses, the converse does not hold. On the other hand, if f is weakly σ-approximately continuous at all points of $\mathbb{R}^n$, then f is σ-approximately continuous on $\mathbb{R}^n$ (cf. Ex. 8.E.1b).

(c) Let f be weakly σ-approximately continuous at x. If $\varepsilon > 0$ and $\{I_k\}$ is a sequence σ-regularly contracting to x, then

$$\lim_{k\to\infty} \frac{1}{\lambda(I_k)} \lambda^*(\{t: |f(t)-f(x)| > \varepsilon\} \cap I_k) = 0.$$

Let F be an additive interval function on $\mathbb{R}^n$. We say that β is the σ-derivative of F at a point $x \in \mathbb{R}^n$ if

$$\beta = \lim_{k\to\infty} \frac{F(I_k)}{\lambda(I_k)}$$

for every sequence of intervals $\{I_k\}$ σ-regularly contracting to x. In this case we denote the common value β of these limits by $F'_\sigma(x)$.

Using (c) of Remark 8.23, a routine argument shows the validity of the next proposition.

8.24 PROPOSITION. Let $f \in \mathcal{L}^1_{loc}(\mathbb{R}^n)$ be locally bounded and weakly σ-approximately continuous at x. If F denotes the indefinite integral of f (i.e. $F(I) = \int_I f$), then $F'_\sigma(x) = f(x)$.

Guided by special situation on $\mathbb{R}$, we say that a function $f \in \mathcal{L}^1_{loc}(\mathbb{R}^n)$ is *σ-upper Lebesgue bounded* at a point $x \in \mathbb{R}^n$ if either $f(x) = +\infty$ or there is $c \in (f(x),+\infty)$ such that

$$\lim_{k\to\infty} \frac{1}{\lambda(I_k)} \int_{I_k} (f(t)-c)^+ dt = 0$$

whenever a sequence $\{I_k\}$ is σ-regular at x. A function f is called *σ-lower Lebesgue bounded* at x if an analogous condition holds. If $f \in \mathcal{L}^1_{loc}(\mathbb{R}^n)$ is weakly σ-approximately continuous and both σ-upper and σ-lower Lebesgue bounded at x and if F denotes the indefinite integral of f, then $F'_\sigma(x) = f(x)$. The proof is almost the same as that given in Lemma 8.8 and we therefore omit it.

We now wish to show that the proof of crucial Lemma 8.9 can be carried on the same way with only immaterial differences. We merely outline the main steps and invite the reader to fill in the details.

8.25 LEMMA. Let $Z \subset \mathbb{R}^n$ be of measure zero and let $X = \mathbb{R}^n \setminus Z$. If f is a Baire one function on a null G_δ-set $M \subset X$, then there are functions $f^* \in B_1(X)$, $g, h \in \mathcal{L}^1_{loc}(X)$ finite on $X \setminus M$ such that h and $-g$ are lower semicontinuous and σ-upper Lebesgue bounded on X and

$$f^* = f \text{ on } M, \quad g \leq f \leq h \text{ on } X.$$

Hint. Let $\tilde{f}$, A and $F_{n,k}$ be as in the proof of Lemma 8.9. Applying Principal Lemma 6.32 for $2^{-m}\sigma(r)$ $r \in \Omega_0$, we get closed set H_m in X such that

$$\lambda(U(x,d) \cap (X \setminus H_m)) < 2^{-m}\sigma(d)d$$

whenever $d > 0$ and $x \in \bigcup_{n,k \leq m} F_{n,k}$. Put again $h(x) = \inf\{m : x \in H_m\}$.

Let $z \in F_{1,1}$ and $d > 0$. Then

$$\int_{U(z,d)} h \leq \lambda(H_1 \cap U(z,d)) + \sum_{m=2}^{\infty} m\, \lambda((H_m \setminus H_{m-1}) \cap U(z,d)) \leq$$
$$\leq \lambda(U(z,d)) + \sum_{m=2}^{\infty} m 2^{1-m} \sigma(d) d < +\infty .$$

Hence $h \in \mathcal{L}^1_{loc}(X)$. Let x and m_0 be as in the proof of Lemma 8.9. Pick a σ-regular sequence $\{I_k\}$ at the point $x \in X$; let $\{r_k\}$ and $c > 0$ correspond to $\{I_k\}$ as in the definition of σ-regularity, and fix $k \in \mathbb{N}$. Put

$$Y_m := I_k \cap \{t \in X: h(t) = m\} \subset U(x,r_k) \cap (X \setminus H_{m-1}),$$

then $\lambda(Y_m) < 2^{1-m} r_k \,\sigma(r_k)$. Hence,

$$\frac{1}{\lambda(I_k)} \int_{I_k} (h(t)-m_o)^+ dt = \frac{\sigma(r_k)}{\lambda(I_k)} \; \frac{1}{\sigma(r_k)} \sum_{m=m_o+1}^{\infty} \int_{Y_m} (m-m_o)\,dt \leq$$

$$\leq \frac{1}{c} \; \frac{1}{\sigma(r_k)} \sum_{m=m_o+1}^{\infty} (m-m_o)\, 2^{1-m} r_k \,\sigma(r_k) =$$

$$= \frac{r_k}{c} \sum_{m=m_o+1}^{\infty} (m-m_o)\, 2^{1-m} ,$$

and consequently

$$\lim_{k\to\infty} \frac{1}{\lambda(I_k)} \int_{I_k} (h(t)-m_o)^+ dt = 0.$$

Using our methods of Section 8.B and taking advantage of the complete Lusin-Menchoff property of the ordinary density topology, we are now going to develop and build upon the following theorem generalizing, in fact, Th. 10 of V. Aversa and M. Laczkovich (1982).

8.26 THEOREM. Let f be a Baire one function, whether finite or infinite, defined on a G_δ-set $M \subset \mathbb{R}^n$ of measure zero. Then f can be extended over the whole $\mathbb{R}^n$ resulting in an (ordinarily) approximately continuous function f^* which is the ordinary derivative of an additive interval function on $\mathbb{R}^n$. Moreover, f^* can be taken to be continuous and finite on $\mathbb{R}^n \setminus \bar{M}$ and upper or lower semicontinuous at those points of M at which f is of this type.

It is clear from the foregoing that also other results of Section 8 could be copied to a more general context. Some of them may be found in the problems for this section. We mention only the following assertion.

8.27 THEOREM. Let $\sigma \in \Omega_o$ and let $A \subset \mathbb{R}^n$ be an F_σ-set of measure zero. Then the restriction of any Baire one function to A can be extended to a σ-approximately continuous σ-derivative of some additive interval function on $\mathbb{R}^n$.

Proof. This assertion follows at once from Theorem 8.22 using Lemma 8.25 and the Lusin-Menchoff property of b_σ (cf. 8.C).

REMARK. Using Ex. 8.D.1, prove that the extended function can be chosen to be continuous on $\mathbb{R}^n \setminus \bar{A}$. On the other hand, it is worth mentioning that we have no extension theorem from F_σ-sets preserving the semicontinuity. In spite of this, we do have some possibilities connected with the conservation of semicontinuity if we extend from closed sets or even from the ambivalent ones (cf. Ex. 8.E.2).

EXERCISES

8.E.1 (a) Let $A \subset \mathbb{R}$ be the set of Example 6.27.f and let $\sigma(t) = t^2$. Show that χ_A is weakly σ-approximately continuous at $x = 0$ but fails to be σ-approximately continuous there.

(b) If a function f is weakly σ-approximately continuous at all points of $\mathbb{R}^n$, then f is σ-approximately continuous on $\mathbb{R}^n$.

Hint. Let the base operator b_σ be as in Section 8.C.1. Select $G \subset \bar{\mathbb{R}}$ open. Show that $f^{-1}(G) \cap b_\sigma(Cf^{-1}(G)) = \emptyset$. Hence, $f^{-1}(G)$ is b_σ-open.

8.E.2 Let $\sigma \in \Omega_0$, $A \subset \mathbb{R}^n$ be an ambivalent set of measure zero and let f be the restriction to A of a Baire one function on $\mathbb{R}^n$. Then there is a weakly σ-approximately continuous function f^* on $\mathbb{R}^n$ which extends f. At the same time, f^* can be chosen to be a σ-derivative of an additive interval function, continuous on $\mathbb{R}^n \setminus A$ and upper or lower semicontinuous at those points where f is.

Hint. Use Ex. 8.C.5 and Lemma 8.25.

REMARK. If A is even closed, you can use the methods of the a.e.-modification (cf. Proposition 8.15 and Corollary 8.16).

8.E.3 Let $\sigma \in \Omega_0$, let $A \subset \mathbb{R}^2$ be an ambivalent set of measure zero and let again f be the restriction to A of a Baire one function. If f is bounded and $M \cap \operatorname{cont}_x A = \emptyset$ for every $x \in \mathbb{R}^2$ (where $M = \{(1,0),(0,1),(-1,0),(0,-1)\}$) then f has an extension f^* to $\mathbb{R}^2$ which is weakly σ-approximately continuous, f^* is σ-derivative and also strong derivative of its indefinite integral, f^* is continuous on $\mathbb{R}^n \setminus \bar{A}$ and preserves the semicontinuity at points of A.

Hint. Use Ex. 8.C.11 (cf. also Ex. 9.D.5). But first of all, state and prove the analogue of Lemma 8.8 for the case of strong differentiability.

REMARK. Since we lack an analogue of Lemma 8.9 for the strong density, we don't know whether the condition of boundedness of f on A can be dropped.

REMARKS AND COMMENTS

As an example illustrating fine topology methods we offer in this section a generalization along with a different proof of a theorem of V. Aversa and M. Laczkovich on the extension to derivatives of additive interval functions on $\mathbb{R}^n$. It would even be possible to go further; we feel that the above examples of exercises are adequate illustrations of our ideas.

9. BOUNDARY BEHAVIOUR OF FUNCTIONS

A. When are boundary functions B_1 ?

B. Jarník - Blumberg method

C. Is any B_1-function a boundary one ?

D. Miscellaneous applications

E. On Bagemihl's and McMillan's result

Let f be a real function on an open subset G of a metric space. If the function f has a limit (a generalized limit) $H(x)$ at each boundary point $x \in \partial G$, we say that H is a *boundary function*. Of course, H is continuous if ordinary limit is in question. But if we deal with some "generalized limits" this is not always the case. In this section we examine the conditions under which boundary functions are of the Baire one class. On the other hand, a question arises whether or not any Baire one function on ∂G is a boundary function.

There is a very extensive literature devoted to the boundary behaviour of functions (cf. E.F. Collingwood and A.J. Lohwater (1966)). Since our point of view is to consider applications of fine topology methods only, we content ourselves here with some special questions connected mainly with the use of the G_δ-insertion method and of the Lusin-Menchoff property. Among important applications we mention the Jarník-Blumberg method which allows us to decide whether some generalized derivatives are of the first class of Baire or not.

To explain the Jarník-Blumberg method, let us remark that some theorems on derivatives can be considered as consequences of theorems concerning boundary behaviour of functions of two variables. Of course, the boundary function of $F(x,y) = \frac{f(x)-f(y)}{x-y}$ at the halfplane $\{(x,y): y < x\}$ is closely related with f'. It seems that this idea is due to V. Jarník (1926). He developed the method by which he proved that any derivative (with finite or infinite values) is a Baire one function. This theorem was rediscovered many times. In fact, in another context the same idea was used, for example, by H. Blumberg in (1930).

9.A When are boundary functions B_1 ?

The question in the title hides a whole series of problems. We start with an illustrative proposition showing one possible approach to them. Notice that this assertion will be the object of further generalization in 9.5 and 9.6.

9.1 PROPOSITION. Let $F(x,y)$ be a function defined on the open halfplane $H:= =\{(x,y) \in \mathbb{R}^2 : y < x\}$. Assume that for each $z \in \mathbb{R}$ there is an approximate limit $f(z)$, whether finite or infinite, of F at the point (z,z) with respect to the Stolz angle

$$U(z) := \{(x,y) : y < z < x\}.$$

Then the boundary function f is in Baire one class.

Proof. For $u,v \in \mathbb{R}^2$ we put

$$\sigma(u,v) = |u-v| + \operatorname{dist}(u, \partial H) + \operatorname{dist}(v, \partial H) \quad (\text{but } 0 \text{ if } u = v)$$

Then the metric σ induces on H the discrete topology and on the boundary ∂H the Euclidean topology. Further, define on $\bar{H}$ the fine topology τ : A set $G \subset \bar{H}$ is τ-open if each point $(z,z) \in G \cap \partial H$ is the density point of G with respect to $U(z)$.

If the extension $\tilde{F}$ of F is defined on ∂H as $\tilde{F}(z,z) = f(z)$, $\tilde{F}$ is τ-continuous. Given a set $A \subset \bar{H}$, we put

$$A^* := (A \cap H) \cup \{(z,z) : \text{for each } n \in \mathbb{N} \text{ there is } m \geq n \text{ such that } \lambda(A \cap U(z) \cap U((z,z),m^{-1})) > \tfrac{1}{2}\lambda(U(z) \cap U((z,z),m^{-1}))\}.$$

Since

$$A_m := \{(z,z) : \lambda(A \cap U(z) \cap U((z,z),m^{-1})) > \tfrac{1}{2}\lambda(U(z) \cap U((z,z),m^{-1}))\}$$

is open in ∂H and $A^* = \bigcap_{n=1}^{\infty} \bigcup_{m=n}^{\infty} A_m \cup (A \cap H)$, A^* is a G_δ-set in $(\bar{H},\sigma)$. Because $\operatorname{int}_\tau A \subset A^* \subset \bar{A}^\tau$, Corollary 2.14 implies that $\tilde{F}$ is a Baire one function. Therefore also f is.

REMARK. We will see later that the consideration of limits w.r.t. Stolz angles applies easily to derivatives.

We see that the establishing of a "boundary topology" seems to be advantageous in the proof of the above proposition. Following this idea we can derive profit from the G_δ-insertion property of a fine boundary topology in many similar examples, too. But we prefer to establish a more general method useful for the investigation of boundary behaviour which we state now as our principal theorem. An alternative more topological approach appears in Ex. 9.A.2 and 9.A.3.

9.2 THEOREM. Let M be a subset of a metric space (P,ρ). Assume that for each $x \in M$ there is a set $A_x \subset P$ which does not contain x. Let f be a function on P, g a function on M such that $g(x) = \lim_{A_x \ni y \to x} f(y)$ for every $x \in M$. Suppose further that the following condition is satisfied.

(W) For each $x \in M$ and for each $\varepsilon > 0$ there is $\delta > 0$ such that $A_x \cap A_y \cap U(x,\varepsilon) \cap U(y,\varepsilon) \neq \emptyset$ whenever $y \in M$ and $\rho(x,y) < \delta$.

Then g is a Baire one function on M.

Proof. Let $a < b$ be reals. Our aim is to find a G_δ-set G such that

$$A := [g \leq a] \subset G \subset [g \leq b]$$

(cf. Theorem 2.12). For $x \in A$ there is $\omega_x > 0$ such that $f(z) < b$ whenever $z \in A_x \cap U(x, \omega_x)$. Let $\delta_{x,n}$ be a positive number from (W) corresponding to x and to $\varepsilon_{x,n} = \min(\omega_x, n^{-1})$. We put

$$G_n := \bigcup_{x \in A} U(x, \delta_{x,n}) \cap M, \qquad G := \bigcap_{n=1}^{\infty} G_n .$$

Obviously, G is a G_δ-subset of M and $A \subset G$. To prove the assertion if suffices to show that $G \subset [g \leq b]$. If $y \in G$ satisfies the inequality $g(y) > b$, then there is $n \in \mathbb{N}$ such that $f(z) > b$ whenever $z \in A_y \cap U(y, n^{-1})$. Since $y \in G$, there is $x \in A$ for which $y \in U(x, \delta_{x,n})$. So we can find a $z \in A_x \cap U(x, \varepsilon_{x,n}) \cap A_y \cap U(y, \varepsilon_{x,n})$. In view of $\varepsilon_{x,n} \leq \omega_x$, we have $f(z) < b$. On the other hand, since $\varepsilon_{x,n} \leq n^{-1}$, we have $f(z) > b$. But this is inconsistent, thereby completing the proof because the insertion of a G_δ-set between $[g \geq b]$ and $[g \geq a]$ is analogous.

Before to proceed we need some necessary notation. We shall deal with the Euclidean space $\mathbb{R}^{2n}$ whose diagonal will be denoted by D, i.e. $D:=\{(x,x): x\in\mathbb{R}^n\}$. Further, let $\mathcal{J}$ denote the collection of all (possibly degenerated) closed intervals in $\mathbb{R}^n$. Every interval $[a_1, b_1]\times\dots\times[a_n, b_n]\in\mathcal{J}$ can be identified with the point $(a_1,\dots,a_n,b_1,\dots,b_n)$ belonging to the set

$$L:=\{(x_1,\dots,x_n,y_1,\dots,y_n): x_i\leq y_i \text{ for each } i=1,\dots,n\}.$$

On $\mathcal{J}$ we consider the usual Hausdorff metric. Obviously, it is equivalent to the usual maximum metric on the set L under idenfication just described.

Hence, the investigation of the limit of an interval function $F(I)$ (we denote it by $\lim\limits_{I\to x} F(I)$) is the same as the investigation of the limit of corresponding function $F^*: L\to\mathbb{R}$ at the point $(x,x)\in\mathbb{R}^{2n}$. In what follows we shall use the language both of interval functions and of point set functions, alternatively.

For each $x\in\mathbb{R}^n$ we put

$$P_x:=\{(y,z)\in\mathbb{R}^{2n}: y=x \text{ or } z=x\},$$

$$P_x^*:=\{I\in\mathcal{J} : x \text{ is a vertex of } I\}.$$

9.3 PROPOSITION. Let f be a function on a set $B\subset\mathbb{R}^n$. If there is a function F on $B\times B\setminus D$ such that

$$\lim_{(B\times B)\cap P_x\ni(y,z)\to(x,x)} F(y,z)=f(x)$$

for every $x\in B$, then $f\in B_1(B)$.

Proof. Let $B^*:=\{(x,x): x\in B\}\subset D$. Extend F on the set B^* arbitrarily. We use Theorem 9.2 putting

$$P=B\times B,\ M=B^*,\ A_x=P_x\cap(B\times B)\setminus\{(x,x)\}.$$

Observe that $(x,x), (y,y)\in M$, $x\neq y$ implies $(x,y)\in A_x\cap A_y$, and, consequently, the condition (W) of Theorem 9.2 is satisfied (it suffices to take $\delta=\varepsilon$). Hence the function f^* defined by $f^*(x,x)=f(x)$ for $x\in B$ is a Baire one function on B^*, which yields that $f\in B_1(B)$.

On the language of interval functions we get the following weakend version (further assertions can be found in Ex. 9.A.5 and 9.A.6).

9.4 PROPOSITION. Let f be a function on a set $B\subset\mathbb{R}^n$. If there is a function F on $\mathcal{J}$ such that

$$\lim_{P_x^* \ni I \to x} F(I) = f(x)$$

for every $x \in B$, then $f \in B_1(B)$.

Now it is just the time to define the notion of a Stolz angle. Let $x \in \mathbb{R}$ and $0 < \theta_1 < \theta_2 < \pi$. The *Stolz angle* with vertex x determined by θ_1 and θ_2, in abbreviation $S_x(\theta_1,\theta_2)$, is defined as the union of all halflines that emanate from the point $(x,0)$ under a direction $\theta \in [\theta_1,\theta_2]$. By the *standard Stolz angle* with vertex x determined by θ $(0 < \theta < \pi/2)$ we understand the (closed) Stolz angle determined by $\theta_1 = \pi/2 - \theta$ and $\theta_2 = \pi/2 + \theta$. We use for it the notation $S_x(\theta)$.

Next, we show how Proposition 9.1 can be derived from our general Theorem 9.2. In fact, a stronger result is valid: The boundary function will be Baire one even on its domain and the approximate limit is replaced by a weak preponderant limit.

Let $H := \{(x,y) \in \mathbb{R}^2,\ y > 0\}$, $z \in \mathbb{R}$ and $\theta \in (0,\ \pi/2)$. Denote $S_z = S_z(\theta)$. We say that b is a *weak preponderant limit* of a function F with respect to the standard Stolz angle S_z if there is a measurable set $M_z \subset S_z$ such that

$$\lim_{M_z \ni t \to z} F(t) = b$$

and the inequality

$$\lambda(M_z \cap U(z,\delta)) > \frac{1}{2}\ \lambda(S_z \cap U(z,\delta))$$

holds for sufficiently small $\delta > 0$. Obviously no generality is lost with the assumption that the last inequality holds for all $\delta > 0$. Of course, if b is an approximate limit of F, then it is a weak preponderant limit as well.

9.5 PROPOSITION. Let $M \subset \mathbb{R}$, $\theta \in (0,\ \pi/2)$. For every $x \in M$ we denote $S_x = S_x(\theta)$. Let F be a function on H having at each point $x \in M$ the weak preponderant limit $f(x)$ with respect to S_x. Then the "preponderant angle boundary function" f is in $B_1(M)$.

Proof. Given $x \in M$, we put $P = \bar{H}$, $A_x = M_x \setminus \{x\}$ ($M_x \subset S_x$ is a measurable set according to the definition of a weak preponderant limit at x) and extend F on $\bar{H}$ arbitrarily in order to prove that (W) of Theorem 9.2 is accomplished. Pick $x \in M$ and $\varepsilon > 0$. There is $c > 0$ such that

$$\lambda(M_x \cap U(x,\varepsilon)) = (1/2 + c)\ \lambda(S_x \cap U(x,\varepsilon)).$$

Apparently there is $\delta > 0$ such that the inequality

$$\lambda((S_x \cap U(x,\varepsilon)) \setminus (S_y \cap U(y,\varepsilon))) < c\ \lambda(S_x \cap U(x,\varepsilon))$$

holds for every $y \in M \cap U(x,\delta)$. Thus, for these y,

$$\lambda((A_x \cap U(x,\varepsilon)) \setminus (A_y \cap U(y,\varepsilon))) \leq \lambda((A_x \cap U(x,\varepsilon)) \setminus (S_y \cap U(y,\varepsilon))) +$$

$$+ \lambda((S_y \setminus A_y) \cap U(y,\varepsilon)) < c\ \lambda(S_x \cap U(x,\varepsilon)) + \frac{1}{2}\ \lambda(S_y \cap U(y,\varepsilon)) =$$

$$= (1/2 + c)\ \lambda(S_x \cap U(x,\varepsilon)) = \lambda(A_x \cap U(x,\varepsilon)).$$

We can see that $A_x \cap U(x,\varepsilon) \cap A_y \cap U(y,\varepsilon) \neq \emptyset$.

9.6 THEOREM. Let $M \subset \mathbb{R}$ and assume that for each $x \in M$ a standard Stolz angle $S_x := S_x(\theta(x))$ $(\theta(x) \in (0, \pi/2))$ is given. Let F be a function on H having at each point $x \in M$ the approximate limit $f(x)$ with respect to S_x. If there is a function $g \in B_1(M)$ such that $0 < g(x) \leq \theta(x)$ for every $x \in M$, then the "approximate boundary function" f is in $B_1(M)$.

Proof. Since $M_n := [g > 1/n]$ are F_σ - subsets of M, we have in view of the previous proposition, that $f \in B_1(M_n)$ for each n. Since

$$\{x \in M: f(x) < a\} = \bigcup_{n=1}^{\infty} \{x \in M_n: f(x) < a\}$$

and analogously for $[f > a]$, we obtain that $f \in B_1(M)$.

REMARK. If $M = \mathbb{R}$, Theorem 9.6 coincides with Mišík's theorem in (1969). L. Mišík proved also that the condition located on function $\theta(x)$ cannot be weakened. The details can be found in Ex. 9.C.2.

We close this section with a general theorem, the converse to it will be examined later on in Section 9.C (Theorem 9.15).

9.7 THEOREM. Let M be a subset of a metric space (P, ρ) and let $\omega \in \Omega_0$ (cf. Section 7.D). For $x \in M$ denote

$$A_x := \{y \in P: \rho(y,M) > \omega(\rho(x,y))\}.$$

If F is a function on $P \setminus \bar{M}$ and f is a function on M such that

$$f(x) = \lim_{A_x \ni y \to x} F(y)$$

for each $x \in M$, then $f \in B_1(M)$.

Proof. Let F be extended to $\bar{M}$ arbitrarily, and let $x \in M$, $x \in \bar{A}_x$. It suffices to verify the condition (W) of Theorem 9.2. Pick again $\varepsilon > 0$, and choose $z \in A_x \cap U(x,\varepsilon)$. Since ρ and ω are continuous functions (which yields $\lim_{y \to x} \omega(\rho(y,z)) = \omega(\rho(x,z))$), there is $\delta > 0$ such that $y \in U(x,\delta) \cap M$ implies $z \in A_y \cap A_x \cap U(y,\varepsilon) \cap U(x,\varepsilon)$.

As a simple corollary of the previous theorem we get immediately the following proposition. Notice that in contrast to Theorem 9.6 on "approximate boundary functions", in the case of "ordinary boundary functions" the standard Stolz angle can vary arbitrarily.

9.8 PROPOSITION. Let $M \subset \mathbb{R}$ and assume that for each $x \in M$ a standard Stolz angle $S_x := S_x(\theta(x))$ is given. Let F be a function on H having at each point $x \in M$ the limit $f(x)$ with respect to S_x. Then the function f belongs to $B_1(M)$.

Proof. Use Theorem 9.7 putting $P = \bar{H}$ and

$$\omega(z) := \begin{cases} z - z^2 & \text{for } z \in [0,1/2), \\ 1/4 & \text{for } z \geq 1/2. \end{cases}$$

EXERCISES

9.A.1 Let τ be the topology of the proof of Proposition 9.1.

(a) Show that τ is a (metric) density topology determined by some differentiation basis.

(b) Prove that τ satisfies the strong essential radius condition (w.r.t. $\mathfrak{S}$) (cf. Ex. 2.D.19).

(c) In light of (b) give an alternative proof of Proposition 9.1 (cf. Ex. 2.D.16).

(d) Let $\mathcal{L}$ be the angle differentiation basis (B_6) in $\mathbb{R}^2$ and let m be the Lebesgue measure. Show that the condition (Bal) of Theorem 6.23 is satisfied. Now, use Ex. 6.C.10 and give a further proof of Proposition 9.1.

9.A.2 We introduce now an idea which is obviously related to the notion of an essential radius condition of Ex. 2.D.16. Let X be a set and let (S,ρ) be a metric space. For each $s \in S$, let $\mathcal{F}_s$ be a filter on X. We say that the *essential radius condition* is satisfied, if for each $s \in S$ and for each $F \in \mathcal{F}_s$ there is an "essential radius" $r(s,F) > 0$ such that

$$\rho(s,t) \leq \min(r(s,F_s), r(t,F_t)) \text{ implies } F_s \cap F_t \neq \emptyset$$

for every $s,t \in S$ and $F_s \in \mathcal{F}_s$, $F_t \in \mathcal{F}_t$.

If the essential radius condition holds, and if f is a function on X, g a function on S such that $g(y) = \lim_{\mathcal{F}_y} f$ for every $y \in S$, then g is a Baire one function on S.

Hint. We may assume $X \cap S = \emptyset$. Put $P = X \cup S$ and define the metric σ on P by $\sigma(x,y) = \dfrac{\rho(x,y)}{1+\rho(x,y)}$ if $x,y \in S$ and $\sigma(x,y) = 1$ $(x \neq y)$ otherwise. Call a subset G of P τ-open if for every $x \in G \cap S$ there is $F \in \mathcal{F}_x$, $F \subset G$. Show that τ satisfies the essential radius condition (w.r.t. σ) and use the result of Ex. 2.D.16.

9.A.3 Let M, (P,ρ), A_x, f and g be as in Theorem 9.2. Suppose that the following condition holds:

(W*) For each $x \in M$ and each $\varepsilon > 0$ there is an "essential radius" $r(x,\varepsilon) > 0$ such that
$\rho(x,y) \leq \min(r(x,\varepsilon_1), r(y,\varepsilon_2))$ implies $A_x \cap A_y \cap U(x,\varepsilon_1) \cap U(x,\varepsilon_2) \neq \emptyset$.

Prove that g is a Baire one function on M.

Hint. Apply Ex. 9.A.2 for $S = M$, $X = P$ and $\mathcal{F}_s$ being the filter having the base $\{A_s \cap U(s,r) : r > 0\}$.

9.A.4 (a) Show that the condition (W) of Theorem 9.2 implies (W*) but, without additional hypotheses, the converse is not true.

Hint. Let H and $U(z)$ be as in Proposition 9.1. Put $P = \bar{H}$, $M = P \setminus H$ and for $b = (z,z) \in M$,

$$A_b = \begin{cases} (U(z) \cap U(b, |z|^2)) \setminus \{z\} & \text{if } z \neq 0, \\ U(z) \setminus \{z\} & \text{if } z = 0. \end{cases}$$

(b) If the system $\{A_x\}$ $(x \in \bar{A}_x \setminus A_x)$ satisfies (W*), then there is a family $\{\tilde{A}_x\}_{x \in M}$ such that $\{\tilde{A}_x\}$ satisfies (W) and for each $x \in M$, $V \cap A_x = V \cap \check{A}_x$ for some neighborhood V of x.

Hint. It does no harm to suppose that $r(x,\varepsilon) < \varepsilon$ for $x \in M$, $\varepsilon > 0$. Put $\tilde{A}_x = A_x \cup (P \setminus U(x, \frac{1}{2} r(x,1)))$. Given $x \in M$ and $\varepsilon > 0$, put $\delta = r(x, \frac{1}{3}\varepsilon)$ and verify (W). For $y \in U(x,\delta)$, distinguish the case $\rho(x,y) < r(y,1)$ from $\rho(x,y) \geq r(y,1)$.

9.A.5 Let $\tilde{\mathfrak{I}} \subset \mathfrak{I}$ be the family of all nondegenerated closed intervals in $\mathbb{R}^n$. Substituting in Proposition 9.4 $\tilde{\mathfrak{I}}$ for $\mathfrak{I}$ and $P^*_x \cap \tilde{\mathfrak{I}}$ for P^*_x show that the more general assertion keeps holding.

Hint. Verify the condition (W).

9.A.6 Let $C \subset \tilde{\mathfrak{I}}$ be the collection of all closed cubes of $\mathbb{R}^n$ and $P^c_x \subset C$ the system of all cubes with center in x. Show that Proposition 9.4 is not valid (even for $B = \mathbb{R}$) replacing $\mathfrak{I}$ by C and P^*_x by P^c_x.

(b) Let $\omega \in \Omega$. Denote

$$P^{\omega}_x = \{I \in C\colon \operatorname{dist}(x,\hat{I}) < \omega(\operatorname{diam} I)\}$$

(where $\hat{I}$ is the center of I). Show that Proposition 9.4 remains true replacing $\mathfrak{I}$ by C and P^*_x by P^{ω}_x.

Hint. Use again (W).

9.A.7 (L. Mišík, 1966)). Let G be an additive interval function in $\mathbb{R}^n$. If G has at any point x of a set $M \subset \mathbb{R}^n$ the ordinary derivative $G'(x)$ (finite or infinite), then $G' \in B_1(M)$.

Hint. Use the previous exercise.

REMARKS AND COMMENTS

It seems that the use of boundary behaviour of arbitrary functions owes its beginnings to V. Jarník (1926) who succeeded to prove by his method that any derivative is in Baire one class. Jarník's theorem (Ex. 9.C.3, (i) $\Leftrightarrow$ (ii)) immediately implies - in terms of interval functions - that any limit of a convergent interval function is in Baire class one which was proved independently by A. Gleyzal (1941) (Proposition

9.4 and Ex. 9.A.5, 9.A.6 - cf. [Bru], Chap. X, Lemma 2.1 - for a more general result see Ex. 9.C.4). Later work in this direction has been done by L.E. Snyder in a collection of papers (1965) - (1967b) who also examined for the first time the boundary behaviour considering approximate limits.

The condition (W) of Theorem 9.2 occupies a place, perhaps, in some papers and it was essentially known, at least for $M = \mathbb{R}$ and $P = \mathbb{R}^2$. A particular case of (W) for $P = \mathbb{R}$ is contained in B.S. Thomson (1982). It should be pointed out that the proof of Theorem 9.2 makes use of the insertion G_δ-method (In-G_δ) so we can prove that certain functions are in Baire class one on an arbitrary subset M of P while other authors use in this situation (generally non-equivalent) conditions (Ba) or (D-P).

A slight alternations of Theorem 9.2 stated in exercises (Ex. 9.A.2, Ex. 9.A.3, Ex. 2.D.16) are sometimes more convenient. Theorem 9.6 is further refinement of Mišík's generalization (1969) of a theorem of L.E. Snyder (1966a).

9.B Jarník-Blumberg method

As we noticed above, the boundary function of $\frac{f(x)-f(y)}{x-y}$ at the halfplane $\{(x,y): y < x\}$ is closely related to the derivative of f. The method using the boundary investigation of functions of two variables (not necessarily of the form indicated above) for obtaining some properties of derivatives is sometimes called the Jarník-Blumberg method. In this section we use this method to prove that certain derivatives (in a generalized sense) are of the Baire one class.

9.9 THEOREM. Let $M \subset \mathbb{R}$ and let f be a function on $\mathbb{R}$. If for each $x \in M$ there is a derivative $f'(x)$, whether finite or infinite, then $f' \in B_1(M)$.

Proof. Let F be an interval function for which

$$F([x,y]) = \frac{f(x)-f(y)}{x-y}$$

whenever $x < y$. Using Proposition 9.4, we get directly $f' \in B_1(M)$.

In the same spirit, using finer Proposition 9.3 we obtain the following theorem on derivative with respect to a subset of $\mathbb{R}$.

9.10 THEOREM. Let $M \subset \mathbb{R}$ and f be a function on M. If for each $x \in M$ there is

$$f'_M(x) := \lim_{M \ni y \to x} \frac{f(x)-f(y)}{x-y} ,$$

then $f'_M \in B_1(M)$.

In the sequel, we use the Jarník-Blumberg method for the investigation of behaviour of approximate derivatives. First of all, we need an obvious geometric lemma:

9.11 LEMMA. Let x be a bilateral accumulation point of a set $M \subset \mathbb{R}$. If

$$\lim_{M \ni y \to x} \frac{f(x)-f(y)}{x-y} = A \text{ (finite or infinite) then}$$

$$\lim_{\substack{M_x \ni (y,z) \to \\ \to (x,x)}} \frac{f(y)-f(z)}{y-z} = A,$$

where $M_x := \{(y,z) \in M \times M : z \lneqq x \leqq y\}$.

9.12 THEOREM. If at every point of a set $M \subset \mathbb{R}$ there is an approximative derivative $f'_{ap}(x)$, whether finite or infinite, then $f'_{ap} \in B_1(M)$.

Proof. Given $x \in M$, there is a measurable set P_x such that x is a density point of P_x and

$$\lim_{P_x \ni y \to x} \frac{f(y)-f(x)}{y-x} = f'_{ap}(x).$$

Put $S_x := \{(y,z) : y \lneqq x \lneqq z\}$, $M_x := S_x \cap (P_x \times P_x)$ (S_x is a Stolz angle with the vertex (x,x) in the halfplane $\{(y,z) : y < z\}$). By Lemma 9.11,

$$f'_{ap}(x) = \lim_{\substack{M_x \ni (y,z) \\ \to (x,x)}} \frac{f(y)-f(z)}{y-z}$$

It is easy to check that the point (x,x) is a density point of M_x with respect to S_x. Hence Proposition 9.5 yields our assertion.

As we have just shown - using the Jarník-Blumberg method - any derivative or any approximate derivative (even getting infinite values) is of Baire class one on its domain. In order to illustrate our methods we show that also any preponderant derivative and any (finite)one-sided approximate derivative are in Baire class one. Since there are more definitions of preponderant derivatives, we adopt the weak one as follows: We say that $f'_{pr}(x) = A$ (finite or infinite) if there is a measurable set P_x such that

$$\lim_{P_x \ni y \to x} \frac{f(y)-f(x)}{y-x} = A$$

and the inequalities

$$\lambda(P_x \cap (x,x+\delta)) > \delta/2, \quad \lambda(P_x \cap (x-\delta,x)) > \delta/2$$

hold for sufficiently small $\delta>0$. Again, as in Section 9.A, we can assume that this inequalities hold for every $\delta>0$.

9.13 THEOREM. Let $M \subset \mathbb{R}$ and let f be a function on $\mathbb{R}$. If for each $x \in M$ there is $f'_{pr}(x)$, whether finite or infinite, then $f'_{pr} \in B_1(M)$.

Proof. Given $x \in M$, let P_x be as furnished by the above definition of $f'_{pr}(x)$. As usual, we put

$$A_{(x,x)} := \{(t,z):\ t < x < z\} \cap (P_x \times P_x)$$

for each $x \in M$. By Lemma 9.11

$$f'_{pr}(x) = \lim_{\substack{M_x \ni (t,z) \to \\ \to (x,x)}} \frac{f(t)-f(z)}{t-z} .$$

To finish the proof, it suffices to verify the condition (W) of Theorem 9.2 for the family of sets $\{A_{(x,x)}\}$. To this end, let $x \in M$ and $\varepsilon>0$ be given. Find $c_x \in (0,\frac{1}{2})$ such that

$$\lambda(P_x \cap (x,x+\tfrac{\varepsilon}{2})) > (\tfrac{1}{2}+c_x)\,\tfrac{\varepsilon}{2},$$

$$\lambda(P_x \cap (x-\tfrac{\varepsilon}{2},x)) > (\tfrac{1}{2}+c_x)\,\tfrac{\varepsilon}{2}.$$

We can consider $\mathbb{R}^2$ equipped with the maximum metric σ. We assert that it is sufficient to put $\delta = \frac{1}{4}\varepsilon c_x$. In fact, let $y \in M$ be such that $\sigma((x,x),(y,y)) < \delta$. Then $|x-y|<\delta$ and we show that the sets

$$M^+ := P_x \cap P_y \cap (x+\delta, x+\varepsilon/2),$$

$$M^- := P_x \cap P_y \cap (x-\varepsilon/2, x-\delta)$$

are nonempty. Indeed, using the obvious inequalities

$$\lambda((x+\delta,x+\varepsilon/2)\setminus(P_x\cap P_y)) \leqq \lambda((x,x+\varepsilon/2)\setminus P_x) + \lambda((y,x+\varepsilon/2)\setminus$$
$$\setminus P_y) \leqq (1/2-c_x)\ \varepsilon/2+\tfrac{1}{2}(\varepsilon/2+\tfrac{1}{4}\varepsilon c_x) =$$
$$= \varepsilon/2-\tfrac{3}{8}\varepsilon c_x < \varepsilon/2-\delta,$$

we see that $M^+ \neq \emptyset$. Similarly, $M^- \neq \emptyset$. Hence, for $z \in M^+$, $t \in M^-$ we have

$$(t,z) \in A_{(x,x)} \cap A_{(y,y)} \cap U((x,x),\varepsilon) \cap U((y,y),\varepsilon).$$

9.14 THEOREM. Let $M \subset \mathbb{R}$ and let f be a function on $\mathbb{R}$. If for each $x \in M$ there is finite $f'_{ap+}(x)$, then $f'_{ap+} \in B_1(M)$.

Proof. We sketch only the main idea, for the details see L. Zajíček (1981). Given $x \in M$, there is a measurable set P_x such that $d_+(x,P_x) = 1$ and

$$\lim_{P_x \ni y \to x} \frac{f(y)-f(x)}{y-x} = f'_{ap+}(x).$$

Put

$$S_x := \{(z,t):\ x \leqq t \leqq z,\quad z-x \geqq 2(t-x)\},$$

$$M_x := S_x \cap (P_x \times P_x)$$

(S_x is a Stolz angle with the vertex (x,x) in the halfplane $\{(z,t):\ z > t\}$).
Then (compare with Lemma 9.11)

$$f'_{ap+}(x) = \lim_{\substack{M_x \ni (t,z) \to \\ \to (x,x)}} \frac{f(t)-f(z)}{t-z}.$$

Further, as it follows from simple calculation, every point (x,x) is a density point of M_x with respect to the Stolz angle S_x. Hence Proposition 9.5 yields the assertion.

REMARK. The example of the well-known Dirichlet function shows that the assumption of finiteness of f'_{ap+} is essential in the above theorem.

EXERCISES

9.B.1 Let f be a real function on $\mathbb{R}$, $x \in \mathbb{R}$. If there is a first category set A such that there is a limit

$$f'_q(x) := \lim_{A \ni y \to x} \frac{f(y)-f(x)}{y-x} ,$$

we say $f'_q(x)$ is the *qualitative derivative* of f at x. If there is $f'_q(x)$ (finite or infinite) for each $x \in M \subset \mathbb{R}$, then f'_q belongs to $B_1(M)$.

Hint. Use the Jarník-Blumberg method taking steps as in the proof of Theorem 9.12.

9.B.2 Define the right qualitative derivative $f'_{q+}(x)$ in the obvious manner and show that $f'_{q+} \in B_1(M)$ provided there is finite $f'_{q+}(x)$ for each $x \in M \subset \mathbb{R}$.

Hint. Use again the Jarník-Blumberg method similarly as in the proof of Theorem 9.14.

REMARKS AND COMMENTS

It stands to reason that any finite derivative is a Baire one function. Using the idea of boundary behaviour of functions of two variables, V. Jarník (1926) proved that any derivative (with finite or infinite values) is in Baire class one.

G. Tolstov (1938) showed that any finite approximate derivative is a B_1-function. The simplified proof of this assertion was given by C. Goffman and C.J.Neugebauer (1960) and L.E. Snyder (1966a) modified Jarník's method and proved this result again. On the other hand, Z. Zahorski (1948) offered an example to produce an approximate derivative not belonging to Baire class one. As shown by D. Preiss (1971b), this example is incorrec and any approximate derivative, even infinite, always belongs to Baire class one. L. Mišík (1972) proved this theorem again using the Jarník-Snyder method. The same idea in regard of the proof together with the use of the boundary and the G_δ-insertion method is described in J. Lukeš and L. Zajíček (1977b) where it is applied also to (possibly infinite) partial approximate derivatives of partially continuous functions (thus simplifying and generalizing Snyder's results (1966b)).

Preponderant notions were introduced as early as in (1915) by A. Denjoy. He showed that a preponderant derivative of a continuous function is in Baire class one. Some properties of preponderant derivatives and open problems are stated in A.M.Bruckner (1977). Using the G_δ-insertion method, another time, K. Pekár (1981) showed that any finite or infinite preponderant derivative on $\mathbb{R}$ is a Baire one function. Independently the same result was obtained by A.M. Bruckner, R.J.O'Malley and B.S.Thomson (1984). Theorem 9.13 extends this result also for an arbitrary subset of $\mathbb{R}$.

Theorem 9.14 (for $M = \mathbb{R}$) concerning one-sided approximate derivatives is due to L. Zajíček (1981).

The qualitative derivative was defined by S. Marcus in (1953). The result of Ex. 9.B.1 is known (see, e.g., L.E. Snyder (1966a), p.422) whereas that of Ex.9.B.2 seems to be new.

9.C Is any B_1-function a boundary function?

In this section we solve the following problem: Given a subset M of a metric space P and a Baire one function f on M, does there exist a (say, continuous) function F on $P\setminus\bar{M}$ for which f is a "boundary function"? The solution of this problem will be based on the following observation: We introduce on P a new "fine" topology (the so-called boundary topology - cf. Section 7.D) which has the Lusin-Menchoff property (Theorem 7.19), and then we use simply the B_1-extension theorem 3.29. Moreover, our theorem will be in a certain sense the converse to Theorem 9.7 (cf. also Ex. 9.C.1).

9.15 THEOREM. Let M be a subset of a metric space (P,ρ) and let $\omega\in\Omega_0$ (cf. Section 7.D). Given a Baire one function f on M, there is a finite continuous function F on $P\setminus\bar{M}$ such that for each $x\in M$,

$$\lim_{n\to\infty} F(y_n) = f(x)$$

whenever $y_n\to x$ and there is $\varepsilon>0$ such that $\rho(y_n,M)>\varepsilon\,\omega(\rho(y_n,x))$. Moreover, if $y_n\to x\in M$ and $y_n\notin\bar{M}$, then

(a) $\limsup F(y_n)\leq f(x)$ provided f is upper semicontinuous at x,

(b) $\liminf F(y_n)\geq f(x)$ provided f is lower semicontinuous at x.

(In particular, if f is continuous at x, then $\lim F(y_n) = f(x)$.)

NOTE. The above condition shows that

$$\lim_{A_x\ni y\to x} F(y) = f(x)$$

provided x is an accumulation point of the set

$$A_x := \{y\in P:\ \rho(y,M)>\varepsilon\,\omega(\rho(y,x))\}.$$

Proof. Putting $X := P\setminus(\bar{M}\setminus M)$, M is a closed subset of X. Let $\tau=\tau(M,\omega)$ be the fine boundary topology (Section 7.D). We know that τ has the Lusin-Menchoff property (Theorem 7.19). It can be easily checked that the assumptions of Theorem 3.30 are satisfied. It now follows that the corresponding extension F has obviously all properties prescribed to it (cf. Ex. 7.D.4).

The following corollary can be easily derived from Theorem 9.15 and for $A = \mathbb{R}\times\{0\}$

it forms a special case of a result of Bagemihl and McMillan which will be discussed later in Section 9.E.

9.16 COROLLARY. Let f be a Baire one function on a set $A \subset \mathbb{R} \times \{0\}$. Then there is a finite continuous function F on the set $H := \mathbb{R} \times (0,+\infty)$ such that

$$\lim_{S_x \ni y \to x} F(y) = f(x)$$

whenever $x \in A$ and S_x is a Stolz angle in H. Moreover, if f is continuous at a point x, then

$$\lim_{H \ni y \to x} F(y) = f(x).$$

Proof. Use Theorem 9.15 putting $P = H$, $M = A$ and ω to be identity. Realize that there is $\varepsilon > 0$ such that $\rho(y_n,A) > \varepsilon\rho(y_n,x)$ provided S_x is a Stolz angle in H with vertex x and $S_x \ni y_n \to x$.

Recall that $\mathcal{J}$ signifies the set of all closed intervals in $\mathbb{R}^n$. Consider on $\mathcal{J}$ the usual Hausdorff metric which is obviously equivalent to the maximum metric σ on the set L (we identify L and $\mathcal{J}$ in a natural way, see the text following Theorem 9.2). Denote $D^* = \{\{x\} : x \in \mathbb{R}^n\}$ and, further, put

$$A_x := \{I \in \mathcal{J} : \text{diam } I > \omega(\rho(x,I))\}$$

(diam I and $\rho(x,I)$ are taken w.r.t. the Euclidean metric) whenever $x \in \mathbb{R}^n$ and $\omega \in \Omega_0$.

9.17 COROLLARY. Let f be a Baire one function on a set $M \subset \mathbb{R}^n$, let $\omega \in \Omega_0$. Then there is a continuous function $F: \mathcal{J} \setminus D^* \to \mathbb{R}$ such that

$$\lim_{A_x \ni I \to \{x\}} F(I) = f(x)$$

for each $x \in M$.

Proof. We use Theorem 9.15 putting $(P,\rho) = (\mathcal{J},\sigma)$. It suffices to find $\tilde{\omega} \in \Omega_0$ for which $\sigma(I,D^*) > \tilde{\omega}(\sigma(\{x\},I))$ for every $I \in A_x$. But we have

$$\sigma(\{x\},I) \leqq \rho(x,I) + \text{diam } I,$$

$$2n^{1/2}\ \sigma(I,D^*) \geqq \text{diam } I.$$

Choosing $I \in A_x$ we get

$$\sigma(\{x\},I) < \omega^{-1}(\text{diam } I) + \text{diam } I$$

and

$$(\omega^{-1} + \text{id})^{-1}(\sigma(\{x\},I)) < \text{diam } I \leq 2n^{1/2}\,\sigma(I,D^*).$$

Hence, we can put

$$\tilde{\omega} := \frac{1}{2n^{1/2}}(\omega^{-1} + \text{id})^{-1}.$$

EXERCISES

9.C.1 Let M be a subset of a metric space (P,ρ) such that each point $x \in M$ is an accumulation point of the set

$$A_x = \{y \in P: \ \rho(y,M) > \omega(\rho(x,y))\}.$$

Show that the following conditions on $U,L \subset M$ and on a function f on M are equivalent:

(i) $f \in B_1(M)$, f is upper semicontinuous at all points of U and lower semicontinuous at all points of L,

(ii) there is a function F on $P \setminus \bar{M}$ such that

(a) $\lim_{A_x \ni y \to x} F(y) = f(x)$ for every $x \in M$,

(b) $\limsup_{\bar{M} \not\ni y \to x} F(y) \leq f(x)$ for every $x \in U$,

(c) $\liminf_{\bar{M} \not\ni y \to x} F(y) \geq f(x)$ for every $x \in L$,

(iii) there is a real continuous function F on $P \setminus \bar{M}$ satisfying (a), (b) and (c).

Formulate also these assertions for special cases $U = L = \emptyset$ and $U = L \neq \emptyset$.

Hint. Use Theorems 9.7 and 9.15.

9.C.2 (a) Let θ be a real function on a complete metric space P. Show that the following assertions are equivalent:

(i) there is $g \in B_1(P)$ such that $0 < g(x) \leq \theta(x)$ for all $x \in P$,

(ii) there are closed sets $F_n \subset P$ and $\varepsilon_n > 0$ such that $P = \bigcup_{n=1}^{\infty} F_n$ and $\theta(x) > \varepsilon_n$ for all $n \in \mathbb{N}$ and $x \in F_n$,

(iii) if $F \neq \emptyset$ is a closed subset of P, then there is an open set G and $\varepsilon > 0$ such that $G \cap F \neq \emptyset$ and $\theta(x) > \varepsilon$ for each $x \in F \cap G$.

(iv) the same as (iii) replacing the closed set F by a perfect one.

(b) (L. Mišík (1969).) Suppose that the condition located on function θ in Proposition 9.6 is not satisfied. Then there is a continuous function F on $H := \mathbb{R} \times (0,+\infty)$ such that F has at any point $(x;0)$ an approximate limit $f(x)$ w.r.t. S_x and $f \notin B_1(\mathbb{R})$.

Hint. By (iv) of (a), find a perfect set D and a sequence $\{r_n\}$ dense in D such that $\theta(r_n) < 2^{-n-1}$. By induction obtain $\delta_n > 0$ in such a way that the sets $S_{r_n}(2\theta(r_n)) \cap U(r_n, \delta_n)$ are pairwise disjoint. Define F such that

$$F(x) = \begin{cases} 0 & \text{if } x \in \bigcup_{n=1}^{\infty} [S_{r_n}(\theta(r_n)) \cap U(r_n, \tfrac{1}{2}\delta_n)] \cap H, \\ 1 & \text{if } x \in H \setminus \bigcup_{n=1}^{\infty} [S_{r_n}(2\theta(r_n)) \cap U(r_n, \delta_n)] . \end{cases}$$

9.C.3 Let f be a function on a set $P \subset \mathbb{R}^n$. Assume that P has not isolated points. Given $x = (x_1,\dots,x_n) \in P$, put

$$A_x = \{(y,z) \in \mathbb{R}^{2n}: y,z \in P,\ y \neq z \text{ and } (y_i-x_i)(z_i-x_i) \leq 0 \text{ for every } i = 1,\dots,n\}.$$

Show that the following assertions are equivalent:

(i) $f \in B_1(P)$,

(ii) (V. Jarník (1926) for perfect $P \subset \mathbb{R}^n$):
there is a function F on $P \times P \setminus \{(y;z) \in \mathbb{R}^{2n}: y = z\}$

such that $\lim_{A_x \ni (y,z) \to (x,x)} F(y,z) = f(x)$ for all $x \in P$,

(iii) (L.E. Snyder (1965)):
there is a real continuous function F as in (ii).

Hint. Use Proposition 9.3 and Corollary 9.17.

9.C.4 Throughout $\mathcal{J}, D^*$ and ω are as in Corollary 9.17. Let A_x have one of the following meanings:

(a) $A_x = \{I \in \mathcal{J} : \text{diam } I > \omega(\rho(x,I))\}$,

(b) $A_x = \{I \in \mathcal{J} : x \in I\}$,

(c) $A_x = \{I \in \mathcal{J} : x \in \text{int } I\}$,

(d) $A_x = \{I \in \mathcal{J} : x \text{ is a vertex of } I\}$,

(e) $A_x = P_x^\omega$ where P_x^ω is as in Ex. 9.A.6.

If f is a function on a set $M \subset \mathbb{R}^n$, then the following assertions are equivalent:

(i) $f \in B_1(M)$,

(ii) there is a function F on $\mathcal{J} \setminus D^*$ such that $\lim_{A_x \ni I \to \{x\}} F(I) = f(x)$ for each $x \in M$,

(iii) there is a real continuous F as in (ii).

REMARKS AND COMMENTS

As early as in the fundamental paper of V. Jarník (1926) it is proved that each B_1 - function on a perfect subset of an Euclidean space is a boundary function of another function. L.E.Snyder (1965) later proved that this function may even be taken as continuous. In a subsequent paper (1967b), L.E.Snyder himself generalizes these results in considering it in a metric space $(P,\rho) \times [0,\infty)$, a boundary behaviour of functions defined on $P \times (0,\infty)$ with respect to "generalized Stolz cones" which are determined by a certain function $\omega \in \Omega_0$.

In this section we are generalizing Snyder's results for the case of functions defined on arbitrary open subsets of a metric space.

The simplicity of the proofs we are suggesting here in contrast to more laborious direct proofs consists in the fundamental thought itself and this is the introduction of the fine boundary topology having the Lusin-Menchoff property. And besides, as a consequence of the theory of fine topologies we obtain an assertion concerning the behaviour in points of semicontinuity or continuity of the boundary function. Corollary 9.17 seems to be a new result.

9.D Miscellaneous applications

In the course of the foregoing sections we have shown some concrete applications of general theorems. It was above all the Jarník-Blumberg method concerning generalized derivatives. The remainder of this section will be devoted to further interesting examples of our procedures. Other related results may be found in the exercises.

D.1 Functions continuous on curves. At about the end of the nineteenth century, R. Baire proved that any separately continuous (i.e. continuous in each variable) function on $\mathbb{R}^2$ is a Baire one function. In fact, using the "strip" method it is not too difficult to construct a sequence of continuous functions converging to it. Another proof of this assertion is based on the insertion of G_δ-sets in the topology of Example 1.1.d (cf. also Example 2.15). Here we give another proof of this fact.

9.18 PROPOSITION. Every separately continuous function on $\mathbb{R}^2$ is in Baire class one.

Proof. For each $x = (x_1,x_2) \in \mathbb{R}^2$ put

$$A_x := \{(t_1,t_2): \ t_1 = x_1 \ \text{ or } \ t_2 = x_2\} \setminus \{x\} .$$

Putting $M = \mathbb{R}^2$ in Theorem 9.2 we can see that the condition (W) of this theorem is obviously satisfied and thus, without much effort, we get the assertion.

The following question occurs to us immediately: What can be said if a function f on $\mathbb{R}^2$ is continuous with respect to any unit circle in the plane? Using the G_δ-insertion method it can be proved that again f must be a Baire one function (cf.Ex. 2.D.10b), but now the direct construction of a sequence of continuous functions converging to f is not evident. On the other hand, Theorem 9.2 yields immediately the result (we can take $A_x = \{(t_1, t_2): (t_1-x_1-1)^2 + (t_2-x_2)^2 = 1 \text{ or } (t_1-x_1+1)^2 + (t_2-x_2)^2 = 1\} \setminus \{x\}$ for any $x = (x_1,x_2) \in \mathbb{R}^2$).

In fact, a more general result is valid. In light of Theorem 9.2 we can deduce easily that any function which is continuous with respect to a family of "translation invariant" triods is in Baire class one (cf. Ex. 2.D.10a).

D.2 A rare function on $\mathbb{R}^2$. In this section we denote by φ *Zahorski's C^∞-function* on $\mathbb{R}$ whose Taylor series at x does not converge on any neighborhood of x whenever $x \in \mathbb{R}$. Its existence is proved in Z. Zahorski (1947).

9.19 PROPOSITION. For every Baire one function f on $\mathbb{R}$ there exists a function F on $\mathbb{R}^2$ such that

(a) $F(x, \varphi(x)) = f(x)$ for each $x \in \mathbb{R}$,

(b) if g is a real analytic function on an open interval I, then $F(x,g(x))$ is continuous on I.

Moreover, F is continuous on the set $\mathbb{R}^2 \setminus \{(x, \varphi(x)) : x \in \mathbb{R}\}$.

Proof. Use Theorems 9.15 putting $X = \mathbb{R}^2$, $M = \{(x, \varphi(x)) : x \in \mathbb{R}\}$ and having $\omega \in \Omega_0$ for which $\lim_{t \to 0+} t^{-n}\omega(t) = 0$ for each $n \in \mathbb{N}$ (such an ω obviously exists!). We obtain a function F on $\mathbb{R}^2 \setminus M$; we complete the definition of F at the points of M in accordance with (a). Since F is continuous on $\mathbb{R}^2 \setminus M$, it remains to show that $F(x,g(x))$ is continuous at x_0 provided g and I are as in (b) and $g(x_0) = \varphi(x_0)$. But in this case there is $k \in \mathbb{N}$ such that $g^{(k)}(x_0) \neq \varphi^{(k)}(x_0)$. It follows that there is $\varepsilon > 0$ such that

$$\rho((x,g(x)), M) > \varepsilon[\rho((x,g(x)), (x_0, \varphi(x_0)))]^k$$

for all x sufficiently near to x_0. In light of the choice of ω, we get

$$\lim_{x \to x_0} F(x,g(x)) = f(x_0) = F(x_0, \varphi(x_0)) = F(x_0, g(x_0)).$$

D.3 Discontinuous strongly approximately continuous functions. Let d_s be a strong density topology on $\mathbb{R}^2$ determined by the interval basis (B_4). Since d_s is deprived of the Lusin-Menchoff property (in fact, d_s is not regular - see C. Goffman, C. J. Neugebauer and T. Nishiura (1961), Th. 5; cf. Ex. 6.D.2), and even the corresponding base operator b_s has not the Lusin-Menchoff property, the construction of discontinuous strongly approximately continuous functions cannot be based on our extension theorems directly. However, we can use the Lusin-Menchoff property of a suitable fine boundary topology as the next proposition shows.

9.20 PROPOSITION. Let $p \subset \mathbb{R}^2$ be a straight line not parallel with the axes. Then d_s is finer than the fine boundary topology $\tau = \tau(p,\omega)$ for $\omega(x) = x$.

Proof. In order to show the assertion, we must prove $\bar{A}^{d_s} \subset \bar{A}^{\tau}$ for every $A \subset \mathbb{R}^2$. It clearly suffices to prove that $x \notin \bar{A}^{d_s}$ provided $x \in p \setminus \bar{A}^{\tau}$. Let $\alpha \in (0,\frac{\pi}{2})$ be the angle between p and the axis y. Pick $\varepsilon > 0$. There is $\gamma > 0$ such that

$$0 < \alpha - \gamma < \alpha < \alpha + \gamma < \frac{\pi}{2}$$

and

$$(*) \qquad (1+\varepsilon) . \mathrm{tg}(\alpha-\gamma) > \mathrm{tg}(\alpha+\gamma) .$$

Put

$$\Gamma_\gamma = \{z \in \mathbb{R}^2 : \rho(z,p) < \rho(z,x) . \sin\gamma\} .$$

Then, as $x \notin \bar{A}^{\tau}$, there is $r > 0$ such that $A \cap U(x,r) \subset \Gamma_\gamma$ (Lemma 7.18). Let I be a closed interval whose sides are parallel with the coordinate axes, $x \in I \subset U(x,r)$. Using Fubini's theorem and $(*)$ we get

$$\lambda_2(I \cap A) \leq \lambda_2(I \cap \Gamma_\gamma) < \varepsilon . \lambda_2(I) ,$$

which implies that $x \notin b_s(A)$. Hence $x \notin \bar{A}^{d_s}$.

9.21 THEOREM. Let $p \subset \mathbb{R}^2$ be a straight line not parallel with the axes. Then any Baire one function on p can be extended over $\mathbb{R}^2$ resulting in a strongly approximately continuous function which is continuous outside p.

REMARK. Compare Theorem 9.21 with the following result of C. Goffman, C.J.Neugebauer and T. Nishiura (1961), Th. 4: If f is strongly approximately continuous on $\mathbb{R}^2$, then for each $x \in \mathbb{R}$ the function $f(x,.)$ is approximately continuous. For a generalization of Theorem 9.21 see Ex. 9.D.5.

D.4. Limits with respect to Stolz angles. In Proposition 9.5 we have shown that the preponderant boundary function with respect to "*conformable standard*" Stolz angles is in Baire class one. On the other hand, Theorem 9.6 stated that the approximate boundary function formed with respect to "*variable standard*" Stolz angles is again in Baire class one provided the change of angles is somehow controlled.

We complete now the boundary investigation by considering ordinary boundary functions formed with respect to "*variable non-standard*" Stolz angles. As usual, we used either G_δ-insertion method or (W) of Theorem 9.2 for proving that a given function f belongs to Baire class one. Sometimes it can happen that the application of both these methods is far from being straightforward. In the next proposition we use in an essential way the completeness of $\mathbb{R}$ to employ (D-P) of Theorem 2.12 together with some category arguments. (A similar method was used recently by L. Larsson (1983) in the proof of the assertion that any symmetric derivative is in Baire class one.)

9.22 PROPOSITION (L.E.Snyder (1965)) Assume that for every $x \in \mathbb{R}$ a Stolz angle $S_x = S_x(\theta_1(x), \theta_2(x))$ is given. Let F be a function on $H := \mathbb{R} \times (0, +\infty)$ having at each point $(x,0)$ a limit $f(x)$ with respect to S_x. If there is a Baire one function g on $\mathbb{R}$ such that $\theta_1 < g < \theta_2$, then f is in $B_1(\mathbb{R})$.

Proof. If f does not belong to $B_1(\mathbb{R})$, (D-P) yields the existence of a closed set $T \subset \mathbb{R}$ and of reals $a < b$ with the property that both the sets $[f \leq a]$ and $[f \geq b]$ are dense in T. A routine argument shows that there are $c, d \in \mathbb{R}$, $a < c < d < b$ such that one of the sets $[f \leq c]$, $[f \geq d]$ is second category in T. Then, as (say) $[f \leq c]$ is a second category set in T, there must be an $n \in \mathbb{N}$ such that

$$A_n := \{x \in T\colon\ f(x) \leq c,\ \ \theta_1(x) < g(x) - \tfrac{1}{n} < g(x) + \tfrac{1}{n} < \theta_2(x)$$

$$\text{and } F(y) < \tfrac{1}{2}(c+d) \text{ for each } y \in S_x \cap U(x, \tfrac{1}{n})\ \}$$

is second category in T.

There is a portion P_1 of T on which A_n is dense (by a portion of T we mean a set of the form $(\alpha, \beta) \cap T \neq \emptyset$). As $g \in B_1(\mathbb{R})$, there is a portion P of T, $P \subset P_1$ such that $\operatorname{osc}_P g < \frac{1}{n}$. Pick $x_0 \in P$ and $\delta > 0$ in such a way that

$$f(x_0) \geq d,\quad F(y) > \tfrac{1}{2}(c+d) \text{ for } y \in S_{x_0} \cap U(x_0, \delta).$$

Since $\operatorname{osc}_P g < \frac{1}{n}$, we get for $x \in P \cap A_n$ sufficiently close to x_0

$$S_{x_0} \cap U(x_0, \delta) \cap S_x \cap U(x, \tfrac{1}{n}) \neq \emptyset.$$

But for y belonging to this set we have $\frac{1}{2}(c+d) < F(y) < \frac{1}{2}(c+d)$. This contradiction yields the required conclusion.

To prove the next result we use, in contrast to the proof of the preceding proposition, the condition (W) of Theorem 9.2.

9.23 PROPOSITION (L.E.Snyder(1965)). Let $\{S_x\}_{x \in \mathbb{R}} = \{S_x(\theta_1(x), \theta_2(x))\}_{x \in \mathbb{R}}$ be a family of Stolz angles. Let $M \subset \mathbb{R}$ and suppose that F is a function on $H := \mathbb{R} \times (0, +\infty)$ having at each point $(x, 0)$, $x \in M$ a limit $f(x)$ with respect to S_x.

Then the function f is honorary Baire two on M.

Proof. Given $x \in \mathbb{R}$, denote $\dot{x} = (x,0)$. For every $x \in M$ and $n \in \mathbb{N}$ there is $\delta_n(x) < \frac{1}{n}$ such that

$$|F(y) - f(x)| < \frac{1}{n}$$

whenever $y \in S_x \cap U(\dot{x}, \delta_n(x))$. Denoting $g_n(x) = (\theta_1(x), \theta_2(x), \delta_n(x))$, the standard argument gives that there is a countable set V_n such that

$$(*) \qquad g_n(x) \in C^+(g_n,x) \cap C^-(g_n,x)$$

for each $x \in M \setminus V_n$. (Here, the right cluster set $C^+(f,x)$ of f at x is defined as the set of all points z for which $f^{-1}(U) \cap (x,x+r) \neq \emptyset$ for each $r > 0$ and each open neighborhood U of z. The left cluster set $C^-(f,x)$ of f at x is defined analogously.)

Our aim is to show that $f \in B_1(A)$ where $A := M \setminus \bigcup_{n=1}^{\infty} V_n$. Denote $T_n(a) = S_a \cap U(\dot{a}, \delta_n(a))$ for $a \in M$, and pick $x \in A$. Making use of $(*)$, there are t_n, z_n such that

$$(**) \qquad x - \frac{1}{n} < t_n < x < z_n < x + \frac{1}{n},$$

$$T_n(x) \cap T_n(z_n) \neq \emptyset, \quad T_n(x) \cap T_n(t_n) \neq \emptyset .$$

Put

$$A_x := (S_x \setminus \{\dot{x}\}) \cup \bigcup_{n=1}^{\infty} (T_n(z_n) \cup T_n(t_n)).$$

For each $y \in T_n(z_n)$ and each $z \in T_n(x) \cap T_n(z_n)$ we have

$$|F(y) - f(x)| < |F(y) - f(z_n)| + |f(z_n) - F(z)|$$

We can see that

$$f(x) = \lim_{A_x \ni y \to x} F(y).$$

Note that, because of $(**)$, the condition (W) of Theorem 9.2 is satisfied. Thus, we obtained the required conclusion.

EXERCISES

9.D.1 Using Theorem 9.2, prove the assertions of Ex. 2.D.5f and 2.D.10.

9.D.2 As an application of the $(\mathcal{S},\omega)$-contingent topology of Ex. 7.D.5, give an alternative proof of Proposition 9.19. To do this, prove the following proposition.

PROPOSITION. There exists a fine topology τ on $\mathbb{R}^2$ which meets the following properties:

(a) τ has the complete Lusin-Menchoff property,

(b) the graph of any Zahorski's $\mathcal{C}^\infty$-function is a τ-isolated set,

(c) if g is a real analytic function on an open interval I and F is τ-continuous on $\mathbb{R}^2$, then $F(x,g(x))$ is continuous on I.

Hint. Let $\omega(t) = \exp(-1/t)$ $(\omega(0) = 0)$ and let $\mathcal{S}_x$ be the family of all graphs of real analytic functions passing through $x \in \mathbb{R}^2$. Consider τ as the $(\mathcal{S},\omega)$-contingent topology of Ex. 7.D.5.

9.D.3 Let $M = \{(1,0), (0,1), (-1,0), (0,-1)\} \subset \mathbb{R}^2$ and let $S \subset \mathbb{R}^2$ be a set for which $\text{cont}(x,S) \cap M = \emptyset$ for every $x \in \mathbb{R}^2$. If f is the restriction to S of a function from $B_1(\mathbb{R}^2)$, then f can be extended to a separately continuous function over $\mathbb{R}^2$.

Hint. Use the M-contingent topology of Ex. 3.B.13 or the boundary topology and the B_1- extension theorem.

9.D.4 Let M be as in the preceding exercise and let $S \subset \mathbb{R}^2$ be a set such that $\text{cont}(x,S) \cap M = \emptyset$ for each $x \in \mathbb{R}^2 \setminus S$. Let f be a function on S which is the restriction of a $B_1(\mathbb{R}^2)$-function such that for every $(x,y) \in S$ we have $\lim f(x_n, y_n) = f(x,y)$ whenever $(x_n-x)/(y_n-y) \to 0$ or $(y_n-y)/(x_n-x) \to 0$, $(x_n;y_n) \in S$, $x_n \to x$, $y_n \to y$. Then f can be extended to a separately continuous function on $\mathbb{R}^2$.

Hint. Use the M-contingent topology and the B_1-extension theorem.

9.D.5 Let M be as in Ex. 9.D.3. Show that the conclusion of Theorem 9.21 remains true if p is an arbitrary closed set of measure zero such that $M \cap \text{cont}(x,p) = \emptyset$ for every $x \in p$.

Hint. Consider again the M-contingent topology τ of Ex. 3.B.13. Show that any both τ-open and density open set is d_s-open. Next use Proposition 8.15.

9.D.6 (L.E. Snyder (1967a)). Say that a set D_x is a <u>bi-arc</u> in $H := \mathbb{R} \times (0,+\infty)$ issuing from a point $(x;0)$ if it is a union of two simple closed arcs D_x^1, D_x^2 issuing from $(x;0)$ which do not intersect outside $(x;0)$ and $D_x \setminus \{(x;0)\} \subset H$.

Let $M \subset \mathbb{R}$ and for $x \in M$, let D_x be a bi-arc in H issuing from $(x;0)$. If F is a function on H and $f(x) = \lim_{D_x \ni t \to (x,0)} F(t)$, $x \in M$, then f is a honorary Baire two function on M.

Hint. It does no harm to suppose that $\sup\{\rho(z,(x;0)) : z \in D_x^i\} > 1$ for $x \in M$, $i = 1,2$. Given $x \in M$ and $n \in \mathbb{N}$, there is $\delta_n(x) < 1/n$ such that $|F(y)-f(x)| < 1/n$ whenever $y \in [D_x \cap U((x;0),\delta_n(x))] \setminus \{(x;0)\}$. Choose $p_i(x) \in D_x^i \cap H \cap U((x;0),\frac{1}{2}\delta_n(x))$, $i = 1,2$, such that any point of D_x^i between points $(x;0)$ and $p_i(x)$ belongs to $U((x;0),\frac{1}{2}\delta_n(x))$. Next, imitate the proof of Proposition 9.23 replacing $\theta_1(x)$, $\theta_2(x)$, S_x by $p_1(x)$, $p_2(x)$, D_x.

9.D.7 (L.E. Snyder (1965)). In the language of the last exercise suppose, in addition, that D_x^1 and D_x^2 are strictly separated by a continuously varying halfline L_x in H. (By this we mean that for each $x \in M$ there is a halfline L_{x_1} in H issuing from $(x;0)$ such that L_x lies "between" D_x^1 and D_x^2, $L_x \cap D_x^1 = L_x \cap D_x^2 = (x;0)$ and the mapping $x \to L_x$ is continuous in an obvious sense.) Show that in this case, f is in $B_1(M)$.

Hint. We may again suppose that $\sup\{\rho(z,(x,0)) : z \in D_x^i\} > 1$ for any $x \in M$ and $i = 1,2$. Now use Theorem 9.2.

REMARKS AND COMMENTS

In this section we give another illustration of the advantage of the use of the condition (W) from Theorem 9.2 and of the use of properties of the fine boundary topology from Theorem 9.15 for certain simple proofs of generalizations of well known theorems (Propositions 9.18 and 9.23) and for new interesting constructions (Proposition 9.19 and Theorem 9.21). Note that in some applications it may be more useful not to use fine boundary topology but to it very closely related contingent topology instead (Ex. 9.D.2-9.D.5).

9.E On Bagemihl's and McMillan's result

In this short section we illustrate our ideas by giving a short proof of Bagemihl's and McMillan's result (1966). The central idea consists in introduction of the fine boundary topology and in essential use of its Lusin-Menchoff property.

Glancing back at Section 7.D, let τ be the fine boundary topology $\tau(F,\omega)$ on $X := \{(x,y): y \geq 0\}$ determined by $\omega(t) = t$ and $F := \mathbb{R} \times \{0\}$. (To keep the notation simple, we identify, as usual, F with $\mathbb{R}$.) Put further $H := X \setminus F$.

Given a real function f defined on a set $D \subset H$, set

$$A_f^o = A_f^o(D) := \{x \in F: \text{there is } y \in \mathbb{R} \text{ such that } \lim_{D \ni z \to x} f(z) = y\},$$

$$A_f = A_f(D) := \{x \in F: \text{there is } y \in \mathbb{R} \text{ such that } \tau\text{-}\lim_{D \ni z \to x} f(z) = y\}.$$

The set A_f^o is called the set of *global convergence* of f, while the set A_f is referred to as the set of *angular convergence* of f. Define, for each $x \in A_f$, the function φ_f as $\varphi_f(x) = \tau\text{-}\lim_{D \ni z \to x} f(z)$.

Then the function φ_f is called a *Stolz angle boundary function* of f.

The following two theorems are proved in F. Bagemihl and J.E. McMillan (1966). The first theorem gives information about the Borel structure of the sets A_f and A_f^o for $D = H$ and of φ_f.

THEOREM A. Let f be a real function on H. Then A_f^o is a G_δ-set, A_f is an $F_{\sigma\delta}$-set, $A_f^o \subset A_f$. Moreover, $A_f \setminus A_f^o$ is a first category set, φ_f is a Baire one function on A_f and it is continuous (relative to A_f) on A_f^o.

The next theorem shows that, in fact, the properties stated in Theorem A characterize the sets of global and angular convergence and no more informations concerning A_f, A_f^o and φ_f can be obtained.

THEOREM B. Let $A^o \subset A \subset F$. If A^o is a G_δ-set, A an $F_{\sigma\delta}$-set such that $A \setminus A^o$ is a first category set, and if φ is a real B_1-function on A which is continuous (relative to A) on A^o, then there is a real continuous function f on H such that

$$A^o = A_f^o, \quad A = A_f \quad \text{and} \quad \varphi = \varphi_f.$$

In what follows, we offer a proof of Theorem B based on our theory which seems to be considerably simpler than the original one. We start with some lemmas.

LEMMA α. Let D be a dense subset of an open set $G \subset H$. Then there is a set $K \subset D$ which is locally finite in H (i.e. each point of H has a neighborhood meeting only a finite subset of K) such that

$$F \cap \operatorname{der} G = \operatorname{der} K, \quad F \cap \operatorname{der}_\tau G = \operatorname{der}_\tau K.$$

Proof is straightforward and we therefore omit it.

LEMMA β. Let P be a dense subset of H. If f is a real continuous function on H, then there is a set $K \subset P$ which is locally finite in H such that $\operatorname{der}_\tau K = F$ and

$$A_f = A_{f \wedge K}, \quad A_f^{\circ} = A_{f \wedge K}^{\circ}.$$

Proof. Let $\{B_n\}_{n=1}^{\infty}$ be a countable base of topology of $[-\infty, +\infty]$. By Lemma α, construct the sets K_n corresponding to $G = f^{-1}(B_n)$ and $D = G \cap P$, and put

$$K := \bigcup_{n=1}^{\infty} K_n \cap (\mathbb{R} \times (0, \tfrac{1}{n}]).$$

LEMMA γ. Let $T \subset F$ be a G_δ-set. Then there is an open set $G \subset H$ such that

$$F \cap \operatorname{der}_\tau G = T \quad \text{and} \quad F \cap \operatorname{der} G = \bar{T}.$$

Proof. It does no harm to suppose that $T \subset (0,1)$ (identifying $\mathbb{R}$ and F !)). Given an open interval $J \subset (0,1)$ and $n \in \mathbb{N}$, set

$$T_J := \{(x,y) : x \in J, \ 0 < y \leqq \rho(x, \mathbb{R} \setminus J)\},$$

$$V_J^n := \{(x,y) : x \in J, \ \frac{1}{2n} \operatorname{dist}^2(x, \mathbb{R} \setminus J) < y < \frac{1}{n} \operatorname{dist}^2(x, \mathbb{R} \setminus J)\}.$$

If $T = \bigcap_{n=1}^{\infty} B_n$ where $(0,1) \supset B_1 \supset B_2 \supset B_3 \supset \dots$ are open sets, put

$$V^n := \bigcup \{V_J^n : J \text{ is a component of } B_n\},$$

$$G := \bigcup_{n=1}^{\infty} V^n \setminus \bigcup \{T_J : J \text{ is a component of } (0,1) \setminus \bar{T}\}.$$

Proof of Theorem B. Let $A^{\circ} = \bigcap_{n=1}^{\infty} U_n$ where $U_1 \supset U_2 \supset \dots$ are open sets, $A = \bigcap_{n=1}^{\infty} H_n^*$

with $H_1^* \supset H_2^* \supset H_3^* \supset \ldots$ F_σ-sets, and let $M \supset A \setminus A^\circ$ be a first category F_σ-set. Put $H_n := U_n \cup (M \cap H_n^*)$. Then $A = \bigcap_{n=1}^{\infty} H_n$ where $H_1 \supset H_2 \supset \ldots$ are F_σ-sets, $U_n \subset H_n$ and $H_n \setminus U_n$ is a first category set. If $T_n := F \setminus H_n$, then $\bar{T}_n = F \setminus U_n$. By Lemma γ, there are open sets $G_n \subset H$ such that

$$F \cap \operatorname{der}_\tau G_n = T_n = F \setminus H_n,\ F \cap \operatorname{der} G_n = \bar{T}_n = F \setminus U_n.$$

The fine boundary topology has the complete Lusin-Menchoff property on X (see Theorem 7.19). Hence τ has the Lusin-Menchoff property on $Y := A \cup H$, and since A is closed in Y, there is a real continuous function g on H such that

$$\tau\text{-}\lim_{z \to x} g(z) = \varphi(x)$$

if $x \in A$ and

$$\lim_{z \to x} g(z) = \varphi(x)$$

if $x \in A_o$ (Corollary 3.31). But it is by no means clear whether $A = A_g$ and $A^\circ = A_g^\circ$, so we proceed as follows: Cite Lemma β to obtain a set $K \subset H$ locally finite in H such that $\operatorname{der}_\tau K = F$ and $A_g = A_{g \wedge K}$, $A_g^\circ = A_{g \wedge K}^\circ$. Find pairwise disjoint sets D_n which do not meet K and are dense in H. Thanks to Lemma α, there are pairwise disjoint sets $K_n \subset H$ which are locally finite in H such that $K_n \cap K = \emptyset$ and

$$(*) \qquad \operatorname{der}_\tau K_n = T_n = F \setminus H_n,\ \operatorname{der} K_n = \bar{T}_n = F \setminus U_n.$$

Set

$$Q := \bigcup_{n=1}^{\infty} (K_n \cap (\mathbb{R} \times (0, \tfrac{1}{n}])),\ S = Q \cup A \cup K$$

and put

$$q(x) := \begin{cases} 0 & \text{if } x \in A \cup K, \\ \frac{1}{n} & \text{if } x \in K_n \cap (\mathbb{R} \times (0, \frac{1}{n}]). \end{cases}$$

Now, S is a closed subset of $Y = A \cup H$ and $(*)$ implies that q is τ-continuous at all points of A and continuous at all points of A°. Obviously, q is continuous at all points of $K \cup Q$. Again, according to Corollary 3.31 there is a real continuous function g^* on H such that $g^* = g$ on $K \cup Q$ and

$$\tau\text{-}\lim_{z\to x} g^*(z) = 0$$

if $x \in A$,

$$\lim_{z\to x} g^*(z) = 0$$

if $x \in A^o$. Finally, put $f = g + g^*$. Then, of course,

$$\tau\text{-}\lim_{z\to x} f(z) = \varphi(x)$$

if $x \in A$,

$$\lim_{z\to x} f(z) = \varphi(x)$$

if $x \in A^o$, and what we must show is that $A \supset A_f$, $A^o \supset A_f^o$. To this end let $x \in F \setminus A$ be given. If $x \in F \setminus A_g$, then $x \notin A_{g \wedge K} = A_{f \wedge K}$ and, therefore, $x \notin A_f$. Assume now that $x \in A_g$. Find $n \in \mathbb{N}$ such that $x \in T_n$ $(x \notin A = \bigcap_{n=1}^{\infty} H_n$. Since $x \in \operatorname{der}_\tau K \cap \operatorname{der}_\tau K_n$, we see that both 0 and $\frac{1}{n}$ are τ-cluster values of g^* at x. Consequently, $x \notin A_{g^*}$, and thus $x \notin A_{g+g^*} = A_f$. The proof of $A_f^o \subset A^o$ is similar.

EXERCISES

9.E.1 When we drop in Theorem B the condition that $A \setminus A^o$ is a first category set, then still there is a real continuous function f on H such that

$$A^o \subset A_f^o, \quad A = A_f \quad \text{and} \quad \varphi = \varphi_f$$

(cf. the original paper by E. Bagemihl and J.E. McMillan (1966), Th. 2).

Hint. It only needs to be observed that if $\varphi \in B_1(A)$ is continuous on A_o, then there exists a G_δ-set A_o^* such that $A_o \subset A_o^* \subset A$, $A \setminus A_o^*$ is a first category set and φ is continuous (relative to A) on A_o^*.

REMARKS AND COMMENTS

We demonstrate the advantage of using the Lusin-Menchoff property of the fine boundary topology again by suggesting a substantially more perspicuous proof of a well known result of F. Bagemihl and J.E. McMillan (1966).

10. FINE TOPOLOGY IN POTENTIAL THEORY

A. Classical fine topology
B. Axiomatic theories
C. Balayage and thinness
D. The fine topology

10.A Classical fine topology

Although it is very difficult to determine where the potential theory exactly originated (by the way, much information in regard of this point can be obtained from many fine lucid articles on this topic), it is obvious that the roots of this theory are closely related to some problems of physics investigated in the 17th and 18th century. It was especially Newton's law of gravitation and Coulomb's law of electrostatics that gave the impulse for the potential theory to come into being. Since the very beginning two fundamental concepts have appeared in the potential theory - harmonic functions (as solutions of the Laplace equation) and potentials (representing the potential energy of a unit mass or charge at a point due to a distribution of mass or charge on the space).

We shall skip a long period of the development of the potential theory and come to a halt only at the concept of the classical fine topology which is the main representative of a wide class of fine topologies the detailed investigation and the use of which in a more general context the next sections are dedicated to.

At the beginning we shall focus our attention on the Euclidean spaces $\mathbb{R}^n$ when $n \geq 3$.

If $z \in \mathbb{R}^n$, we define the *fundamental harmonic function* u_z with pole z on $\mathbb{R}^n$ by

$$u_z(x) = \begin{cases} |x-z|^{2-n} & \text{for } x \neq z, \\ +\infty & \text{for } x = z. \end{cases}$$

If μ is a Radon measure on $\mathbb{R}^n$, the *Newtonian potential* N^μ is defined on $\mathbb{R}^n$ by

$$N\mu(x) = \int_{\mathbb{R}^n} u_z(x)\, d\mu(z), \quad x \in \mathbb{R}^n.$$

It follows immediately that $N\mu$ is nonnegative and lower semicontinuous (use Fatou's lemma).

Define the *fine topology* on $\mathbb{R}^n$ as the smallest topology on $\mathbb{R}^n$ for which all Newtonian potentials are continuous, i.e. the fine topology is the weak topology in the sense of Section 2.B induced on $\mathbb{R}^n$ by the family $\mathcal{N} := \{ N\mu : \mu \text{ is a Radon measure on } \mathbb{R}^n \}$.

Before moving on to some properties of the fine topology which typify its importance, it is convenient to recall some definitions. A real function h on an open set $G \subset \mathbb{R}^n$ having continuous second partial derivatives is called a *harmonic function* if $\Delta h = 0$ on G. Notice that the fundamental harmonic function with pole z is obviously harmonic on $\mathbb{R}^n \setminus \{z\}$.

It is one of the remarkable and fruitful properties of harmonic functions that they satisfy the *averaging principle*: If h is harmonic on G and if $\overline{U(z,R)} \subset G$, then the value $h(z)$ is equal to the average of h over the boundary of the ball $U(z,R)$. Indeed, denote

$$\Phi(r) = \frac{r^{1-n}}{\sigma_n} \int_{\partial U_r} h \, d\sigma_{z,r}$$

(where $U_r := U(z,r)$, $\sigma_{z,r}$ denotes the Hausdorff $(n-1)$-dimensional measure on ∂U_r and σ_n is the surface area of a sphere of radius 1), we have

$$\Phi(s) = \frac{r^{1-n}}{\sigma_n} \int_{\partial U_r} h\left(\frac{s}{r} y\right) d\sigma_{z,r}(y)$$

for every $s, r \in (0,R]$. Green's formula now provides

$$\Phi'(r) = \frac{1}{r^n \sigma_n} \int_{\partial U_r} \frac{\partial h}{\partial n} \, d\sigma_{z,r} = \frac{1}{r^n \cdot \sigma_n} \int_{U_r} \Delta h \, d\lambda_n = 0$$

(where $\frac{\partial h}{\partial n}$ is just the directional derivative of h in the direction of the outer unit normal to the $\partial U_{z,r}$). Hence, Φ must be constant on $(0,R]$ and $h(z) = \lim_{r\to 0_+} \Phi(r) = \Phi(R)$.

We now consider a large family of functions, where the requirement that they satisfy the averaging principle is relaxed to require that they have the "superaveraging property". If f is a function on ∂U_r, define

$$L_f(z,r) = \frac{r^{1-n}}{\sigma_n} \int^*_{\partial U_r} f \, d\, \sigma_{z,r}.$$

A function f on G is said to be *super-mean-valued* on G if $f(x) \geq L_f(x,r)$ whenever $\overline{U(x,r)} \subset G$.

10.1 LEMMA. The fundamental harmonic function with pole z is super-mean-valued on $\mathbb{R}^n$.

Proof. Let $U := U(x,r)$. Since everything holds if $x = z$, or $|x-z| > r$ (then u_z is a harmonic function on an open set containing $\bar{U}$), assume first $|x-z| = r$. Put $z_k := z + \frac{1}{k}(z-x)$. Then $|x-z_k| > r$ and $u_{z_k} \uparrow u_z$. Hence the assertion easily follows.

The way we can prove the assertion for $0 < |x-z| < r$ is by using a certain kind of symmetry. Let $y \in \partial U_r$. Putting

$$z^* = x + \frac{r^2}{|z-x|^2}(z-x),$$

we have

$$|z^*-x| > r \quad \text{and} \quad |y-z| = \frac{|z-x|}{r} \, |y-z^*|$$

(cf. [He], p.14). Hence

$$u_z(y) = \left(\frac{|z-x|}{r}\right)^{2-n} u_{z^*}(y),$$

and therefore

$$L_{u_z}(x,r) = \left(\frac{|z-x|}{r}\right)^{2-n} u_{z^*}(x) \leq u_z(x).$$

(We use again the fact that u_{z^*} is harmonic on an open set containing $\overline{U(x,r)}$.)

10.2 PROPOSITION. Any Newtonian potential is a super-mean-valued function on $\mathbb{R}^n$.

Proof. Let μ be a Radon measure on $\mathbb{R}^n$. Choose $z \in \mathbb{R}^n$ and $r > 0$. Using Fubini's theorem and Lemma 10.1 we have

$$L_{N^\mu}(z,r) = \frac{1}{\sigma_n r^{n-1}} \int_{\partial U(z,r)} N^\mu \, d\sigma_{z,r} = \int_{R^n} \frac{1}{\sigma_n r^{n-1}} N^{\sigma_{z,r}} d\mu \leq$$

$$\leq \int_{R^n} u_z \, d\mu = N^\mu(z).$$

10.3 COROLLARY. If μ is a Radon measure on $\mathbb{R}^n$, $z \in \mathbb{R}^n$, and $r > 0$, then

$$N^\mu(z) \geq \frac{n}{\sigma_n r^n} \int_{U(z,r)} N^\mu \, d\lambda_n.$$

Proof. Denote $f := N^\mu$. Given $R \in (0,r)$, we have $f(z) \geq L_f(z,R)$ by Proposition 10.1. Therefore

$$\int_{U(z,r)} f \, d\lambda_n = \int_0^r \Big(\int_{\partial U(z,R)} f(y) \, d\sigma_{z,R}(y) \Big) dR =$$

$$= \int_0^r \sigma_n R^{n-1} L_f(z,R) \, dR \leq f(z) \int_0^r \sigma_n R^{n-1} \, dR =$$

$$= \frac{\sigma_n r^n}{n} f(z).$$

10.4 EXAMPLE. We will construct discontinuous Newtonian potentials. So, let z be an accumulation point of a sequence $\{x_k\}$ of points in $\mathbb{R}^n$, $z \neq x_k$, and let α_k be positive constants.

If $\mu = \sum \alpha_k \varepsilon_{x_k}$, then $N^\mu(z) = \sum \alpha_k |z-x_k|^{2-n}$ while $N^\mu(x_k) = +\infty$. Choosing constants α_k such that $\sum \alpha_k |z-x_k|^{2-n} < +\infty$ we easily produce the required example.

Note that there are even bounded discontinuous Newtonian potentials (see Ex. 10.A.1).

10.5 THEOREM. Every Newtonian potential on $\mathbb{R}^n$ is approximately continuous.

Proof. Let u be a Newtonian potential on $\mathbb{R}^n$. Since u is lower semicontinuous, it remains to show that u is upper semicontinuous in the density topology. Pick $z \in \mathbb{R}^n$ and assume $u(z) < +\infty$. Given $\varepsilon > 0$ and $b > u(z)$, there is $a < u(z)$ with $u(z)-a < \varepsilon(b-a)$. Using the lower semicontinuity of u find $\delta > 0$ such that $u > a$ on $U(z,\delta)$. If $0 < r < \delta$ and $U := U(z,r)$, then by Corollary 10.3 we have

$$\lambda_n(U \cap [u \geq b]) \leq \frac{1}{b-a} \int_U (u-a)\,d\lambda_n \leq \frac{u(z)-a}{b-a} \lambda_n U \leq$$

$$\leq \varepsilon \lambda_n U.$$

10.6 COROLLARY. The fine topology on $\mathbb{R}^n$ is coarser than the ordinary density topology on $\mathbb{R}^n$.

10.7 PROPERTIES OF THE FINE TOPOLOGY. Some simple properties of the fine topology could be derived directly from the definition and from the above considerations. We mention some of them.

(a) The fine topology is finer than the Euclidean topology. Indeed, since $U(z,r) = \{x \in \mathbb{R}^n : N^{\varepsilon_z}(x) > r^{2-n}\}$, each ball in $\mathbb{R}^n$ is finely open. The existence of discontinuous Newtonian potentials implies that the fine topology is properly finer.

(b) By (a) and Theorem 2.3, the fine topology is Hausdorff and completely regular.

(c) Any nonempty finely open set must be uncountable in light of Corollary 10.6. In particular, the fine topology has no isolated points.

(d) Any countable set $S \subset \mathbb{R}^n$ is finely closed, even finely isolated. Assume, in order to obtain a contradiction, that there is a fine accumulation point z of S; we may assume that $z \notin S$. According to Example 10.4, there is a Radon measure μ such that $N^{\mu}(z) < +\infty$ while $N^{\mu} = +\infty$ on S. The set $\{x \in \mathbb{R}^n : N^{\mu}(x) < N^{\mu}(z) + 1\}$ is a fine neighborhood of z which does not meet S.

(e) It now follows from Theorem 2.1 that finely compact sets are just the finite sets and that the fine topology is not metrizable.

REMARK. Using logarithmic potentials, a similar introduction of the fine topology on $\mathbb{R}^2$ may be found in the exercises for this section. It would even be possible to go further, replacing "fundamental harmonic functions" by more general suitable kernels. The next example of potential theory for the heat equation is also very important.

Define the function Γ on $\mathbb{R}^{n+1}$ by

$$\Gamma(x,t) := \begin{cases} (4\pi t)^{-\frac{n}{2}} e^{-\frac{|x|^2}{4t}} & \text{on } \{(x,t) \in \mathbb{R}^{n+1}: t > 0\}, \\ 0 & \text{elsewhere.} \end{cases}$$

As Γ is a solution of the heat equation

$$\Omega f = 0$$

on $\mathbb{R}^{n+1} \setminus \{0\}$ where the differential operator Ω equals $\Delta - \frac{\partial}{\partial t}$,

Γ plays a role of the *fundamental caloric function* for $\mathbb{R}^{n+1}$ with pole 0. The interested reader could find the way to develop the theory of *heat potentials*

$$(x,t) \longmapsto \int_{\mathbb{R}^{n+1}} \Gamma(x-y,\ t-s)\ d\mu(y,s)$$

of Radon measures and the corresponding fine topology on $\mathbb{R}^{n+1}$. Since the behaviour of the caloric functions as the solutions of the heat equation is similar to that of harmonic functions, this analogy led to introducing a more general setting of harmonic spaces.

EXERCISES

10.A.1 (a) In a similar fashion as in the proof of Lemma 10.1 deduce that

$$N^{\sigma_{z,r}}(y) = \sigma_n r^{n-1} L_{u_z}(y,r) = \begin{cases} \sigma_n r^{n-1} u_z(y) & \text{for } |y-x| > r, \\ \sigma_n r & \text{for } |y-x| \leq r. \end{cases}$$

(b) Let $S = \{x_1, x_2, \dots\}$ be a countable subset of $\mathbb{R}^n$ $(n \geq 3)$, and let $z \in \mathbb{R}^n \setminus S$. Construct a bounded Newtonian potential u such that $u \geq 2$ on S and $u(z) \leq 1$.

Hint. Construct by induction positive reals r_n, a_n such that the Newtonian potentials $u_n := N^{\mu_n}$ where $\mu_n = \sigma_{x_1,r_1} + \dots + \sigma_{x_n,r_n}$ satisfy.

$$u_n(x_n) \geq 2,\ u_n \leq 2+1+\dots+2^{2-n} \text{ on a neighborhood } U \text{ of } x_n,$$

$$N^{\sigma_{x_n,r_n}}(y) \leq 2^{-n} \text{ for } y \in \{z\} \cup (\mathbb{R}^n \setminus U).$$

Now, consider the Newtonian potential of the measure $\sum_{n=1}^{\infty} \varepsilon_{x_n, r_n}$.

10.A.2 The Lebesgue spine (H. Lebesgue (1913), cf. [He], p. 175). Let $D = \{x \in \mathbb{R}^3 : x_1 \in [0,1],\ x_2 = x_3 = 0\}$ and let μ be the Radon measure given by the formula

$$\mu(A) = \int_{A \cap D} x_1 d\lambda_1(x_1).$$

Put

$$U := \{x \in \mathbb{R}^3 : |x| < 1 \quad \text{and} \quad N^{\mu}(x) < 2\}$$

and prove the following assertions:

(a) $N^{\mu}(0) = 1$, $N^{\mu} = +\infty$ on $D \setminus \{0\}$, N^{μ} is continuous on $\mathbb{R}^3 \setminus \{0\}$ and harmonic outside D (= the support of μ),

(b) 0 is a boundary point of U and an accumulation point of ∂U,

(c) $\{0\} \cup (\mathbb{R}^3 \setminus U)$ is a fine neighborhood of 0,

(d) U is a bounded open subset of $\mathbb{R}^3$ having a connected complement.

10.A.3 Logarithmic potentials on $\mathbb{R}^2$. The fundamental harmonic function u_z with pole z is defined on $\mathbb{R}^2$ by putting

$$u_z(x) = \begin{cases} -\log|x-z| & \text{if } x \neq z \\ +\infty & \text{if } x = z. \end{cases}$$

The *logarithmic potential* L^{μ} of a measure μ on $\mathbb{R}^2$ having a compact support is defined by

$$L^{\mu}(x) = \int u_z(x)\, d\mu(z) \quad \left(= \int u_z(x)\, d\mu(x)\right).$$

Show that

(a) any logarithmic potential is lower semicontinuous and super-mean-valued (note that the logarithmic potential need not be nonnegative and therefore Fubini's theorem is not directly applicable, cf. [He], p.100),

(b) there are discontinuous logarithmic potentials,

(c) any logarithmic potential is approximately continuous.

10.A.4 Hyperharmonic and superharmonic functions. A function u is said to be *hyperharmonic* on an open set $U \subset \mathbb{R}^n$ $(n \geq 2)$ if u is lower semicontinuous, lower finite and super-mean-valued on U.

(a) Let u be a hyperharmonic function on U. It can be proved that the following assertions are equivalent:

(i) u is not identically $+\infty$ on any component of U,

(ii) u is finite a.e. on U (relative to Lebesgue measure),

(iii) u is finite on a dense subset of U.

Hint. The implications (ii) $\Rightarrow$ (iii) $\Rightarrow$ (i) are clear, concerning (i) $\Rightarrow$ (ii) see [He] , Th. 4.10.

A hyperharmonic function which satisfies (i), and hence all equivalent conditions above, is termed *superharmonic*.

(b) Show that any Newtonian potential is either superharmonic on $\mathbb{R}^n$ or identically $+\infty$, and every logarithmic potential on $\mathbb{R}^2$ is superharmonic.

(c) Let u be a lower semicontinuous, lower finite function on an open set $U \subset \mathbb{R}^n$. The following assertions are equivalent:

(i) u is hyperharmonic on U,

(ii) for every $x \in U$ there is $r_x > 0$ such that

$L_u(x,r) \leq u(x)$ whenever $r < r_x$ (u is *locally* super-mean-valued,

(iii) if $W \subset \overline{W} \subset U$ is a relatively compact open set and if $f \in \mathcal{C}(\overline{W})$ is harmonic on W and $f \leq u$ on ∂W, then $f \leq u$ on W.

Hint. Consult [He] , Chap. 4.

(d) A function h is harmonic on an open set $U \subset \mathbb{R}^n$ if and only if h and $-h$ are both superharmonic on U.

Hint. Use the fact that h is harmonic on U if and only if it is finite, continuous on U and satisfies the averaging principle.

(e) (Riesz decomposition theorem, F. Riesz (1930).) Let u be a superharmonic function on an open set U, and let $W \subset \overline{W} \subset U$ be a relatively compact open set. Then there is a unique Radon measure μ on W and a harmonic function h on W such that

$$u = \begin{cases} N^{\mu} + h \text{ on } W & \text{if } n \geq 3, \\ L^{\mu} + h \text{ on } W & \text{if } n = 2. \end{cases}$$

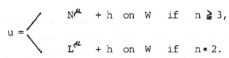

For the proof, see [He], § 6.4 or N.S. Landkof (1972), § 1.5. Show that any nonnegative superharmonic function s on $\mathbb{R}^n$ $(n \geq 3)$ can be uniquely expressed as $s = N^{\mu} + h$, where μ is a Radon measure and $h \geq 0$ is the greatest harmonic minorant of s on $\mathbb{R}^n$.

10.A.5 Show that the fine topology on $\mathbb{R}^n$ $(n \geq 3)$ is generated by the following families of functions:

(i) the family of all Newtonian potentials of Radon measures with compact support,

(ii) the family of all nonnegative superharmonic functions on $\mathbb{R}^n$.

Further show that any hyperharmonic function on any open subset of $\mathbb{R}^n$ $(n \geq 2)$ is finely continuous.

Hint. Use (e) of the previous exercise.

10.A.6 The fine topology on $\mathbb{R}^2$. The fine topology on $\mathbb{R}^2$ is defined as the coarsest topology on $\mathbb{R}^2$ making all logarithmic potentials continuous.

(a) Show that the fine topology on $\mathbb{R}^2$ has similar properties as the fine topology on $\mathbb{R}^n$ in the $n \geq 3$ case. In particular, Corollary 10.6 and all properties of 10.7 remain valid.

(b) An interesting property of the fine topology on $\mathbb{R}^2$ can be stated as follows: If V is a fine neighborhood of a point $z \in \mathbb{R}^2$, then there is a sequence of real numbers $r_k \searrow 0$ such that $\partial U(z,r_k) \subset V$. (See [He], Th. 10.14, or M. Brelot (1971), Prop. IX.6.)

(c) Let z be an endpoint of a line segment L in $\mathbb{R}^2$. It is clear from foregoing that $\{z\} \cup (\mathbb{R}^2 \setminus L)$ is not a fine neighborhood of z. For contrast, consider now the case $n \geq 3$ and appealing to Ex. 10.A.2c show that $\{z\} \cup (\mathbb{R}^n \setminus L)$ is a fine neighborhood of z whenever L is a line segment in $\mathbb{R}^n$ and z is its endpoint.

10.A.7 Connectivity properties of the fine topology.

(a) Every open connected subset of $\mathbb{R}^n$ $(n \geq 2)$ is finely connected.

Hint. Use Corollary 10.6 or Ex. 10.A.6a and the results of Section 6.C (e.g. Ex. 6.C.16). Cf. also Theorem 11.13.

(b) Show that every finely open connected subset of $\mathbb{R}^2$ is finely connected.

Hint. In light of Ex. 10.A.6b verify (ii) of Proposition 5.4. (Cf. also T.W.Gamelin and T.J. Lyons (1983), Lemma 2.1.)

Further interesting properties of the fine topology on $\mathbb{R}^n (n \geq 2)$ can be summarized as follows:

(c) The fine topology on $\mathbb{R}^n$ is locally connected (B. Fuglede (1971b) and (1972), cf. Ex. 11.D.7).

(d) Every finely open finely connected subset of $\mathbb{R}^n$ is arcwise connected (Xuan-Loc Nguyen and T. Watanabe (1972), cf. B. Fuglede (1975b).Even, any two points of a finely open finely connected set U can be joined by a polygonal path in U (B. Fuglede (1975b) for the n=2 case, probabilistic proof - T.J. Lyons (1980), analytic proof - B. Fuglede (1980)).

REMARKS AND COMMENTS

We have already mentioned that the roots of potential theory could be traced to physical problems of 17th century. Their development needs to be associated with the names of L. Euler, P.S.Laplace, J. Lagrange, J. Fourier, G. Green, C.F.Gauss, S.D. Poisson, W. Thomson (= Lord Kelvin), and potential theory was intensively studied in the course of the last century by authorities like L. Dirichlet, B. Riemann, K.Weierstrass, H.A. Schwartz, C. Neumann, H. Poincaré or A. Harnack. Many outstanding mathematicians continued the research in this interesting field. Among them we would like to recall at least D. Hilbert, H. Lebesgue, P. Fatou, G.C. Evans, I. Fredholm, G. Herglotz, F. Riesz, Ch. de la Vallée Poussin, N. Wiener, O. Perron.

At this stage we reach a period of modern potential theory which today forms a nonnegligible part of analysis. For the details of the whole story, the interested reader can consult various interesting contributions (e.g. H. Lebesgue (1907), O.D. Kellog (1926), L. Gårding (1979), M. Brelot (1952), (1972), H. Bauer (1984 a), A.F. Monna (1975)).

Investigating irregular points, M. Brelot (1939) originated the study of a classical notion of thinness. In 1940, H. Cartan called attention to the fact that the complements of sets which are thin at a given point form a local base for a certain topology and this one can be characterized as the coarsest topology making continuous each element of the family of all superharmonic functions. Later on, H. Cartan labelled this topology as the fine topology and studied it in (1946).

Notice also that superharmonic functions having continuous second partial derivatives were introduced in the 19th century, but lower semicontinuous superharmonic functions were first defined by F. Riesz (1926). An interesting paper clarifying relations between harmonic functions and averaging principles is that of I. Netuka (1975).

Remarkable results concerning connectivity and local connectivity results of the fine topology were recently reached by B. Fuglede. They are used to advantage in investigations concerning finely holomorphic functions.

Although it was known that thinness is closely related to the notion of a dispersion point, it was B. Fuglede (1971a) who explicitely noted that the fine topology is coarser than the density topology according to Wiener's test.

Related paper : J. Král and I. Netuka (1985).

10.B Axiomatic theories

Speaking about important concepts of the classical potential theory we must not omit *harmonic* and *hyperharmonic functions*. Recall that in the classical framework a hyperharmonic function on an open set U is defined as a lower semicontinuous, lower finite super-mean-valued function on U. Looking at the concepts of harmonic and hyperharmonic functions we find that one can be defined by the use of the other. Their study, together with the introduction and examination of other concepts as well (like potentials or the Dirichlet problem) creates the contents of the classical theory even at the present time. It is easy to comprehend that in the course of time these concepts have been generalized and investigated in various other respects. New theories have arisen: studying potentials it was possible to investigate more general kernels than the Newtonian one; solutions of the Laplace equation have been replaced by the study of solutions of more general classes of partial differential equations; the probabilistic aspects of potential theory have recently influenced many other directions of the development of the theory of harmonic and hyperharmonic functions.

Let us concentrate on those abstract theories which are covered today under the name of axiomatic potential theory where the introduction of the family of harmonic or hyperharmonic functions is given by certain system of axioms.

At present two most important axiomatic theories covering both elliptic and parabolic cases are used. The axiomatic system of C. Constantinescu and A. Cornea developed in [C & C] is more general and in some sense more self-evident for verification of axioms in parabolic cases. (Starting by the notion of hyperharmonic functions they defined a function h to be harmonic if both h and $-h$ are hyperharmonic.) The axiomatic system of H. Bauer is indisputably more simple and also more imaginative for non-specialist in the field.

So, we shall briefly introduce Bauer's axiomatic system to brief the reader on basic notions of this theory. It is not difficult to transfer all definitions and theorems to the axiomatic system of C. Constantinescu and A. Cornea. In fact, all assertions remain true if we deal with a $\mathfrak{P}$-harmonic space with a countable base in their axiomatic system.

We consider a locally compact Hausdorff space X with a countable base which is equipped with a sheaf $\mathcal{H}$ associating to any open subset $U \neq \emptyset$ of X a vector space $\mathcal{H}(U)$ of real continuous functions on U called *harmonic functions* on U. A relatively compact open set $U \subset X$ is termed *regular* if the *Dirichlet problem* is solvable in the following sense: For every function $f \in \mathcal{C}(\partial U)$, there is a unique continuous extension F_f^U of f to $\bar{U}$ such that F_f^U is harmonic on U. Besides, F_f^U must be nonnegative provided f is . The couple $(X,\mathcal{H})$ is termed a *$\mathfrak{P}$-harmonic space* if the following axioms are satisfied:

Sheaf axiom: If h is a harmonic function on an open set U, then the restriction of h to any open subset V of U is harmonic on V. If $\{U_\alpha\}$ is a collection of open sets and h is a function on $U := \bigcup_\alpha U_\alpha$ which is harmonic on each U_α, then h is harmonic on U.

Base axiom: The topology of X has a base consisting of regular sets.

Doob's convergence axiom: The limit of any increasing sequence of harmonic functions on any open set is a harmonic function whenever it is finite on a dense set.

Positivity axiom: For any point $x \in X$ there exists a harmonic function defined in an open neighborhood of x that does not vanish at x.

Separation axiom: The set of all potentials separates the points of X linearly, i. e. for each couple of points $x \neq y$ there exist potentials p,q on X such that $p(x)q(y) \neq p(y)q(x)$.

It remains to explain the notion of a potential.
A *hyperharmonic function* on an open subset U of X is a lower finite, lower semicontinuous function u which satisfies the following condition: Whenever V is a regular set, $\bar{V} \subset U$, and f is a continuous function on ∂V, $f \leq u$ on ∂V, then $F_f^V \leq u$ on V. A hyperharmonic function which is finite on a dense subset of U is called *superharmonic* on U. A nonnegative superharmonic function on U for which any of its nonnegative harmonic minorants vanishes identically is termed a *potential* on U. It can be proved that the families of all potentials or all superharmonic functions on an open set U form convex cones of lower semicontinuous functions.

The sheaf of continuous solutions of the Laplace equation on $\mathbb{R}^n$ or the heat equation on $\mathbb{R}^{n+1}$ satisfies these axioms (in case of the Laplace equation for $n = 1$ or $n = 2$ we must restrict the space X to be a bounded open subset of $\mathbb{R}^n$). Further nice examples of harmonic spaces, especially those of solutions of linear partial differential equations, are described in H. Bauer (1984b), or in [C&C] .

Along with the theory of harmonic spaces, it was approximately in the fifties that the research began to include even more abstract cones of functions that saved some properties of cones of superharmonic functions or potentials. Let us recall, for example, Choquet theory of convex cones of concave functions on compact convex sets or, more generally, convex cones of lower semicontinuous functions on (locally) compact spaces; potential cones of G. Mokobodzki and D. Sibony; examinations of abstract cones of lower semicontinuous functions given by M. Brelot in (1971), balayage theory on convex cones of continuous functions on Baire spaces as given in Chapter 4 of [C&C] ; investigations of J. Bliedtner and W. Hansen on balayage spaces. Since the seventies the theory of H-cones of N. Boboc, Gh. Bucur and A. Cornea has also developed successfully. In their theory a large part of balayage theory turns out to be valid in certain lattice cones. This theory includes in itself as a concrete model even the axiomatic theory of harmonic spaces and its greater generality consists mainly in removing the sheaf axiom and omitting the assumption of the given space being locally compact. Its advantage is that it permits to develop the duality theory and also includes much more probabilistic models. Nevertheless, our opinion is that from the point of view of applications to the solutions of partial differential equations the theory of harmonic spaces still remains more convenient.

Let now S be a convex cone of nonnegative functions on a topological space X. A family $F \subset S$ is said to be *increasing* to $s \in S$ if F is upper directed and $\sup F = s$. Here, of course, we consider on S the *natural order* which is defined pointwise: $s \leqq t$ if $s(x) \leqq t(x)$ for all $x \in X$. We shall use the notation $\wedge F$, $\vee F$, $s \wedge t$, $s \vee t$, for the natural infima and suprema in $(S, \leqq)$ provided they exist while $\inf F$, $\sup F$, $\inf(s,t)$, $\sup(s,t)$ denotes, of course, the pointwise operations (e.g. $\inf F\,(x) = \inf\{f(x) : f \in F\}$, etc.).

Let u be a positive function on X. An element $s \in S$ is termed *u-continuous* if for every $\varepsilon > 0$ and for every family $F \subset S$ increasing to s there exists $t \in F$ such that $s \leqq t + \varepsilon u$. An element $s \in S$ is called *universally continuous* if it is u-continuous for each strictly positive function $u \in S$. The set of all universally continuous elements of S will be denoted by S_o.

The convex cone S is called a *standard H-cone of functions on* X if the following axioms are satisfied:

(S_1) For any $s,t,u \in S$ such that $s+u \leqq t+u$, we have $s \leqq t$.

(S_2) *Axiom of upper directed sets:* If $s \in S$ and if $F \subset S$ is an upper directed family such that $\sup F \leqq s$, then the function $\sup F$ belongs to S.

(S_3) *Axiom of lower semicontinuous regularization:* For any nonempty family F of S, there exists $\wedge F$ and, furthermore,

$$\wedge(F + s) = s + \wedge F$$

for each $s \in S$.

(S_4) *The Riesz decomposition property:* For any $p,q,r \in S$ such that $p+q \geqq r$

there exist $s,t \in S$ such that $s \preccurlyeq p$, $t \preccurlyeq q$ and $s+t = r$.

(S_5) *Separation axiom:* $\mathbf{1} \in S$ and S separates the points of X (here, $\mathbf{1}$ denotes the constant function equal 1 on X).

(S_6) *Axiom of standardness:* There exists a countable subset of S_o which is increasingly dense. (A set $D \subset S$ is *increasingly dense* if the set $\{d \in D: d \preccurlyeq s\}$ is increasing to s for each $s \in S$.)

(S_7) The topology of X is the weak topology induced on X by S_o.

According to [BBC] , pp. 114-115, X is a topological subspace of a compact metrizable space. By (S_6) and (S_7), any element of the cone S is lower semicontinuous on X. Further, from (S_2) and (S_3) it follows that the ordered cone $(S,\preccurlyeq)$ is Dedekind complete (i.e. each subset of S having an upper bound has a supremum and each nonempty subset has an infimum).

Finally, notice that every $\mathbf{1}$-continuous element s of S is bounded and continuous since the family $\{\min(d,n): d \in S_o, d \preccurlyeq s, n \in \mathbb{N}\}$ which is increasing to s has a uniformly convergent subsequence converging to s.

We have already introduced the concept of superharmonic functions in abstract harmonic spaces.The convex cone S of all nonnegative superharmonic functions on a $\mathfrak{P}$-harmonic space X with a countable base provides for us the most important standard H-cones of functions if assuming that the constant functions are superharmonic. (Remark that no essential generality is lost with this assumption; for if not we may consider the cone $\tilde{S} = \{ \frac{s}{s_o} : s \in S\}$ where s_o is a fixed positive continuous function from S.)

The framework of standard H-cones of functions also includes standard balayage spaces of J. Bliedtner and W. Hansen and various cones of excessive functions (cf. [BBC]). Let us give two illustrative examples of standard H-cones.

EXAMPLES: (a) Let S be the cone of all nonnegative concave functions on a bounded interval $I \subset \mathbb{R}$. Then S is a standard H- cone of functions on I. If the interval I is open, then the sheaf of all locally linear functions on open subsets of I forms Bauer's $\mathfrak{P}$-harmonic space and S is the set of all nonnegative superharmonic functions with respect to it. If I is not open,then a correspondence between S and the family of superharmonic functions of a harmonic space does not exist (cf. Ex. 11.D.1).

(b) Let S be the cone of all nonnegative lower semicontinuous increasing functions on an interval $I \subset \mathbb{R}$. Apparently, S is again a standard H-cone of functions on I. If I does not contain its left endpoint, then S is associated with the set of superharmonic functions on a $\mathfrak{P}$-harmonic space in the sense of [C & C] . This space has no base of regular sets, and therefore is not covered up by Bauer's axiomatic system. The weak topology induced on $I = \mathbb{R}$ by S (i.e. the fine topology) is the Sorgenfrey topology of Ex. 1.A.16.

EXERCISES

10.B.1 (H. Bauer). Let X be a harmonic space with a countable base not necessarily satisfying the separation axiom. The following assertions are equivalent:

(i) X is a $\mathfrak{P}$-harmonic space,

(ii) for any $x \in X$ there is $p \in \mathcal{P}$ with $p(x) > 0$,

(iii) there is a positive real continuous potential on X,

(iv) there is a positive potential on X.

For the proofs see [Bau], Satz 2.5.3, Kor. 2.7.3, or [C&C], Prop. 2.3.2 and Prop. 7.2.1.

REMARKS AND COMMENTS

As it was pointed out by [C&C], "by the end of the last century three basic principles of potential theory, namely the Dirichlet problem, the minimum principle and the convergence property were established. Gradually it became clear that these properties of the Laplace equation are also shared by other partial differential equations ...... it was remarked that a large part of the results of potential theory could be obtained using only the above three principles. It seems therefore quite natural to develop an axiomatic system which would unify these theories and extend potential theory to these partial differential equations. This axiomatic theory was constructed around the 1950's by G. Tautz, J.L.Doob, M. Brelot and H. Bauer."

In the sixties, N. Boboc, C. Constantinescu and A. Cornea in a series of papers built up a further more general axiomatic theory, in fact a localized version of Bauer's axiomatic system. One more step was done by J. Köhn (1968) who left the axiom of the existence of a base of regular sets, and the effort culminated by the appearance of the monograph of C. Constantinescu and A. Cornea in (1972) which closes this period.

Probabilistic aspects of potential theory and an interplay between classical potential theory and Brownian motion led to further generalizations. In (1978), J. Bliedtner and W. Hansen introduced balayage spaces which generalize the notion of harmonic spaces. In a forthcoming publication by these authors, a potential theory on balayage spaces has been developed in details.

In the seventies, another generalization of harmonic spaces appeared. N. Boboc and A. Cornea in (1970a), (1970b) and (1970c), as well as N. Boboc, Gh. Bucur and A.Cornea in (1975) began their new theory of H-cones. The monograph [BBC] by the same authors forms a summary of this new part of potential theory.

Among many survey articles on axiomatic theories, the papers by M. Brelot (1972), H. Bauer (1975), (1981), (1984a), (1984b), C. Constantinescu (1966), J. Bliedtner (1980), J. Veselý (1983) could be recommended.

10.C Balayage and thinness

To save repetition, we assume for the rest of this chapter that S denotes a *standard H-cone of functions* on a metrizable topological space X with a countable base. The reader who is primarily interested in the theory of harmonic spaces can realize S as the cone of all nonnegative superharmonic functions on a $\mathfrak{P}$-harmonic space X with a countable base assuming the positive constant functions are in S.

The *fine topology* f is defined as the coarsest topology on X making all functions from S continuous. It is clear that f is finer than the original topology on X in light of the axiom (S_7). According to Theorems 3.1.3 and 4.3.2 of [BBC] (cf. [C&C], Prop. 5.1.4 and Th. 5.1.1)

$$\wedge F = \widehat{\inf F}^{f} = \widehat{\inf F}$$

for any nonempty subset F of S.

Given $s \in S$ and a subset A of X, we define the reduced function or the *reduite* of s on A by

$$R_s^A := \inf \{ t \in S: \ t \geqq s \ \text{ on } A \}$$

and the swept-out function, or the *balayage* of s on A by

$$\hat{R}_s^A := \wedge \{ t \in S : \ t \geqq s \ \text{ on } A \}.$$

Obviously, $\hat{R}_s^A = \widehat{R_s^A} = \widehat{R_s^A}^{f} \leqq R_s^A \leqq s$, $R_s^A = s$ on $\overline{A}^f$ and $\hat{R}_s^A = R_s^A = s$ on $\text{int}_f A$.

If A is finely open, then $\hat{R}_s^A \in \{ t \in S : t \geq s \text{ on } A \}$, and therefore $\hat{R}_s^A = R_s^A$ on X.

Routine arguments show that $R_s^A \leqq R_t^B$ and $\hat{R}_s^A \leqq \hat{R}_t^B$ provided $A \subset B$ and $s \leqq t$. Also, the Riesz decomposition property (S_4) yields $R_{t+s}^A = R_t^A + R_s^A$ and $\hat{R}_{t+s}^A = \hat{R}_t^A + \hat{R}_s^A$ for any $s, t \in S$ and $A \subset X$ (cf. Th. 4.2.1 of [C&C], or Prop. 3.2.1 of [BBC]).

Notice that in harmonic spaces we define R_s^A and $\hat{R}_s^A$ by means of the cone of all nonnegative hyperharmonic functions on X instead of S.

Next we define the notion of *thinness* which is one of the major themes of potential theory. Historically, it is most closely allied to the notion of fine topology;

in the forties, its key role in potential theory was established by M.Brelot and H.Cartan. A set $A \subset X$ is called *thin* at a point $x \in X$ if there is $s \in S$ such that $\hat{R}_s^A(x) < s(x)$. The set of all points of X where A is not thin is called the *base* of A and it is denoted by bA.

Conserving our earlier terminology of Section 1.B, a set $Z \subset X$ is termed *totally thin* if bZ is empty. Every union of a countable collection of totally thin sets is called a *semipolar set*.

EXAMPLES. Each one point set in $\mathbb{R}^n$ $(n \geq 2)$ is totally thin for the Laplace equation. Any line segment is totally thin for the Laplace equation in $\mathbb{R}^n$ $(n \geq 3)$ but not in the $n = 2$ case (see Ex.10.A.6c). Also any countable subset of $\mathbb{R}^n$ $(n \geq 2)$ is totally thin: it is always finely isolated but can be dense in $\mathbb{R}^n$.

Consider now the harmonic space of the solutions of the heat equation. If $c \in \mathbb{R}$ and $H := \{(x,t) \in \mathbb{R}^{n+1} : t \geq c\}$, then the halfspace H is thin at any point (x,c). This can be obtained from the observation that $\hat{R}_1^H(x,t) = 0$ for every $t \leq c$. It follows that any hyperplane $\{(x,t) \in \mathbb{R}^{n+1} : t = c\}$ is totally thin. Hence the set $T := \{(x,t) \in \mathbb{R}^{n+1} : t \in \mathbb{Q}\}$ is semipolar. It is remarkable that T is finely dense in $\mathbb{R}^{n+1}$, and that, according to Ex. 11.D.3, this phenomenon cannot occur in case of the Laplace equation.

Now we present some useful properties of these notions, sometimes without complete proofs, together with the background necessary to make the following sections self-contained for the reader's convenience.

10.7 PROPOSITION. Let $\{A_n\}$ be an increasing sequence of subsets of X, $A := \bigcup A_n$ and let $s, s_n \in S$, $s_n \nearrow s$ on A. Then

$$R_{s_n}^{A_n} \longrightarrow R_s^A, \quad \hat{R}_{s_n}^{A_n} \longrightarrow \hat{R}_s^A.$$

Proof. See [C&C], § 4.2, or [BBC], Th. 3.2.5 and 4.3.5.

10.8 PROPOSITION. The mapping $b : A \longrightarrow bA$ is a base operator and $bX = X$.

Proof. Obviously, $bX = X$ and $b(A \cup B) \supset bA \cup bB$. Assume that $x \notin bA \cup bB$. Then there are $s,t \in S$ such that

$$\hat{R}_s^A(x) < s(x), \quad \hat{R}_t^B(x) < t(x).$$

Put $u = s+t$. Then $u \in S$ and

$$\max(\hat{R}_u^A(x), \hat{R}_u^B(x)) < u(x).$$

To complete the proof it suffices to find $v \in S$ for which $\hat{R}_v^{A \cup B}(x) < v(x)$. The way to prove this is exactly the same as that one given in the proof of Prop. 6.3.1 of [C&C] .

10.9 REMARK. Before treating further properties of bases let us note that the additivity of b means nothing but that the union $A \cup B$ is thin at a point x if and only if A and B are thin at x.

By definition, $bA \subset \bigcap_{s \in S} [\hat{R}_s^A = s]$. Our aim now is to manufacture at least one element $p \in S$ guaranteeing the equality $bA = [\hat{R}_p^A = p]$.

A function $p \in S$ is called a *generator* if for every $q \in S$ there are sequences $\{p_n\}$, $\{q_n\}$ of elements of S and $\{\alpha_n\}$ of positive constants such that $q_n \nearrow q$ and $p_n + q_n = \alpha_n p$ for every $n \in \mathbb{N}$.

There exists always a bounded continuous generator. Indeed, let $\{d_n\} \subset S_o$ be a countable increasingly dense set ordered into a sequence. Find real constants $a_n > 0$ such that $a_n d_n \leq 2^{-n}$ for each $n \in \mathbb{N}$. Obviously, the function $\sum a_n d_n$ has all required properties.

10.10 PROPOSITION. Let $p \in S$ be a generator. Then $bA = [\hat{R}_p^A = p]$ for any $A \subset X$.

Proof. Let $A \subset X$. Obviously $bA \subset [\hat{R}_p^A = p]$. Pick $x \in [\hat{R}_p^A = p]$. Given $q \in S$, find sequences $\{p_n\}$, $\{q_n\}$ of elements of S and $\{\alpha_n\}$ of positive constants such that $q_n \nearrow q$ and $p_n + q_n = \alpha_n p$ for every $n \in \mathbb{N}$. It should be clear that $\hat{R}_{q_n}^A(x) = q_n(x)$ and hence, using Proposition 10.7, $\hat{R}_q^A(x) = q(x)$. It follows $x \in bA$.

10.11 COROLLARY. For any subset A of X, the base bA is a finely closed set of type G_δ.

Proof. This is so because $bA = [\hat{R}_p^A = p]$ for a continuous generator $p \in S$ and the function $x \mapsto \hat{R}_p^A(x)$ is lower semicontinuous and finely continuous.

We turn next to the principal lemma in the present section. Its key role will be securely established.It has many interesting and important applications, as we shall see several times in the sequel.

10.12 KEY LEMMA. Let $p \in S$ be bounded, let K be compact, and $K \subset U \subset X$. Then for every $\varepsilon > 0$ there is $q \in S$ such that $q = p$ on CU and $q < \hat{R}_p^{CU} + \varepsilon$ on K.

(If X is a harmonic space, then the Key Lemma still holds even for unbounded p).

Proof. For harmonic spaces see [C&C], Cor. 5.3.2. Moreover, Key lemma carries over to any standard H-cone without essential change in the mechanics of the proof. What is necessary only to show is the fact that $\hat{R}_p^A$ is continuous on $X \setminus \bar{A}$ for any subset A of X and for any bounded $p \in S$. But this assertion is true in light of Prop. 5.6.14 of [BBC]. Notice only that the assumption of saturation of X in this proposition is superfluous and can be easily overcome.

10.13 COROLLARY. If $p \in S$ and U is a subset of X, then $R_p^{CU} = \hat{R}_p^{CU}$ on U.

Proof. Pick $x \in U$. In the case when p is bounded, use Key Lemma for $K = \{x\}$. Next, if $p \in S$ is arbitrary, look at the functions $p_n = p \wedge n$ $(n \in \mathbb{N})$. Appealing to Proposition 10.7,

$$R_p^{CU}(x) = \lim R_{p_n}^{CU}(x) = \lim \hat{R}_{p_n}^{CU}(x) = \hat{R}_p^{CU}(x).$$

10.14 PROPOSITION. Let $A \subset X$ and $x \in X$. The following assertions are equivalent:

(i) A is thin at x,

(ii) there is an open neighborhood V of x such that $A \cap V$ is thin at x,

(iii) there is a fine neighborhood V of x such that $A \cap V$ is thin at x.

Proof. It is that (i) $\Rightarrow$ (ii) $\Rightarrow$ (iii). Assume that (iii) holds. Appealing to Theorem 2.3 there are $s, t \in S$ such that $x \in [s < t] \subset V$. Hence $\hat{R}_t^{CV}(x) \leqq R_t^{CV}(x) \leqq s(x) < t(x)$, and therefore CV is thin at x. By assumption, $A \cap V$ is also thin at x, so $A \subset (A \cap V) \cup CV$ is thin at x in light of Remark 10.9.

10.15 PROPOSITION. Let $A \subset X$ and $x \in X \setminus A$. The following assertions are equivalent:

(i) A is thin at x,

(ii) CA is a fine neighborhood of x,

(iii) there is $u \in S$ such that

$$u(x) < \liminf_{A \ni y \to x} u(y).$$

Proof. The equivalence (ii)$\Leftrightarrow$(iii) is Cartan's theorem 2.5. (i)$\Rightarrow$(ii): There exists $s \in S$ such that $\hat{R}_s^A(x) < s(x)$. Thanks to Corollary 10.13 there is $u \in S$, $u \geq s$ on A and $u(x) < s(x)$. Now, the set $[u < s]$ is a fine neighborhood of x contained in CA.

Next, assume that (iii) holds. There is $a \in \mathbb{R}$ and an open neighborhood V of x such that $u(x) < a$ and $u > a$ on $V \cap A$. Hence

$$\hat{R}_a^{A \cap V}(x) \leq R_a^{A \cap V}(x) \leq u(x) < a.$$

Thus (iii) implies that $A \cap V$ is thin at x, and it suffices to cite the previous proposition.

10.16 PROPOSITION. The mapping $b: A \to bA$ is a strong base operator (in the sense of Section 1.A) and the fine topology coincides with the b-topology. Moreover, $\bar{A}^f = A \cup bA$ and $\mathrm{der}_f A \subset bA \subset \bar{A}^f$ for any subset A of X.

Proof. It immediately follows from (i)$\Leftrightarrow$(ii) of the previous proposition that $\bar{A}^f = A \cup bA$, which is what we wanted. Indeed, this implies that a set A is finely closed if and only if $bA \subset A$, which is the case if and only if A is b-closed. Now, $b(bA) \subset bA$ since bA is always finely closed (Corollary 10.11). Finally, Theorem 1.2 yields $\mathrm{der}_f A = bA \cup \{x \in A: x \in b\{x\}\}$, and therefore $\mathrm{der}_f A \subset bA$.

10.17 THEOREM. (Convergence theorem). Let $F \subset S$ be a nonempty family. Then the set $[\wedge F < \inf F]$ is semipolar.

Proof. A quite simple proof in the framework of H-cones appears in [BBC], Th. 3.3.7.

A subset V of X is termed *finely regular* if $V = CbCV$. According to Corollary 10.11, any finely regular set is finely open and of type F_σ. Notice that $V \subset CbCV$ for any finely open set, and also remark that a relatively compact open set V (in a harmonic space) is finely regular if and only if it is regular.

Given a set $U \subset X$, denote by $I(U)$ the collection of all finely regular sets such that $\bar{V} \subset U$.

A subset M of X is called a *sweeping kernel* of U if $M \subset U$ and $\hat{R}_p^{CM} = \hat{R}_p^{CU}$ for every $p \in S$.

10.18 PROPOSITION. For a finely open subset V of X, the following conditions are equivalent:

(i) V is finely regular,

(ii) $\hat{R}_p^{CV} = p$ on CV for every $p \in S$,

(iii) $R_p^{CV} = \hat{R}_p^{CV}$ for every $p \in S$,

(iv) $R_p^{CV} \in S$ for every $p \in S$.

Proof. We already know that

$$\bigcap_{p \in S} [\hat{R}_p^{CV} = p] = bCV \subset CV \subset \bigcap_{p \in S} [R_p^{CV} = p] .$$

Hence, the implications (i) $\Rightarrow$ (ii) $\Rightarrow$ (iii) $\Rightarrow$ (iv) $\Rightarrow$ (iii) $\Rightarrow$ (i) are easily verified.

10.19 PROPOSITION. If $F \subset X$ is a finely closed set, then the fine interior of F is a finely regular set.

Proof. Put $U := \mathrm{int}_f F$. Then

$$CU = \overline{CF}^f = CF \cup bCF \subset bCF \subset bCU \subset CU.$$

10.20 PROPOSITION. Let s be a real continuous function of S and let $U \subset X$. Then

$$R_s^{CU} = \inf \{R_s^{CV} : V \in I(U)\} , \quad \hat{R}_s^{CU} = \bigwedge \{\hat{R}_s^{CV} : V \in I(U)\} .$$

Proof. Let $u \in S$, $\varepsilon > 0$ such that $u \geqq s$ on CU. If $V := \mathrm{int}_f [u+\varepsilon \leqq s]$, then V is finely regular in view of the previous proposition, $\overline{V}^f \subset [u + \varepsilon \leqq s] \subset U$ and $u + \varepsilon \geqq s$ on CV. Hence

$$R_s^{CU} = \inf \{u+\varepsilon : u \in S, u \geqq s \text{ on } CU, \varepsilon > 0\} \geqq$$

$$\geqq \inf \{R_s^{CV} : V = \mathrm{inf}_f [u+\varepsilon \leqq s], \ u \in S, u \geqq s \text{ on } CU, \ \varepsilon > 0\} \geqq$$

$$\geqq \inf \{R_s^{CV} : V \in I(U)\} .$$

The reverse inequality is obvious. The second assertion now easily follows. Namely, we have

$$\hat{R}_s^{CU} = \widehat{R_s^{CU}} = \widehat{\inf\{R_s^{CV} : V \in I(U)\}} = \wedge\{R_s^{CV} : V \in I(U)\} \geqq$$

$$\geqq \wedge\{\hat{R}_s^{CV} : V \in I(U)\} \geqq \hat{R}_s^{CU}.$$

We now draw an important conclusion from the above proposition. But before the theorem we make some observations.

REMARKS. Let X be a topological space with a countable base.

(a) If $\mathcal{F}$ is a family of lower semicontinuous functions on X, then there is a countable subfamily $\mathcal{F}_o \subset \mathcal{F}$ such that $\sup \mathcal{F}_o = \sup \mathcal{F}$. If, moreover, $\mathcal{F}$ is upper directed, then $\mathcal{F}_o$ can be taken as an increasing sequence.

(b) If $\mathcal{Y}$ is a family of functions on X, then there is a countable subfamily $\mathcal{Y}_o \subset \mathcal{Y}$ such that $\widehat{\inf \mathcal{Y}_o} = \widehat{\inf \mathcal{Y}}$.

Since any topological space with a countable base is Lindelöf,

(a) follows from Theorem 1.4 putting $\mathcal{Z} = \{\emptyset\}$. The assertion

(b) is known as *Choquet's lemma*. In order to prove it, for every set U of a countable base $\mathcal{L}$ of X associate a countable family $\mathcal{Y}_U \subset \mathcal{Y}$ such that

$$\inf\{f(x) : x \in U,\ f \in \mathcal{Y}\} = \inf\{f(x) : x \in U,\ f \in \mathcal{Y}_U\}$$

and set $\mathcal{Y}_o := \bigcup\{\mathcal{Y}_U : U \in \mathcal{L}\}$.

10.21 THEOREM. Let U be a subset of X. Then there is a sweeping kernel of U which is an F_σ-set. Moreover, if U is finely open, then each sweeping kernel of U is finely open as well.

Proof. Let D be a countable subset of S_o which is increasingly dense. Let $p \in D$. By Proposition 10.20 and thanks to Choquet's lemma, there is a sequence $\{V_n^p\}_n \subset I(U)$ such that $\hat{R}_p^{CU} = \bigwedge_n \hat{R}_p^{CV_n^p}$. Set $V^p = \bigcup_n V_n^p$. Then $V^p \subset U$ is an F_σ-set and

$$\hat{R}_p^{CU} \leqq \hat{R}_p^{CV^p} \leqq \bigwedge_n \hat{R}_p^{CV_n^p} = \hat{R}_p^{CU}.$$

If $V := \bigcup_{p \in D} V^p$, then $V \subset U$ is again an F_σ-set and, for any $s \in D$

$$\hat{R}_s^{CU} \leqq \hat{R}_s^{CV} \leqq \hat{R}_s^{CV^s} = \hat{R}_s^{CU}.$$

According to the previous remark (a) and Proposition 10.7, $\hat{R}_s^{CU} = \hat{R}_s^{CV}$ for every $s \in S$.

Let now p be a bounded continuous generator. Then by Proposition 10.10

$$V \subset U \subset CbCU = [\hat{R}_p^{CU} < p] = [\hat{R}_p^{CV} < p] = CbCV,$$

and thus V is finely open.

EXERCISES

10.C.1 (Cf. [BBC], Lemma 4.3.7.) For any positive element $u \in S$ there exists a u-continuous generator.

Hint. Use the following observation ([BBC], Prop. 4.1.2d): If $s \in S$ and $\{s_n\}$ is a sequence of u-continuous functions such that $\frac{s_n}{u}$ converges uniformly to $\frac{s}{u}$, then s is u-continuous.

10.C.2 Every element of S is finite on a finely dense subset of X.

Hint. Use Theorem 2.3 and axiom (S_1).

10.C.3 Polar sets. A subset A of X is termed *polar* if $\hat{R}_1^A = 0$.

(a) Any polar set is totally thin. In particular, polar sets are finely closed, finely isolated and finely nowhere dense.

(b) Let $s \in S^{\uparrow}$. The following conditions on $A = [s = +\infty]$ are equivalent:

(i) A is polar,

(ii) CA is finely dense,

(iii) CA is dense.

Hint. Suppose (iii). Then $\hat{R}_1^A \leq \bigwedge_n \{\min(1, \frac{1}{n} s)\} = 0$. (Observe that $\min(1,t) \in S$ whenever $t \in S^{\uparrow}$.)

(c) If $s \in S$ then $[s = +\infty]$ is polar.

Hint. Use (b) and Ex. 10.C.2.

(d) The family of all polar sets is a σ-ideal.

Hint. Use Proposition 10.7.

(e) The following conditions on a set A are equivalent:

(i) A is polar,

(ii) there is $t \in S^{\uparrow}$ such that t is finite on a finely dense set and $t = +\infty$ on A,

(iii) $\hat{R}_s^A = 0$ for every $s \in S$.

Hint. Assume that (i) holds and fix a countable dense set $Z \subset [R_1^A = 0]$. Find $t_n \in S$ such that $t_n \geq 1$ on A and $\sum t_n(x) < +\infty$ for every $x \in Z$. Set $t = \sum t_n$ and use (b) to prove (ii). For the proof of (ii)$\Rightarrow$(iii) use $\hat{R}_s^A \leq \bigwedge_n \{\min(s, \frac{1}{n}t)\}$.

(f) Every polar set is contained in a polar set of type G_δ.

(g) If $F \subset S$, $\wedge F = 0$, then $[\inf F > 0]$ is a polar set (cf. Proposition 12.10).

Hint. $[\inf F > 0] = \bigcup_{n \in \mathbb{N}} [\inf F \geq \frac{1}{n}]$.

10.C.4 Let A be a relatively compact polar subset of X, and let f be a real nonnegative lower semicontinuous function on X. Then given any $\varepsilon > 0$, there is $s \in S^{\uparrow}$ such that $|s-f| < \varepsilon$ on A.

Hint. No generality is lost with the assumption that $\varepsilon > 1$. By virtue of Key Lemma 10.17 there are $s_n \in S$ such that $s_n = 1$ on $A \setminus K_n$ and $s_n < 2^{-n}$ on K_n where $K_n := \{x \in \bar{A}: f(x) \leq n\}$. Now put $s = \sum s_n$.

REMARKS AND COMMENTS

In this section we collected main results on balayage and thinness. Even if the roots of the balayage method can be traced back to C.F. Gauss historically, only H. Poincaré's "méthods de balayage" (1894) was devised to solve the problem of the equilibrium distribution. Poincaré's work stimulated further development of the balayage theory. The notions of the reduite and the balayage of superharmonic functions as well as their fundamental properties are examined by M. Brelot (1945).
The key lemma is attributed to N. Boboc, A. Cornea and C. Constantinescu (1965).

Concerning the convergence theorem, E. Szpilrajn and T. Rado (see T. Rado (1937)) showed that in the classical theory of harmonic functions the limit of a decreasing sequence of positive superharmonic functions differs from a superharmonic function on a set of Lebesgue measure zero. M. Brelot (1938b) proved that this exceptional set is of inner capacity zero (all of its compact subsets are polar). H. Cartan (1942),

(1945) improved this result replacing sequences by directed families and showing that this set itself is polar (of outer capacity zero). Later on, the convergence theorem entered into axiomatic theories (J.L. Doob (1956), M. Brelot (1962), H. Bauer (1965), C. Constantinescu and A. Cornea [C&C] (1972)) and was generalized in various ways (cf. M. Brelot (1966)). A quite simple proof in the framework of H-cones appeared in [BBC], Th. 3.3.7 and is recommended for interested readers.

Polar sets were introduced into classical potential theory by M. Brelot (1941), and it was H. Cartan (1945) who proved that they coincide with the sets of capacity zero.

H. Wallin (1963) states that a compact subset of $\mathbb{R}^n$, $n \geq 3$ has capacity zero if and only if every positive continuous function on it is the restriction of a positive Newtonian potential. This result was recently generalized by R. Jesuraj (1985) to compact polar subsets of a harmonic Brelot space. In Ex. 6.3.6 of [C&C], a similar result on polar and semipolar subsets of $\mathfrak{P}$-harmonic spaces using approximation "up to ε" can be found. An assertion of Ex. 10.C.4 is a simple consequence of Key lemma and it concerns an approximation of lower semicontinuous functions on relatively compact polar sets.

Related papers: W. Hansen (1971), I. Netuka (1974).

10.D The fine topology

In this section we establish important features of the fine topology on X. All the properties here have been chosen on the basis of their usefulness and intrinsic interest.

10.22 PROPOSITION. In the fine topology, X is a completely regular Hausdorff space. In particular, each point of X has a neighborhood base consisting of finely closed sets.

Proof. Since the fine topology is defined as the weak topology induced on X by a family of functions, the assertion follows from Theorem 2.3.

Recall that a *cozero set* in a topological space X is a set of the form $[f > 0]$ for some real continuous function f on X.

10.23 LEMMA. For a subset C of X, the following are equivalent:

(i) C is a fine cozero set,

(ii) there is a sequence $\{V_n\}$ of finely regular sets such that $C = \bigcup_n V_n$,

(iii) there is an upper semicontinuous, finely continuous function $g \geq 0$ such that $C = [g > 0]$,

(iv) $C = \bigcup_n W_n$ where $W_n \in I(C)$.

Proof. (i)$\Rightarrow$(ii): Let $C = [f > 0]$ for a finely continuous $f \geq 0$. Put $V_n :=$ $= \mathrm{int}_f [f \geq \frac{1}{n}]$. Taking Proposition 10.18 into account, we know that V_n are finely regular. We can see also that $C = \bigcup_n V_n$ by noting that $[f > \frac{1}{n}] \subset V_n \subset [f \geq \frac{1}{n}]$.

(ii)$\Rightarrow$(iii): Let p be a bounded continuous generator. According to Proposition 10.10, $V_n = \mathrm{CbCV}_n = [\hat{R}_p^{CV_n} < p]$ for every $n \in \mathbb{N}$. Now, we can put $g := \sum 2^{-n}(p - \hat{R}_p^{CV_n})$.

(iii)$\Rightarrow$(iv): Set again $W_n = \mathrm{int}_f [g \geq \frac{1}{n}]$. Then $\bar{W}_n \subset [g \geq \frac{1}{n}] \subset C$ and $\bigcup_n W_n = C$.

The implications (iv)$\Rightarrow$(ii) and (iii)$\Rightarrow$(i) are obvious.

10.24 THEOREM. The σ-topology of all fine cozero sets has the Lusin-Menchoff property with respect to the original topology of X (see Ex. 3.B.20).

Proof. What we must show is that given a closed set F and a fine cozero set C containing F, there is a fine cozero set W such that $F \subset W \subset \bar{W} \subset C$. To this end let $f \geq 0$ be upper semicontinuous and finely continuous, $g \geq 0$ continuous, such that $C = [f > 0]$, $F = [g = 0]$. Put $W = [g < f]$. Then W is a fine cozero set and

$$F \subset W \subset \bar{W} \subset [g \leq f] \subset C.$$

10.25 THEOREM. Let U be a finely open subset of X, and let $F \subset U$ be a closed σ-compact set. Then there is a finely regular set V such that $F \subset V \subset \bar{V} \subset U$.

Proof. Let $\{K_n\}$ be an increasing sequence of compact sets, $F = \bigcup_n K_n$. Let p be a bounded continuous generator. By Proposition 10.10, $\hat{R}_p^{CU} < p$ on $\mathrm{CbCU} \supset U$. It now follows from the Key Lemma 10.12 that there are $q_n \in S$ such that $q_n \geq p$ on CU and $q_n < \hat{R}_p^{CU} + \frac{1}{n}$ on K_n. Set

$$C := \bigcup_n [q_n < p].$$

Obviously, C is a fine cozero set and $F \subset C \subset U$. According to the previous theorem, there is a finely open set W such that $F \subset W \subset \bar{W} \subset C$. Thus if we put $V := \mathrm{int}_f \bar{W}$

we have achieved a finely regular set with the property $F \subset W \subset V \subset \bar{W} \subset C \subset U$.

10.26 COROLLARY. (The Lusin-Menchoff property of the fine topology.) Let X be a σ-compact space. Then the fine topology has the complete Lusin-Menchoff property.

Proof. Since every closed subset of X is σ-compact, the fine topology has the Lusin-Menchoff property by the preceding theorem. Furthermore, for every $M \subset X$ the base bM is a G_δ -set (Corollary 10.11) and $\bar{M}^f = M \cup bM$ (Corollary 10.15). Theorem 3.19 now shows that the fine topology has even the complete Lusin-Menchoff property.

10.27 COROLLARY. The fine topology of X is cometrizable.

Proof. It follows immediately from Theorem 10.25 that each point of X has a fine neighborhood base the elements of which are closed. But even simpler arguments yield this assertion (cf. Ex. 10.D.2).

10.28 THEOREM. If X is completely metrizable, then X endowed with the fine topology is a strong Baire space.

Proof. To prove the assertion it suffices to use the preceding corollary, Corollary 10.11 and Corollary 4.5 (cf. also Ex. 4.A.12 and Proposition 10.16).

10.29 THEOREM. Every totally thin subset of X is finely closed, finely discrete and finely nowhere dense. Any semipolar set is first category in the fine topology. Hence, if (X,f) is a Baire space, there are no finely open nonempty semipolar subsets of X.

Proof. Let $M \subset X$ be totally thin. Then $\mathrm{der}_f M = \emptyset \subset M$. Thus, M is finely closed and finely discrete. Further,

$$\mathrm{int}_f M \subset CbCM = CX = \emptyset .$$

10.30 THEOREM. The fine topology on X has the G_δ-insertion property. Any finely continuous function on X (or, more generally, on any G_δ-set which is not thin at any of its points) belongs to Baire class one.

Proof. As $\mathrm{int}_f A \subset bA \subset \bar{A}^f$ for any $A \subset X$ and bA is always a G_δ-set, the first assertion follows.

Let now M be a G_δ -set, $M \subset bM$. Denote $\tau = f \wedge M$ and select $A \subset M$. There is a finely open set $V \subset X$ such that $\mathrm{int}_\tau A = V \cap M$. Since $M \setminus A \subset CV$, we have $V \subset$

$\subset CbCV \subset Cb(M \setminus A)$. Therefore

$$V \cap M \subset Cb(M \setminus A) \cap [bA \cup b(M \setminus A)] \subset bA.$$

Finally,

$$\text{int}_\tau A = V \cap M \subset M \cap bA \subset M \cap \bar{A}^f = \bar{A}^\tau$$

and we can apply Corollary 2.14 (note that $M \cap bA$ is a G_δ-subset of M).

10.31 THEOREM. (Quasi-Lindelöf property of the fine topology.) Let $\mathcal{V}$ be a family of finely open subsets of X. Then there is a countable subfamily of $\mathcal{V}$ whose union differs from the union of the whole family by a semipolar and finely nowhere dense

Proof. Let p be a bounded continuous generator. Denote

$$\mathcal{F} := \{u \in S\colon \text{there is } U \in \mathcal{V} \text{ such that } u \geqq p \text{ on } CU\}.$$

By Choquet's lemma (cf. Remark following Proposition 10.20) there are countable families $\{u_n\} \subset \mathcal{F}$, $\{U_n\} \subset \mathcal{V}$ such that

$$\widehat{\inf u_n} = \widehat{\inf \mathcal{F}} \quad \text{and} \quad u_n \geqq p \text{ on } CU_n.$$

Pick $x \in M := \bigcup_{V \in \mathcal{V}} V \setminus \bigcup_n U_n$. If $G \in \mathcal{V}$ contains x, then

$$\widehat{\inf u_n}(x) = \widehat{\inf \mathcal{F}}(x) \leqq \hat{R}_p^{CG}(x) < p(x) \leqq \inf u_n(x)$$

Hence M is a subset of $[\widehat{\inf u_n} < \inf u_n]$ which is semipolar in light of Convergence theorem 10.6. Let now W be a finely open subset of M. According to Proposition 10.10, $\widehat{\inf \mathcal{F}} < p$ on W. On the other hand, $\widehat{\inf u_n} \geqq p$ on W. Thus W must be empty. Now, it easily follows that $\bar{M}^f$ cannot contain a finely interior point.

REMARK. If X equipped with the fine topology is a Baire space, then any finely open subset of $\bar{M}^f$ must be empty in light of Theorem 10.29.

The next Berg's theorem characterize standard H-cones of functions whose fine topology is normal. On the other hand, let us underline that the most important examples of harmonic spaces derived from the solutions of partial differential equations belong to spaces for which the fine topology is not normal.

10.32 THEOREM (Ch. Berg (1971) for harmonic spaces). The following assertions are equivalent:

(i) every totally thin set is countable,

(ii) every semipolar set is countable,

(iii) the fine topology is Lindelöf,

(iv) the fine topology is paracompact,

(v) the fine topology is normal.

Proof. The proof of the implication (i)$\Longrightarrow$(ii) is straightforward, the implications (iii)$\Longrightarrow$(iv)$\Longrightarrow$(v) are true in any regular topological space.
(ii)$\Longrightarrow$(iii): Let $\mathcal{V}$ be a family of finely open sets. By quasi-Lindelöf property (Theorem 10.31) there is a countable subcollection $\mathcal{V}_o$ of $\mathcal{V}$ whose union differs from the union of all $\mathcal{V}$ by a semipolar set Z. Since Z is countable, there is a countable covering $\mathcal{V}_1 \subset \mathcal{V}$ of Z. Obviously, $\mathcal{V}_o \cup \mathcal{V}_1$ is a countable subcover of $\mathcal{V}$.
(v)$\Longrightarrow$(i): Let T be an uncountable totally thin set. According to Theorem 10.21 there is a totally thin G_δ-set B containing T. The cardinality of B must be c. Now, we can use either Theorem 2.2 or we can proceed as follows: Assume that the fine topology is normal. For any $A \subset B$, the sets A and $B \setminus A$ are both finely closed. Therefore, there exists a finely continuous function f on X such that $f = 0$ on A and $f = 1$ on $B \setminus A$. It easily follows that the cardinality of the set of all finely continuous functions on X is at least 2^c. This contradicts the fact that any finely continuous function on X belongs to the Baire class one (Theorem 10.30) whose cardinality is $c < 2^c$.

EXERCISES

10.D.1 (M. Brelot). (a) Let $x \in X$. Assume

(Thi) for every subset A of X which is thin at x $(x \in \overline{A} \setminus A)$ there is $s \in S$ such that $s(x) < +\infty = \lim_{A \ni y \to x} s(y)$.

Then a function f on X is finely continuous at x (finely lower semicontinuous, resp.) if and only if there is a fine neighborhood V of x such that

$$f(x) = \lim_{V \ni y \to x} f(y) \quad (f(x) \leq \liminf_{V \ni y \to x} f(y), \text{ resp.}).$$

Hint. According to Theorem 2.8 we need only to show that for any sequence $\{V_n\}$ of fine neighborhoods of x there is a decreasing sequence $\{W_n\}$ of neighborhoods of x such that $V := X \setminus \bigcup_n (W_n \setminus V_n)$ is a fine neighborhood of x. Find $u_n \in S$ such that $u_n(x) = 1$ and

$$\liminf_{CV_n \ni y \to x} u_n(y) = +\infty$$ and a decreasing sequence $\{W_n\}$ of neighborhoods

of z such that $u_n > 3^n$ on $W_n \setminus V_n$. Put

$$V := X \setminus \bigcup_n (W_n \setminus V_n), \quad u := \min(2, \sum_n 2^{-n} u_n).$$

Then $u(x) = 1 < 2 = \liminf_{CV \ni y \to x} u(y)$.

REMARK. The assumption (Thi) is fulfilled provided X is a $\mathfrak{P}$-harmonic space with a countable base and $\{x\}$ is a polar set (classical result of M. Brelot (1944), cf. R.-M. Hervé (1962) and [C&C], Ex.8.2.2).

(b) A number $c \in \bar{\mathbb{R}}$ belongs to the fine cluster of a function g on X at a point $x \in X$ if and only if there is a set $M \subset X$ such that $x \in bM \setminus M$ and $c = \lim_{M\, y \to x} f(y)$.

Hint. By virtue of Theorem 2.9 we need to prove that for any sequence $\{A_n\}$ of sets satisfying $x \in bA_n \setminus A_n$ there is a decreasing sequence $\{W_n\}$ of neighborhoods of x for which the set $A := \bigcup_n (A_n \setminus W_n)$ is not thin at x. Let us fix a bounded generator p on X and find a decreasing sequence $\{W_n\}$ in such a way that

$$\hat{R}_p^{A_n \setminus W_n}(x) > p(x) - \frac{1}{n}.$$

In this case $\hat{R}_p^A(x) = p(x)$ and this means $x \in bA$. The construction of $\{W_n\}$ is straightforward: If $\{U_n\}$ is a decreasing sequence of neighborhoods of x for which $\bigcap_n U_n = \{x\}$, by Proposition 10.7 $R_p^{A_j \setminus U}(x) \nearrow R_p^{A_j}(x) = p(x)$ for every j fixed (note that $x \notin A_n$).

REMARK. These assertions can be found in M. Brelot (1971) even in a very abstract context.

10.0.2 Let τ be a weak topology on a topological space (Y,ρ) induced by a convex cone of lower semicontinuous functions, and let $x \in X$. Show that

(a) τ has a local base at x consisting of closed sets,

(b) if, moreover, Y is locally compact and τ is finer than ρ, then $\mathcal{V}_\tau(x)$ has a base which is formed by compact sets.

REMARKS AND COMMENTS

This section enumerates the most important properties of the fine topology in standard H-cones. The Lusin-Menchoff property of the fine topology was introduced in potential theory by J. Lukeš (1977) and established in harmonic spaces was discovered axiom of polarity. J. Malý (1980) proved the Lusin-Menchoff property without any restriction. Notice that a similar property of quasi-normality as well as an extension theorem of Tietze's type of special type are examined in [Fug].

The result that a harmonic space endowed with the fine topology is a Baire space, attributed to A. Cornea in 1966 can be found in [C&C] or, in a more general form in M. Brelot (1971).

Since the classical fine topology in $\mathbb{R}^n$ is coarser than the density topology, it follows that any finely continuous function is in the Baire class one. This was observed by B. Fuglede (1971a). The same result was proved by I. Netuka and L. Zajíček (1974) for the case of the heat equation. Finally, B. Fuglede (1974) showed that finely continuous functions are B_1 in any axiomatic theory of harmonic functions. The simplified proof of J. Lukeš and L. Zajíček of this assertion in (1977a) and (1977b) suggested a new method of "insertion of G_δ-sets" (cf. Section 2.D).

The quasi-Lindelöf property of the fine topology in harmonic spaces was discovered by J.L. Doob (1966) (cf. Section 1.B).

11. HARMONIC SPACES

We bring together in this chapter most of the material that will be relied on in subsequent chapters. We content ourselves with the background necessary to make our exposition completely self-contained for the reader's convenience.

Throughout this chapter, $(X,\mathcal{H})$ denotes a $\mathfrak{P}$-harmonic space with a countable base. We denote by $\mathcal{H}^*(U)$ the set of all hyperharmonic functions on an open set $U \subset X$. Further, S stands for the set of all nonnegative superharmonic functions on X.

In Sections 10.C and 10.D we have achieved some results in standard H-cones of functions assuming in accordance with [BBC] that the positive constants are in S. Dealing as we are with harmonic spaces rather than more general standard H-cones, it is important to know that all results remain valid in this framework even if the constant functions are not superharmonic. All notions and assumptions can be immediately transferred into this general situation. For this purpose it is convenient to fix a real positive continuous superharmonic function s_o on X and the phrases like "p is bounded" should be translated as "p/s_o is bounded".

11.A Potentials

Recall that a nonnegative superharmonic function p is called a *potential* if the greatest harmonic minorant of p is zero. Denote by $\mathcal{P}$ the set of all potentials on X. Further, $\mathcal{P}^c$ ($\mathcal{P}^f$, $\mathcal{P}^b$, resp.) will be the set of all real continuous (real, locally bounded, resp.) potentials on X. We say that a function f is *lower $\mathcal{P}^c$-bounded* (lower $\mathcal{P}^f$-bounded,...,resp.) on a set $U \subset X$ if there is $q \in \mathcal{P}^c$ ($q \in \mathcal{P}^f$,...,resp.) such that $-q \lneq f$ on U.

11.1 PROPOSITION. Any nonnegative hyperharmonic function on X is the supremum of an increasing sequence of elements of $\mathcal{P}^c$ each being harmonic outside a compact subset of X.

Proof. Originally, H. Bauer (1963), Cf. [Bau], Satz 2.5.8, or [C&C], Cor. 2.3.1.

By an *Evans function* of a superharmonic function s on X we understand any nonnegative hyperharmonic function u on X such that for every positive $\alpha \in \mathbb{R}$ the set $[u < \alpha s]$ is relatively compact.

11.2 PROPOSITION. Let p be a nonnegative superharmonic function on X and

$$h = \inf\{R_p^{CK} : K \text{ is a compact subset of } X\}.$$

Then

(a) h is the greatest harmonic minorant of p,

(b) the following assertions are equivalent:

(i) p is a potential on X,

(ii) $h = 0$,

(iii) $\bigwedge\{\hat{R}_p^{CV} : V \text{ is a relatively compact finely open subset of } X\} = 0$,

(iv) for every $x \in X$ there is an Evans function of p finite at x.

Proof. (a) Let G be a relatively compact open subst of X. By the remark preceding Theorem 10.21, there is a decreasing sequence $\{K_n\}$ of compact subsets of CG such that

$$-h = \sup_n (-R_p^{CK_n}) \quad \text{on } G.$$

(Notice that R_p^{CK} is always harmonic on int K for any compact set K ([C&C] , Prop. 5.3.1).) It follows from the convergence axiom that h is harmonic on G. Appealing to a sheaf axiom we see that h is harmonic on X. If h' is a harmonic minorant of p, then thanks to Bauer's minimum principle 11.9, $h' \leq h$.

(b) According to (a), (i)$\Longleftrightarrow$(ii). The equivalence (ii)$\Longleftrightarrow$(iii) is obvious.

(ii)$\Longrightarrow$(iv): Select $x \in X$. Find a sequence $\{K_j\}$ of compact subsets of X such that $R_p^{XK_j}(x) < +\infty$. Now put

$$q := \sum \hat{R}_p^{CK_j}.$$

Pick $k \in \mathbb{N}$ and denote $K := \bigcup_{j \leq k} K_j$. Then $kp \leqq \sum_{j=1}^{k} \hat{R}_p^{CK_j} \leqq q$ on CK which shows that q is an Evans function for p.

(iv)$\Longrightarrow$(ii): Choose $x \in X$. let q be an Evans function of p finite at x. For any $\varepsilon > 0$, the function $-h + \varepsilon q$ is nonnegative outside a compact subset of X. From Bauer's minimum principle 11.9 we deduce that $-h + \varepsilon q \geqq 0$ on X. Letting $\varepsilon \to 0$ we obtain $-h(x) \geqq 0$.

REMARK. It can be even proved that for every potential p on X there is an Evans function q such that q is a potential, q is finite whenever p is finite and q is continuous at any point where p is continuous ([C&C], Prop. 2.2.4).

11.3 PROPOSITION. If $A \subset X$ is a Borel set and if u is a nonnegative hyperharmonic function on X, then

$$R_u^A = \sup \{ R_u^K : K \subset A \text{ is compact} \},$$

$$\hat{R}_u^A = \sup \{ \hat{R}_u^K : K \subset A \text{ is compact} \}.$$

Proof follows from results on capacitability of Borel subsets of locally compact spaces with countable base. Cf. [C&C], Cor. 5.3.3. In Ex. 11.B.1 we present another proof of a similar assertion in the context of standard H-cones.

11.4 PROPOSITION. For every function $f \in \mathcal{K}(X)$ with compact support K and for every $\varepsilon > 0$ there are $p, q \in \mathcal{P}^c$ such that $p = q$ on CK, p,q are harmonic on CK and $|p-q-f| < \varepsilon$ on K. In particular, the vector space $\mathcal{P}^c - \mathcal{P}^c$ is dense in $\mathcal{K}(X)$.

Proof. See [Bau], Satz 2.7.4, or [C&C], Th. 2.3.1.

REMARK. We see now that any two Radon measures coincide provided they agree on $\mathcal{P}^c$.

11.5 PROPOSITION. (a) Each real continuous potential which is harmonic outside a compact set is universally continuous (in the sense of standard H-cones - see Section 10.B).

(b) Each universally continuous function s belong to $\mathcal{P}^c$.

Proof. (a) The assertion is an easy consequence of Dini's theorem and Bauer's minimum principle 11.9.

Concerning (b), let $p_n \in \mathcal{P}^c$, $p_n \nearrow s$ (Proposition 11.1). Let $q \in \mathcal{P}^c$ be a positive potential. By definition, for every $\varepsilon > 0$ there is $n \in \mathbb{N}$ such that $s \leqq p_n + \varepsilon q$. Hence, s is continuous, finite, and a potential as well.

REMARK. It may be worth while to note that Proposition 11.5 together with Proposition 11.1 and Th. 5.11 of [C&C] give arguments to conclude that the set S of all nonnegative superharmonic functions on a $\mathfrak{P}$-harmonic space is actually a standard H-cone of functions (provided the positive constants are superharmonic).

A potential p on X is called *strict* provided any two Radon measures μ, ν on X coincide if $\mu(p) = \nu(p) < +\infty$ and if $\mu(u) \leqq \nu(u)$ for each $u \in \mathcal{H}^*_+(X)$. There exists always a real continuous strict potential on X ([C&C], Prop. 7.2.1, cf. Remark of Ex. 11.A.2), and for any real strict potential p on X and for any $A \subset X$ we have $bA = [\hat{R}^A_p = p]$ ([C&C], Prop. 7.2.2).

The notion of a strict potential has been employed for a number of years in the theory of harmonic spaces. In the framework of standard H-cones it was replaced by a similar notion of a generator. Keeping a tradition, in what follows we shall prefer to use strict potentials.

EXERCISES

11.A.1 Let $(\mathbb{R}^n, \mathcal{H})$ $(n \geqq 3)$ be the harmonic space associated with the Laplace equation. Then a function p is a potential on $\mathbb{R}^n$ if and only if p is a Newtonian potential and p is finite somewhere.

Hint. Use Riesz decomposition theorem of Ex. 10.A.4.

11.A.2 Let s be a generator on a harmonic space X.

(a) Show that $s > 0$ on X.

(b) Let $0 \leq h \leq s$ be a harmonic function. Then $p := s - h$ is also a generator.

Hint. Let $q \in S$ and $n \in N$. There are sequences $\{q_j^n\}$, $\{t_j^n\} \subset S$ and $\{\alpha_j^n\} \subset \mathbb{R}_+$ such that $n \leq \alpha_j^n$, $q_j^n \nearrow q \wedge np$, $q_j^n + t_j^n = \alpha_j^n s$. Put $p_j^n := t_j^n - \alpha_j^n h$. Then $p_j^n \in S$ and $q_j^n + p_j^n = \alpha_j^n s$. Set $q_k := \bigvee_{n \leq k} q_k^n$. Obviously $q_k \nearrow q$. Using Cor. 2.1.3 of [BBC] or Prop. 4.1.8 of [C&C] there are p_k, α_k such that $p_k + q_k = \alpha_k p$.

(c) Let h be the greatest harmonic minorant of s. Using (b) show that $s-h$ is a strict potential on X.

(d) If u is a positive potential and s is a u-continuous generator (its existence is stated in Ex. 10.C.1), then s is a strict potential.

REMARK. In light of (c) or (d) the existence of a strict potential $p \in \mathcal{P}^c$ can be derived by knowing the existence of a real continuous generator.

REMARKS AND COMMENTS

A convenient definition of potentials in axiomatic theories originated from M.Brelot (1958).The proof of Proposition 11.4 on approximation ensued from the classical theory of harmonic functions using a uniform approximation of continuous functions by $\mathcal{C}^2$-smooth functions. The idea of the proof based on Stone's approximation lemma is due to J. Deny and it is employed by R.-M. Hervé (1959).

11.B Balayage of measures

Let A be a subset of X and $x \in X$. Then there is a unique Radon measure ε_x^A, called the *balayaged measure* of ε_x on A, such that $\hat{R}_u^A(x) = \varepsilon_x^A(u)$ for every function $u \in \mathcal{L}_+^*(X)$. Moreover, if $U \subset X$ is a finely open set and $x \in U$, then the measure ε_x^{CU} is carried by any Borel subset of X containing $\bar{U} \setminus U$ ([C&C] - Prop. 7.1.2, Cor. 7.1.2, Prop. 7.1.3 and Th. 7.2.1 ; cf. also Ex. 11.B.1c).

If f is a function on a subset of X, define $f^A : x \longmapsto \varepsilon_x^A(f)$ on the set of all $x \in X$ for which the integral $\varepsilon_x^A(f)$ is defined (finite or infinite).

In what follows, we shall need the following important properties of balayaged measures.

11.6 PROPOSITION. Let $A \subset X$ and let $f \in \mathcal{K}^{\uparrow}(X)$. Then f^A is a finely lower semicontinuous, Borel function on X.

Proof. We know that $p^A = \hat{R}_p^A$ is a finely continuous and Borel function whenever $p \in \mathcal{P}^c$. Since $\mathcal{P}^c - \mathcal{P}^c$ is uniformly dense in $\mathcal{K}(X)$, the same assertion holds for any $f \in \mathcal{K}(X)$. If $f \in \mathcal{K}^{\uparrow}(X)$, then there are $g_n \in \mathcal{K}(X)$ such that $g_n \nearrow f$. By the Lebesgue monotone convergence theorem we deduce that the function f^A is a limit of an increasing sequence $\{(g_n)^A\}$ of finely continuous Borel functions.

11.7 PROPOSITION. Let U,V be finely open subsets of X, U being a Borel set, $x \in V \subset U$ and let $f \geq 0$ be a Borel function on $\bar{U}$. Then

$$\varepsilon_x^{CU}(f) = \int_{CU} f \, d\varepsilon_x^{CV} + \int_U f^{CU} \, d\, \varepsilon_x^{CV}.$$

Proof. The usual maneuvers involving integration techniques show that it suffices to demonstrate the proposition for $p \in \mathcal{P}^c$. Our aim is to show $\varepsilon_x^{CV}(R_p^{CU}) = R_p^{CU}(x)$. To this end, let $\varepsilon > 0$ be given. Find $u \in \mathcal{H}_+^*(X)$, $u \geq p$ on CU, $u(x) < R_p^{CU}(x) + \varepsilon$. Then

$$\varepsilon_x^{CV}(R_p^{CU}) \leq \varepsilon_x^{CV}(u) \leq u(x) < R_p^{CU}(x) + \varepsilon .$$

Conversely, there is a compact set $K \subset U$ such that $\int_{U \setminus K} p \, d\varepsilon_x^{CV} < \varepsilon$. Key Lemma 10.12 now provides $q \in \mathcal{H}_+^*(X)$ such that $q \leq p$ on X, $q = p$ on CU, $q < R_p^{CU} + \varepsilon$ on K. Then

$$\varepsilon_x^{CV}(R_p^{CU}) + 2\varepsilon \geq \varepsilon_x^{CV}(q) = \hat{R}_q^{CV}(x) \geq \hat{R}_q^{CU}(x) \geq \hat{R}_p^{CU}(x) = R_p^{CU}(x),$$

and therewith the assertion is established.

11.8 PROPOSITION. Let $A \subset X$ and $x \in \bar{A}$. The following assertions are equivalent:

(i) $x \in bA$,

(ii) $\varepsilon_x^A = \varepsilon_x$,

(iii) $\varepsilon_x = \text{w-lim}_{CA \ni y \to x} \varepsilon_y^A$.

Proof. The equivalence (i)$\iff$(ii) is trivial.

(ii)$\Longrightarrow$(iii): Pick $p \in \mathcal{P}^c$. We have

$$\limsup_{CA \ni y \to x} \varepsilon_x^A(p) \leq p(x) = \varepsilon_x^A(p) = \hat{R}_p^A(x) \leq \liminf_{CA \ni y \to x} \hat{R}_p^A(y) =$$

$$= \liminf_{CA \ni y \to x} \varepsilon_y^A(p).$$

Since $\mathcal{P}^c - \mathcal{P}^c$ is dense in $\mathcal{K}(X)$, we get (iii).

(iii)$\Longrightarrow$(ii): Since any $p \in \mathcal{P}^c$ is a limit of an increasing sequence of functions from $\mathcal{K}(X)$, we get

$$p(x) \leq \liminf_{CA \ni y \to x} \varepsilon_y^A(p) = \liminf_{CA \ni y \to x} \hat{R}_p^A(y) = \hat{R}_p^A(x).$$

Therefore, $\varepsilon_x^A = \varepsilon_x$ on $\mathcal{P}^c$.

EXERCISES

11.B.1 The measure sweeping property. A standard H-cone S of functions on X has the *measure sweeping property* if for any subset A of X and $x \in X$ there is a measure m on the σ-algebra $\mathcal{B}$ of all Borel subsets of X such that m is inner regular (i.e. $m(B) = \sup\{m(K): K \subset B \text{ compact}\}$ for all $B \in \mathcal{B}$) and $m(s) = \hat{R}_s^A(x)$ for each $s \in S$. The completion of m is denoted by ε_x^A and called the *balayage* (or *sweeping*) of ε_x onto A.

(a) Realize that the measure sweeping property holds on any $\mathfrak{P}$-harmonic space with a countable base.

(b) For any standard H-cone S of functions on a set X there is a standard H-cone S_1 of functions on a set X_1 and a mapping $\mathcal{X}: X \to X_1$ such that

(b_1) $\mathcal{X}$ is a homeomorphism of X onto $\mathcal{X}(X)$ in both natural and fine topologies,

(b_2) $\mathcal{X}(X)$ is finely dense in X_1,

(b_3) for every $s \in S$ there is $\bar{s} \in S_1$ such that $\bar{s} * \mathcal{X} = s$ on X, and any element of S_1 is of this form, $s \mapsto \bar{s}$ is an algebraic and lattice isomorphism,

(b_4) S_1 has the measure sweeping property on X_1,

(b_5) X_1 is completely metrizable.

For the proof see [BBC], pp. 113-115.

REMARK. The "completion" X_1 of (b) is not uniquely determined by the properties (b_1) - (b_5). The construction mentioned in [BBC] is in a certain sense maximal, and X is termed *saturated* if it coincides with this extension.

Let now S be a standard H-cone of functions on X having the measure sweeping property.

(c) For any $p \in S$ and any finely closed set $A \subset X$ which is ε_x^A-measurable for each $x \in X$ we have $\hat{R}_p^A = \sup\{\hat{R}_p^K : K \text{ compact}, K \subset A\}$.

Hint. Let $x \in X$, $p(x) < +\infty$. Then

$$\hat{R}_p^A(x) = \varepsilon_x^A(p) = \sup\left\{\int_K p \, d\varepsilon_x^A : K \subset A \text{ compact}\right\}.$$

Let $K \subset A$ be compact, $u \in S$, $u \geqq p$ on A. Then

$$\int_K p \, d\varepsilon_x^A \leqq \int_K u \, d\varepsilon_x^A \leqq \varepsilon_x^A(u) \leqq u(x).$$

Hence

$$\int_K p \, d\varepsilon_x^A \leqq R_p^A(x).$$

We see that $\hat{R}_p^A$ and $\sup\{\hat{R}_p^K : K \text{ compact}, K \subset A\}$ coincide on a finely dense set (namely $[p < +\infty] \cap [\hat{R}_p^A = R_p^A]$). Conclude the assertion using the fine continuity of both functions.

(d) If $U \subset X$ is a Borel finely open set and $x \in U$, then $\varepsilon_x^{CU}(U) = 0$.

Hint. Assume $\varepsilon_x^{CU}(K) > 0$ for some compact set $K \subset U$. By Theorem 10.25 there is a finely regular set $V \subset X$ such that $K \subset V \subset U$. Let $p \in S$ be a bounded continuous generator. Then $\hat{R}_p^{CU} = \hat{R}_q^{CU}$ where $q = R_p^{CV}$, so $\varepsilon_x^{CU}(p-q) = 0$ and therefore $\varepsilon_x^{CU}(V) = 0$, which is a contradiction.

REMARKS AND COMMENTS

An important notion of balayaged measures was introduced formerly in connection with balayages of Newtonian potentials. Our definition expressing a duality between balayages of measure and balayages of superharmonic functions appeared as a theorem in M. Brelot (1945). The existence of balayaged measures in axiomatic theories was established by R.-M. Hervé (1962), N. Boboc, C. Constantinescu and A. Cornea (1965).

Studying Riesz potentials, O. Frostman (1938) characterizes regular boundary points by means of a balayage of Dirac measures.

Proposition 11.7 is a powerful result of J. Bliedtner and W. Hansen (1976) and can be stated in various degrees of generality.

11.C The Dirichlet problem

Let $U \subset X$ be a regular set. Then, for any $f \in \mathcal{C}(\partial U)$ there is a continuous function F_f^U on $\bar{U}$ which is harmonic on U and coincides with f on ∂U. Moreover, $F_f^U \geqq 0$ if $f \geqq 0$. Hence, the mapping $\mu_x^U: f \longmapsto F_f^U(x)$ is for any $x \in U$ a Radon measure on ∂U, the so-called *harmonic measure* on U at x.

It is known that not every open set is regular. Nevertheless, given an arbitrary open set $U \subset X$, we can assign to any continuous function $f \in \mathcal{K}(\partial U)$ a harmonic function on U in a "reasonable" way. There are several methods how to construct this "generalized solution". In what follows, we will describe the *Perron method.*

Let $U \subset X$ be an open set and let f be a function on ∂U. We denote by $\bar{\mathcal{U}}_f^U$ the set of all hyperharmonic functions u on U which are lower bounded, nonnegative outside the intersection with U of a compact subset of X, and such that for any $y \in \partial U$

$$\liminf_{U \ni x \to y} u(x) \geqq f(y).$$

We denote

$$\bar{H}_f^U = \inf \bar{\mathcal{U}}_f^U , \quad \underline{H}_f^U = -\bar{H}_{-f}^U.$$

The following Bauer's minimum principle for hyperharmonic functions guarentees that $\underline{H}_f^U \leqq \bar{H}_f^U$.

11.9 BAUER'S MINIMUM PRINCIPLE. Any hyperharmonic function u on an open set $U \subset X$ must be nonnegative provided $u \geqq 0$ outside the intersection with U of some compact subset of X and

$$\liminf_{U \ni x \longrightarrow y} u(x) \geqq 0 \quad \text{for every } y \in \partial U.$$

Proof. See [Bau] , Kor. 1.3.6 or [C&C], Th. 1.3.1.

A function f on U is called *resolutive* if $\underline{H}_f^U = \bar{H}_f^U$ on U and the common value, denoted in this case simply by H_f^U, is harmonic on U. A set U is termed *resolutive* if any function of $\mathcal{K}(\partial U)$ is resolutive. It is easy to see that the map

$$\mu_x^U: f \longmapsto H_f^U(x)$$

is for every $x \in U$ a Radon measure on ∂U. This measure is again called the *harmonic measure* on U at x. Of course, any regular set is resolutive. Moreover, $H_f^U = F_f^U$, whenever U is regular.

Important properties of the Perron solutions are summarized in the next theorem.

11.10 THEOREM. (a) Any open subset of X is resolutive and

$$\mu_x^U = \varepsilon_x^{CU} \quad \text{for every } x \in U.$$

(b) For any function f on ∂U and any $x \in U$,

$$\bar{H}_f^U(x) = \int^* f \, d\mu_x^U$$

(c) A function f on ∂U is resolutive if and only if f is μ_x^U-integrable for any $x \in U$ (Doob's convergence axiom is essential for (c)).

Proof. See [Bau], IV. §1, or [C&C], Th. 2.4.2, Th. 2.4.4, Th. 7.1.2, Ex. 1.2.6.

REMARK. This theorem admits an important corollary. Let v be a lower semicontinuous, lower finite function on an open subset U of X. Then $v \in \mathcal{H}^*(U)$ if and only if $\varepsilon_x^{CW}(v) \leqq v(x)$ whenever W is a relatively compact open subset of X and $x \in W \subset \bar{W} \subset U$.

EXERCISES

11.C.1 An abstract version of the Dirichlet problem. Let $\mathcal{U}$ be a convex cone of lower finite functions on an abstract set Y, and let $U \subset Y$. Assume that the following two axioms are satisfied:

AD 1. If $u_n \in \mathcal{U}_+$, then $\sum u_n \in \mathcal{U}$.

AD 2. If $u \in \mathcal{U}$, $u \geqq 0$ on $Y \setminus U$, then $u \geqq 0$.

Let f be a function on Y. Define

$$\mathcal{U}_f = \{u \in \mathcal{U} : u \geqq f \text{ on } Y \setminus U\},$$

$$\bar{H}_f = \inf \mathcal{U}_f, \quad \underline{H}_f = -\bar{H}_{-f}.$$

Then $\underline{H}_f \leqq \bar{H}_f$. In the event that $\underline{H}_f = \bar{H}_f$, we denote their common value by H_f. Denote by $\mathcal{R}$ the family of all finite functions on Y for which $\underline{H}_f = \bar{H}_f$ and H_f is finite ("resolutive functions"). Further let $\mathcal{K} \subset \mathcal{R}$ be a vector lattice of functions on Y such that $\min(f,1) \in \mathcal{K}$ whenever $f \in \mathcal{K}$. Finally, denote by Σ the σ-algebra of subsets of $Y \setminus U$ generated by the family of functions $\{f \wedge Y \setminus U : f \in \mathcal{K}\}$.

(a) Show that $\mathcal{R}$ is a vector space and the map $f \mapsto H_f$ is a positive linear operator on $\mathcal{R}$.

(b) If $h \in \mathcal{U} \cap -\mathcal{U}$, then $h \in \mathcal{R}$ and $H_h = h$.

(c) If $f_k \in \mathcal{R}$, $f_k \nearrow f$, then $\bar{H}_f = \lim H_{f_n}$.

Hint. The heart of the proof lies in the observation that $\sum v_n \in \mathcal{U}_{f-f_1}$ provided $v_n \in \mathcal{U}_{f_{n+1}-f_n}$ (cf. proof of Lemma 13.5).

(d) Given $x \in U$ there is a measure μ_x ("harmonic measure") on Σ such that $\mu_x(f) = H_f(x)$ for every $f \in \mathcal{K}$.

Hint. Use the Stone-Daniell representation theorem taking into account (c).

(e) If $x \in U$ and $f \in \mathcal{K}^{\uparrow}$, then $\mu_x(f) = \bar{H}_f(x)$.

(f) If $\mathcal{U} \subset \mathcal{K}^{\uparrow}$ and if $x \in U$, then

$$\bar{H}_f(x) = \int^* f \, d\mu_x$$

for every function f on Y (cf. Theorem 13.6).

Hint. Show that

$$\int^* f \, d\mu_x = \inf\left\{\int^* s \, d\mu_x : s \in \mathcal{K}^{\uparrow},\ s \geqq f\right\} = \inf\left\{\bar{H}_s(x) : s \in \mathcal{K}^{\uparrow},\ s \geqq f\right\} = \bar{H}_f.$$

REMARK. The equality of (f) was proved in axiomatic theory of harmonic functions by H. Bauer (1962).

REMARKS AND COMMENTS

Extensive comments on the Dirichlet problem are shifted to Sections 13.A and 13.B. An important tool of potential theory on harmonic spaces is Bauer's minimum principle. Its beginning goes back to the sixties (H. Bauer (1960), (1962)) when the study of concave lower semicontinuous functions on compact convex sets was developed. In Ex. 12.C.1 we state a more general version of Bauer's minimum principle.

11.D The fine topology

We have already seen that the fine topology of standard H-cones of functions possesses many properties which seem to be interesting even to topologists. The aim of this section is to continue the examination of the fine topology in the framework of harmonic spaces. Thus, as our next step, we consider further properties of the fine topology imposing on a given "space" new requirements. In fact, we derive profit mainly from the local compactness of our space and sheaf properties of harmonic or hyperharmonic functions.

11.11 PROPOSITION. Every nonempty finely open set is uncountable. In particular, there are no isolated points in the fine topology.

Proof. Let V be a nonempty finely open countable subset of X. Select $x \in V$ and find a strict potential $p \in \mathcal{P}^c$. By means of Key Lemma 10.12 there is $q \in \mathcal{P}$ such that

$$q \lneqq p \text{ on } X,\ q = p \text{ on } CV \text{ and } q(x) < p(x).$$

From the positivity axiom it follows the existence of a harmonic function h defined on a neighborhood U of x such that $q(x) < h(x) < p(x)$. Further, there is a relatively compact open neighborhood W of x with the following properties:

$$h < p \text{ on } \bar{W},\ \bar{W} \subset U \text{ and } V \cap \partial W = \emptyset.$$

Indeed, denoting $W_r = [g < r]$, where g is a nonnegative continuous function on X such that $\{x\} = [g = 0]$, the set

$$\{r > 0:\ [g = r] \cap V \neq \emptyset,\ \bar{W}_r \subset U,\ h < p \text{ on } \bar{W}_r\}$$

is countable. Hence

$$\varepsilon_x^{CW}(q) \lneqq q(x) < h(x) = \varepsilon_x^{CW}(h) \leqq \varepsilon_x^{CW}(p) = \varepsilon_x^{CW}(q)$$

which is a contradiction. (Keep in mind that ε_x^{CW} is supported by $\partial W \subset CV$ and $q = p$ on CV !)

11.12 PROPOSITION. Every point of X has a fine local base consisting of fine neighborhoods which are compact and connected (in the initial topology on X).

Proof. Let V be a fine neighborhood of a given point $x \in X$. Making use of Ex. 11.D.2b there is a compact fine neighborhood K of x, $K \subset V$. Let $\mathcal{D}$ be the family of all both closed and open (in K) subsets of K containing x, and let $T := \bigcap \mathcal{D}$. Since the family $\mathcal{D}$ is closed under finite intersections, T is compact (obvious) and connected. (Indeed, assume that $T = F_1 \cup F_2$ where F_1, F_2 are disjoint nonempty closed subsets of T, and $x \in F_1$. Since F_1 and F_2 are also closed in K, they can be separated by open sets G_1 and G_2 in K. Compactness arguments yields $D \in \mathcal{D}$, $D \subset G_1 \cup G_2$. Since $D \cap G_1 \in \mathcal{D}$, we have arrived at a contradiction.) To conclude the proof, it suffices to check that T is a fine neighborhood of x. To this end let $p \in \mathcal{P}^c$ be a strict potential. In light of Key Lemma 10.12, there is a potential q such that

$$q \lneq p \text{ on } X, \quad q = p \text{ on } CK \text{ and } q(x) < p(x).$$

For every $D \in \mathcal{D}$ define the function q_D as follows:

$$q_D := \begin{cases} q & \text{on } C(K \setminus D), \\ p & \text{on } CD \end{cases}$$

(note that $q = p$ on $C(K \setminus D) \setminus D = CK$). Since the sets $C(K \setminus D)$ and CD are open, it follows from the sheaf property of hyperharmonic functions that $q_D \in \mathcal{H}^*_+(X)$. As pointed out, the family $\mathcal{D}$ is directed by inclusion, and thus $\{q_D\}$ is upper directed. Therefore $t := \sup q_D \in \mathcal{H}^*_+(X)$ and the set $[t < p]$ is a fine neighborhood of x contained in T.

The rest of this section is devoted to a result which brings out both the topological properties of a harmonic space X and the nature closely related to the ellipticity of X.

A set F is termed *absorbent* in an open set $U \subset X$ if $F \subset U$ and there is $u \in \mathcal{E}^*_+(U)$ such that $F = \{x \in U: u(x) = 0\}$. It is known that a set is absorbent if and only if it is closed in U and finely open (Ch. Berg (1971), cf. [C&C], Ex. 7.2.11).

A harmonic space is said to be *elliptic* if it has a basis of relatively compact resolutive sets G such that for every $x \in G$, the corresponding harmonic measure μ_x^G has a support all of ∂G.

Before treating the next theorem, let us note a useful fact: Any $\mathfrak{P}$-harmonic space is locally connected ([Bau], Satz 1.1.10, [C&C], Th. 1.1.1).

11.13 THEOREM. The following assertions are equivalent:

(i) X is an elliptic harmonic space,

(ii) any absorbent set of any open connected subset U of X is either empty or equal to U,

(iii) fine closure of any open set U of X is dense in the boundary of U,

(iv) any connected open subset of X is finely connected.

Proof. The equivalence (i)$\Leftrightarrow$(ii) is proved in [C&C], Prop. 6.1.3. Assume that X is elliptic. Let U be an open subset of X such that the set $N := bU \cap \partial U$ is not dense in ∂U. Thus, there is $x \in \partial U$ and its open connected neighborhood V such that $V \cap N = \emptyset$. Then $V \setminus (V \cap U)$ is a nonempty absorbent set in V different from V. This proves that (ii) implies (iii). Conversely, suppose that there is an open connected set $U \subset X$ and an absorbent set F in U with $\emptyset \neq F \neq U$. Referring to the space U, the fine closure of the open (and finely closed) nonempty set $U \setminus F$ cannot be dense in the boundary of $U \setminus F$. This shows that (iii)$\Rightarrow$(i). It remains to prove (iii)$\Leftrightarrow$(iv). This will follow from general Corollary 5.2 taking into account that the fine topology on X has the G_δ-insertion property (Theorem 10.30).

A harmonic space X satisfies the *domination axiom D* if for any relatively compact open subset U of X and for any $x \in U$, the set of all points of ∂U at which U is thin is of μ_x^U-measure zero.

It is known that the harmonic space associated to the Laplace equation satisfies the domination axiom D, while the heat equation harmonic space does not.

11.14 THEOREM. (J. Köhn and M. Sieveking (1967).) If the domination axiom D holds on X, then X is an elliptic harmonic space.

Proof. Let U be a relatively compact open subset of X. Suppose that $\partial U \cap bU$ is not dense in ∂U. Then there is $z \in \partial U$ and its relatively compact open neighborhood V such that $V \cap \partial U \subset CbU$. Let $f \in \mathcal{K}_+(\partial U)$, $f(z) > 0$, $f = 0$ outside V. Since $\mu_x^U(\partial U \setminus bU) = 0$ for every x (axiom D !), we have $H_f^U = 0$ on U (note that $H_f^U(x) = \varepsilon_x^{CU}(f) = \mu_x^U(f)$). On the other hand, CU is not thin at z (by Proposition 10.8, $bX = X$!) and (v) of Proposition 11.8 yields $f(z) = 0$.

EXERCISES

11.D.1 Show that there is no $\mathfrak{P}$-harmonic space on $[0,1)$ for which the set of all nonnegative superharmonic functions coincides with the standard H-cone of functions of Example (a) in Section 10.B.

Hint. Show that the one-point set $\{0\}$ is finely open.

11.D.2 Consider the harmonic space determined by the heat equation on $\mathbb{R}^2$. Prove the following assertions:

(a) Let $c \in \mathbb{R}$ and $H := \{(x;t) \in \mathbb{R}^2 : t \leq c\}$. Then H is an absorbent set and $V \cup \{(x;\ c)\}$ is a fine neighborhood of $(x;c)$ for any $x \in \mathbb{R}$.

(b) The set $\{(x;t) \in \mathbb{R}^2 : t = c\}$ is finely isolated.

(c) The fine topology is not locally connected. Acutally (b) gives a stronger result: The fine topology is totally disconnected, i. e. the only nonempty finely connected subsets of $\mathbb{R}^2$ are the one-point sets.

11.D.3 In classical potential theory on $\mathbb{R}^n$ $(n \geq 2)$ determined by solutions of the Laplace equation the following results are valid:

(a) For any subset A of $\mathbb{R}^n$, $A \setminus bA$ is a polar set. (M. Brelot (1944); cf. [He], Th. 10.9.)

(b) Any totally thin set is polar. Consequently, any semipolar set is polar and finely nowhere dense.

Hint. Use (a).

11.D.4 Fine topology induced by the band $\mathcal{M}$. Let $\mathcal{M}$ be the family of all potentials on X which are of the form $\sum p_n$ with $p_n \in \mathcal{P}^c$. (In the classical potential theory, every locally bounded potential belongs to $\mathcal{M}$. The band $\mathcal{M}$ was considered by N. Boboc, C. Constantinescu and A. Cornea (1965) since potentials of $\mathcal{M}$ conserve many useful properties of classical potentials like Frostman's domination principle or the Evans-Vasilescu continuity principle.)

Let τ_m be the weak topology on X induced by $\mathcal{M}$. Let A be a subset of a $\mathfrak{P}$-harmonic space X with countable base thin at a polar point $z \in X \setminus A$. If $\bar{A} \subset A \cup \{z\}$, then $X \setminus A$ is τ_m-open.

Hint. See J. Malý (1983).

11.D.5 (R. Wittmann, unpublished). If every point of X is polar, then the following two assertions are equivalent:

(i) every semipolar subset of X is polar (axiom of polarity),

(ii) the τ_m-topology has the quasi-Lindelöf property w.r.t. the σ-ideal of polar sets.

Hint. Trivially (i) implies (ii). Assume now that there exists a totally thin set A which is not polar; we may suppose that A is compact. Ex. 11.D.4 states that $U_x := \{x\} \cup CA$ is τ_m-finely open for every $x \in X$. Then $\{U_x : x \in A\}$ is a covering of X. The assumption then leads to the conclusion that A must be countable, and therefore polar.

11.D.6 (Cf. Ex. 3.1.7 of [C&C].) Let $X = (-1,1)$ be equipped with a harmonic sheaf consisting for an open subset U of X of all continuous functions h on U such that h is locally linear on $U \setminus \{0\}$ and if $0 \in U$ then h is constant on some interval $(-\varepsilon, 0]$. Show that X is a $\mathfrak{P}$-harmonic space and

(a) the interval $(-1,0]$ is an absorbent subset of X,

(b) the domination axiom D does not hold on X,

(c) X is locally connected in the fine topology.

11.D.7 Local connectedness of the fine topology. Let X be a $\mathfrak{P}$-harmonic space with countable base. The fine topology on X has the *Brelot property* if for every $z \in X$ and $u \in \mathcal{H}^*(X)$ there is a fine neighborhood of z on which u is continuous. (It was shown by M. Brelot (1967) that in the presence of the domination axiom D given a finely lower semicontinuous function f on a finely open subset U of X, for every $z \in U$ there is a fine neighborhood V of z on which f is lower semicontinuous. More general examination is contained in M. Brelot (1971).)

Given a subset A of X, we denote by Clop A the collection of all finely open-closed subsets of A.

(a) Suppose that for each finely regular subset U of X, the family Clop U is a σ-algebra. Use the quasi-Lindelöf property of the fine topology and Theorem 10.29 to show that the fine topology is locally connected.

Hint. First of all, realize that each point z has a local base of finely regular sets (Propositions 10.22 and 10.19). Pick a finely regular set U containing z. Set $V := \bigcap \{A: \; z \in A \in \text{Clop } U\}$. Show, as indicated, that there are $A_n \in$ Clop U containing z such that $V = \bigcap_{n=1}^{\infty} A_n$. By assumption, $V \in$ Clop U. It follows easily that V is a finely connected fine neighborhood of z.

(b) The fine topology on a $\mathfrak{P}$-harmonic space is locally connected provided it has the Brelot property.

Hint. Let U be a finely regular subset of X. We want to show that Clop U is a σ-algebra. Select $V_n \in$ Clop U and pick $z \in \bigcap V_n$. Let $p \in \mathcal{P}^c$ be strict and

$$f = \sum 2^{-n} \left(\hat{R}_p^{V_n} + \hat{R}_p^{U \setminus V_n}\right).$$

There is a compact connected fine neighborhood W of z (Proposition 11.12) on which f is continuous. Now, n being fixed, $\hat{R}_p^{CV_n}$ is continuous on W (and V_n is finely regular!), so that

$$V_n \cap W = W \cap [\hat{R}_p^{CV_n} < p]$$

is open in W. Similarly, $W \setminus K_n$ is open in W. Since W is connected, we get $W \subset V_n$. We conclude that $\bigcap V_n$ is a fine neighborhood of z.

(c) Let X be a $\mathfrak{P}$-harmonic space which is locally connected in the fine topology. Then X satisfies the following *axiom of broken ellipticity*: For any open subset U of X and any $x \in U$ there is an open neighborhood V of x, $V \subset U$ such that for any absorbent set A in U, either $x \in A$ or $A \cap V = \emptyset$.

Hint. Since every open subset of X is a "subspace", reduce the general case to the case when $U = X$. Proceed via the following steps:

(α) Let $x \in X$ and $p \in \mathcal{P}^c$ positive on X. Further, let $\{V_n\}$ be a local base of open neighborhoods of x. Assume that B_n are absorbent sets (in X) such that $x \in X \setminus B_n$ and $B_n \cap V_n \neq \emptyset$. Put $A_n = A_{n-1} \cup B_n$, $A = \overline{\bigcup A_n}$. Show that $x \in A \setminus \bigcup A_n$. By Prop. 6.1.2 of [C&C], A and A_n are absorbent sets.

(β) Since A is an absorbent set, it is finely open. Set $f := \sum \chi_{CA_n} \cdot p$. Then f is hyperharmonic on X.

(γ) Let M be the fine component of U containing x. Obviously, $M \subset A \setminus \bigcup A_n$. It follows that $\hat{R}_p^M = 0$ on every A_n, and hence $\hat{R}_p^M(x) = 0$. Thus M cannot be a fine neighborhood of x.

REMARK. Each elliptic harmonic space satisfies the axiom of broken ellipticity. Ex. 11.D.6 is a typical example of a non-elliptic harmonic space satisfying the axiom of broken ellipticity.

11.D.8 The fine topology on any $\mathfrak{P}$-harmonic space with countable base has the complete Lusin-Menchoff property. Furthermore, if $K \subset U$, K compact, U finely open, then there is a relatively compact finely regular set V such that $K \subset V \subset \overline{V} \subset U$.

Hint. Use Corollary 10.26. If K is compact, find a relatively compact open set G containing K and use Theorem 10.25 for $K \subset U \cap G$.

11.D.9 Assume that any point of a $\mathfrak{P}$-harmonic space X is polar. Then

(a) any countable subset of X is finely closed,

(b) finely compact sets are exactly the finite sets,

(c) if a fine limit of a sequence $\{x_n\}$ is x, then the set $\{n \in \mathbb{N}: x_n \neq x\}$ is finite,

(d) (X,f) is not separable,

(e) no point of X has a countable fundamental system of fine neighborhoods,

(f) the fine topology is neither normal, nor paracompact, nor Lindelöf,

(g) (J. Lukeš and L. Zajíček (1976)) assuming the continum hypothesis, (X,f) is not a Blumberg space (although it is a strong Baire space).

Hints. (a): Use Ex. 10.C.3.

(b)-(e): Use Theorem 2.1 together with Proposition 11.11.

(f): See Berg's theorem 10.32.

(g): Use Theorem 4.7 where $\mathcal{M}$ stands for the family of all G_δ -sets.

11.D.10 Let X be a $\mathfrak{P}$-harmonic space. Show that there is a Baire two function on X which is not a limit of a sequence of finely continuous functions.

Hint. Let p be a strict continuous potential, $U \subset X$ an open set and $h > 0$ a harmonic function on U. Put $D := \{x \in U: p(x)h^{-1}(x) \in \mathbb{Q}\}$. Establish that D and $U \setminus D$ are finely dense in U. (If, say, D is not finely dense in U, then there is $z \in U$ and its open and relatively compact neighborhood V such that $\overline{V} \subset U$ (Proposition 11.12) and $\hat{R}_p^{D \cup CV}(z) < p(z)$. Find $r \in \mathbb{Q}$ such that $\hat{R}_p^{D \cup CV}(z) < r\,h(z) < p(z)$. Now, apply Bauer's minimum principle to $V \cap \{x \in U: p(x) > r\,h(x)\}$.) Since X endowed with the fine topology is a Baire space and $U \setminus D$ is a fine-G_δ-set, D cannot be a fine-G_δ-set. Now cite Theorem 3.32.

11.D.11 Let X be a $\mathfrak{P}$-harmonic space. Then there is no Radon measure m on X such that the fine topology on X is an abstract density topology with respect to m. (Cf. Section 6.E, in particular Theorem 6.39.)

Hint. Assume that such a Radon measure m exists. Analogously as in Theorem 6.19 show that any F_σ-set is fine-G_δ. Now, use Ex. 11.D.10 and Theorem 3.32.
Another argument: Pick $z \in X$ and a continuous function $f \geqq 0$ for which $\{z\} = [f = 0]$. The set $L := \{r \in \mathbb{R}: m([f = r]) > 0\}$ being countable, there are $r_n \searrow 0$, $r_n \notin L$. Put $A := \bigcup_{n=1}^{\infty} [f = r_n]$. Since $m(A) = 0$, A is τ-closed. Pick a fine neighborhood V of x. By Proposition 11.12 there is a connected fine neighborhood

$W \subset V$ of x. By virtue of Proposition 11.11, W is uncountable. Since $f(W)$ is connected and $0 \in f(W)$, it must be $W \cap A = \emptyset$. Hence $z \in \overline{A}^{\tau} = A$ and this contradiction yields the required conclusion.

11.D.12 Consider the harmonic space associated with the heat equation on $\mathbb{R}^{n+1}$. Let $U \subset \mathbb{R}^{n+1}$ be an open set and $z = (y;s) \in U$. Denote

$$A := \{(x;t) \in U: \quad t \lneqq s\}.$$

Then A is an absorbent set in U and the harmonic measure μ_z^U is supported by $\overline{A} \cap \partial U$.

REMARKS AND COMMENTS

We bring together in this section further properties of the fine topology in harmonic spaces. The results are not new, but we present here some alternative proofs.

Proposition 11.11 is conform to Ex. 5.1.5 and Cor. 5.1.2 of [C&C]. Proposition 11.12 is due to C. Constantinescu and A. Cornea (1963).

The question on connectivity of the fine topology was raised by Ch. Berg and was settled by B. Fuglede in (1971b) and [Fug] for harmonic spaces satisfying the domination axiom D. As any harmonic space fulfilling this axiom is elliptic (the result of J. Köhn and M. Sieverking (1967); cf. [C&C], Prop. 9.2.1), Theorem 11.13 generalizes Fuglede's theorem. Its proof is taken from J. Lukeš and L. Zajíček (1977c). Likewise the result of Ex. 11.D.7 slightly generalizes another Fuglede's theorem of (1971b) and [Fug] which says that the fine topology is locally connected provided the domination axiom D is assumed.

An interesting property of fine topologies of Ex. 11.D.4 is due to J. Malý (1983), the result of Ex. 11.D.5 is a private communication of R. Wittmann.

Related paper : G. Wildenhain (1983).

12. FINELY HYPERHARMONIC FUNCTIONS

A. Fundamental properties

B. Properties of the cone $\mathcal{H}_+^*(U)$

C. Lsc pointwise finely hyperharmonic functions

In the seventies, B. Fuglede has developed the theory of finely harmonic functions on harmonic spaces in the presence of the domination axiom D. He defined (finely) harmonic and hyperharmonic functions on finely open sets which notions were defined previously on open sets only. Having the family of finely harmonic functions he succeeded to solve the generalized Dirichlet problem on finely open sets. This new builded fine potential theory appeared to be useful and under its influence some new results were obtained even in case of (initially) open sets.

The central role in the fine potential theory is played, perhaps, by the notion of finely hyperharmonic functions. It seems to be several ways how to define this concept in general harmonic spaces, all these definitions being equivalent in the spaces satisfying the domination axiom D. However, only some of them lead to a good theory in the framework of spaces without the domination axiom D. If we imitate Fuglede's definition, e.g., in the harmonic space associated with the heat equation, we get open sets on which the set of all "finely hyperharmonic functions" is far from the set of all hyperharmonic functions thereon. In this case, the set of all "finely hyperharmonic functions" does not form even a cone.

In this chapter we make use of Fuglede's characterization of finely hyperharmonic functions to introduce this concept in general harmonic spaces. In the definition we cannot use the language of the fine topology in its pure form since we deal unavoidably with closures in the initial topology. We prove that the family of all finely hyperharmonic functions defined in such a way has the Riesz decomposition property as well as the suitable lattice properties. Moreover, we prove the fine continuity of finely hyperharmonic functions. Under the assumption concerning lower boundedness the notions of finely hyperharmonic functions and hyperharmonic functions coincide on open sets.

Later on we shall prove a minimum principle for finely hyperharmonic functions which plays the crucial role in investigation of the Dirichlet problem.

Notice that, omitting the domination axiom D in fine potential theory, we lose some pleasant properties of finely hyperharmonic functions as, e.g., the fine sheaf property. Though we shall neither need nor use the domination axiom D, it may be worth while to note that some theorems and assertions on finely hyperharmonic functions in general harmonic spaces are proved under stronger assumptions than those ones in Fuglede's theory (e.g. we need the lower semicontinuity of considered functions instead of Fuglede's fine lower semicontinuity). To overcome this defect we use in Section 14 quasi-topological methods in order to derive results covering up both theories.

A final remark should be added. The primary aim of the third part is to introduce a fine potential theory on harmonic spaces. As many results have their counterpart within the theory of standard H-cones we have hesitated how to arrange the material of each section. On the one hand, it would be nice to include results as general as possible and, on the other hand, it should be kept in mind that the theory in this case is apparently more complicated. While trying to maintain the balance just described, we finally decided to present the interpretation in the context of harmonic spaces. For this reason, we have included a selection of exercises and the Appendix which provide extensions of the theory presented in the text to the more general framework of standard H-cones of functions.

12.A Fundamental properties

Before moving on to our main definitions, it is convenient to introduce some notation. Given a finely open subset U of a harmonic space X, $\mathcal{B}(U)$ denotes the collection of all relatively compact, finely open sets V with $\overline{V} \subset U$. Moreover, $\mathcal{B}_o(U)$ is reserved for the family of all Borel sets in $\mathcal{B}(U)$.

A function u is said to be *finely hyperharmonic* on a finely open set U if it is lower finite, finely lower semicontinuous, if u^{CV} is defined and if

$$-\infty < u^{CV} \leqq u$$

holds on V for every $V \in \mathcal{B}(U)$.

A function h is said to be *finely harmonic* on U if both the functions h and $-h$ are finely hyperharmonic on U. The set of all nonnegative finely hyperharmonic functions is denoted by $\mathcal{H}^*_+(U)$. Further, we denote by $\mathcal{S}(U)$ the set of all $s \in \mathcal{H}^*_+(U)$ which are finite on a finely dense subset of U. The functions from $\mathcal{S}(U)$ are called *finely superharmonic* on U.

Both $\mathcal{H}^*_+(U)$ and $\mathcal{S}(U)$ are min-stable convex cones of functions on U.

A function f is said to be *pointwise finely hyperharmonic* on a finely open set U provided it is lower finite, finely lower semicontinuous on U and if for each $x \in U$ the collection

$$\left\{V \in \mathcal{B}(U):\ \int^* f \, d\varepsilon_x^{CV} \leqq f(x)\right\}$$

forms a fine local base at x.

12.1 REMARKS. (a) In the presence of the domination axiom D, B. Fuglede defines the family of all finely hyperharmonic functions in a little different manner. Apart from some necessary minor changes (cf. [Fug], p.68) it is possible to transfer this definition to general harmonic spaces as follows: A lower finite, finely lower semicontinuous function u on a finely open set U is *finely hyperharmonic in Fuglede's sense* if the collection of all $V \in \mathcal{B}(U)$ such that u is lower bounded on $\overline{V}$ and

$$\int^* u \, d\varepsilon_x^{CV} \leqq u(x)$$

holds for all $x \in V$, forms a base for the fine topology on U. The family of all finely hyperharmonic functions in Fuglede's sense is denoted by $\mathcal{H}_F^*(U)$.

Of course, each finely hyperharmonic function belongs to $\mathcal{H}_F^*(U)$. If the domination axiom D holds on a harmonic space, then $\mathcal{H}_+^*(U) = (\mathcal{H}_F^*)_+(U)$ but, in general, these classes can differ. Hence, in harmonic spaces without axiom D the family $\mathcal{H}_F^*(U)$ lacks of the good global behaviour and we use a narrower class of finely hyperharmonic functions.

Later on, we shall show that $\mathcal{H}_F^*(U)$ is close to the family of all pointwise finely hyperharmonic functions (cf. Ex. 12.A.6, or Proposition 14.14). At this point notice only that fine hyperharmonicity is not a finely local concept: There exist functions which are finely hyperharmonic on a fine neighborhood of each point of U without being finely hyperharmonic on U (cf. Ex. 12.A.2 or Ex. 13.A.1). These problems and related questions are discussed in J. Lukeš and J. Malý (1982).

(b) The inequality $\int u \, d\varepsilon_x^{CV} \leqq u(x)$ in the definition of fine hyperharmonicity suffices to be verified for Borel sets $V \in \mathcal{B}_o(U)$ only. Indeed, given a finely open set $V \in \mathcal{B}(U)$ and $x \in V$, there is an F_σ- sweeping kernel $W \subset V$ (Theorem 10.21). Put $B = W \cup \{x\}$. Then $W \subset B \subset V$ and hence B is also an F_σ-sweeping kernel of V, and B is a finely open fine neighborhood of x. Therefore,

$$\int u \, d\varepsilon_x^{CV} \leqq u(x), \text{ provided } \int u \, d\varepsilon_x^{CB} \leqq u(x).$$

(c) It will follow from Proposition 12.3 that a nonnegative function u on U is finely hyperharmonic on U provided it is finely lower semicontinuous on U and

$$\int^* u \, d\varepsilon_x^{CV} \leqq u(x)$$

whenever $x \in V \in \mathcal{B}(U)$ (or, $x \in V \in \mathcal{B}_o(U)$).

(d) In what follows, we restrict ourselves (for the sake of simplicity) sometimes to nonnegative finely hyperharmonic functions. It would even be possible to go further and to generalize some results to the case of functions attaining somewhere negative values, the examples in Exercises being adequate illustrations of these ideas. The straightforward but tedious improvements require as a rule clumsy additional assumptions concerning certain lower boundedness conditions. We do not feel that added generality justifies the effort involved.

12.2 THEOREM. Let u be a locally lower bounded function on an open set $U \subset X$. Then the following assertions are equivalent:

(i) u is finely hyperharmonic on U,

(ii) u is hyperharmonic on U.

Proof. (i) $\Rightarrow$ (ii): It easily follows from [C&C], Prop. 5.1.4. (ii) $\Rightarrow$ (i): Pick $x \in V \in \mathcal{B}_o(U)$. By Proposition 11.3

$$\hat{R}_p^{CV}(x) = \sup \{ \hat{R}_p^K(x) : K \subset CV, K \text{ compact} \}$$

for every $p \in \mathcal{P}^c$. It easily follows that

$$\hat{R}_p^{CV}(x) = \lim_{W \in \mathcal{V}} \hat{R}_p^{CW}(x)$$

where W runs through the net $\mathcal{V}$ of all relatively compact open neighborhoods of V with $\overline{W} \subset U$. From the density of $\mathcal{P}^c - \mathcal{P}^c$ in $\mathcal{K}(X)$ we deduce

$$\varepsilon_x^{CV} = w\text{-}\lim_{W \in \mathcal{V}} \varepsilon_x^{CW}.$$

Taking into account the lower semicontinuity of every hyperharmonic function and Remark following Theorem 11.10, we obtain

$$\varepsilon_x^{CV}(u) = \liminf_{W \in \mathcal{V}} \varepsilon_x^{CW}(u) \leqq u(x).$$

12.3 PROPOSITION. Every pointwise finely hyperharmonic function u on a finely open set U is a Borel function on U.

Proof. Choose $c, d \in \mathbb{R}$, $c < d$. The fine lower semicontinuity of u implies that the set $[u \leqq d]$ is finely closed in U and hence

$$U \cap b[u \leqq d] \subset [u \leqq d].$$

Let $z \in U \setminus b[u \leqq d]$. Then $W := \{z\} \cup [u > d]$ is a fine neighborhood of z. Let h be a harmonic function on an open neighborhood G of z such that $c < h < d$ on G (its existence follows from the positivity axiom). We can find $V \in \mathcal{B}_o(U)$ such that

$$z \in V \subset \overline{V} \subset W \cap G, \quad u^{CV}(z) \leqq u(z)$$

and u is lower bounded on $\overline{V}$. Thus, by Theorem 12.2

$$c < h(z) = h^{CV}(z) \leqq u^{CV}(z) \leqq u(z).$$

It proves that

$$U \setminus b\,[u \leqq d] \subset [u > c].$$

We have shown that

$$[u \leqq c] \subset U \cap b\,[u \leqq d] \subset [u \leqq d].$$

Since bA is a Borel set for any $A \subset X$ (see Corollary 10.11), the assertion follows now easily (cf. proof of Theorem 2.13).

12.4 LEMMA. Let U, V be finely open subsets of X. Assume that $\bar{V} \subset U$ and, moreover, V is a Borel set. If u, v are nonnegative functions on X, u hyperharmonic on X, v finely hyperharmonic on U, $v \leqq u$ on X and $v = u$ on $X \setminus V$, then v is hyperharmonic on X.

Proof. In view of Theorem 12.2, it is enough to show that v is finely hyperharmonic on X. Apparently, v is finely lower semicontinuous on X. Pick $W \in \mathcal{B}_o(X)$ and let us show that $v^{CW} \leqq v$ on W. (Notice that v is a Borel function on $X = (X\setminus V) \cup V$ using Proposition 12.3.) If $x \in W \setminus V$, then using the fine hyperharmonicity of u on X,

$$\varepsilon_x^{CW}(v) \leqq \varepsilon_x^{CW}(u) \leqq u(x) = v(x).$$

Choose now $x \in W \cap V$ and denote $M = X \setminus (W \cap V)$. Since $\varepsilon_x^M(W \cap V) = 0$ it follows that

$$v^{CW} \leqq v \quad \varepsilon_x^M\text{- almost everywhere on } W.$$

By Proposition 11.7 we get

$$v^{CW}(x) = \int v\, d\, \varepsilon_x^{CW} = \int_W v^{CW}\, d\, \varepsilon_x^M + \int_{CW} v\, d\varepsilon_x^M \leqq$$

$$\leqq \int_W v\, d\, \varepsilon_x^M + \int_{CW} v\, d\, \varepsilon_x^M = \varepsilon_x^M(v) \leqq v(x)$$

(we use that $v \in \mathcal{H}_+^*(U)$ and $W \cap V \in \mathcal{B}_o(U)$).

REMARK. Another proof can be found in the Appendix.

The key for proving the main properties of finely hyperharmonic functions is the next crucial lemma.

12.5 LEMMA. Let $u \in \mathcal{H}_+^*(U)$, $p \in \mathcal{P}^c$, where U is a finely open subset of X.

Then there is a unique $q \in \mathcal{P}^b$ such that

$$q = \min(p^{CU} + u,\ p) \quad \text{on} \quad U$$

and

$$\hat{R}_p^{CU} \leqq q \leqq p \quad \text{on} \quad X.$$

Proof. Let $p' \in \mathcal{P}^c$ be strictly positive. Pick $\varepsilon > 0$ and $v \in \mathcal{H}_+^*(X)$, $v \geqq p$ on CU. Set $V := [\varepsilon p' + v < p]$. Then V is a finely open Borel set and

$$\bar{V} \subset [\varepsilon p' + v \leqq p] \subset U.$$

Denote

$$f_{V,\varepsilon} := \begin{cases} p & \text{on } X \setminus V \\ \min\ (p,\ v + \varepsilon p' + u) & \text{on } V. \end{cases}$$

By Lemma 12.4, $f_{V,\varepsilon} \in \mathcal{H}_+^*(X)$. Finally, we put

$$q := \wedge \{ f_{V,\varepsilon} \ : \varepsilon > 0,\ v \in \mathcal{H}_+^*(X),\ v \geqq p \ \text{on} \ CU \}.$$

Since $q \leqq p$, we get $q \in \mathcal{P}^b$. We know also, that $R_p^{CU} = \hat{R}_p^{CU}$ on U (cf. Corollary 10.13), which yields $q = \min(p, \hat{R}_p^{CU} + u)$ on U. The inequality $\hat{R}_p^{CU} \leqq q \leqq p$ on X is obvious. Since the semipolar set $[\hat{R}_p^{CU} < R_p^{CU}]$ equals $\{x \in CU: \hat{R}_p^{CU}(x) < p(x)\}$ (cf. Corollary 10.13 and Theorem 10.17) and since q is finely continuous, the uniqueness of q follows from Theorems 10.29 and 10.28.

12.6 THEOREM. Every finely hyperharmonic function u on a finely open set U is finely continuous on U.

Proof. Choose $z \in U$. Assuming first $u(z) > 0$, find $V \in \mathcal{B}(U)$ such that $z \in V$ and $u > 0$ on V. The function u is finely lower semicontinuous by definition. It is enough to consider $u(z) < +\infty$. In this case there is $p \in \mathcal{P}^c$ such that

$$\hat{R}_p^{CV}(z) + u(z) < p(z).$$

By Lemma 12.5 there is $q \in \mathcal{P}^b$ such that

$$q = \min(u + \hat{R}_p^{CV},\ p)$$

on V. The set

$$W := \{x \in V: \ q(x) < p(x)\}$$

is a fine neighborhood of z. Since $u = q - \hat{R}_p^{CU}$ on W and the function $q - \hat{R}_p^{CU}$ is finely continuous, it follows that u is finely continuous at z. Now, if $u(z) \lneqq 0$, we may replace u by $u+s$ where s is a superharmonic function on X with $-u(z) < s(z) < +\infty$.

12.7 LEMMA. Let U be finely open and let $p \in \mathcal{P}^f$. Then p^{CU} is finely harmonic on U.

Proof. Pick $x \in V \in \mathcal{B}_o(U)$ and let U' be an F_σ - sweeping kernel of U. Put $W := U' \cup \overline{V}$. Then W is an F_σ - sweeping kernel of U and $V \in \mathcal{B}_o(W)$. By Proposition 11.7 we have

$$p^{CU}(x) = p^{CW}(x) = \varepsilon_x^{CV}(R_p^{CW}) = \varepsilon_x^{CV}(p^{CW}) = \varepsilon_x^{CV}(p^{CU}).$$

Since $p^{CU} = \hat{R}_p^{CU}$, we see that p^{CU} is finely continuous on U and the proof is complete.

REMARK. A different proof of Lemma 12.7 is outlined in Ex. 12.A.7. Notice also that in Corollary 12.13 we derive a more general result.

12.8 PROPOSITION. If a function f on a finely open set U is finely hyperharmonic on each set of $\mathcal{B}_o(U)$, then f is finely hyperharmonic on U.

Proof. Given $V \in \mathcal{B}_o(U)$, in light of the Lusin-Menchoff property there is a set $W \in \mathcal{B}_o(U)$ (cf. Ex. 11.D.8) such that $\overline{V} \subset W$. Since f is finely hyperharmonic on W by assumption, the assertion follows easily by the definition of fine hyperharmonicity.

EXERCISES

12.A.1 Let u be a finely hyperharmonic function on an open set $U \subset X$. Using Doob's convergence axiom show that u is locally bounded (and hence hyperharmonic) on U.

Hint. Let $W \in \mathcal{B}_o(U)$ be open. Then

$$-\infty < u^{CW} = \lim (\max(u,-n))^{CW} \leqq u.$$

Since u^{CW} is harmonic on W (cf. Theorem 11.10.c), u is locally lower bounded.

12.A.2 Example. (J.Lukeš and J. Malý (1982)). Let $X = [0,1)$ and $T = \{0\} \cup \{2^{-n}: n \in \mathbb{N}\}$. A function h is said to be harmonic on an open set $U \subset X$ if h is continuous on U, locally linear on $U \setminus T$ and "right constant" at any point of $U \cap T$. Prove the following assertions:

(a) The sheaf of harmonic functions on X satisfies Bauer's axioms.

(b) If $f = \chi_{[1/2,1)}$, then f is a nonnegative function being finely hyperharmonic on a fine neighborhood of each point of X. Show that f is not finely hyperharmonic on X.

(c) If $f = 0$ on each interval $[2^{-2j}, 2^{-2j+1})$, $f = 1$ on each interval $[2^{-2j+1}, 2^{-2j+2})$ and $f(0) = 0$, then f is pointwise hyperharmonic on X and it is not finely continuous at 0.

(d) Show that the family of all nonnegative pointwise finely hyperharmonic functions on X is not a convex set.

Find also similar counterexamples on the space of locally increasing functions (Ex. (b) of Section 10.B).

12.A.3 Consider a harmonic space of the heat equation on $\mathbb{R}^2$. Find a strictly increasing sequence $\{a_k\}$ such that $a_k \nearrow 0$, $\varepsilon_0^{CM_k}(f_k) \leqq 1$ for any k even where $M_k = \{(x;t): k|x| < 1, \ t \in (a_k, 0)\}$,

$$f_k(x,t) = \begin{cases} 0 & \text{for } t \in (a_{k-1}, a_{k+1}] \\ 2 & \text{for } t \geqq 0 \\ 5 & \text{elsewhere} \end{cases}$$

Set $u := \inf\{f_{4n}: n \in \mathbb{N}\}$, $v = \inf\{f_{4n+2}: n \in \mathbb{N}\}$ and show that u, v are pointwise finely hyperharmonic functions on $\mathbb{R}^2$ but u+v is not.

12.A.4 Let u be a pointwise finely hyperharmonic function on a finely open set U, let $x \in U$. Show that

$$u(x) = \liminf_{\mathcal{V}_f(x)} \varepsilon_x^{CV}(u)$$

where V runs through the filter $\mathcal{V}_f(x)$ of all fine neighborhoods of x.

Hint. The way to prove the inequality $u(x) \leqq \liminf \varepsilon_x^{CV}(u)$ is by using the fine lower semicontinuity of u at x together with the positivity axiom. The reverse inequality is obvious by definition of pointwise hyperharmonicity.

REMARK. The assertion remains valid if we suppose instead of pointwise fine hyperharmonicity that u is finely continuous at x. It is also not too hard to see that the hypothesis can be weakened by assuming only that u is "pointwise finely hyperharmonic at x".

12.A.5 Let u, U and x be as in the previous exercise. Show that

$$u(x) = \text{fine-}\liminf_{y \to x} u(y).$$

Hint. Use Ex. 12.A.4.

12.A.6 (J. Lukeš and J. Malý (1982)). Let U be a finely open subset of X. Prove that every finite pointwise finely hyperharmonic function belongs to $\mathcal{X}_F^*(U)$ by showing the next observation: (*)

Let $V \subset X$ be a relatively compact finely open set and $f < +\infty$ a lower bounded, finely lower semicontinuous Borel function on $\bar{V}$. If $z \in V$ and $\varepsilon_z^{CV}(f) \leqq f(z)$, then there is a finely open set $W \subset V$ such that $z \in W$ and $f^{CW} \leqq f$ on W.

Hint. Set $A_k := CV \cup \{x \in V\colon\ f(x) \leqq \varepsilon_x^{CV}(f) - \frac{1}{k}\}$, $A = \bigcup_{k=1}^{\infty} A_k$. If $\varepsilon_z^{A_k}(f) < f(z)$ for some k, put $W = \{x \in CA_k\colon\ \varepsilon_x^{A_k}(f) < f(x)\}$. If $\varepsilon_z^{A_k}(f) \geqq f(z)$ for all k, put $W = C\bar{A}^f$. Now, a verification of the required inequality makes substantial use of the Bliedtner-Hansen theorem 11.7.

12.A.7 Establish Lemma 12.7 proceeding via the following steps:

(a) Pick $V \in \mathcal{B}_o(U)$ and find a Borel sweeping kernel W of U, $V \subset W \subset U$.

(b) By Proposition 11.3, find an increasing sequence $\{K_n\}$ of compact sets such that $\hat{R}_p^{K_n} \nearrow \hat{R}_p^{CW} = \hat{R}_p^{CU} = p^{CU} \leqq p$.

(c) Functions $\hat{R}_p^{K_n}$ are harmonic on CK_n ([C&C] , Prop. 5.3.1), therefore finely harmonic on W.

(d) Use the Lebesgue monotone convergence theorem, (b) and (c) to show that $(p^{CU})^{CV} = p^{CU}$ on V.

(e) N. Boboc and Gh. Bucur have kindly informed the authors that the assumption of finiteness of f is essential in this proposition therefore putting in order Theorem 8 and its Corollary 9 of paper mentioned.

REMARKS AND COMMENTS

It is only now that remarkable progress has been made in the fine potential theory. This has been due to the fact of introducing new topological techniques into the field of abstract harmonic spaces. Although the introduction of the fine topology goes back to the forties, much of the recent development in the area is due to B. Fuglede in the 1970's. He started his theory from consideration of harmonic spaces fulfilling the domination axiom D. There is a number of his papers, cf. (1970), (1971b), (1974), (1975a), investigating mainly the properties both of finely hyperharmonic functions and of finely harmonic ones. Almost all results are collected in [Fug] where the whole theory is presented.

The case when axiom D is not assumed was investigated by J. Bliedtner and W. Hansen (1975), (1978) and by J. Lukeš and J. Malý (1982). Recently, the framework of harmonic spaces is often left out and fine hyperharmonicity is studied in various contexts. Let us mention only the localization theory on standard H-cones of N. Boboc, Gh. Bucur and A. Cornea (1975), [BBC], N. Boboc and Gh. Bucur (1985a), (1985b), (1985c)).

Under the axiom D, B. Fuglede himself proved that each function from $\mathcal{H}_F^*(U)$ is finely hyperharmonic and raised the question ([Fug], p. 69) of whether it is true that any pointwise finely hyperharmonic function is finely hyperharmonic. An affirmative answer was given by T. Lyons (1982). He used the Choquet theory and the proof is based on the Brelot property. J. Lukes and J. Malý (1982) proved that every finite pointwise finely hyperharmonic function belongs to $\mathcal{H}_F^*$, even in general harmonic spaces (cf. Ex. 12.A.6). Using quasi-topological methods, we present in Section 14.C (Proposition 14.14) another proof in spaces with axiom D.

Related papers: B. Fuglede (1982), J. Lukeš (1984).

12.B Properties of the cone $\mathcal{H}_+^*(U)$

In this section we investigate the lattice properties of the cone $\mathcal{H}_+^*(U)$. As a consequence, we obtain a convergence property of finely harmonic functions. We prove also fine harmonicity and fine continuity of functions of the type "f^{CU}" and, finally, we introduce a further important topology on finely open sets.

We suppose throughout this section that U is a finely open subset of X, and that p is a fixed finite continuous strict potential on X. Denote $e = p - p^{CU}$.

12.9 PROPOSITION. (Property of upper directed sets.) Let $\mathcal{W}$ be an upper directed set of finely hyperharmonic functions on U. Then $f := \sup \mathcal{W}$ is finely hyperharmonic.

Proof. According to the quasi-Lindelöf property (Theorem 10.31 and 1.4) there is an increasing sequence $\{f_n\} \subset \mathcal{W}$ such that the set

$$M := \{x \in U : \lim f_n(x) < f(x)\}$$

is semipolar. By a straightforward use of the Lebesgue monotone convergence theorem, it can be easily shown that $\lim f_n$ is finely hyperharmonic on U. Hence $\lim f_n$ is finely continuous on U, and thus M is a finely open set. As M is semipolar, it must be empty (Theorem 10.29). We see that $f = \lim f_n$ is finely hyperharmonic on U.

Given $u \in \mathcal{H}^*_+(U)$ and $n \in \mathbb{N}$, we denote $u^{(n)} = \min(u, ne)$. Obviously, $u^{(n)}$ is finely hyperharmonic on U. Further, $\tilde{u}^{(n)} \in \mathcal{P}^b$ will denote the unique potential on X having the properties:

$$\tilde{u}^{(n)} = u^{(n)} + np^{CU} \quad \text{on } U,$$

$$\tilde{u}^{(n)} = np \quad \text{on } [p = \hat{R}_p^{CU}]$$

(cf. Lemma 12.5).

12.10 PROPOSITION. (Property of fine lower semicontinuous regularization). Let $\mathcal{W} \subset \mathcal{H}^*_+(U)$. Then $g := \widehat{\inf \mathcal{W}}^f \in \mathcal{H}^*_+(U)$. Moreover, g differs from $\inf \mathcal{W}$ in a semipolar set.

Proof. Denote

$$\mathcal{W}_n := \{v^{(n)} : v \in \mathcal{W}\}, \quad \tilde{\mathcal{W}}_n := \{\tilde{v}^{(n)} : v \in \mathcal{W}\}.$$

By [C&C], Th. 5.1.1, $\widehat{\inf \tilde{\mathcal{W}}_n}^f \in \mathcal{H}^*_+(X)$. Since $-p^{CU} \in \mathcal{H}^*_+(U)$ (Lemma 12.7) and

$$\widehat{\inf \mathcal{W}_n}^f = -np^{CU} + \widehat{\inf \tilde{\mathcal{W}}_n}^f$$

on U, we get $\widehat{\inf \mathcal{W}_n}^f \in \mathcal{H}^*_+(U)$. Further,

$$\widehat{\inf \mathcal{W}_n}^f \nearrow g.$$

Taking into account Proposition 12.9 we obtain $g \in \mathcal{H}^*_+(U)$. The set $[g < \inf \mathcal{W}]$ is contained in

$$\bigcup_{n=1}^{\infty} [\widehat{\inf \tilde{\mathcal{W}}_n}^f < \inf \tilde{\mathcal{W}}_n],$$

and therefore it is semipolar by Theorem 10.17.

REMARKS. (a) We see that the convex cone $\mathcal{H}^*_+(U)$ is a complete lattice with respect to its *natural order* (cf. Section 10.B). For any finite subset $\mathcal{K} \subset \mathcal{H}^*_+(U)$, $\wedge\mathcal{K}$ exists and equals to $\inf\mathcal{K}$.

If $\mathcal{W} \subset \mathcal{H}^*_+(U)$, then $\wedge\mathcal{W} = \widehat{\inf \mathcal{W}}^f$ and

$$v + \wedge\mathcal{W} = \wedge(v + \mathcal{W})$$

for any $v \in \mathcal{H}^*_+(U)$. Further

$$\vee\mathcal{W} = \wedge\{v \in \mathcal{H}^*_+(U) : v \succcurlyeq w \text{ for any } w \in \mathcal{W}\}.$$

If $\mathcal{W}$ is upper directed, then $\vee\mathcal{W} = \sup\mathcal{W}$.

(b) We have shown that $\mathcal{H}^*_+(U)$ possesses the property of upper directed sets and of fine lower semicontinuous regularization. Hence, $\mathcal{H}^*_+(U)$ satisfies the axioms given in Chapter 4.1 of [C&C]. In particular, all propositions, which can be derived from these axioms are valid.

Now, we prove that $\mathcal{H}^*_+(U)$ has also the Riesz decomposition property, therefore completing the above mentioned list of axioms. Taking into account Remark (a) we see that $\mathcal{S}(U)$ forms an H-cone of functions provided it contains positive constants. (It is even a standard H-cone of functions, see Appendix.)

12.11 PROPOSITION. (The Riesz decomposition property.) If $v, w, u \in \mathcal{H}^*_+(U)$, $v+w \succcurlyeq u$, then there exist $f, g \in \mathcal{H}^*_+(U)$ such that

$$f \preccurlyeq w,\ g \preccurlyeq v \quad \text{and} \quad f + g = u.$$

Proof. Fix $n \in \mathbb{N}$ and denote

$$f'_n = \wedge\{s \in \mathcal{H}^*_+(X) : s \geqq n\hat{R}_p^{CU} + \tilde{u}^{(n)} - \tilde{v}^{(n)}\}.$$

As

$$n\hat{R}_p^{CU} + \tilde{u}^{(n)} \leqq \tilde{v}^{(n)} + f'_n,$$

the Riesz decomposition property of $\mathcal{H}^*_+(X)$ yields the existence of g'_n, $f''_n \in$ $\in \mathcal{H}^*_+(X)$ such that

$$n\hat{R}_p^{CU} + \tilde{u}^{(n)} = g'_n + f''_n \quad \text{and} \quad g'_n \leqq \tilde{v}^{(n)}, \quad f''_n \leqq f'_n.$$

Hence $n\hat{R}_p^{CU} + \tilde{u}^{(n)} \leqq \tilde{v}^{(n)} + f''_n$, and we deduce that $f''_n = f'_n$. Define the following functions on U:

$$f_n := f'_n - np^{CU}, \; h_n = g'_n - np^{CU}, \; g_n = \wedge\{h_j : j \geqq n\}.$$

Obviously,

$$f_n \leqq w^{(n)}, \; h_n \leqq v^{(n)},$$

$$f_n + h_n = f'_n + g'_n - 2np^{CU} = f''_n + g'_n - 2np^{CU} = u^{(n)}$$

on U. (Since $\tilde{w}^{(n)} = np$ on $[p=\hat{R}_p^{CU}]$ and $\tilde{w}^{(n)} \geqq \tilde{u}^{(n)} - \tilde{v}^{(n)} + np^{CU}$ on U, we deduce in light of Theorems 10.17 and 10.29 that $\tilde{w}^{(n)} \geqq n\hat{R}_p^{CU} + \tilde{u}^{(n)} - \tilde{v}^{(n)}$ on X. Hence $f_n \leqq w^{(n)}$.) Apparently, the sequence $\{g_n\}$ is increasing on U. It can be checked that the same holds for $\{f_n\}$. (Pick $s \in \mathcal{H}^*_+(X)$, $s \geqq (n+1)\hat{R}_p^{CU} + \tilde{u}^{(n+1)} - \tilde{v}^{(n+1)}$. Since

$$s - p^{CU} \geqq np^{CU} + \tilde{u}^{(n)} - \tilde{v}^{(n)} \quad \text{on} \quad [u \geqq v],$$

$$s - p^{CU} \geqq np^{CU} \geqq np^{CU} + \tilde{u}^{(n)} - \tilde{v}^{(n)} \quad \text{on} \quad [u < v],$$

we get $f'_{n+1} - p^{CU} \geqq f'_n$ on U.) Set $f := \lim f_n$, $g := \lim g_n$. Then $f, g \in \mathcal{H}^*_+(U)$ and for $j \geqq n$ we have

$$f_j + g_n \leqq f_j + h_j = u^{(j)} \leqq u,$$

$$f + h_j \geqq f_j + h_j = u^{(j)} \geqq u^{(n)}.$$

Hence

$$u^{(n)} \leqq \inf\{f + h_j : j \geqq n\} = f + g_n = \lim_{j\to\infty} (f_j + g_n) \leqq u.$$

Letting $n \to \infty$, we obtain $f + g = u$. Since

$$f_n \leq w^{(n)} \leqq w, \quad g_n \leqq h_n \leqq v^{(n)} \leqq v,$$

we have $f \leqq w$, $g \leqq v$.

12.12 THEOREM. Let $\mathcal{V}$ be an upper directed set of finely harmonic functions on U, $h := \sup \mathcal{V} < +\infty$ on U. Then h is finely harmonic on U.

Proof. It follows from Proposition 12.9 that h is finely hyperharmonic. In particular, h is finely continuous on U. Pick $x \in V \in \mathcal{B}_o(U)$. Then

$$h(x) = \sup \{ f(x) : f \in \mathcal{V} \} = \sup \{ f^{CV}(x) : f \in \mathcal{V} \} \leqq h^{CV}(x)$$

(notice that h is a Borel function on U in light of Proposition 12.3).

12.13 COROLLARY. Let v be a function on X such that $\int^* \min(v,0)\, d\,\varepsilon_x^{CU} > -\infty$ for each $x \in U$. Then the function

$$h: x \longmapsto \int^* v \; d\,\varepsilon_x^{CU}$$

is finely hyperharmonic on U. If, moreover, $h < +\infty$ on U, then h is finely harmonic on U.

Proof. Let $n \in \mathbb{N}$ and set $v_n := \min(v, np)$,

$$\mathcal{V}_n := \{ s : s \text{ is lower semicontinuous and lower finite on } X, \; s \geqq 0 \text{ outside a compact set}, \; v_n \leqq s \leqq np \}.$$

Pick $s \in \mathcal{V}_n$ and find $f_k \in \mathcal{K}(X)$ such that $f_k \nearrow s$. Then f_k^{CU} are finely harmonic on U. (According to Lemma 12.7, q^{CU} is finely harmonic on U for every $q \in \mathcal{P}^C$. Now, it only needs to be observed that $\mathcal{P}^C - \mathcal{P}^C$ is dense in $\mathcal{K}(X)$.) Since $f_k^{CU} \nearrow s^{CU}$ and $s^{CU} \leq np^{CU} \leq np < +\infty$, Theorem 12.12 yields that s^{CU} is finely harmonic on U. The same reason leads to the conclusion that

$$h_n = \inf \{ s^{CU} : s \in \mathcal{V}_n \}$$

(where $h_n(x) := \int^* v_n \; d\,\varepsilon_x^{CU}$) is finely harmonic on U. (Notice that the set $\{s^{CU} : s \in \mathcal{V}_n\}$ is lower directed and that

$$-\infty < \int^* \min(v,0) \; d\,\varepsilon_x^{CU} \leq \int^* v_n \; d\,\varepsilon_x^{CU} = h_n(x).)$$

Now, in accordance with the monotone convergence theorem for upper integrals we get that $h = \lim h_n$. Proposition 12.9 implies that h is finely hyperharmonic on U and, if $h < +\infty$, h is finely harmonic on U in view of Theorem 12.12.

REMARK. If v is lower $\mathcal{P}^f$- bounded on $\bar{U}$, then $\int^* \min(v,0)\; d\varepsilon_x^{CU} > -\infty$ for all $x \in U$. Notice also that h is finite provided v is $\mathcal{P}^f$-bounded on $\bar{U}$.

Now, we introduce a new topology on U which has pleasant and useful properties and fills the gap between the fine topology and the initial one.

The *topology* $\nu(U)$ is defined as the coarsest topology on U which is finer than the initial one and which makes all functions from $\mathcal{H}^*_+(U)$ lower semicontinuos. Notice that $\nu(U)$ coincides with the natural topology with respect to $S'(U)$ in the sense of [BBC] (cf. Appendix).

12.14 THEOREM. The following topologies on U coincide:

(a) $\tau_1 = \nu(U)$,

(b) the weak topology τ_2 induced on U by the family $\mathcal{C}(U) \cup \{q^{CU}: q \in \mathcal{P}^c\}$,

(c) the weak topology τ_3 induced on U by the family

$$\mathcal{P}^c \cup \{p^{CU}\} .$$

Proof. Obviously τ_2 is finer than τ_3. For every $q \in \mathcal{P}^c$, the function $q - q^{CU}$ belongs to $\mathcal{H}^*_+(U)$. Hence, the topology τ_1 makes all functions q^{CU} continuous. Thus τ_1 is finer than τ_2. If $u \in \mathcal{H}^*_+(U)$, then $u^{(n)} = \tilde{u}^{(n)} - np^{CU}$ is τ_3-lower semicontinuous, and therefore τ_3 is finer than τ_1.

REMARK. It follows immediately from Theorem 12.14.c that U endowed with $\nu(U)$ is a metrizable space with a countable base.

We conclude this section by examining the sheaf properties of the family of finely hyperharmonic functions. The starting point provides the next lemma.

12.15 LEMMA. Assume that the system $\mathcal{H}^*_+(X)$ has the following property: If U_1, U_2 are finely open, $X = U_1 \cup U_2$, f is finely hyperharmonic and nonnegative on both U_1 and U_2, then f is finely hyperharmonic on X. Then for any finely open set $U \subset X$, the set of all points of CU where U is thin is of ε_x^{CU}-measure zero for every $x \in U$.

Proof. Let U be a finely open set. Suppose that $v \in \mathcal{H}_+^*(X)$, $v \geq \hat{R}_p^U$ on CU. Fix $\varepsilon > 0$ and put $U_\varepsilon := [v + \varepsilon p > p]$, $G := \mathrm{int}_f(CU)$. Then U and G are finely open sets, and $X = G \cup (U \cup U_\varepsilon)$. Indeed, if $z \in X \setminus (G \cup U)$, then $z \in bU$. But in this case $v(z) \geq p(z) = \hat{R}_p^U(z)$. Define a function u as follows:

$$u := \begin{cases} p & \text{on } G \\ \min(p, v + \varepsilon p) & \text{on } U \cup U_\varepsilon \end{cases}.$$

By assumption, u is finely hyperharmonic on X, hence it is hyperharmonic on X (Theorem 12.2). Since $u = p$ on CU, we have $u \geq R_p^{CU}$. We know (Corollary 10.13) that $R_p^{CU} = \hat{R}_p^{CU}$ on U, hence $u \geq \hat{R}_p^{CU}$ on U. It easily follows (letting $\varepsilon \to 0$) that $v \geq \hat{R}_p^{CU}$ on U. We see that, for any $x \in U$,

$$\hat{R}^{CU}_{\hat{R}_p^U}(x) = R^{CU}_{\hat{R}_p^U}(x) \geq \hat{R}_p^{CU}(x),$$

hence $\varepsilon_x^{CU}(\hat{R}_p^U) = \varepsilon_x^{CU}(p)$. Thus $\varepsilon_x^{CU}[\hat{R}_p^U < p] = 0$.

We say that the system of all finely hyperharmonic functions has the *fine sheaf property*, or that the fine sheaf property holds on X, if

(a) for each couple $V \subset U$ of finely open sets, if $f \in \mathcal{H}_+^*(U)$, then $f \restriction V$ is finely hyperharmonic on V,

(b) a nonnegative function defined on the union U of finely open sets U_α is finely hyperharmonic on U if and only if, for any α, the restriction of f to U_α is finely hyperharmonic.

12.16 THEOREM. The following assertions are equivalent:

(i) the domination axiom D holds on X,

(ii) the fine sheaf property holds on X,

(iii) for any finely open set U, the set of all points where U is thin is of ε_x^{CU}-measure zero for any $x \in U$.

Proof. (i)$\Rightarrow$(ii), As we will see shortly, any nonnegative pointwise finely hyperharmonic function is finely hyperharmonic if the domination axiom D holds on X (Proposition 14.14). The assertion that the family of all pointwise finely hyperharmonic functions has the fine sheaf property is an easy exercise.

(ii)$\Rightarrow$(iii): Lemma 12.15.

(iii)$\Rightarrow$(i): Obvious by the definition.

EXERCISES

12.B.1 Let $\mathcal{V} \subset \mathcal{H}^*_+(U)$, $\wedge\mathcal{V} = 0$. Then $\varepsilon_x^{CV}(\inf \mathcal{V}) = 0$ whenever $x \in V \in \mathcal{B}(U)$.

Hint. Make use of the proof (v)$\Rightarrow$(i) of Theorem 14.8.

REMARK. We know that the set $W := \{y \in U: \inf \mathcal{V} > 0\}$ is semipolar (Proposition 12.10). In light of Ex. 10.C.3g and 14.B.5, W is even U-polar.

12.B.2 Let $\mathcal{V} \subset \mathcal{H}^*(U)$. Suppose that $\inf \mathcal{V}$ is lower $\mathcal{P}^f$-bounded on $\overline{V}$ for every $V \in \mathcal{B}(U)$. Show that $\widehat{\inf \mathcal{V}}^f \in \mathcal{H}^*(U)$.

Hint. Use directly the definition and Corollary 12.13.

12.B.3 Let U be an arbitrary Borel finely open subset of X. A convex subcone $\mathcal{F}$ of $\mathcal{H}^*_+(U)$ is all of $\mathcal{H}^*_+(U)$ if and only if:

(a) $\wedge\mathcal{A} \in \mathcal{F}$ whenever $\mathcal{A} \subset \mathcal{F}$,

(b) $\sup s_n \in \mathcal{F}$ whenever $\{s_n\} \subset \mathcal{F}$ is an increasing sequence,

(c) $u/U \in \mathcal{F}$ whenever $u \geq 0$ is a hyperharmonic function on an open set containing U.

Hint. Let $u \in \mathcal{H}^*_+(U)$. Then $u = \sup_n \min(u, ne)$. Fix $n \in \mathbb{N}$, there is a potential q such that $\min(u,ne) = q - p^{CU}$ on X and $(q - p^{CU}) / U = \bigwedge_K (q - p^K) / U$ where K runs through all compact subsets of CU (notice also that $q - p^K \in \mathcal{H}^*_+(CK)$).

12.B.4 Let W be a sweeping kernel of a finely open set $U \subset X$. Then the mapping $u \mapsto u/W$ is an algebraic and lattice isomorphism of $\mathcal{H}^*_+(U)$ onto $\mathcal{H}^*_+(W)$.

12.B.5 Let U be a finely open subset of X. Show that generally the fine topology on U can differ from the $\nu(U)$-topology.

Hint. Consider a harmonic space having all points polar. Then any nonempty finely open set is uncountable and the fine topology on U is not metrizable (cf. Ex. 11.D.2f). Now compare the Remark following Theorem 12.14.

12.B.6 Give an example of a finely open set U for which the $\nu(U)$-topology differs on U from the original topology.

Hint. Consider the harmonic space of the Laplace equation on $\mathbb{R}^3$. Let $\{B_n\}$ be a sequence of pairwise disjoint closed balls, B_n having centre at z_n, such that $z_n \to 0$ and $\hat{R}_2^B(0) < 1$ where $B = \bigcup_n B_n$. Put $U = \mathbb{R}^3 \setminus B$, $V = \{x \in U: \hat{R}_2^B(x) < 1\}$. Then V is $\nu(U)$-open but it is not open in U (there are $x_n \in U$, $x_n \to 0$ such that $\hat{R}_2^B(x_n) \to 2$).

12.B.7 Let $V \subset U$ be finely open sets.

(a) Show that the $\nu(V)$-topology is finer than the $\nu(U) \wedge V$-topology.

(b) If V is $\nu(U)$-open, then $\nu(V) = \nu(U) \wedge V$.

Hint. The collection of all sets of the form

$$U \cap [p^{CU} < \beta p] \cap W$$

where $\beta < 1$ and W is open, is a base for $\nu(U)$. Denote $V_1 := U \cap [p^{CU} < \beta p]$. Then $\beta p^{CV_1} = p^{CU}$ on V_1, thus $\nu(V_1) = \nu(U) \wedge V_1$. Let $V_2 = V_1 \cap W$. If $u \in \mathcal{H}_+^*(V_2)$ there are $q_n \in \mathcal{H}_+^*(V_1)$ such that $q_n = \min(np, np^{CW}+u)$ on V_2 (similarly as in the proof of Lemma 12.5) and the assertion follows.

12.B.8 Let u be a finely hyperharmonic function on a finely open subset of a $\mathfrak{P}$-harmonic space X. Suppose that u is locally lower bounded on U in the $\nu(U)$-topology. Prove that u is $\nu(U)$-lower semicontinuous.

Hint. Let $V \subset U$ be a $\nu(U)$-open set on which u is lower $\mathcal{P}^c$-bounded. Then $u+p \geq 0$ is finely hyperharmonic on V for some $p \in \mathcal{P}^c$. Hence $u+p$ is $\nu(V)$-lower semicontinuous. But now Ex. 12.B.7b allows to conclude that u is $\nu(U)$-lower semicontinuous on V.

12.B.9 Let U be a finely open subset of X.

(a) If x_n, $z \in U$ and $x_n \to z$ in the $\nu(U)$-topology, then there is $B \in \mathcal{B}(U)$

such that $x_n, z \in V$ and $x_n \to z$ in $\nu(V)$.

Hint. Let $p \in \mathcal{P}^c$ be a strict potential, and let $\{H_n\}$ be a decreasing $\nu(U)$--local base of $\nu(U)$-open neighborhoods of z. Denote $S := \{z\} \cup \bigcup_n \{x_n\}$, and put $W_0 = U$. Proceed by induction. Given $n \geq 1$, by Key lemma 10.12 and the Lusin-Menchoff property there is $V_n \in \mathcal{B}(W_{n-1})$ such that $S \subset V_n$ and $\hat{R}_p^{CW_n} < \hat{R}_p^{CW_{n-1}} + 2^{-n}$ on S. Set $W_n := V_1 \cup \dots \cup V_n \cup H_n$, $V = \bigcup V_n$. Obviously, $W_n \searrow V \in \mathcal{B}(U)$. Select $m \in \mathbb{N}$. By Ex. 12.B.7, $\lim \hat{R}_p^{CW_{m-1}}(x_n) = \hat{R}_p^{CW_{m-1}}(z)$ and

$$\limsup \hat{R}_p^{CV}(x_n) \leq \limsup \hat{R}_p^{CV_m}(x_n) \leq \lim \hat{R}_p^{CW_{m-1}}(x_n) + 2^{-m} =$$

$$= \hat{R}_p^{CW_{m-1}}(z) + 2^{-m} \leq \hat{R}_p^{CV}(z) + 2^{-m}.$$

(b) Deduce by (a) the following corollary: If $G \subset X$, and if $G \cap V$ is $\nu(V)$--open for every $V \in \mathcal{B}(U)$, then $G \cap U$ is $\nu(U)$-open.

REMARKS AND COMMENTS

It was the purpose of this section to establish the main properties of the cone of all nonnegative hyperharmonic functions.

If S is a standard H-cone of functions on X and $U \subset X$ is a finely open set, then S induces on U in a reasonable way a standard H-cone $S'(U)$ of functions (N. Boboc and Gh. Bucur (1985a), essentially also N. Boboc, Gh. Bucur and A. Cornea (1975) and [BBC]). It follows from results of this section or from N.Boboc and Gh. Bucur (1985c) that in the case of a harmonic space, $S'(U)$ is exactly the set of all finely superharmonic functions on U. The properties of $\mathcal{H}_+^*(U)$ examined in this section form a basis of various axiomatic systems of potential theory (Chap. IV of [C&C] , balayage spaces, H-cones or hyperharmonic cones).

Several sheaf properties are discussed in N. Boboc and Gh. Bucur (1985a) and (1985b).

Natural topology was introduced by N. Boboc, Gh. Bucur and A. Cornea [BBC] in the framework of standard H-cones. We make use of a different but equivalent definition (cf.Appendix).

12.C Lsc pointwise finely hyperharmonic functions

It is well-known that a locally lower bounded function f is hyperharmonic on an open set U if and only if f is"pointwise hyperharmonic" and lower semicontinuous on U (cf. [Bau] , Satz 1.3.8). Guided by this situation, we consider its analogue in a more general setting of fine hyperharmonicity. We prove that in case of the lower semicontinuity assumption, the local and global definitions coincide. From this fact we draw the theorem on removable singularities of finely hyperharmonic functions.

The basic tool of this section is the minimum principle for the class of all lower semicontinuous "generalized" pointwise finely hyperharmonic functions, which is a special case of Bauer's minimum principle of Ex. 12.C.1. We utilize a countable base argument (through the concept of a strict potential) to obtain a quite elementary proof. Almost all results of this section will be improved in Section 14.B.

Define the system $\mathcal{H}^{\#}(U)$ on a finely open set U as the set of all lower finite lower semicontinuous functions f on U which have the following property: For every $x \in U$ and for every $V \in \mathcal{B}(U)$, $x \in V$, there is $W \in \mathcal{B}(U)$ such that $W \subset V$, CW is thin at x and $\varepsilon_x^{CW}(f) \leqq f(x)$.

It is important to notice that $\mathcal{H}^{\#}(U)$ contains all lower semicontinuous pointwise finely hyperharmonic functions on U.

12.17 THEOREM. (Minimum principle for $\mathcal{H}^{\#}(U)$. Let U be a finely open set. If $f \in \mathcal{H}^{\#}(U)$ is nonnegative outside the intersection of U with some compact set and if $\liminf_{U \ni x \to y} f(x) \geqq 0$ for every $y \in \overline{U} \setminus U$, then $f \geqq 0$ on U.

Proof. Suppose that there is $y \in U$ such that $f(y) < 0$. Let p be a strict potential on X. Under these assumptions there is a smallest $m > 0$ such that $f + mp \geqq 0$ on U. Then the set

$$M := \{ x \in U : f(x) + mp(x) = 0 \}$$

is non-empty and, of course, $M \subset U$. Pick $x \in M$. There is a finely open set W satisfying $\overline{W} \subset U$, $\varepsilon_x^{CW} \neq \varepsilon_x$ and $\varepsilon_x^{CW}(f) \leqq f(x)$. We have

$$0 \leqq \varepsilon_x^{CW}(f+mp) < f(x) + mp(x) = 0,$$

which is a contradiction.

12.18 LEMMA. Let U be a finely open relatively compact set. If f is a lower finite, lower semicontinuous and finely continuous function on $\overline{U}$ which is finely hyperharmonic on U, then $\varepsilon_x^{CU}(f) \leqq f(x)$ for all x from the fine closure of U.

Proof. Assume $x \in U$. According to Propositions 10.20 and 11.2 (cf. also Section

13.B),

$$\varepsilon_x^{CU} = \text{w-lim}_{V \in \mathcal{B}(U)} \varepsilon_x^{CV}.$$

Hence

$$\varepsilon_x^{CU}(f) \leqq \liminf_{V \in \mathcal{B}(U)} \varepsilon_x^{CV}(f) \leqq f(x).$$

Since the function $f^{CU} - f$ is finely lower semicontinuous on the fine closure of U (cf. Proposition 11.6), $f^{CU} - f \leqq 0$ on $\overline{U}^f$.

12.19 THEOREM. Any function from $\mathcal{H}^{\#}(U)$ (in particular, any lower semicontinuous pointwise finely hyperharmonic function) is finely hyperharmonic on U.

Proof. Let $u \in \mathcal{H}^{\#}(U)$. Choose $V \in \mathcal{B}_o(U)$ and $x \in V$. By Theorem 13.12.a (whose proof does not depend on this theorem and its consequences) or by Prop. 3.3 of J. Bliedtner and W. Hansen (1978) there is a set $\mathcal{T}$ of lower finite lower semicontinuous functions on $\overline{V}$ such that each $t \in \mathcal{T}$ is finely hyperharmonic on V, $t \geqq -u$ on $\overline{V} \setminus V$ and

$$\varepsilon_x^{CV}(u) = \sup\{-t(x) : t \in \mathcal{T}\}.$$

Given $t \in \mathcal{T}$, we have $t+u \in \mathcal{H}^{\#}(V)$ in light of Lemma 12.18. Minimum principle 12.17 implies $t+u \geqq 0$ on V. Thus

$$\varepsilon_x^{CV}(u) = \sup\{-t(x) : t \in \mathcal{T}\} \leqq u(x).$$

12.20 THEOREM. (Removable singularities for finely hyperharmonic functions.) Let $G \subset U$ be finely open sets and let f be a lower finite, lower semicontinuous and finely continuous function on U. If f/G is finely hyperharmonic on G, then f is finely hyperharmonic on U provided the set $U \setminus G$ is semipolar.

Proof. If $A \subset X$ is finely closed and $f \in \mathcal{H}^{\#}(U \setminus A)$, then $f \in \mathcal{H}^{\#}(U \setminus bA)$. Indeed, given $z \in V \in \mathcal{B}(U \setminus bA)$, put $W = V \setminus A$. According to Lemma 12.18, $\varepsilon_x^{CW}(f) \leqq f(x)$ for every x from the fine closure of W, in particular for $x = z$. Further, $CW = CV \cup A$ is thin at z. Now, let B be the smallest finely closed subset of $X \setminus G$ such that $f \in \mathcal{H}^{\#}(U \setminus B)$. It follows $B \cap U \subset (bB) \cap U \subset b(B \cap U)$, which yields

$$B \subset \beta(CG) = ib(CG) = i(CG) \subset CU$$

(cf. Corollary 1.17). We conclude that $f \in \mathcal{H}^{\#}(U)$. In light of Theorem 12.19, f is finely hyperharmonic on U.

EXERCISES

12.C.1 Bauer's minimum principle. Let F be a family of lower semicontinuous, lower finite functions on a metrizable compact space X. Assume that there is a min-stable convex cone $F^c \subset F$ of nonnegative continuous functions linearly seperating the points of X. Then $F^c - F^c$ is dense in $\mathcal{C}(X)$. Let $\{s_n - t_n : s_n, t_n \in F^c\}$ be a countable dense subset of $\mathcal{C}(X)$ and put

$$p := \sum a_n (s_n + t_n),$$

where $a_n > 0$ are constants, $a_n \max_{x \in X} (s_n(x) + t_n(x)) < 2^{-n}$. (Compare with constructions of strict potentials.) Further define the *Choquet boundary* $\mathrm{Ch}_F X$ of X with respect to F as the set of all $x \in X$ such that the Dirac measure ε_x is the only Radon measure μ on X for which $\mu(s) \leqq s(x)$ for all $s \in F$. Show that the following Bauer's minimum principle holds: If $s \in F$ and $s \geqq 0$ on $\mathrm{Ch}_F X$, then $s \geqq 0$ everywhere on X.

Hint. Let $z \in X$ and $c > 0$ be such that

$$\frac{s(z)}{p(z)} = \min_X \frac{s}{p} = -c.$$

If μ is a Radon measure on X for which $\mu(s) \leqq s(z)$ for all $s \in F$, then $\mu(p) = p(z)$ if and only $\mu = \varepsilon_z$. Since

$$cp(z) = -s(z) \leqq \mu(-s) \leqq \mu(cp) \leqq cp(z),$$

it must be $\mu = \varepsilon_z$.

REMARKS AND COMMENTS

The coincidence of pointwise finely hyperharmonic functions and finely hyperharmonic functions in the class of lower semicontinuous functions was proved by J. Malý (1980) and by L. Stoica (1983). Analogous results concerning ordinary hyperharmonic functions are due to G. Tautz (1943), M. Brelot (1959) and H. Bauer (1962).

Concerning removable singularities, there is an extensive number of papers, both for harmonic and hyperharmonic functions starting with H.A.Schwarz (1872) in classical potential theory (one point sets are removable for harmonic functions).

M. Brelot (1941) extends the result for closed polar sets and the class of hyperharmonic functions. This result is easily transferable to axiomatic theories (cf. [C&C], Th. 6.2.1). I. Netuka and J. Veselý (1978) deal with removability of closed semipolar sets for the class of harmonic and superharmonic functions. In (1978), J. Bliedtner and W. Hansen showed that every finely closed semipolar set is removable in the class of continuous finely superharmonic functions. Notice that our Theorem 12.20 as well as Theorem 12.19 will be further generalized in Section 14.C.

H. Bauer in (1960) obtained an abstract minimum principle which he used in (1962) to prove a boundary minimum principle in the framework of harmonic spaces. Later on a series of results concerning minimum principles appeared generalizing more or less Bauer's principle. Among a list of papers let us mention at least B. Fuchsteiner (1971) and J. Bliedtner and K. Janssen (1974). R. Wittmann (1983) presents an "elementary" proof without the use of the Hahn-Banach theorem or Zorn's lemma. In Ex. 12.C.1 we present a short proof in metrizable case which is sufficient for our purposes. A similar simple idea is concealed in L. Stoica (1983).

Related papers: R. Harvey and J.C.Polking (1972), J. Hyvönen (1979)

13. THE FINE DIRICHLET PROBLEM

A. The Perron method.
B. The Wiener method.

13.A The Perron method

As stated above, the classical Dirichlet problem for a continuous function f on the boundary of an open bounded set U consists in finding a continuous extension of f to the whole closure $\bar{U}$ of U which is harmonic in U. We know that the classical Dirichlet problem is not always solvable. However, there are methods allowing to assign to any continuous function on the boundary of U a function harmonic on U in a reasonable way (the Wiener method) or working even also for discontinuous functions (the Perron method).

In (1950), M. Brelot succeeded in a modified Perron method using the fine topology. More precisely, first of all he proved the fine minimum principle for hyperharmonic functions on open sets:

"If f is a lower bounded hyperharmonic function on an open relatively compact set U and if the fine lower limit of f at any point of the fine boundary of U is nonnegative, then $f \geq 0$ on U."

Then he defined the upper and lower Perron solution using the fine limits of hyperharmonic functions at the fine boundary and he proved that these solutions coincide with the usual Perron ones. All this was given for the classical case of harmonic functions derived from the Laplace equation. In (1960), M. Brelot himself extends these results for a general axiomatic including the domination axiom D. In the presence of the same axiom, B. Fuglede in the seventies develops his theory of finely harmonic functions including Perron solution. He works with finely hyperharmonic functions on finely open sets using fine limits on the fine boundary.

In our theory without the domination axiom D there is no chance to prove the boundary minimum principle for the fine boundary of U. This one is too small for solving the Dirichlet problem on it. This can be easily illustrated in case of the heat equation. (Consider any open rectangle in $\mathbb{R}^2$ whose sides are parallel with the axes.) Moreover, we know that the balayaged measure need not be carried by the fine boundary. This is the reason why we consider the "*bitopological*" boundary $\bar{U}\setminus U$ (which coincides with the initial one in case of an open set U) and we take ordinary limits on it.

Another approach to the Dirichlet problem using a weaker concept of limits is developed in Section 14.C.

As usual, U will denote a finely open subset of a $\mathfrak{P}$-harmonic space X. Further, denote by $\mathcal{H}_1^*(U)$ the set of all finely hyperharmonic functions on U whose lower limits at any point of $\bar{U}\setminus U$ are $>-\infty$.

13.1 THEOREM. (Minimum principle for finely hyperharmonic functions.) Let f be a finely hyperharmonic lower $\mathcal{P}^f$-bounded function on U. If $\liminf_{U\ni x\to y} f(x) \geq 0$ for every $y\in\bar{U}\setminus U$, then $f\geq 0$ on U.

Proof. Let $f\geq -p$ on U for $p\in\mathcal{P}^f$. Let $q\in\mathcal{P}^f$ be a positive Evans function of p (cf. Proposition 11.2). Choose $\varepsilon>0$. Since the set $[p-\varepsilon' q>0]$ is relatively compact for each $\varepsilon'>0$, there is a compact set $L\subset X$ such that $f+\varepsilon q$ is (strictly) positive outside $L\cap U$ and $\liminf_{U\ni x\to y}(f(x)+\varepsilon q(x))>0$ for each $y\in\bar{U}\setminus U$. Putting K to be the closure of the set $\{x\in U: f(x)+\varepsilon q(x)<0\}$, K is a compact set and, of course, $K\subset U$. Making use of the Lusin-Menchoff property of the fine topology, there is a finely open set V with compact closure such that $K\subset V\subset\bar{V}\subset U$. Let $z\in K$. Since

$$-\infty < -p(z)+\varepsilon q(z)\leq \varepsilon_z^{CV}(-p+\varepsilon q)\leq \varepsilon_z^{CV}(f+\varepsilon q),$$

and since ε_z^{CV} is carried by $\bar{V}\setminus K$, we get immediately

$$f(z)+\varepsilon q(z)\geq \varepsilon_z^{CV}(f+\varepsilon q)\geq 0.$$

It follows that $f+\varepsilon q \geq 0$ on U. Since $\varepsilon > 0$ was arbitrary, the Theorem is proved.

If f denotes an arbitrary function defined at least on the bitopological boundary $\bar{U} \setminus U$ of U, by a *fine superfunction* for f we mean a function u defined on U such that

(a) $u \in \mathcal{H}_1^*(U)$,

(b) u is lower $\mathcal{P}^f$-bounded on U,

(c) $\liminf_{U \ni x \to y} u(x) \geq f(y)$ for every $y \in \bar{U} \setminus U$.

The family of all fine superfunctions for f will be denoted by $\bar{\mathcal{U}}_f(U)$. Further we define

$$\bar{H}_f^U = \inf \bar{\mathcal{U}}_f(U)$$

and, analogously, put

$$\underline{\mathcal{U}}_f(U) = -\bar{\mathcal{U}}_{-f}(U), \quad \underline{H}_f^U = -\bar{H}_{-f}^U .$$

It follows from the minimum principle 13.1 that $\underline{H}_f^U \leq \bar{H}_f^U$ on U. The function f is termed *finely resolutive* for U if $\underline{H}_f^U = \bar{H}_f^U$ and if this common value, denoted simply by H_f^U, is a finely harmonic function on U.

13.2 PROPOSITION. The set $\mathrm{Res}(U)$ of all real finely resolutive functions on $\bar{U} \setminus U$ is a vector space and the map $f \mapsto H_f^U$ is a positive linear operator on $\mathrm{Res}(U)$.

If $f_n \in \mathrm{Res}(U)$, $f_n \to f$ uniformly and if there exists a compact subset K of X such that $f=f_1=f_2=\dots$ on $(\bar{U} \setminus U) \setminus K$, then $f \in \mathrm{Res}(U)$ and $H_f^U = \lim H_{f_n}^U$.

Proof. Notice that the set of all finely harmonic functions on U is a vector space closed under the locally uniform convergence. It is not difficult to check that

$$\underline{H}_f^U + \underline{H}_g^U \leq \underline{H}_{f+g}^U \leq \bar{H}_{f+g}^U \leq \bar{H}_f^U + \bar{H}_g^U$$

for any finite f,g on $\bar{U} \setminus U$. Hence, $\mathrm{Res}(U)$ is a vector space and the mapping $f \mapsto H_f^U$ is a positive linear mapping on it.

Choose a strictly positive potential $p \in \mathcal{P}^c$. For every $\varepsilon > 0$ there is $m \in \mathbb{N}$ such that $f \leq f_n + \varepsilon p$ on $\overline{U} \setminus U$ whenever $n \geq m$. If u is a fine superfunction for f_n, then $u + \varepsilon p$ is a fine superfunction for f, hence

$$\bar{H}_f^U = H_{f_n}^U + \varepsilon p \quad \text{on } U.$$

The similar inequality holds for the lower solutions, thus

$$\bar{H}_f^U = \underline{H}_f^U = \lim H_{f_n}^U$$

and the convergence is uniform on compact subsets of U (since p is bounded on them). As mentioned above, H_f^U is finely harmonic on U.

13.3 THEOREM. Any function $f \in \mathcal{K}(\partial U)$ is finely resolutive and $H_f^U(x) = \varepsilon_x^{CU}(f)$ for any $x \in U$.

Proof. Since $\mathcal{P}^c - \mathcal{P}^c$ is dense in $\mathcal{K}(\partial U)$, according to Proposition 13.2 it is sufficient to prove the theorem for every $p \in \mathcal{P}^c$. Suppose that u is a hyperharmonic function on X, $u \geq p$ on CU. Then also $u \in \overline{\mathcal{U}}_p(U)$, and thus $\hat{R}_p^{CU} = R_p^{CU} \geq H_p^U$ on U. By Lemma 12.7, $\hat{R}_p^{CU} \in \underline{\mathcal{U}}_p(U)$. Finally, we obtain

$$\hat{R}_p^{CU} \leq \underline{H}_p^U \leq \bar{H}_p^U \leq \hat{R}_p^{CU} \quad \text{on } U.$$

Again by Lemma 12.7, $\hat{R}_p^{CU}$ is finely harmonic on U and thus the proof is complete.

13.4 COROLLARY. If p is a continuous potential on X, then $H_p^U = \hat{R}_p^{CU}$ on U.

13.5 LEMMA. If f is nonnegative outside a compact subset of ∂U and it is lower semicontinuous, lower finite, then $\bar{H}_f^U(x) = \varepsilon_x^{CU}(f)$ for any $x \in U$.

Proof. Let $f_n \in \mathcal{K}(\partial U)$, $f_n \nearrow f$. Using the Lebesgue monotone convergence theorem,

$$\varepsilon_x^{CU}(f) = \lim \varepsilon_x^{CU}(f_n) = \lim H_{f_n}^U(x) \leq \bar{H}_f^U(x).$$

For the reverse, fix $\varepsilon > 0$ and $x \in U$. There is a sequence $\{u_n\}$ such that

$$u_1 \in \mathcal{U}_{f_1}(U), \quad u_1(x) < \varepsilon_x^{CU}(f_1) + \tfrac{1}{2}\varepsilon,$$

$$u_n \in \overline{\mathcal{U}}_{f_n - f_{n-1}}(U)\ , \ u_n(x) < \varepsilon_x^{CU}(f_n - f_{n-1}) + 2^{-n}\varepsilon$$

for every $n > 1$. Setting $u = \sum_{n=1}^{\infty} u_n$, we have $u \in \overline{\mathcal{U}}_f(U)$. Again, using the Lebesgue theorem,

$$\bar{H}_f^U(x) \leq u(x) = \sum_{n=1}^{\infty} u_n(x) \leq \varepsilon_x^{CU}(f_1) + \sum_{n=2}^{\infty} \varepsilon_x^{CU}(f_n - f_{n-1}) + \varepsilon =$$

$$= \varepsilon_x^{CU}(f) + \varepsilon \ .$$

13.6 THEOREM. Let f be a function on ∂U. Then, for any $x \in U$,

$$\bar{H}_f^U(x) \leq \int^* f \, d\, \varepsilon_x^{CU}.$$

Moreover, if U is a Borel set,

$$\bar{H}_f^U(x) = \int^* f \, d\,\varepsilon_x^{CU} \ .$$

Proof. Given an arbitrary $k > \int^* f \, d\,\varepsilon_x^{CU}$, there is a lower semicontinuous, lower finite function s on ∂U nonnegative outside a compact subset of ∂U such that $s \geq f$ on ∂U and $\varepsilon_x^{CU}(s) \leq k$. Hence

$$\bar{H}_f^U(x) \leq \bar{H}_s^U(x) = \varepsilon_x^{CU}(s) \leq k.$$

We deduce

$$\bar{H}_f^U(x) \leq \int^* f \, d\,\varepsilon_x^{CU}.$$

Now, let U be a Borel finely open set, $x \in U$. Assume that $\bar{H}_f^U(x) < +\infty$. Choosing $m > \bar{H}_f^U$, there is $u \in \overline{\mathcal{U}}_f(U)$ such that $u(x) < m$. Define the function v on ∂U:

$$v(y) = \liminf_{U \ni x \to y} u(x) \ , \ y \in \partial U.$$

Then v is lower semicontinuous and nonnegative outside a compact subset of ∂U. (If not, we add a suitable potential to u, cf. the proof of Theorem 13.1). Obviously, $u \in \overline{\mathcal{U}}_v(U)$. Since ε_x^{CU} is carried by $\bar{U} \setminus U$, $f \leq v$ ε_x^{CU}-almost everywhere. Finally, we obtain

$$\varepsilon_x^{CU}(f) \leqq \varepsilon_x^{CU}(v) = \bar{H}_v^U(x) \leqq u(x) < m.$$

13.7 THEOREM. Let f be a function on ∂U.

(a) If f is ε_x^{CU}-integrable for every $x \in U$, then f is finely resolutive.

(b) If U is a Borel finely open set, and if f is finely resolutive, then f is ε_x^{CU}-integrable for every $x \in U$.

(c) If the assumptions of (a) or (b) are satisfied, then

$$\varepsilon_x^{CU}(f) = H_f^U(x) \qquad \text{for every } x \in U.$$

Proof. (a) If f is ε_x^{CU}-integrable, then $\underline{H}_f^U(x) = \bar{H}_f^U(x) = \varepsilon_x^{CU}(f)$ (Theorem 13.6). Taking into account Corollary 12.13, we get that H_f^U is finely harmonic on U.

(b) It follows immediately from Theorem 13.6.

(c) Obvious.

A point $z \in \partial U$ is termed *regular* if $\lim\limits_{x \to z} H_f^U(x) = f(z)$ for any $f \in \mathcal{K}(\partial U)$.

13.8 THEOREM. For $z \in \partial U$, the following assertions are equivalent:

(i) z is a regular point,

(ii) for any continuous potential $p \in \mathcal{P}^c$ we have

$$\lim_{U \ni x \to z} \hat{R}_p^{CU}(x) = p(z),$$

(iii) there is a strict continuous potential $q \in \mathcal{P}^c$ such that

$$\lim_{U \ni x \to z} \hat{R}_p^{CU}(x) = q(z),$$

(iv) $\varepsilon_z^{CU} = \varepsilon_z$,

(v) $z \in bCU$.

Proof. (i)$\Rightarrow$(ii). Let $f_n \in \mathcal{K}(\partial U)$, $f_n \nearrow p$ on ∂U. Then

$$p(z) = \lim_{n\to\infty} f_n(z) = \lim_{n\to\infty} (\lim_{x\to z} H^U_{f_n}(x)) \leqq \liminf_{x\to z} H^U_p \leqq$$

$$\leqq \limsup_{x\to z} H^U_p(x) = \limsup_{U\ni x\to z} R^{CU}_p(x) \leqq \limsup_{x\to z} p(x) = p(z).$$

(ii)$\Rightarrow$(iii): Obvious.

(iii)$\Rightarrow$(iv). Since $\hat{R}^{CU}_q(z) = \liminf_{U\ni x\to z} \hat{R}^{CU}_q(x)$,

we obtain $\hat{R}^{CU}_q(z) = q(z)$ and, in view of the strictness of q, $\varepsilon^{CU}_z = \varepsilon_z$.

(iv) $\Rightarrow$ (v). If q is a strict potential, then

$$bCU = [\hat{R}^{CU}_q = q] = \{x \in X: \ \varepsilon^{CU}_x(q) = \varepsilon_x(q)\}.$$

Hence, $\varepsilon^{CU}_z = \varepsilon_z$ implies $z \in bCU$.

(v)$\Rightarrow$(i). If $z \in bCU$, then

$$p(z) = \hat{R}^{CU}_p(z) = \liminf_{U\ni x\to z} \hat{R}^{CU}_p(x) \leqq \limsup_{U\ni x\to z} \hat{R}^{CU}_p(x) \leqq p(z)$$

for any $p \in \mathcal{P}^c$. Since $\mathcal{P}^c - \mathcal{P}^c$ is dense in $\mathcal{K}(\partial U)$, (i) follows easily.

For a while, we could call a finely open set $U \subset X$ to be "*regular*" if for any $f \in \mathcal{K}(\partial U)$ there is a continuous extension of f to $\overline{U}$ which is finely harmonic on U.

It is easy to see that U is "regular" if and only if each point $z \in \partial U$ is regular which is the case if and only if $\partial U \subset bCU$. Of course, if U is open and relatively compact, the notions "regular" and regular coincide.

Let U be finely open and let f be a real function on $\overline{U} \setminus U$. A function h on U is said to be a *finely classical solution* for f if h is $\mathcal{P}^f$-bounded, finely harmonic on U and if $\lim_{U\ni x\to z} h(x) = f(z)$ for every $z \in \overline{U} \setminus U$.

13.9 THEOREM. (a) Any "regular" set U is open.

(b) The finely classical solution is unique and equals H^U_f provided it exists.

(c) A set U is finely regular if and only if for any $f \in \mathcal{K}(\partial U)$ there is a finely classical solution for f.

Proof. (a) Obviously, $\partial U \subset bCU \subset CU$. Hence U is open.

(b) If h is a finely classical solution for f, then h and $-h$ are fine superfunctions for f.

(c) Obvious.

In the remainder of this section we prove that the Perron solution H_f^U can be also constructed using the family of lower semicontinuous fine superfunctions. In the sequel, $U \subset X$ will denote always a Borel finely open set.

Given a function f defined at least on $\bar{U} \setminus U$, we denote

$$\tilde{\mathcal{U}}_f(U) = \{v \in \bar{\mathcal{U}}_f(U) : v \text{ is lower semicontinuous on } U\},$$
$$\underset{\sim}{\mathcal{U}}_f(U) = -\tilde{\mathcal{U}}_{-f}(U),$$
$$\tilde{H}_f^U = \inf \tilde{\mathcal{U}}_f(U), \quad \underset{\sim}{H}_f^U = \sup \underset{\sim}{\mathcal{U}}_f(U).$$

Clearly,

$$\underset{\sim}{H}_f^U \leqq \underline{H}_f^U \leqq \bar{H}_f^U \leqq \tilde{H}_f^U \quad \text{on } U.$$

In what follows, we prove that any function $f \in \mathcal{K}(\partial U)$ is again "resolutive" in the sense that $\underset{\sim}{H}_f^U = \tilde{H}_f^U$. In fact, we prove a stronger result: for any function f on ∂U the equality $\tilde{H}_f^U = \bar{H}_f^U$ holds. Thus, in general, we cannot expect that lower semicontinuous upper functions yield a continuous solution.

13.10 THEOREM. If $f \in \mathcal{K}(\partial U)$, then $\underset{\sim}{H}_f^U(x) = \tilde{H}_f^U(x) = \varepsilon_x^{CU}$ for every $x \in U$.

Proof. It is enough to show that $\underset{\sim}{H}_p^U = \tilde{H}_p^U = \hat{R}_p^{CU}$ on U for every $p \in \mathcal{P}^c$ (compare with the proof of Theorem 13.3). Obviously, $\tilde{H}_p^U \leqq R_p^{CU} = \hat{R}_p^{CU}$ on U. On the other hand,

$$\underset{\sim}{H}_p^U \geqq \sup\{(\sup \underset{\sim}{\mathcal{U}}_p(CK)) \wedge U : K \subset CU, K \text{ compact}\} =$$
$$= \sup\{\hat{R}_p^K : K \subset CU, K \text{ compact}\} = \hat{R}_p^{CU}$$

on U (use Proposition 11.3).

13.11 THEOREM. Let f be a lower semicontinuous function on ∂U which is nonnegative outside a compact subset of ∂U. Then

$$\tilde{H}_f^U(x) = \varepsilon_x^{CU}(f)$$

for every $x \in U$.

Proof. Step by step the same as the proof of Lemma 13.5 (cf. also Ex. 11.C.1c).

13.12 THEOREM. (a) For any function f on ∂U and any $x \in U$,

$$\tilde{H}_f^U(x) = \int^* f \, d\, \varepsilon_x^{CU}.$$

(b) For each $x \in U$, $\underset{\sim}{H}_f^U(x) = \tilde{H}_f^U(x)$ if and only if $\underline{H}_f^U(x) = \bar{H}_f^U(x)$. (In particular, if f is ε_x^{CU}-integrable.)

Proof. Using Theorem 13.11, the proof of the equality $\tilde{H}_f^U(x) = \int^* f \, d\varepsilon_x^{CU}$ is the same as the proof of Theorem 13.6. Of course, (b) follows from (a).

EXERCISES

13.A.1 Consider the harmonic space introduced in Ex. 11.D.6. Let $f(x) = (2x-1)\cdot \chi_{(0,1)}(x)$. Then f is finely hyperharmonic on a fine neighborhood of each point of X (in particular, f is pointwise finely hyperharmonic on X), $f \geqq 0$ outside a compact subset of X but f is not nonnegative. Hence, minimum principle 13.1 has no immediate analogy for pointwise hyperharmonic functions.

13.A.2 Let U be a Borel finely open subset of X, and let $V \subset U$ be a finely open set. Let f be a finely resolutive function for U. If $g = H_f^U$ on U and $g = f$ on $\bar{U} \setminus U$, then g is a finely resolutive function for V and $H_g^V = H_f^U$ on U.

Hint. Operate with lower semicontinuous fine superfunctions on U.

REMARK. Realize that this assertion is in fact a reformulation of Bliedtner and Hansen's Proposition 11.7.

13.A.3 Boundary behaviour of the Perron solution. Let U be a finely open subset of a $\mathfrak{P}$-harmonic space X.

(a) Let $z \in \bar{U}$ and $x_n \in U$, $x_n \to z$, and let μ be a Radon measure on X. Then the following conditions are equivalent:

(i) $\mu = \text{w-lim}\ \varepsilon_{x_n}^{CU}$,

(ii) $\mu(p) = \lim \hat{R}_p^{CU}(x_n)$ for every $p \in \mathcal{P}^c$.

(b) Let $z \in \bar{U}$. Denote $\Lambda(z,U) = \{\text{w-lim}\ \varepsilon_{x_n}^{CU} : \{x_n\}$ is a sequence of points of U, $x_n \to z\}$.

If $\lambda \in \Lambda(z,U)$, then there is $\alpha \in [0,1]$ such that

$$\lambda = \alpha \varepsilon_z + (1-\alpha) \varepsilon_z^{CU}.$$

Hint. By [C&C], Theorem 7.2.3, λ lies on the segment between $\varepsilon_z^{CU\setminus\{z\}}$ and ε_z. Since

$$\varepsilon_z^{CU}(p) = \liminf_{U \ni x \to z} \varepsilon_x^{CU}(p) \leq \lambda(p) \leq p(z)$$

for every $p \in \mathcal{P}^c$, λ must be between ε_z^{CU} and ε_z.

(c) Let $z \in \bar{U}$. A sequence $\{x_n\}$, $x_n \in U$, $x_n \to z$ is termed *regular* (for U) if $\text{w-lim}\ \varepsilon_{x_n}^{CU} = \varepsilon_z$, and semiregular (for U) if $\text{w-lim}\ \varepsilon_{x_n}^{CU} = \varepsilon_z^{CU}$. We say that z is semiregular (for U) if $\text{w-lim}_{U \ni x \to z}\ \varepsilon_x^{CU} = \varepsilon_z^{CU}$, and strictly semiregular (for U) if moreover $\varepsilon_z^{CU} \neq \varepsilon_z$. Points of $\bar{U}$ which are not regular are called irregular.

(c_1) A point z is regular (semiregular, resp.), if and only if any sequence $\{x_n\}$ of points of U converging to z is regular (semiregular, resp.).

(c_2) There exists always a semiregular sequences converging to z.

Hint. If $p \in \mathcal{P}^c$, then

$$\liminf_{U \ni x \to z} \hat{R}_p^{CU}(x) = \hat{R}_p^{CU}(z).$$

Now, use a strict potential $p \in \mathcal{P}^c$.

(d) Let $x_n \in U$, $x_n \to z$. Then the following assertions are equivalent:

(i) $\{x_n\}$ is regular,

(ii) $\{x_n\}$ is regular for $G \cap U$ whenever G is an open neighborhood of z containing $\{x_n\}$,

(iii) if f is lower semicontinuous at z, ε_x^{CU}-integrable for every $x \in U$ and if $\liminf_{U \ni x \to z} H_f^U(x) > -\infty$, then $f(z) \leq \limsup H_f^U(x_n)$.

Hint. Use Bliedtner and Hansen's Proposition 11.7 for a suitable Borel sweeping kernel of U.

(e) Let $z \in \bar{U}$. The following assertions are equivalent:

(i) z is strictly semiregular,

(ii) there is an open neighborhood G of z such that

$$\varepsilon_x^{CU}(G) = 0 \quad \text{for any} \quad x \in U,$$

(iii) there is an open neighborhood G of z such that

$$\varepsilon_x^{CU} = \varepsilon_x^{C(U \cup G)} \quad \text{for any} \quad x \in U,$$

(iv) there is $f \in \mathcal{K}(X)$ such that

$$\limsup_{U \ni x \to z} H_f^U < f(z).$$

Hint. (i)$\Longrightarrow$(ii): Let $p \in \mathcal{P}^c$ be a strict potential. Find an open neighborhood G of z and a continuous function g on X such that $0 \lneq g \lneq p$, $g = p$ on CG, $g < p$ on G and $\hat{R}_p^{CU} < g$ on $G \cap U$. Then $\varepsilon_x^{CU}(p-g) = H_{p-g}^U(x) = 0$ for any $x \in U$.

(ii)$\Longrightarrow$(iii): Use again Bliedtner and Hansen's Proposition 11.7 for a suitable Borel sweeping kernel of $U \cup G$.

(iii)$\Longrightarrow$(iv): Find $f \in \mathcal{K}(X)$ such that $f \geq 0$, $f(z) > 0$ and $f = 0$ and CG.

(iv)$\Longrightarrow$(ii) is similar as (i)$\Longrightarrow$(ii), (iii)$\Longrightarrow$(i) is obvious.

(f) If a point $z \in \bar{U}$ is not semiregular, then there is a regular sequence of points of U converging to z.

(g) The set M of all strictly semiregular points for U is open and $\varepsilon_x^{CU}(M) = 0$ for every $x \in U$. If $G \subset X$ is open, $G \cap \bar{U} \subset M$, then $\varepsilon_x^{CU} = \varepsilon_x^{C(U \cup G)}$ for every $x \in U$.

(h) Let G be an open neighborhood of z. A point z is semiregular for U if and only if z is semiregular for $G \cap U$.

Hint. Use (f) and (d).

(j) Let z be an irregular point for U. Then the following assertions are equivalent:

(i) z is semiregular,

(ii) if h is a nonnegative finely harmonic function on U such that $\limsup_{U \ni x \to z} h(x) < +\infty$, then there is a harmonic function g on a neighborhood G of z such that $h = g$ on $G \cap U$.

Hint. Let $p \in \mathcal{P}^c$ be a strict potential such that

$$\limsup_{U\ni x\to z} \hat{R}_p^{CU}(x) + h(x) < p(z) - 1.$$

Find an open neighborhood G of z such that $\hat{R}_p^{CU} + h < p - 1$ on $\bar{G} \cap \bar{U}$. There is $q \in \mathcal{P}$ such that $\hat{R}_p^{CU} \leq q \leq p$ and $q = \min(\hat{R}_p^{CU} + h, p)$ on U. We have $\hat{R}_q^{C(U\cap G)} = q$. By (h) and (g), $\hat{R}_q^{CG} = \hat{R}_q^{C(G\cap U)}$ on $G \cap U$. Now, put

$$g = (\hat{R}_q^{CG} - \hat{R}_p^{C(G\cup U)}) \wedge G.$$

(k) Let $z \in \bar{U}$. Then there are four possibilities only:

(α) z is a regular point for U,

(β) z is a strictly semiregular point for U,

(γ) $\varepsilon_z \neq \varepsilon_z^{CU}$, $\Lambda(z,U) = \{\varepsilon_z, \varepsilon_z^{CU}\}$,

(δ) $\varepsilon_z \neq \varepsilon_z^{CU}$, $\Lambda(z,U) = \{\alpha\varepsilon_z + (1-\alpha)\,\varepsilon_z^{CU} : \alpha \in [0,1]\}$

Hint. Assume that neither of (α), (β), (δ) occurs. Let $p \in \mathcal{P}^c$ be a strict potential, $p(z) = 1$. Then there is $\alpha \in (0,1)$ with the following property: If $c :=$ $= \alpha p(z) + (1-\alpha)\hat{R}_p^{CU}(z)$, then there is an open neighborhood G of z and an interval (a,b) containing c such that $[\hat{R}_p^{CU} < ap] \cup [\hat{R}_p^{CU} > bp] \supset U \cap G$. Denote

$$V_s = \{x \in U \cap G : \hat{R}_p^{CU} < ap\},$$

$$V_r = \{x \in U \cap G : \hat{R}_p^{CU} > bp\}.$$

Now, show that

$$\lim_{V_s \ni x \to z} \hat{R}_p^{CU}(x) = R_p^{CU}(z), \quad \lim_{V_r \ni x \to z} \hat{R}_p^{CU}(x) = p(z).$$

(m) (W. Hansen (1983)) Let A be now an arbitrary subset of X, $z \in X$, and let $\Lambda^A(z)$ denote the set of all Radon measures λ on X such that $\lambda = \text{w-lim}\, \varepsilon_{x_n}^A$ for some sequence $\{x_n\}$ converging to z. Show that either $\Lambda^A(z) \subset \{\varepsilon_z, \varepsilon_z^A\}$ or $\Lambda^A(z) = \{\alpha\varepsilon_z + (1-\alpha)\,\varepsilon_z^A : \alpha \in [0,1]\}$.

Hint: Use the preceding exercise for $U = \text{int}_f CA$.

13.A.4 Let U be a finely regular subset of X, and let g be a finely continuous $\mathcal{P}^f$-bounded function on X. Prove that there is a finely continuous function h on $\bar{U}$ which is finely harmonic on U such that $g=h$ on CU.

Hint. First, show that g is finely resolutive (Theorems 10.30 and 13.7). Put $h = H_g^U$ on U, $h = g$ on CU. Obviously, h is finely continuous on $X \setminus \partial_f U$. Fix $z \in \partial_f U$, it remains to show $g(z) = \text{fine-}\lim_{U \ni x \to z} h(x)$. We may assume that $g \gtrsim 0$ (otherwise consider $g = (g+p) - p$ for $p \in \mathcal{P}^f$). Given $\varepsilon > 0$, put $A = [g < g(z) - \varepsilon]$. Since A is thin at z, there is $w \in \mathcal{H}_+^*(X)$ such that w is $\mathcal{P}^f$-bounded and

$$w(z) + g(z) \leqq \liminf_{A \ni x \to z} w(x).$$

Then

$$w(z) + g(z) \leqq \liminf_{x \to z} (w(x) + g(x)) + \varepsilon$$

and form Ex. 13.A.3 we easily deduce (z is regular !) that

$$\liminf_{U \ni x \to z} H_{w+g}^U \geqq w(z) + g(z) - \varepsilon .$$

Since $\hat{R}_w^{CU}$ is finely continuous, we have

$$\text{fine-}\liminf_{x \to z} h(x) \geqq g(z) - \varepsilon .$$

13.A.5 Incomplete fine Dirichlet problem. Let U be a finely regular set and let $F \subset X \setminus U$ be a finely closed set. If f is the restriction to F of a $\mathcal{P}^f$-bounded Baire one function on X and $f \wedge F$ is finely continuous, then there exists a finely continuous extension f^* of f to $\bar{U}$ which is finely harmonic on U. Moreover, if t, s are lower semicontinuous functions on $\bar{U}$ which are finely hyperharmonic on U and $-t \leqq s$ on $\bar{U}$, $-t \leqq f \leqq s$ on F, then we can select f^* such that $-t \leqq f^* \leqq s$ on $\bar{U}$.

Hint. Use the previous exercise and the B_1-extension theorem 3.29.

REMARKS. (a) Notice that $B_1(X) \wedge F = B_1(F)$ provided F is a G_δ-set.

(b) Since the fine boundary $\partial_f U$ of a finely regular set U is always a G_δ-set ($\partial_f U = bU \cap bCU$), the previous remark and exercise provide a solution of the fine

Dirichlet problem on the fine boundary: Given a finely continuous $\mathcal{P}^f$-bounded function g on $\partial_f U$, there is a finely continuous function F_g on $\bar{U}$ such that $F_g = g$ on $\partial_f U$ and F_g is finely harmonic on U. Of course, if the domination axiom D does not hold on X, the uniqueness of F_g is lost. Cf. also Proposition 14.17.

13.A.6 Generalized superfunctions. Let U be a finely open set and f a function on $\bar{U} \setminus U$. A function $u \in \mathcal{H}_l^*(U)$ is termed a *reg-superfunction* for f if u is lower $\mathcal{P}^f$-bounded and $\liminf u(x_n) \geq f(z)$ whenever $\{x_n\}$ is a regular sequence of points of U converging to a point $z \in \bar{U} \setminus U$.

(a) Minimum principle. If u is a reg-superfunction for 0, then $u \geq 0$.

Hint. Fix $x \in U$ and $\varepsilon > 0$. Find a strict potential $p \in \mathcal{P}^c$ and a sequence $\{w_n\}$ of functions from $\mathcal{H}_+^*(X)$ such that $w_n \geq p$ on CU and $\sum (w_n(x) - \hat{R}_p^{CU}(x)) < \varepsilon$. Now use Theorem 13.1 for $u + \sum (w_n - \hat{R}_p^{CU})$.

(b) Perron-type solution. If f is a finely resolutive function for U, then H_f^U is the infimum of all reg-superfunctions for f.

(c) Let f be a continuous $\mathcal{P}^c$-bounded function on X, and let h be finely harmonic and $\mathcal{P}^c$-bounded on U. Then the following assertions are equivalent:

(i) $h = H_f^U$,

(ii) if $z \in \bar{U} \setminus U$ and $\{x_n\}$ is a regular sequence in U converging to z, then $\lim h(x_n) = f(z)$.

REMARKS AND COMMENTS

The problem of finding a harmonic function on an open bounded set U in $\mathbb{R}^n$ taking on preassigned continuous boundary values is known as the Dirichlet problem. It is as old as the potential theory itself and it appeared in gravitation problems, electrostatics, heat conduction and elasticity theory. The development of the Dirichlet problem in the nineteenth century which is associated with the names of S.D.Poisson, G.Green, C.F.Gauss, B.Riemann, K.Weierstrass, H.A.Schwarz and C. Neumann is outlined in L.Gårding (1979).

Since that time, various methods for solving the Dirichlet problem were discovered. Let us mention at least methods of the calculus of variations connected with the Dirichlet integral, the alternating method of H.A.Schwarz, the method of the aritmetic mean of C.Neumann and "méthode de balayage" of H.Poincaré, or the method of integral equations of I.Fredholm. Each of these methods, however, had its limitations on the shape of the domain or on the properties of the boundary values.

Growing efforts in solving the Dirichlet problem for a wide class of domains brought about a requirement of generalized solutions. As a matter of fact, H.A.Schwarz (1872) and M.Bôcher (1903) proved that an isolated singularity was removable for bounded harmonic functions. As remarked by S.Zaremba (1911), this meant the existence of open sets for which the classical Dirichlet problem was not solvable for all continuous boundary data. More specifically, Zaremba pointed out that domains with isolated boundary points are not regular for the Dirichlet problem. An even more striking example of a non-regular domain was given for three-dimensional space by H. Lebesgue in (1913).

It was H. Lebesgue who explicitely proposed to separate the investigation of the Dirichlet problem into two parts: Firstly to produce a harmonic function depending in a way on the given boundary condition and then investigate the boundary behaviour of the resulting candidate for a solution.

Some of the old methods for a construction of such a harmonic function were mentioned above, but none of them applied to the case of general domains. On the other hand, in the twenties, two completely different new methods without any limitation on the region were proposed. O.Perron (1923) and R.Remak (1924) considered an arbitrary bounded function f on the boundary of a general bounded open set U and defined the upper class for f as the set of all continuous superharmonic functions on U whose lower limit dominates f at each boundary point of U. The infimum $\overline{H}_f^U$ of such a class was shown to be a harmonic function called the upper solution. The lower solution $\underline{H}_f^U$ was defined similarly and the relation $\underline{H}_f^U \leq \overline{H}_f^U$ was established. The equality of the upper and lower solutions for a continuous f was verified by N.Wiener (1925).

Notice that one year before, N.Wiener defined in (1924) another type of solution of the Dirichlet problem.

A more general treatment of Perron's method was given by M. Brelot (1938a).He considered for an arbitrary numerical function on ∂U the Perron upper class formed by lower semicontinuous superharmonic functions and characterized functions for which upper and lower solutions coincide.

Perron's method was extended by W. Sternberg (1929) to the heat equation and later by many authors to abstract theories of harmonic spaces (cf.M.Brelot (1966), [Bau], [C&C]).

As already pointed out, the concept of the Dirichlet problem is central to the study of potential theory. As the harmonic functions needed for its solution were derived from more and more general partial differential equations, the need for a common basis in the form of abstract harmonic spaces which should be applicable as much as possible became acute. Moreover, in this situation even the fine potential theory on harmonic spaces was recently built. Having a good theory of finely (hyper) harmonic functions, it is only a short step to go to the investigation of the fine Dirichlet problem.

The basis for it is the validity of the minimum principle. In the presence of the axiom D, B.Fuglede proved the fine minimum principle by means of the quasi-Lindelöf property of the fine topology. By contrast, the proof of the minimum principle 13.1 which holds in general harmonic spaces is based on the Lusin-Menchoff property of the fine topology. Notice that the both above mentioned minimum principles are not comparable. Their unification is the object of the next chapter.

It needs to be observed that J. Bliedtner and W.Hansen (1978) use the Choquet theory to derive the minimum principle for some convex cones of continuous finely hyperharmonic functions. Their Prop. 3.3 is, in fact, a Perron type solution of the fine Dirichlet problem for continuous boundary functions.

The notion of semiregular points was introduced as a local analogy of Bauer's semiregular sets (H. Bauer (1962)) by J. Lukeš and J. Malý (1981). There, most of the material of Ex. 13.A.3 is presented in the context of open sets. The existence of semiregular sequences at each boundary point of an open set was proved by J. Köhn and M. Sieveking (1967); the existence of a regular sequence at all non-semiregular points is one of the results of J. Lukeš and J. Malý (1981).

In this paper the classification of irregular boundary points of open sets is given generalizing the classical result by O.Frostman (1939) and improving investigation of N. Boboc and A. Cornea (1967). Further study of cluster sets of balayaged measures of arbitrary sets examined by W. Hansen (1983) yields further generalization of the mentioned results (cf. Ex. 13.A.3). T. Ikegami (1984) established similar results for resolutive compactifications of harmonic spaces.

Related paper : J. Hyvönen (1979).

13.B The Wiener method

In (1924), N. Wiener defined in the classical case a completely different type of the solution of the Dirichlet problem, the idea of which was based on the observation described in Prologue. The Wiener type solution has not been studied in the abstract potential theory. The reason lies in the fact that, in general, one cannot exhaust a given domain by a sequence of regular sets. It means that one cannot insert a regular set V between U and an arbitrarily chosen compact set $K \subset U$ in the sense that $K \subset V \subset \overline{V} \subset U$. This was observed by H. Bauer in [Bau] ,p. 147. On the other hand, in any elliptic $\mathfrak{P}$-harmonic space, the existence of an exhaustion formed by regular sets is guaranteed by a result of R.-M. Hervé (1962) (Prof. 7.1, cf. [C&C] , ex. 3.1.14), and in fact Wiener's method may be repeated without changes in the frame of elliptic spaces. Some conditions for the insertion of regular sets can be found in E. Čermáková (1978).

Let us mention that in J. Lukeš and I. Netuka (1976) some modified Wiener procedure is described. The exhaustion is realized by the so-called Keldych sets (see Ex.

13.B.2) on which the Perron solution is the unique reasonable solution of the Dirichlet problem.

An upper directed system $\mathcal{R}$ of finely open sets is a *fine exhaustion* of a finely open set U if for any compact set $K \subset U$ there is $V \in \mathcal{R}$ such that $K \subset V \subset \overline{V} \subset U$.

From Theorem 10.25 it follows that there is always a fine exhaustion of a given finely open set formed by finely regular sets(cf. Ex. 11.D.8). Let us remark that an exhaustion of an open set by open sets is also a fine exhaustion. While it is not always possible to exhaust an open set by open *regular* sets, there is a fine exhaustion consisting of (relatively compact) *finely regular* sets.

Let $\mathcal{R}$ be a fine exhaustion of U formed by relatively compact finely regular sets. Let $p \in \mathcal{P}^c$. If $V \subset \overline{V} \subset U$ is a finely regular set and K is a compact subset of X, then

$$\hat{R}_p^{CK} + \hat{R}_p^{CU} \geq p \quad \text{outside a set } W \in \mathcal{R} .$$

Hence from Propositions 11.2 and 10.20 we obtain

$$\hat{R}_p^{CU} = \bigwedge \{ \hat{R}_p^{CV} : V \in I(U) \} + \bigwedge \{ \hat{R}_p^{CK} : K \text{ is a compact subset of } X \} =$$

$$= \bigwedge \{ \hat{R}_p^{CV} : V \in \mathcal{R} \} .$$

It follows

$$\varepsilon_x^{CU} = \operatorname*{w-lim}_{V \in \mathcal{R}} \varepsilon_x^{CV}$$

for any $x \in U$. In other words, if g is a continuous extension of a given function $f \in \mathcal{K}(\partial U)$ to X, then

$$f^{CU} = \lim_{V \in \mathcal{R}} h_g^V \quad \text{on } U$$

where h_g^V are the finely classical solutions for g on the sets $V \in \mathcal{R}$. The last assertion is nothing else than the promised *Wiener method* of solving of the (fine) Dirichlet problem.

EXERCISES

13.B.1 A finely open set U can be finely exhausted by a *sequence* of (finely regular) sets if and only if it is locally compact.

Hint. Using the Lusin-Menchoff property of the fine topology, a set U can be exhausted by a sequence of finely regular sets if and only if there is a sequence $\{K_n\}$ of compact sets in U such that for every compact set $K \subset U$ there is $n \in \mathbb{N}$ such that $K \subset K_n$. For U being locally compact this condition is obviously satisfied. Let us suppose that U is not locally compact and there is a sequence with the properties just described. Then there is $z \in U$ such that no K_n is a neighborhood of z. Assume that $\{V_n\}$ is a local base at z, $\bigcap_{n=1}^{\infty} V_n = \{z\}$, $V_1 \supset V_2 \supset V_3 \supset \dots$ If $x_n \in V_n \setminus K_n$, then the set

$$K := \{z\} \cup \{x_n : n \in \mathbb{N}\}$$

is a compact set which is not contained in any K_n.

13.B.2 Keldych operators. Let U be a relatively compact finely open subset of X. Suppose that the vector space $H(U) = \{f \in \mathcal{C}(\bar{U}) : f \text{ is finely harmonic on } U\}$ contains constant functions and separates the points of $\bar{U}$. A *Keldych operator* on U is any positive linear operator L associating with every $f \in \mathcal{C}(\partial U)$ a finely harmonic function Lf on U such that $L(h \restriction \partial U) = h \restriction U$ whenever $h \in H(U)$.

(a) Show that $H^U : f \mapsto H_f^U$ is a Keldych operator.

(b) An outstanding theorem of J. Bliedtner and W. Hansen (1975) says that $H(U)$ is *simplicial* and $\varepsilon_x^{\beta CU}$ is a unique minimal *representing measure* for each $x \in U$ (β is the essential base operator of Section 1.D).

(c) Using (b) show that $D^U : f \mapsto f^{\beta CU} \restriction U$ is again a Keldych operator (you can also profit from Theorem 12.20) and $D^U w \leqq Lw \leqq H^U w$ whenever L is a Keldych operator on U and $w = \min(h_1, \dots, h_n)$, $h_j \in H(U)$.

(d) It can be proved that $\hat{R}^{\beta CU} = \sup\{q \in \mathcal{P}^c : q \leqq p,\ q = \hat{R}_q^{CU}\}$ for every $p \in \mathcal{P}^c$ (J.Bliedtner and W.Hansen (1975)). Analogously, put

$$P_p^{CU} = \sup\{q \in \mathcal{P} : q \leqq p,\ \hat{R}_q^{CU} = q\},\quad p \in \mathcal{P}^c .$$

For every $x \in U$ there is a unique Radon measure ν_x^U such that $\nu_x^U(p) = P_p^{CU}$ for any $p \in \mathcal{P}^c$. Prove that $T^U : f \mapsto \nu_x^U(f)$, $x \in U$ is a Keldych operator (In J.Lukeš (1973), (1974) this Keldych operator is called a *prinicipal solution* of the Dirichlet problem).

(e) We say that U is a *finely Keldych set* if there is a unique Keldych operator on U. The following assertions are equivalent:

(i) U is a finely Keldych set.

(ii) $D^U = H^U$,

(iii) $T^U = H^U$,

(iv) the set of all irregular points for U is of ε_x^{CU}-measure zero for any $x \in U$,

(v) U is a sweeping kernel of a finely regular set.

Hint. Use methods indicated in references concerning Keldych operators. Especially, if (iv) holds, then $bCU = \beta CU$ and $\varepsilon_x^{CU} = \varepsilon_x^{bCU}$ for every $x \in U$. By virtue of (c), (ii) implies (i) and (iii). For the proof of (iii)$\Longrightarrow$(iv) use the fact that $P_p^{CU} \leqq \hat{R}_q^{CU} \leqq q$ where $q = \hat{R}_p^{CU}$, and hence $\varepsilon_x^{CU}[\hat{R}_p^{CU} < p] = 0$ for any $p \in \mathcal{P}^c$.

(f) Any finely regular subset of U is a finely Keldych set. In particular, given a compact set $K \subset U$, there is always a finely Keldych set V such that $K \subset V \subset \overline{V} \subset U$.

REMARKS AND COMMENTS

As pointed out, the Wiener procedure of solving the Dirichlet problem is not transferable to general harmonic spaces. It is somewhat surprising that the methods of the fine topology yield the possibility to use Wiener's approach. Modified Wiener's method, as explained in J.Lukeš and I.Netuka (1976), uses the properties of the so-called Keldych operators (cf. Ex. 13.B.2).

A.F.Monna noticed in (1938) and (1939) (see A.F.Monna (1971) where relevant references and interesting comments on the subject may be found) that the methods of Perron and Wiener are special constructions only and investigated the unicity of the Dirichlet problem from the functional analysis point of view. He asked whether an operator of the Dirichlet problem (submitted to certain natural conditions) was uniquely determined. A similar question was posed and solved by M.V.Keldych (1941a), (1941b) by proving that there is exactly one positive linear operator sending continuous functions on ∂U into harmonic functions on U such that its value is the classical solution, if it exists. An operator possessing these properties is termed a *Keldych operator*. For the Laplace equation there is therefore the only one "reasonable" solution of the Dirichlet problem and Perron's or Wiener's methods appear thus as its special constructions. The same unicity result holds for a wide class of partial differential equations of elliptic type as shown by M.Brelot (1961).

In the general situation of harmonic spaces, uniqueness questions turn out to be more delicate. In contrast to the Laplace case, there is e.g. for some open sets $U \subset \mathbb{R}^{n+1}$ more than one reasonable solution of the Dirichlet problem for the heat equation, so that the Keldych theorem fails. Consequently, one can ask what is in fact the "right solution" of the Dirichlet problem. An attempt to answer this question is given by J. Lukeš and I. Netuka (1979) out of which we also use the historical remarks and comments.

Concerning Keldych operators, their investigation goes back to papers of J.Lukeš (1973), (1974) which were followed by a series of papers of I.Netuka (1980a), (1980b), (1982a), (1982b), (1982c).

An attempt to modify Wiener's procedure for the heat equation and to more general parabolic equations is to be ascribed to E.M. Landis (1969), (1971).

Exercise 13.B.1 is due to P.Pyrih (unpublished).

Related papers concerning Keldych operators: E.M.J. Bertin (1978), T.Ikegami (1981), (1983), J.Lukeš (1977b), H.Schirmeier and U.Schirmeier (1978), W.Hansen (1985).

14. QUASI-TOPOLOGICAL METHODS

A. Abstract quasi-continuity
B. Finely hyperharmonic functions revised
C. Quasi-solution of the fine Dirichlet problem

In the previous sections, a fine potential theory on harmonic spaces without axiom D was developed. As we have seen, there is an unmistakable distinction admitting of the axiom D, or not. For example, the family of all pointwise finely hyperharmonic functions does not coincide with the set of "globally" finely hyperharmonic functions. This phenomena cannot occur in Fuglede's theory with the axiom D where all kinds of definitions of hyperharmonic functions lead to the same theory. In similar fashion, omitting the axiom D, the Dirichlet problem cannot be uniquely solved by considering fine limits on fine boundary only.

To remedy this deficiency, we develop in the next sections the quasi-topological methods to answer some of the natural questions. We now consider families of hyperharmonic functions, where the requirement of lower semicontinuity is relaxed to require that they are only quasi-l.s.c. The reason lies in the fact that, in the presence of axiom D, any finely lower semicontinuous function is quasi-l.s.c. Thus, starting from the additional assumption of quasi-l.s.c., we can characterize finely hyperharmonic functions. We will see that a nonnegligible role is reserved for the $\mathcal{V}(U)$-topology. The results of Section 14.B have important ramifications later when we show the Dirichlet problem in a new light.

14.A Abstract quasi-continuity

Let τ be a fine topology on a Hausdorff topological space (Y,ρ) and assume that

(Y,τ) is a Baire space. Let Φ be a convex cone of nonnegative τ-continuous functions on Y which contains the function $+\infty$. We consider the natural order on Φ given by the pointwise ordering of functions of Φ. We use the notation $\wedge \mathcal{W}$ and $\vee \mathcal{W}$ for the natural infimum and supremum of a set $\mathcal{W} \subset \Phi$.

We impose upon Φ the following two axioms of C. Constantinescu and A. Cornea ([C&C] , § 4.1):

Axiom of upper directed sets. If $\mathcal{W} \subset \Phi$ is a nonempty upper directed set, then $\vee \mathcal{W} \in \Phi$.

Axiom of τ-lower semicontinuous regularization. If $\mathcal{W} \subset \Phi$ is a nonempty set, then the τ-lower semicontinuous regularization $\widehat{\inf \mathcal{W}}^{\tau}$ of $\inf \mathcal{W}$ belongs to Φ.

We see that $\wedge \mathcal{W} = \widehat{\inf \mathcal{W}}^{\tau}$ and $\vee \mathcal{V} = \sup \mathcal{V}$ provided $\mathcal{V}$ is upper directed. The following important proposition is a special case of Prop. 4.1.3 from [C&C] .

14.1 PROPOSITION. Let $\{\mathcal{V}_n\}$ be a sequence of subsets of Φ such that $\wedge \mathcal{V}_n = 0$ for each $n \in \mathbb{N}$. Let $\mathcal{V}$ be the set of all sums $\sum v_n$ where $v_n \in \mathcal{V}_n$. Then $\wedge \mathcal{V} = 0$.

Proof is straightforward and uses the fact that $[\widehat{\inf \mathcal{W}}^{\tau} < \inf \mathcal{W}]$ is a τ-first category set and that (Y,τ) is a Baire space.

We denote by $\Phi^{\#}$ the set of all lower semicontinuous elements of Φ. For any $A \subset Y$ we define the *capacity* of A by

$$c(A) = \wedge \{ u \in \Phi^{\#} : u \geq 1 \text{ on } A \}.$$

Notice that this capacity has values in the convex cone Φ but it can happen that $c(A) \notin \Phi^{\#}$. If A is τ-open, then $c(A) \geq 1$ on A. Obviously, the capacity c is a monotone map, and in view of Proposition 14.1, it is countable subadditive.

A family $\mathcal{H}$ of subsets of Y is called *evanescent* if

$$\wedge \{ c(A) : A \in \mathcal{H} \} = 0.$$

A function f on Y is said to be *quasi-l.s.c.* on Y (with respect to Φ) if there is an evanescent family $\mathcal{H}$ of open sets such that $f \restriction Y \setminus G$ is lower semicontinuous for every $G \in \mathcal{H}$. Similarly we define *quasi-u.s.c.* and *quasi-continuous* functions on Y.

14.2 PROPOSITION. Let $\{\mathcal{H}_n : n \in \mathbb{N}\}$ be a system of evanescent families. Then the family of all unions $\bigcup G_n$ where $G_n \in \mathcal{H}_n$ for each $n \in \mathbb{N}$ is again evanescent.

Proof. Obvious application of Proposition 14.1.

14.3. PROPOSITION. (a) Quasi-l.s.c. functions on Y form a convex cone containing $\inf(f,g)$ and $\sup(f,g)$ with every f,g.

(b) The limit of a sequence $\{f_n\}$ of quasi-l.s.c. functions is quasi-l.s.c. provided either $\{f_n\}$ converges locally uniformly or $\{f_n\}$ is increasing.

(c) A function f is quasi-continuous if and only if it is quasi-l.s.c. and quasi-u.s.c.

Proof. It is a routine application of Proposition 14.2 in combination with elementary properties of lower semicontinuous functions.

14.4 THEOREM. Consider the following conditions on a function f on Y.

(i) f is quasi-l.s.c.,

(ii) there is $\mathcal{V} \subset \Phi^{\#}$ such that $\wedge \mathcal{V} = 0$ and $f+v$ is lower semicontinuous for every $v \in \mathcal{V}$,

(iii) there is $\mathcal{V} \subset \Phi^{\#}$ such that $\wedge \mathcal{V} = 0$ and for every $v \in \mathcal{V}$ there is a lower semicontinuous function g with $f \leqq g \leqq f+v$.

Then (ii) $\Rightarrow$ (iii) $\Rightarrow$ (i). If, moreover, $f \geqq 0$, then all these conditions are equivalent.

Proof. (i) $\Rightarrow$ (ii) ($f \geqq 0$). Put

$$\mathcal{V}' := \bigcup \{\{u \in \Phi^{\#} : u \geqq 1 \text{ on } G\} : G \text{ is open, } f \text{ is lower semicontinuous on } Y \setminus G\},$$

$$\mathcal{V} := \{\textstyle\sum n v_n : v_n \in \mathcal{V}'\}.$$

By Proposition 14.1, $\wedge \mathcal{V} = 0$. Pick $\{v_n\} \subset \mathcal{V}'$ and denote $v = \sum n v_n$. For $k \in \mathbb{N}$ denote $g_k = \inf(k, f+v)$. Choose $x \in Y$ and $k \in \mathbb{N}$. Denote

$$G := \{y \in Y: v_k(y) > 1\}.$$

We have

$$g_k(x) \leq k \leq \inf (k, \liminf_{G\ni y\to x} kv_k(y)) \leq \liminf_{G\ni y\to x} g_k(y).$$

Obviously,

$$g_k(x) \leq \liminf_{G\not\ni y\to x} g_k(y).$$

We see that g_k is lower semicontinuous for every $k \in \mathbb{N}$ and hence $f+v = \sup g_k$ is lower semicontinuous.

(ii)$\Rightarrow$(iii): Trivial.

(iii)$\Rightarrow$(i). Put

$$\mathcal{V}^* := \{\sum nv_n : v_n \in \mathcal{V}\}$$

and

$$\mathcal{Y} := \{[v > 1] : v \in \mathcal{V}^*\}.$$

Then $\mathcal{Y}$ is a family of open subsets of Y and

$$\wedge\{c(G) : G \in \mathcal{Y}\} \leq \wedge \mathcal{V}^* = 0$$

by Proposition 14.1. Fix a sequence $\{v_n\} \subset \mathcal{V}$. Denote $v := \sum nv_n$ and $G := [v > 1]$. For every $n \in \mathbb{N}$ there is a lower semicontinuous function g_n such that

$$f \leq g_n \leq f+v_n \leq f + \frac{v}{n} \leq f + \frac{1}{n}$$

on $Y \setminus G$. Since a uniform limit of a sequence of lower semicontinuous functions is also lower semicontinuous, we conclude f is lower semicontinuous on $Y \setminus G$.

14.5 REMARKS. (a) Concerning the implication (i)$\Rightarrow$(ii), the assumption of nonnegativity of f can be weakened, see Ex. 14.A.2.

(b) Assuming that $u \in \Phi$ is quasi-l.s.c., we obtain

$$u = \wedge\{v \in \Phi^{\#} : v \geq u\}.$$

If all elements of Φ are quasi-l.s.c., then

$$c(A) = \wedge\{v \in \Phi : v \geq 1 \text{ on } A\}$$

for every $A \subset U$. Cf. Corollary 14.10.

EXERCISES

14.A.1 A subset A of Y is called *quasi-open* if there is an evanescent family $\mathcal{H}$ of open sets such that $A \cup G$ is open for all $G \in \mathcal{H}$. Show that:

(a) the family of all quasi-open sets forms a σ-topology on Y (cf. Ex.3.B.5),

(b) a function f on Y is quasi-l.s.c. if and only if any set $[f > c]$, $c \in \mathbb{R}$ is quasi-open.

14.A.2 Let f be a lower finite quasi-l.s.c. function on Y. Assume that there is $\mathcal{W} \subset \Phi^{\#}$ such that $\wedge \mathcal{W} = 0$ and $f+w$ is locally lower bounded for every $w \in \mathcal{W}$. Then there is $\mathcal{V} \subset \Phi^{\#}$ such that $\wedge \mathcal{V} = 0$ and $f+v$ is lower semicontinuous for every $v \in \mathcal{V}$ (cf.Theorem 14.4).

14.A.3 A function f is quasi-l.s.c. on Y if and only if there is an evanescent family $\mathcal{H}$ of arbitrary subsets of Y such that $f \wedge Y \setminus N$ is lower semicontinuous for every $N \in \mathcal{H}$.

REMARKS AND COMMENTS

It was N. Wiener (1924) who introduced the mathematical concept of capacity in accordance with previously defined physical capacity. There followed a series of papers by Ch. de la Vallée Poussin, O. Frostman, A.F. Monna, M. Brelot culminating by famous Choquet's theorem on capacitability of Borel sets.

It was shown by H. Cartan (1945) that the potential-theoretic counterpart of Lusin's theorem is valid: Every superharmonic function is quasi-continuous, i.e. its restriction is continuous everywhere but a set of arbitrarily small capacity. Since that time the study of quasi-topological notions has widely spread. The axiomatic theory of capacities was built up mainly by M. Brelot and G. Choquet. A systematic research of an abstract capacity and quasi-topological properties was done by B.Fuglede (1971a). An expansion and the applications of the theory of capacities in several branches of analysis in quite recent period seem to be noteworthly.

A notion of evanescent families was introduced by M. Brelot (1965).

Related paper :D. Feyel (1981b)

14.B Finely hyperharmonic functions revised

Come back to fine potential theory on harmonic spaces. Throughout this section we assume again that U is a fixed finely open subset of a $\mathfrak{P}$-harmonic space X.

Recall that the convex cone $\mathcal{H}^*_+(U)$ of all nonnegative finely hyperharmonic functions on U satisfies the axiom of upper directed sets and the axiom of fine lower semicontinuous regularization (Propositions 12.9 and 12.10). Further, U endowed with the fine topology is a Baire space (Theorem 10.28). Thus we can consider $\mathcal{H}^*_+(U)$ as Φ of the previous section.

Notice that $\mathcal{H}^\#_+(U)$ is nothing else than the family of all lower semicontinuous elements of $\mathcal{H}^*_+(U)$. All quasi-notions are defined by means of the capacity with values in $\mathcal{H}^*_+(U)$ and by evanescent families of open sets in U. We emphasize this fact by the prefix "U-quasi".

Recall also that $\nu(U)$ is the coarsest topology on U finer than the initial topology and making all functions from $\mathcal{H}^*_+(U)$ lower semicontinuous.

The next proposition should be compared with Lemma 12.7 and its proof in Ex. 12.A.7.

14.6 PROPOSITION. Let U be a Borel finely open set and let $p \in \mathcal{P}^f$. Then the function p^{CU} is U-quasi-continuous on U.

Proof. By Proposition 11.3,

$$\hat{R}_p^{CU} = \sup \{ \hat{R}_p^K : K \subset CU, K \text{ compact} \} .$$

Put

$$\mathcal{V} = \{ (\hat{R}_p^{CU} - \hat{R}_p^K) \wedge U : K \subset CU, \ K \text{ compact} \} .$$

Then $\mathcal{V} \subset \mathcal{H}^\#_+(U)$, $\wedge \mathcal{V} = 0$ and $-\hat{R}_p^{CU} + v$ is continuous on U for every $v \in \mathcal{V}$. Hence $\hat{R}_p^{CU}$ is U-quasi-u.s.c. on U by Theorem 14.4. Of course, $\hat{R}_p^{CU}$ is lower semicontinuous.

14.7 COROLLARY. Let f be a function on a Borel finely open set U. If f is a lower semicontinuous function with respect to $\nu(U)$, then f is U-quasi-l.s.c. on U.

Proof. Let $p \in \mathcal{P}^c$ be strict. If $G \subset X$ is open and if p^{CU} is continuous on $U \setminus G$, then f is lower semicontinuous on $U \setminus G$ by Theorem 12.14. To conclude the proof it suffices to use Proposition 14.6.

The previous series of results, giving necessary conditions and sufficient conditions for quasi-continuity notions, culminates now with a weighty characterization of finely hyperharmonic functions.

14.8 THEOREM. Let $U \subset X$ be a Borel finely open set. Let u be a nonnegative pointwise finely hyperharmonic function on U. Then the following assertions are equivalent:

(i) u is finely hyperharmonic on U,

(ii) u is $\nu(U)$-lower semicontinuous,

(iii) u is U-quasi-l.s.c. on U,

(iv) there is $\mathcal{V} \subset \mathcal{H}_+^{\#}(U)$ such that $u+v \in \mathcal{H}_+^{\#}(U)$ for every $v \in \mathcal{V}$ and $\wedge \mathcal{V} = 0$,

(v) there is $\mathcal{V} \subset \mathcal{H}_+^{*}(U)$ such that $u+v \in \mathcal{H}_+^{*}(U)$ for every $v \in \mathcal{V}$ and $\wedge \mathcal{V} = 0$.

Proof. (i) $\Longrightarrow$ (ii). According to Lemma 12.5, u is the limit of an increasing sequence of functions of the form $q - p^{CU}$, where q, p are finite potentials on X, p is continuous and $\hat{R}_p^{CU} \leqq q \leqq p$ (indeed, $u^{(n)} \nearrow u$ on U where $u^{(n)}$ were introduced in Section 12.B). Now the assertion follows from Theorem 12.14.
(ii) $\Longrightarrow$ (iii). See Corollary 14.7.
(iii) $\Longrightarrow$ (iv). It follows from Theorem 14.4 and Theorem 12.19.
(iv) $\Longrightarrow$ (v). It is obvious.
(v) $\Longrightarrow$ (i). Put

$$f := \wedge \{ u+v : v \in \mathcal{V} \}.$$

Then f is finely hyperharmonic and we want to prove that $f = u$. Obviously $u \leqq f$. Pick $x \in U$ and choose $r < f(x)$. Let h be a harmonic function defined on a neighborhood of x such that $r < h(x) < f(x)$. There is a fine neighborhood $V \in \mathcal{B}(U)$ of x such that $\overline{V}$ is contained in the domain of $h, h < f$ on $\overline{V}$ and $\varepsilon_x^{CV}(u) \leqq u(x)$. By Corollary 12.13, the function $g: x \longrightarrow \int^* \inf \mathcal{V} \, d\varepsilon_x^{CV}$ is finely continuous on V. Since $g \leqq \inf \mathcal{V}$ on V, $g = 0$ except a semipolar set (Proposition 12.10). Hence $g = 0$ on V and we deduce

$$r < h(x) = \varepsilon_x^{CV}(h) \leqq \varepsilon_x^{CV}(f) \leqq \int^* \inf(u + \mathcal{V}) \, d\varepsilon_x^{CV} =$$
$$= \int^* (\inf \mathcal{V} + u) \, d\varepsilon_x^{CV} \leqq \varepsilon_x^{CV}(u) \leqq u(x).$$

14.9 COROLLARY. Let u be a pointwise finely hyperharmonic function on a finely open set U. Assume that for every $V \in \mathcal{B}(U)$ there is $p \in \mathcal{P}^f$ such that $-p^{CV} \leqq u$ on V. Then u is finely hyperharmonic on U if and only if it is V-quasi-l.s.c. on every set $V \in \mathcal{B}(U)$ (or, on every $V \in \mathcal{B}_o(U)$).

Proof. It follows from Theorem 14.8, Proposition 14.6 and from Proposition 12.8 (cf. also its proof).

14.10 COROLLARY. Let $U \subset X$ be a Borel finely open set.

(a) If $u \in \mathcal{H}^*_+(U)$, then

$$u = \wedge \left\{ v \in \mathcal{H}^{\#}_+(U) : v \geq u \right\}.$$

(b) If $A \subset U$, then

$$c(A) = \wedge \left\{ v \in \mathcal{H}^*_+(U) : v \geqq 1 \text{ on } A \right\}.$$

Proof. These are obvious consequences of Theorem 14.8.

At the end of this section, we return to Theorem 12.20 on removable singularities for finely hyperharmonic functions. Acutally, it is possible to give its more general version where the requirement that the function is lower semicontinuous is relaxed to require that it is only quasi-l.s.c.

14.11 THEOREM. (Removable singularities for finely hyperharmonic functions). Let u be a finely continuous function on a finely open set U. Suppose that for each $V \in \mathcal{B}_o(U)$ there is $p \in \mathcal{P}^f$ such that $-p^{CV} \leqq u$ on V and u is V-quasi-l.s.c. on V. If there is a finely open set $G \subset U$ such that $U \setminus G$ is semipolar and u is finely hyperharmonic on G, then u is finely hyperharmonic on U.

Proof. Let $V \in \mathcal{B}_o(U)$, $p \in \mathcal{P}^f$ and $-p^{CV} \leqq u$ on V. According to Theorem 14.4, there is $\mathcal{V} \subset \mathcal{H}^*_+(V)$ such that $\wedge\mathcal{V} = 0$ and $u+v+p^{CV}$ is lower semicontinuous for every $v \in \mathcal{V}$. Pick $v \in \mathcal{V}$. Then $u+v+p^{CV} \in \mathcal{H}^*_+(V)$ by Theorem 12.20. Since the functions $u+p^{CV}$ and $\wedge(u+p^{CV}+\mathcal{V})$ are finely continuous on V and equal except a semipolar set, they coincide on V. Thus u is finely hyperharmonic on V, and also on U in light of Proposition 12.8.

EXERCISES

14.B.1 Let X be a harmonic space of Ex. 11.D.6. If

$$f = \chi_{(0,1]}, \quad g = \chi_{(0,1)},$$

then f is a nonnegative pointwise finely hyperharmonic function on X (even it is finely hyperharmonic on a fine neighborhood of each point of X). Both f and g are (finely) hyperharmonic outside the semipolar set $\{0\}$. However, neither f nor g is (finely) hyperharmonic on X. Compare with Theorems 14.8 and 14.11 and show that

(a) f is not X-quasi-l.s.c.,

(b) g is not finely continuous.

14.B.2 Natural sheaf property. Let u be a nonnegative function on a finely open set U. If u is finely hyperharmonic on some $\nu(U)$-neighborhood of each point of U, then u is finely hyperharmonic on U.

Hint. By Ex. 12.B.7b, u is $\nu(U)$-lower semicontinuous on U and it suffices to use Theorem 14.8.

14.B.3 Using the quasi-Lindelöf property of the fine topology or Choquet's lemma for $\nu(U)$ show that any evanescent family of subsets of U contains a countable evanescent subfamily.

14.B.4 Let U be a finely open subset of a harmonic space X on which the function 1 is superharmonic. Show that there is a real-valued "capacity" γ (= countably subadditive increasing nonnegative set function, $\gamma\emptyset = 0$) such that a family $\mathcal{K}$ is evanescent if and only if $\inf\{\gamma A: A \in \mathcal{K}\} = 0$.

Hint. By Ex. 8.3.9 of [C&C] there is a Radon measure μ on X such that every finely open set is μ-measurable and of positive measure and every semipolar set is of μ-measure zero. Obviously μ can be constructed to be finite. Put $\gamma A = \int_U c(A)\, d\mu$.

14.B.5 A subset A of a Borel finely open set $U \subset X$ is termed *U-polar* if $c(A) = 0$.

(a) Using the results given in Appendix show that U-polar sets coincide with polar sets determined by the standard H-cone $S'(U)$. In particular, all assertions stated in Ex. 10.C.3 can be stated in the language of U-polar sets.

(b) Consider the harmonic space of the heat equation on $\mathbb{R}^2$. Show that the set $\{(x;t) \in \mathbb{R}^2: t = 0\}$ is U-polar if $U = \{(x;t): t \lneq 0\}$ but it is not polar.

14.B.6 Quasi-continuity on non-Borel sets. If A is a subset of a finely open set $U \subset X$, put

$$c'(A) = \wedge\{v \in \mathcal{X}^*_+(U): \ v \geq 1 \ \text{ on } A\}$$

(cf. Corollary 14.10.b). Given $p \in \mathcal{P}^f$, there is a family $\mathcal{H}$ of subsets of U such that $\wedge\{c'(A): A \in \mathcal{H}\} = 0$ and $\hat{R}_p^{CU}$ is continuous on $U \setminus A$ for every $A \in \mathcal{H}$.

Hint. Let W be a Borel sweeping kernel of U. There is an evanescent family $\mathfrak{M}$ for W such that $\hat{R}_p^{CU} = \hat{R}_p^{CW}$ is continuous on $W \setminus G$ for every $G \in \mathfrak{M}$. Now, put $\mathcal{H} := \{G \cup (U \setminus W): \ G \in \mathfrak{M}\}$ and use Ex. 12.B.4.

REMARK. It would be possible to define a weaker concept of "quasi-continuity" by means of the "capacity" c'. Notice only that the implication (i)$\Longrightarrow$(ii) of Theorem 14.4 has no simple analogy in this more general setting.

14.B.7 Let f be a finely lower semicontinuous function on a Borel finely open set U. Assume that f is U-quasi-l.s.c., lower $\mathcal{P}^f$-bounded on U and $E(U)$-lower bounded (cf. Ex. 14.C.2). If for every open set $G \subset X$ and $x \in G \cap U$ there is $V \in \mathcal{B}(G \cap U)$ such that $x \in V$ and

$$\varepsilon_x^{CV}(f) \leq f(x)$$

then f is finely hyperharmonic on U.

Hint. Proceed via the following steps:

(1) Suppose $f \geq 0$, f is lower semicontinuous. Let $p \in \mathcal{P}^c$ be strict. Define

$$g := \begin{cases} f & \text{on } U \\ 0 & \text{on } CU. \end{cases}$$

Using Theorem 11.10 and Bauer's minimum principle of Ex. 12.C.1, prove that

$$u_n := \bigwedge_{V \in I(U)} \min(\hat{R}_{np}^{CV} + g, \ np) \in \mathcal{P}$$

(cf. Lemma 12.5). Now,

$$u_n - \hat{R}_{np}^{CU} \nearrow f$$

on U.

(2) Let f be lower semicontinuous, $f \geq -q \in \mathcal{P}^c$, and let u be an Evans function of q. Then $f + u \geq 0$ outside an intersection of U with a compact subset K of $\mathbf{X}$. Let g' be a lower semicontinuous function on K such that

$$-q \leq g' \text{ on } K \text{ and } g' = f + u \text{ on } U \cap K.$$

Use again Bauer's minimum principle of Ex. 12.C.1 for $\mathcal{F} = \mathcal{P}^f \cup (\mathcal{P}^f + g') \wedge K$ to show that $-\hat{R}_q^{CU} \leq f$ on U. Now, use (1) and the decomposition $f = (f+\hat{R}_q^{CU}) - \hat{R}_q^{CU}$.

(3) Now, in general case appealing to Corollary 14.10.a and Ex. 14.A.2 there is a family $\mathcal{V} \subset \mathcal{X}_+^{\#}(U)$ such that $\wedge\mathcal{V} = 0$ and $f+v$ is lower semicontinuous and lower $\mathcal{P}^c$-bounded for each $v \in \mathcal{V}$. Denote

$$w = \widehat{f + \inf \mathcal{V}^f}.$$

By Ex. 12.B.2, w is finely hyperharmonic on U. Let $q \in \mathcal{P}^f$, $-q \leq w$ on U. Then

$$-\bigwedge\{\hat{R}_q^{CW} : W \in \mathcal{B}(U)\}$$

is a finely harmonic minorant of w on U, so we may assume that $w \geq 0$. Realize that by Ex. 12.B.1 $\varepsilon_x^{CW}(w) = \varepsilon_x^{CW}(f + \inf \mathcal{V}) = \varepsilon_x^{CW}(f)$ for every $W \in \mathcal{B}(U)$. Hence also $f \geq 0$. Fix $x \in U$ and a decreasing local base $\{G_n\}$ of open neighborhoods of x. For every $n \in \mathbb{N}$ find $V_n \in \mathcal{B}(U \cap G_n)$ such that $x \in V_n$ and $\varepsilon_x^{CV_n}(f) \leq f(x)$. From Proposition 10.7 deduce that $\varepsilon_x = \text{w-lim}\, \varepsilon_x^{CV_n}$ and using Lemma 12.5

$$w(x) \leq \liminf \varepsilon_x^{CV_n}(w) = \liminf \varepsilon_x^{CV_n}(f) \leq f(x).$$

Hence $w = f$ and we are through.

REMARK. The assumption on the $E(U)$-lower boundedness cannot be dropped (consider $f = -u_o$, where u_o is the fundamental caloric function at 0, cf. the final remark of Section 10.A).

REMARKS AND COMMENTS

Although the finely hyperharmonic functions we are dealing with are not lower semicontinuous, results of this section provide conclusions which seem to be typical for the behaviour of lower semicontinuous functions. The root of it all lies in the observation that by performing some quasi-topological procedures (like the addition of an appropriate "small function" or omitting a "small set") we get lower semicontinuous functions on compact sets where Bauer's minimum principle is applicable.

The results of this section appeared in a preliminary text of J. Malý (1984) and were generalized by N. Boboc and Gh. Bucur (1985c) in the framework of standard H-cones for families of nonnegative functions. It is to be noted that similar ideas were originated by T.J. Lyons in (1982) and further developed in the probabilistic framework (private communication). Since in any harmonic space satisfying the domination axiom D every finely lower semicontinuous function is quasi-l.s.c., the issues referred-to are common generalizations of results of B. Fuglede, T.J. Lyons and the outcome of Section 12.C. Further relevant investigation can be found in Appendix.

Related paper : Y. Le Jan (1983)

14.C Quasi-solution of the Dirichlet problem

The Perron method solving the Dirichlet problem gives an answer to a natural query how to introduce a solution of the Dirichlet problem in case when the classical solution does not exist. Immediately a further question arises: To what extent the Perron solution h for f on a (finely) open set U conserves the property $\lim\limits_{U \ni x \to z} h(x) = f(z)$ which is required on the classical solution ? We already know that the notion of regular points plays an important role in the investigation of behavior of the Perron solution.

Now, another idea is at hands. Removing a set G of small capacity from U we can ask for the equality $\lim\limits_{x \to z} h(x) = f(z)$ for $x \in U \setminus G$ only. It should be pointed out the striking analogue with the Lusin theorem: A real measurable function becomes continuous removing an (open) set of arbitrarily small measure.

So, we introduce an alternative approach to the generalized (fine) Dirichlet problem weakening the classical concept of solution. Since the role of small sets is played by elements of evanescent families, our method is in fact quasi-topological. In comparison with the Perron method we need not repeat the whole procedure using the notion of (fine) superfunction. However, it is worth pointing out that considering families of quasi-superfunctions it is useful to compare the quasi-solution with the Perron solution. The key to understanding this procedure we speak about is a careful study of Ex. 14.C.2

Finally, notice that the results concerning the quasi-solution of the Dirichlet problem are nice applications of our ideas contained in previous sections.

Let U be a finely open set and let g be a function on $\overline{U} \setminus U$. Denote $(\mathcal{P}^f)^{CU} = \{ p^{CU} : p \in \mathcal{P}^f \}$. A function h on U is termed a *quasi-solution* for g if

(a) h is finely harmonic on U,

(b) h is $(\mathcal{P}^f)^{CU}$-bounded on U,

(c) there is an evanescent family $\mathcal{H}$ of sets open in U such that $\lim\limits_{U\setminus G\ni x\to z} h(x) = g(z)$ for every $G\in\mathcal{H}$ and $z\in\overline{U\setminus G}\setminus U$.

Similarly, a function s is termed a *quasi-superfunction* for g if s is finely hyperharmonic and lower $(\mathcal{P}^f)^{CU}$ - bounded on U and if there is an evanescent family $\mathcal{H}$ of sets open in U such that $g(z) \leqq \liminf\limits_{U\setminus G\ni x\to z} s(x)$ for all $G\in\mathcal{H}$ and $z\in\overline{U\setminus G}\setminus U$.

14.12 THEOREM. Let g be a function on $\overline{U}\setminus U$ such that g is $\mathcal{P}^f$-bounded on $\overline{U}\setminus U$ and ε_x^{CU}-integrable for every $x\in U$. Then the following two assertions hold:

(a) The function H_g^U is a quasi-solution for g.

(b) If s is a quasi-superfunction for g, then $s \geqq H_g^U$.

In particular, H_g^U is the unique quasi-solution for g.

Proof. Define the topology

$$\rho' := \{G\subset\overline{U} : G \text{ is finely open, } G\setminus U\subset \operatorname{int}_{\rho_0} G\}$$

where ρ_0 is the restriction of the initial topology on $\overline{U}$.

Denote by Φ the convex cone of all nonnegative functions on $\overline{U}$ which are finely hyperharmonic on U and by $\Phi^{\#}$ the set of all lower semicontinuous elements of Φ. It is simple to verify that Φ satisfies the axioms of Section 14.A (the axiom of upper directed set and the axiom of τ-lower semicontinuous regularization) provided we define the topology τ on $\overline{U}$ to be the fine topology on U and the discrete topology on $\overline{U}\setminus U$. We observe that a finely harmonic $(\mathcal{P}^f)^{CU}$-bounded function h on U is a quasisolution for a $\mathcal{P}^f$-bounded function g on $\overline{U}\setminus U$ if and only if the function which equals h on U and g on $\overline{U}\setminus U$ is quasi-ρ'-continuous on $\overline{U}$ w.r.t. Φ.

Denote now

$$\overline{g} := \begin{cases} g^{CU} & \text{on } U \\ g & \text{on } \overline{U}\setminus U. \end{cases}$$

(a) Let Υ be the set of all sums $s+t$ where s,t are lower finite, ρ'-lower semicontinuous on $\bar{U}$, finely hyperharmonic on U and $-t \lneq g \lneq s$ on $\bar{U} \setminus U$. We see by Theorem 13.7 that $\Upsilon \subset \Phi^{\#}$ and $\wedge\Upsilon = 0$. According to Theorem 14.4, $\bar{g}$ is Φ-quasi-ρ'-l.s.c. on $\bar{U}$. We deduce that H_g^U is a quasi-superfunction for g. Similarly, $-H_g^U$ is a quasi-superfunction for $-g$.

(b) Let s be a function on $\bar{U}$ such that s is a quasi-superfunction for g on U and $s = f$ on $\bar{U} \setminus U$. Find $p \in \mathcal{P}^f$ such that $s \geqq -p$ on $\bar{U}$ and $s \geqq -p^{CU}$ on U. Denote

$$\bar{p} := \begin{cases} p & \text{on } \bar{U} \setminus U \\ p^{CU} & \text{on } U. \end{cases}$$

According to (a), $\bar{p}$ is Φ-quasi-ρ'-continuous on $\bar{U}$. Hence, $s+\bar{p}$ is a nonnegative Φ-quasi-ρ'-l.s.c. function on $\bar{U}$. By Theorem 14.4 there is a set $\Upsilon \subset \Phi^{\#}$ such that $\wedge\Upsilon = 0$ and $s+\bar{p}+v$ is ρ'-lower semicontinuous for every $v \in \Upsilon$. Choose $v \in \Upsilon$. Since $s+\bar{p}+v$ is a fine superfunction to $g+p$, we have

$$s+\bar{p}+v \geqq \bar{g}+\bar{p} \quad \text{on} \quad \bar{U} \setminus U,$$

$$s+\bar{p}+v \geqq g^{CU} + p^{CU} = \bar{g}+\bar{p} \quad \text{on} \quad U.$$

Thus, $s+v \geqq \bar{g}$ on $\bar{U}$. Hence

$$s = \bigwedge_{\Phi} \{ s+v : v \in \Upsilon \} \geqq \bar{g} \quad \text{on} \quad \bar{U}.$$

(Indeed, these relations hold on the union $\bar{U} \setminus U$ with a finely dense subset of U, and the functions in considerations are finely continuous on U.)

14.13 COROLLARY. (The quasi-minimum principle). Let u be a lower $(\mathcal{P}^f)^{CU}$-bounded finely hyperharmonic function on U. If there is an evanescent family $\mathcal{H}$ of sets open in U such that

$$\liminf_{U\setminus G \ni x \to z} u(x) \geqq 0$$

for every $G \in \mathcal{H}$ and every $z \in \overline{U \setminus G} \setminus U$, then $u \geqq 0$ on U.

Proof. Apparently, u is a quasi-superfunction to 0 on U.

We close this section by some observations concerning harmonic spaces X satisfying the axiom D. Nowadays, the following theorem ranks among almost classical results: Any function on X which is finely lower semicontinuous on X except a polar set, is quasi-l.s.c. (cf. [Fug] , 5.5).

14.14 PROPOSITION. Assume X satisfies the axiom D. Let u be a pointwise finely hyperharmonic function on a finely open set U. If for every $V \in \mathcal{B}(U)$ there is $p \in \mathcal{P}^f$ such that $-p^{CV} \leqq u$ on V, then u is finely hyperharmonic on U.

Proof. Define $u_V = u$ on the fine closure $\bar{V}^f$ of V and $u_V = +\infty$ elsewhere. Then u_V is finely lower semicontinuous on X, and therefore it is quasi-l.s.c. on X. By Corollary 14.9 one reaches the conclusion.

14.15 COROLLARY. If a harmonic space satisfies the domination axiom D, then the fine sheaf property holds on X.

14.16 PROPOSITION. Assume that X satisfies the domination axiom D. Let s be a finely hyperharmonic function on a finely open set $U \subset X$. Suppose there is a potential $p \in \mathcal{P}^f$ such that $s \geqq -p^{CU}$. If $\text{fine-}\liminf_{x \to y} s(x) \geqq 0$ for every $y \in \partial_f U$ except a polar set, then $s \geqq 0$ on U.

Proof. Set

$$f := \begin{cases} \min(s,0) & \text{on } U, \\ 0 & \text{on } CU. \end{cases}$$

Then f is finely lower semicontinuous on X, except a polar set, hence it is quasi-l.s.c. and we can use Corollary 14.13 to deduce $s \geqq f \geqq 0$ on U.

REMARK. This is a slight modification of Fuglede's fine boundary minimum principle ([Fug] , Th. 9.1., cf.B. Fuglede (1974); the difference concerning the lower boundedness assumptions will be weakened in Ex. 14.C.2). Now, following the same lines as in [Fug], Section 14, the investigation of the fine Dirichlet problem on the fine boundary can be developed. We mention only an immediate consequence of Theorem 14.12.

14.17 PROPOSITION. Assume that X satisfies the domination axiom D. Let U be a finely open set and let f be a function on its fine boundary. Assume that f is ε_x^{CU}- integrable for every $x \in U$ and there is a potential $p \in \mathcal{P}^f$ such that $|f| \leqq p$ on $\partial_f U$. Then there is a unique function h finely harmonic on U such that

$|h| \leqq p^{CU}$ and $\operatorname{fine-lim}_{x \to y} h(x) = f(y)$ for all $y \in \partial_f U$ except a set of ε_x^{CU}-measure zero for each $x \in U$.

EXERCISES

14.C.1 Let X be the harmonic space of Ex. 11.D.6. If $u(x) = 2x-1$ for $x \in U = (0,\frac{1}{2})$, then u is a (finely) harmonic function on a relatively compact open subset of X, $\operatorname{fine-lim}_{U \ni x \to y} u(x) = 0$ for every $y \in \partial_f U$ but $u < 0$ on U. Show directly that u is not a quasi-superfunction for 0.

14.C.2 This exercise is an improvement of Theorem 14.12 by relaxing the assumption of lower boundedness. A function f on a finely open set U is said to be *E(U)-lower bounded* if there is a family $\mathcal{V} \subset \mathcal{H}_+^*(U)$ such that $\wedge\mathcal{V} = 0$ and $f+v$ is lower $\mathcal{P}^C$-bounded on U for every $v \in \mathcal{V}$. In particular, if $f \geqq -q$ for some $q \in \mathcal{M}$ (cf. Ex. 11.D.4), then f is $E(U)$-lower bounded.

The following two assertions on a finely resolutive function f for U and a finely harmonic function h on U are equivalent:

(i) $h = H_f^U$,

(ii) h is $E(U)$-bounded and there is an evanescent family $\mathcal{H}$ of open sets in U such that $\lim_{U \setminus G \ni x \to z} h(x) = f(z)$ whenever $G \in \mathcal{H}$ and $z \in \overline{U \setminus G} \setminus U$.

14.C.3 A function h defined on a Borel finely open set U is termed *U-quasi-uniformly continuous* if there is an evanescent family $\mathcal{H}$ of open sets in U closed under finite intersections such that h has a real continuous extension to $\overline{U \setminus G}$ for every $G \in \mathcal{H}$.

(a) Let f be a finely resolutive function for U. Show that there is an evanescent family $\mathcal{H}$ of open sets in U which is closed under finite intersections such that the function

$$g = \begin{cases} f & \text{on } \overline{U} \setminus U \\ H_f^U & \text{on } U \end{cases}$$

is continuous on $\overline{U \setminus G}$ for every $G \in \mathcal{H}$. In particular, H_f^U is U-quasi-uniformly continuous.

(b) Conversely, let h be a U-quasi-uniformly continuous and finely harmonic function on U. If h is $E(U)$-bounded, then there is a function f on $\overline{U} \setminus U$

such that $h = H_f^U$.

Hint. Let $\mathcal{H}$ be an evanescent family from the definition of the U-quasi-uniform continuity of h. Show that

$$\lim_{U \setminus G_1 \ni x \to z} h(x) = \lim_{U \setminus G_2 \ni x \to z} h(x)$$

whenever $G_1, G_2 \in \mathcal{H}$ and $z \in \overline{U \setminus G_1} \cap \overline{U \setminus G_2}$. Define $f(z)$ to be this limit for $z \in \bigcup_{G \in \mathcal{H}} \overline{U \setminus G}$ and $f(z) = 0$ elsewhere.

REMARKS AND COMMENTS

In this section, a quite new approach to the Dirichlet problem is suggested. We employ the construction just accomplished to provide one of various ways of solving the Dirichlet problem. The method of quasi-solutions serves a possibility how to cover both Fuglede's approach and the results of Section 13.A.

APPENDIX: Fine hyperharmonicity and the Dirichlet problem in standard H-cones

A. Localization

B. The Dirichlet problem

C. Pointwise fine hyperharmonicity

The object of this Appendix is to present fine potential theory in the framework of standard H-cones. Thus, throughout S is a standard H-cone of functions on X having the measure sweeping property of Ex. 11.B.1, and U is a finely open subset of X. Let S^f stands for the family of all finite elements of S and $S^c = S^f \cap \mathcal{C}(X)$.

We fix a family $\mathbb{B}$ of finely open subsets of U and denote $U_{\mathbb{B}} = \bigcup \mathbb{B}$.

If V is a subset of X, let $\Sigma(V)$ be the σ-algebra of those subsets of X which are ε_x^{CV}-measurable for all $x \in V$. Put $\Sigma(\mathbb{B}) = \bigcap\{\Sigma(V) : V \in \mathbb{B}\}$.

A lower finite function u on X is said to be <u>$\mathbb{B}$-finely hyperharmonic</u> on U if it is finely lower semicontinuous, $\Sigma(\mathbb{B})$-measurable and if

$$\varepsilon_x^{CV}(u) \leqq u(x)$$

for every $V \in \mathbb{B}$ and $x \in V$. A function h on X is called <u>$\mathbb{B}$-finely harmonic</u> on U if both h and $-h$ are $\mathbb{B}$-finely hyperharmonic on U.

Consider now a further fine topology t on X which is coarser than the fine topology f, and denote

$$\mathcal{P}' = \{p \in S^f : p \text{ is t-continuous and } \bigwedge\{\hat{R}_p^{W \setminus V} : V \in \mathbb{B}\} = 0 \text{ for every finely open set } W \text{ with } \overline{W}^t \subset U\}.$$

A function g on X is said to be <u>lower $\mathcal{P}'$-bounded</u> if there is $p \in \mathcal{P}'$ such that $-p \leqq g$ on X.

We say that $X = (X, S, U, \mathbb{B}, t)$ is a <u>$\mathcal{P}'$-space</u> if $\mathcal{P}'$ contains a positive element.

<u>EXAMPLES</u>. The most important examples of the topology t which we bear in mind above all are the choices where

- t is the initial topology,
- t is the fine topology,
- t is the weak topology induced on X by the family

$$\{s : s \in S_0\} \cup \{R_s^{CU} : s \in S_0\}.$$

(To gain some insight into the significance of the last topology t, we note that this one induces on U the natural topology $\nu(U)$ w.r.t. $S'(U)$ (cf. Theorem A-10) and describes the convergence along regular filters at the points of $\overline{U} \setminus U$ (cf. Ex. 13.A.6).)

Similarly as in the framework of harmonic spaces we can consider $\mathbb{B}$ to be the collection of all relatively compact finely open sets V with $\overline{V}^t \subset U$. Then there is a narrow analogy with the definition of the fine hyperharmonicity of Chapter 12 provided t equals the initial topology. There is a difference, but only a small one: When considering standard H-cones, all functions in question are defined on the whole space X. On the other hand, the choice of t being the fine topology enables us to investigate the Dirichlet problem using fine limits on the fine boundary of U. (Note that in case of harmonic spaces without the domination axiom D such a choice of $\mathbb{B}$ is artificial since we must then break the convention of U being the domain of finely hyperharmonic functions in order to have finely hyperharmonic functions defined ε_x^{CV}-almost everywhere for all $V \in \mathbb{B}$ and $x \in V$.)

We will introduce here two more examples showing that the above mentioned choice of $\mathbb{B}$ consisting of relatively compact sets may have serious disadvantages.

Let S be the standard H-cone of all nonnegative, lower semicontinuous and finite functions s on $X := (0,1] \cup (2,3]$ such that the function $x \mapsto xs(x)$ is increasing on X, and let t be the Euclidean (=initial) topology on X. If $U = X$ and $\mathbb{B}$ is the collection of all relatively compact finely open sets V with $\overline{V} \subset U$, then every element of $\mathcal{P}'$ vanishes on $(2,3]$ and X is not a $\mathcal{P}'$-space.

Let now

$$X = \bigcup_{n=1}^{\infty} (1-\frac{1}{2n}, 1-\frac{1}{2n+1}] \cup \{1\} ,$$

and let S be again the standard H-cone of all nonnegative, lower semicontinuous and finite functions s on X such that $x \mapsto xs(x)$ is increasing on X. Then X is saturated and it is easy to see that the point 1 has no compact fine neighborhood. Thus, $X_{\mathbb{B}} \neq X$ for $\mathbb{B}$ as in the preceding example.

Let X, S, t and U be given. Choosing $\mathbb{B}$ as the family of all finely open sets V with $\overline{V}^t \subset U$, the function $\mathbb{1}$ is obviously a positive element of $\mathcal{P}'$ and $U_{\mathbb{B}} = U$. Thus we have always the means at hand to fill up $\mathbb{B}$ in order to satisfy the requirements that X is a $\mathcal{P}'$-space and $U_{\mathbb{B}} = U$. Before proceeding, a few words are in order. The choice of $\mathbb{B}$ just indicated does not guarantee the coincidence between $\mathbb{B}$-finely hyperharmonic and finely hyperharmonic functions in case of harmonic spaces.

A Localization

Assume that the family $\mathbb{B}$ is upper directed when ordered by inclusion, and that $\mathbb{B}$ contains any finely regular set which is a subset of some element of $\mathbb{B}$.

Let $\mathcal{D}(U)$ be the family of all differences $s - R_s^{CU}$ where $s \in S^f$; $\mathcal{D}^{\uparrow}(U)$ be, as usually, the family of all limits of increasing sequences from $\mathcal{D}(U)$, and finally let

$$S'(U) := \{ f/\wedge U : f \in \mathcal{D}^{\uparrow}(U) \text{ and } f < +\infty \text{ on a finely dense subset of } U \}.$$

Following up N. Boboc and A. Cornea (1975), N. Boboc and Gh. Bucur recently proved in (1985a) that $S'(U)$ is always a standard H-cone of functions on the (properly topologized) set U and clarified in (1985c) the relation between $S'(U)$ and the family of all $\mathbb{B}$-finely hyperharmonic functions for a suitable family $\mathbb{B}$. Proofs of their results make substantial use of the theory of resolvents. Notice also that a similar result was obtained by B. Fuglede (1982) in the framework of $\mathfrak{P}$-harmonic spaces satisfying the domination axiom D. Our aim is to present analytic proofs the heart of which lies in the use of a principal operator. (Notice that in another context a similar idea is used studying Keldych operators, cf. Ex. 13.B.2.)

A-1 Bliedtner-Hansen lemma. Let V be a $\Sigma(V)$-measurable finely open subset of U. Then

$$\varepsilon_x^{CV}(R_s^{CU}) = R_s^{CU}(x)$$

whenever $s \in S^f$ and $x \in U$.

Proof. Use the idea of the proof of Proposition 11.7. Cf. also Gh. Bucur and W. Hansen (1984).

A-2 COROLLARY. If $U \in \Sigma(\mathbb{B})$ and $p \in S^f$, then R_p^{CU} is $\mathbb{B}$-finely harmonic on U (cf. Lemma 12.7).

A-3 COROLLARY. If $U \in \Sigma(\mathbb{B})$ and $s \in \mathcal{D}^{\uparrow}(U)$, then s is $\mathbb{B}$-finely hyperharmonic on U.

For the further relations between $\mathcal{D}^{\uparrow}(U)$ and the family of all $\mathbb{B}$-finely hyperharmonic functions on U we proceed in the following steps.

A-4 LEMMA. For any $p \in \mathcal{P}'$,

$$\bigwedge_{V \in \mathbb{B}} \hat{R}_p^{CV} = \hat{R}_p^{CU} .$$

Proof. Use the idea of the proof of Proposition 10.20 and the inequality $\hat{R}_p^{CV} \leq \hat{R}_p^{W \setminus V} + \hat{R}_p^{CW}$ where W is as in the definition of $\mathcal{P}'$.

A-5 LEMMA. Let $p \in S^f$. If $W \in \mathbb{B}$ is finely regular and $u \geq 0$ is a $\mathbb{B}$-finely hyperharmonic function on U, then the function $q := \min(R_p^{CW} + u, p)$ belongs to S.

Proof. Since $R_p^{CW} = \hat{R}_p^{CW}$ and $q = p$ on CW, q is finely lower semicontinuous on X. Let $\mathcal{F} := \{f\colon f$ is a finely continuous function on X, $R_p^{CW} \leq f \leq q$, $\sup(f - R_p^{CW}) < +\infty\}$. Put

$$r = r_f = \bigwedge\{s \in S\colon s \geq f\}.$$

The proof that $q \in S$ is achieved by showing that $f \leq r_f \leq q$. Granting this it will follow that

$$q = \bigvee\{r_f : f \in \mathcal{F}\}$$

since $q = \sup \mathcal{F}$ in view of the complete regularity of the fine topology.

Take now any $f \in \mathcal{F}$ and any $\varepsilon > 0$. By Proposition 10.19 there is a finely regular set V such that

$$[f+\varepsilon < r] \subset V \subset [f+\varepsilon \leqq r] (\subset [f < r] \subset [R_p^{CW} < p] \subset W).$$

It follows that $V \in \mathbb{B}$. We now wish to show that $R_r^{CV} = r$. Suppose $c := \sup(r - R_r^{CV}) > 0$. Then $R_r^{CV} + b \geqq f$ for a suitable $b > 0$, $c - \varepsilon < b < c$. Hence $R_r^{CV} + b \geqq r$ and $b \geqq c$, which yields the required conclusion. Since $r \leqq f+\varepsilon \leqq q$ on CV, we get

$$r = R_r^{CV} \leqq (f+\varepsilon)^{CV} \leqq f+\varepsilon$$

on V. Letting $\varepsilon \to 0$, we thus have $r \leqq f$.

Guided by the definition of principal balayage of Ex. 13.B.2, we consider now its analogue in a more general setting of standard H-cones. If $p \in S$ and f is a function on U dominated by some element of S, put

$$A(p,f) := \bigwedge \{t \in S : t \geqq \hat{R}_p^{CU} \text{ on } X \text{ and } t \geqq \hat{R}_p^{CU} + f \text{ on } U\}$$

and define the <u>principal modification of A</u> by

$$P(p,f) := \bigvee \{q \in S : q \leqq \min(p, A(q,f))\}.$$

A routine argument shows that $A(P(p,f),f) = P(p,f)$ whenever $\hat{R}_p^{CU} + f \leqq p$ on U.

A-6 <u>THEOREM</u>. Let $p \in \mathcal{P}'$ and let $u \geqq 0$ be a $\mathbb{B}$-finely hyperharmonic function on U. If u is finely continuous at all points of $U \setminus U_{\mathbb{B}}$, then

(a) there is $q \in S$ such that $\hat{R}_p^{CU} \leqq q \leqq p$ on X and $q = \min(p, \hat{R}_p^{CU} + u)$ on U,

(b) there is $d \in \mathcal{D}(U)$ such that $d = \min(p - \hat{R}_p^{CU}, u)$ on U.

<u>Proof</u>. (a) Put $q = A(p, \min(p - \hat{R}_p^{CU}, u))$ and proceed as in the proof of Lemma 12.5 making substantial use of Lemmas A-4 and A-5.

(b) Denote $f = \min(p - \hat{R}_p^{CU}, u)$ on U, $s = P(p,f)$ and put $d = s - \hat{R}_s^{CU}$. Then $s \in \mathcal{P}'$, $d \in \mathcal{D}(U)$ and $\hat{R}_p^{CU} + f \leqq p$ on U. Hence

$$s = A(s,f) \geqq \hat{R}_s^{CU} + f$$

on U, and thus by (a) there is $q \in S$ such that $\hat{R}_s^{CU} \leqq q \leqq s$ on X and $q = \min(\hat{R}_s^{CU} + f, s) = \hat{R}_s^{CU} + f$ on U. Therefore $q \geqq A(s,f)$. It follows that

$$\hat{R}_s^{CU} + f = q \geqq A(s,f) = s \geqq \hat{R}_s^{CU} + f$$

on U, and we have $s - \hat{R}_s^{CU} = f$ on U, as needed.

A-7 <u>COROLLARY</u>. If u is a $\mathbb{B}$-finely hyperharmonic function on U, then u is finely continuous on $U_{\mathbb{B}}$.

Proof. It suffices to imitate the proof of Theorem 12.6 using (a) of the previous theorem.

If $p > 0$ is a positive element of $\mathcal{P}'$, it should be clear that every min $(p,\mathbb{1})$-continuous generator is an element of $\mathcal{P}'$ being a $\mathbb{1}$-continuous generator. Granting this the next theorem easily follows from (b) of Theorem A-6.

A-8 THEOREM. Let X be a $\mathcal{P}'$-space. Any function on X which is finely continuous and $\mathbb{B}$-finely hyperharmonic on U and vanishes on CU belongs to $\mathcal{D}^{\uparrow}(U)$. In particular, if $U \in \Sigma(\mathbb{B})$ and $U = U_{\mathbb{B}}$, then $\mathcal{D}^{\uparrow}(U)$ is exactly the family of all nonnegative $\mathbb{B}$-finely hyperharmonic functions vanishing on CU.

In similar fashion as in Section 12.B define the topology $\nu(U)$ as the coarsest topology on U which is finer than the initial one and which makes all functions from $S'(U)$ lower semicontinuous. This topology is the weak topology induced on U by the family $S^c \cup \{R_s^{CU} : s \in S^c\}$ (cf. Theorem 12.4).

A-9 LEMMA. Any function $s \in S'(U)$ is the limit of an increasing sequence of functions from $S'(U)$ which are $\nu(U)$-continuous.

Proof. We can express s as the limit of an increasing sequence of functions of the form $A(p, d-\hat{R}_p^{CU}) \wedge U$, where $p, d \in S^c$. Let now $p, d \in S^c$ and fix $z \in U$. Consider $u \in S$, $\max(\hat{R}_p^{CU}, d) \leqq u \leqq p$. We conclude the proof by showing that for every such a function u and every $\varepsilon > 0$ we can find $v \in S$ such that v is $\nu(U)$-continuous at z and

$$\max(\hat{R}_p^{CU}, d) - \varepsilon \leqq v \leqq u.$$

According to Proposition 10.7,

$$u = \sup\{\hat{R}_u^{CW \cup \{z\}} : W \text{ is a } \nu(U)\text{-neighborhood of } z\}.$$

There is a $\nu(U)$-neighborhood W of z such that

$$\hat{R}_u^{CW \cup \{z\}}(z) \geqq d(z) - \varepsilon$$

and a finely open $\nu(U)$-neighborhood $V \subset W$ of z such that

$$\hat{R}_u^{CV \cup \{z\}} \geqq \hat{R}_u^{CW \cup \{z\}} \geqq d - \varepsilon$$

on $\overline{V}^f$. (Of course, $\hat{R}_u^{CV \cup \{z\}} \geqq u \geqq d - \varepsilon$ on $X \setminus \overline{V}^f = bCV$.) Put

$$v = \min(u, \hat{R}_u^{CV} + u(z) - \hat{R}_u^{CV}(z)).$$

Analogously as in Ex. 12.B.7b, we have $\nu(U) \wedge V = \nu(V)$ and hence v is $\nu(U)$--continuous at z. We have $v \leqq u$ on $CV \cup \{z\}$, and thus

$$v \geqq \hat{R}_u^{CV \cup \{z\}} \geqq d - \varepsilon.$$

A-10 THEOREM. $S'(U)$ is a standard H-cone of functions on $(U, \nu(U))$.

Proof. According to the existence of a Borel sweeping kernel of U and Ex. 12.B.4

we reduce the general case to the case when U is a Borel set. Put $\mathbb{B} = I(U)$ and let t be the initial topology on U. Then $U = U_{\mathbb{B}}$, and so $S'(U)$ is exactly the set of restrictions to U of all $\mathbb{B}$-finely hyperharmonic functions on U which are finite on finely dense subsets of U. We can easily check that $S'(U)$ satisfies the axiom (S_2) of upper directed sets, the axiom of fine lower semicontinuous regularization (from which the axiom (S_3) follows) and has the Riesz decomposition property. The proofs are almost the same as those given in Section 12.B and we therefore omit them.

Let now p be a $\mathbb{1}$-continuous generator and D be a countable increasingly dense subset of S_0. Put

$$E := \{[A(np, d-\hat{R}^{CU}_{np}) - \hat{R}^{CU}_{np}] \wedge U:\ n \in \mathbb{N},\ d \in D\}.$$

Then E is countable and $s = \bigvee\{t \in E:\ t \leqq s\}$ for every $s \in S'(U)$. As $E \subset S'(U)$, the axiom (S_6) of standardness will follow by Prop. 4.2.2 of [BBC], provided we can prove that each element of E is $\mathbb{1}$-continuous. So let $T \subset S'(U)$ be increasing to $q := A(np, d-\hat{R}^{CU}_{np})-\hat{R}^{CU}_{np}$ where $n \in \mathbb{N}$, $d \in D$. For each $t \in T$ there is $s_t \in S$ such that

$$\hat{R}^{CU}_{np} \leqq s_t \leqq np \quad \text{on } X \quad \text{and} \quad s_t = \hat{R}^{CU}_{np} + t \quad \text{on } U.$$

Then the set $\{s_t \wedge d:\ t \in T\}$ is increasing to d. Since $d \in S_0$, given $\varepsilon > 0$ there is $t_\varepsilon \in T$ such that $(s_{t_\varepsilon} \wedge d) + \varepsilon \geqq d$ on X. It can be easily checked that $t_\varepsilon + \varepsilon \geqq q$ on U.

It remains to show that the topology $\nu(U)$ is the weak topology ν induced on U by the family $(S'(U))_0$ (which is called the <u>natural topology</u> on U w.r.t. $S'(U)$). We infer to Lemma A-9 that any element of $(S'(U))_0$ is $\nu(U)$-continuous. Let now $s \in S_0$, $s \leqq 1$. Since the restrictions of $\hat{R}^{CU}_s$ and $\mathbb{1}-\hat{R}^{CU}_s$ belong to $S'(U)$, they are ν-lower semicontinuous. It follows from the previous considerations that the functions $\hat{R}^{CU}_s$ and

$$s-\hat{R}^{CU}_s = A(s, s-\hat{R}^{CU}_s) - \hat{R}^{CU}_s$$

e. $\mathbb{1}$-continuous. Hence s is ν-continuous, and therefore $\nu = \nu(U)$.

A-11 <u>THEOREM</u>. If U is a Borel set, then $S'(U)$ has the measure sweeping property.

<u>Proof</u>. Let $p \in S^f$. According to Theorems A-1 and A-6, given $A \subset U$ and $x \in U \setminus (A \setminus bA)$,

$$\bigwedge\{t \in S'(U):\ t \geqq p-R^{CU}_p \ \text{ on } A\}(x) = R^{CU\cup A}_p(x) - R^{CU}_p(x) =$$
$$= \varepsilon^{CU\cup A}_x(p) - \varepsilon^{CU\cup A}_x(R^{CU}_p) = (\varepsilon^{CU\cup A}_x \wedge U)(p-R^{CU}_p).$$

It follows that

$$\bigwedge\{t \in S'(U):\ t \geqq s \ \text{ on } A\}(x) = (\varepsilon^{CU\cup A}_x \wedge U)(s)$$

for any $s \in S'(U)$.

B The Dirichlet problem

Throughout X is a $\mathcal{P}'$-space, $U \in \Sigma(\mathbb{B})$, $\mathbb{B}$ satisfies the assumptions formulated in Section A and

$$\hat{R}_p^{CU} = \sup\{\hat{R}_p^{CW} : U \subset W \text{ and } \hat{R}_p^{CW} \text{ is t-continuous on } U\}$$

for any $p \in S^c$. (The reader should realize that the last "resolutivity assumption" is satisfied if, for example, t is finer than $\mathcal{V}(U)$ or if $U \in \Sigma(U)$ - cf. Ex. 11.B.1c.)

We will take the time now to present a Perron type solution of the Dirichlet problem. Again, the following minimum principle is a toe in the door for the application of Ex. 11.C.1.

A-12 The minimum principle. Let u be a lower $\mathcal{P}'$-bounded and t-lower semicontinuous function on X which is $\mathbb{B}$-finely hyperharmonic and finely continuous on U. If $u \geqq 0$ on CU, then $u \geqq 0$ everywhere on X.

Proof. Let $p \in \mathcal{P}'$ be such that $-p \leqq u$ on X. Pick $x \in U$ and $\varepsilon > 0$. The set $M := [u+\varepsilon < 0]$ is finely open since u is finely continuous and, of course, $\overline{M}^t \subset [u+\varepsilon \leqq 0] \subset U$. Hence, $\bigwedge_{V \in \mathbb{B}} \hat{R}_p^{M \setminus V} = 0$. If $V \in \mathbb{B}$ and $v \in S$ are chosen so that $v \geqq p$ on $M \setminus V$, then $u+v+\varepsilon \geqq 0$ on CV, and therefore

$$0 \leqq \varepsilon_x^{CV}(u+v+\varepsilon) \leqq u(x)+v(x)+\varepsilon .$$

We see that $u(x) + \hat{R}_p^{M \setminus V}(x) + \varepsilon \geqq 0$. According to the definition of $\mathcal{P}'$ we get $u(x) \geqq 0$.

We turn now to an important application of Ex. 11.C.1 to obtain the Perron type solution of the Dirichlet problem even in standard H-cones of functions. To begin with, let $\mathcal{U}$ be the family of all t-lower semicontinuous and $\mathcal{P}'$-lower bounded functions on X which are $\mathbb{B}$-finely hyperharmonic and finely continuous on U. If $u_n \in \mathcal{U}$, $u_n \geqq 0$, then $u_n \wedge U \in S'(U)$, and therefore $\sum u_n \in S'(U)$. Hence, the function $\sum u_n$ is finely continuous on U and the Lebesgue monotone convergence theorem implies $\sum u_n \in \mathcal{U}$. Since also the minimum principle holds for the functions of $\mathcal{U}$, we can apply all methods and results developed in Ex. 11.C.1 and in Section 13.A to the present situation. In particular, if f is a function on X and

$$\overline{H}_f^U = \inf\{u \in \mathcal{U} : u \geqq f \text{ on } CU\}, \quad \underline{H}_f^U = -\overline{H}_{-f}^U,$$

we have the next basic result.

A-13 THEOREM. If a function f is ε_x^{CU}-integrable for all $x \in U$, then f is "resolutive" and

$$\underline{H}_f^U(x) = \overline{H}_f^U(x) = \varepsilon_x^{CU}(f)$$

for every $x \in U$.

Now, we would like to apply some results of Section 14.A to the convex cone $S'(U)$. The concept of a capacity there introduced is convenient for the examination of properties of $S'(U)$. For some other purposes, like for the investigation of the Dirichlet problem, we need to define a more complicated notion of a capacity. Thus, although we cannot apply the results of Section 14.A directly, we can proceed analogously in the following situation:

- Φ will be the convex cone $S'(U)$ (Φ obviously satisfies the axioms of upper directed sets and fine - lsc. regularization),
- $\Phi^\#$ will be the family of all t-lower semicontinuous functions ≥ 0 on X which are $\mathbb{B}$-finely hyperharmonic on U,
- <u>t-capacity</u> of a set $A \subset X$ (not only $A \subset U$!) is defined as

$$c(A,U,t) = \bigwedge_\Phi \{t \wedge U:\ t \in \Phi^\#,\ t \geq 1 \text{ on } A\},$$

- a family $\mathcal{G}$ of subsets of X is termed <u>t-evanescent</u> if $\bigwedge\{c(G,U,t):\ G \in \mathcal{G}\} = 0$,
- a function f on X is termed <u>U-quasi-t-lower semicontinuous</u> if there is a t-evanescent family $\mathcal{G}$ of t-open subsets of X such that $f/\complement G$ is t-lower semicontinuous whenever $G \in \mathcal{G}$,
- a function h on X is <u>$E'(U)$-bounded</u> if

$$\bigwedge_\Phi \{t \wedge U:\ t \in \Phi^\#,\ -|h|+t \text{ is lower } \mathcal{P}'\text{-bounded}\} = 0$$

(analogously $E'(U)$-lower bounded).

Having these definitions in mind, one of the main results is the following.

A-14 <u>THEOREM</u>. Let f be a real function on X which is ε_x^{CU}-integrable for each $x \in U$. Then there is a unique $E'(U)$-bounded function h on X such that

(a) h is U-quasi-t-continuous on X,

(b) h is $\mathbb{B}$-finely harmonic on U,

(c) $h = f$ on CU.

For the proof of this theorem proceed as in Ex. 14.C.2 and 14.C.3. The details are left to the reader.

A-15 <u>COROLLARY</u>. Assume $U \in \mathbb{B}$. Then any $\mathbb{B}$-finely hyperharmonic function u on U is U-quasi-t-lower semicontinuous.

<u>Proof</u>. Put $h = \overline{H}_u^U$. By the preceding theorem, h is U-quasi-t-continuous and $u-h \in S'(U)$ is U-quasi-t-lower semicontinuous by an analogous argument as in Theorem 14.8.

A-16 REMARK. Let $U \in \mathbb{B}$ and $W \subset U$. Then the following assertions are equivalent:

(i) $c(U \setminus W, U, t) = 0$ (i.e. $U \setminus W$ is (U,t)-polar),

(ii) W is a sweeping kernel of U,

(iii) $U \setminus W$ is polar with respect to $S'(U)$.

It follows (analogously as in Ex. 12.B.4) that $U \setminus W$ is a removable singularity: Any function of $S'(W)$ is a restriction of some function from $S'(U)$.

C Pointwise fine hyperharmonicity

Throughout we assume that t is the initial topology and $\mathbb{B}$ is formed by all finely open sets whose closure is compact and lies in U.

If X is a $\mathfrak{P}$-harmonic space, then a function f is $\mathbb{B}$-finely hyperharmonic on U exactly when f is finely hyperharmonic on U in the sense of Section 12.A.

It would be possible to state some results analogous to those of Section 14.B. Instead, we present one result of this type only leaving other variations to the reader.

A-17 THEOREM. Let X be a $\mathcal{P}'$-space, $U \in \Sigma(U)$ and let f be a U-quasi-l.s.c., $\Sigma(U)$-measurable and $E'(U)$-bounded function on X. If for each $x \in U$ there is a finely open set V, $x \in V \subset U$ such that $\varepsilon_x^{CV}(f)$ is defined and $\varepsilon_x^{CV}(f) \leqq f(x)$, then $\int^* f \, d\varepsilon_z^{CU} \leqq f(z)$ for every $z \in U$.

Proof. Similarly as in Ex. 14.B.7 there is $\mathcal{V} \subset \Phi^\#$ such that

$$\bigwedge \{v \wedge U : v \in \mathcal{V}\} = 0$$

and for any $v \in \mathcal{V}$ the function $f+v$ is lower semicontinuous and lower $\mathcal{P}'$-bounded. We may even assume that for each $v \in \mathcal{V}$ there is a fine cozero set U_v such that $v = +\infty$ on $U \setminus U_v$. (Indeed, having $\mathcal{V}$ with properties ascribed, we can add up to functions of $\mathcal{V}$ functions of the form

$$\sum_{n=1}^{\infty} (\hat{R}_p^{CV_n} - R_p^{K_n})$$

where $p \in S^c$ is a generator, V_n are finely regular, $\overline{V}_n \subset U$ and $K_n \subset CU$ are closed.) Fix $z \in U$ and a finely open set $W \subset U$ such that $z \in W$ and $\varepsilon_z^{CW}(f) \leqq f(z)$. Choose $v \in \mathcal{V}$. By Theorem A-13 we have

$$\varepsilon_z^{CU}(f+v) - \sup\{-t(z) : t \in \mathcal{T}\}$$

where $\mathcal{T}$ is the set of all lower semicontinuous lower $\mathcal{P}'$-bounded functions which are $\mathbb{B}$-finely hyperharmonic on U. Let $t \in \mathcal{T}$ be fixed. Then t is a supersolution to itself on W, and hence $\varepsilon_z^{CW}(t) \leq t(z)$. The set

$$M := [f+v+t+\varepsilon \leqq 0]$$

is closed and $M \subset U_v$. By the "Lusin-Menchoff property" of Theorem 10.24 there is a finely regular set G such that $M \subset G \subset \overline{G} \subset U_v$. Find $p \in \mathcal{P}'$ such that $-p \leqq t+f+v$. Then

$$\wedge\{\hat{R}_p^{G \setminus K} : K \subset X \text{ is compact}\} = 0.$$

Fix a compact set $K \subset X$ and $u \in S$, $u \geqq p$ on $G \setminus K$ (so $t+f+v+u+\varepsilon \geqq 0$ on CK). Then using Bauer's minimum principle of Ex. 12.C.1 for K we deduce that $t+f+v+u+\varepsilon \geqq 0$ on X. We have

$$-t(z) \leqq \varepsilon_z^{CW}(-t) \leqq \varepsilon_z^{CW}(f+v+u+\varepsilon).$$

Similarly as in Ex. 12.B.1 we get

$$\varepsilon_z^{CW}(\inf\{R_p^{G \setminus K} : K \subset X \text{ is compact}\}) = 0,$$

so that

$$\int^* f \, d\varepsilon_z^{CU} = \varepsilon_z^{CU}(f+v) = \sup\{-t(z) : t \in \mathcal{T}\} \leqq \varepsilon_z^{CW}(f+v).$$

Taking the infimum over all $v \in \mathcal{V}$ gives

$$\int^* f \, d\varepsilon_z^{CU} = \varepsilon_z^{CW}(f) \leqq f(z).$$

THE END

BIBLIOGRAPHY

Books

BAUER, H.

[Bau] Harmonische Räume und ihre Potentialtheorie
In: Lecture Notes in Mathematics 22, Springer 1966

BOBOC, N., BUCUR, Gh., and CORNEA, A.

[BBC] Order and convexity in potential theory: H-cones
Lecture Notes in Mathematics 853, Springer 1981

BOURBAKI, N.

[Bour] General topology, Part 1 (transl.)
Addison-Wesley, Reading (1966)

BRUCKNER, A. M.

[Bru] Differentiation of real functions
Lecture Notes in Mathematics 659, Springer 1978

CONSTANTINESCU, C., and CORNEA, A.

[C&C] Potential theory on harmonic spaces
Berlin, Heidelberg, New York: Springer 1972

ČECH, E.

[Čech] Topological spaces, Rev. Ed. (transl.)
Interscience, New York (1966)

ENGELKING, R.

[Eng] General topology
Monografie matematyczne, Warszawa (1977)

FUGLEDE, B.

[Fug] Finely harmonic functions
In: Lecture Notes in Mathematics 289, Springer 1972

de GUZMÁN, M.

[Guz] Differentiation of integrals in $\mathbb{R}^n$
Lecture Notes in Mathematics 481, Springer 1975

HAUSDORFF, F.

[Haus] Set theory, 3rd Ed. (transl.)
Chelsea, New York (1957)

HELMS, L. L.

[He] Introduction to potential theory
New York-London-Sydney-Toronto: Wiley 1969

KELLEY, J. L.

[Kel] General topology
Van Nostrand, Princeton, New York 1955

KURATOWSKI, K.

[Kur] Topology, Vol. I (transl.)
Academic Press, New York (1966)

SAKS, S.
[Saks] Theory of the integral
Monografie matematyczne 7, New York (1937)

WILLARD, S.
[Will] General topology
Addison-Wesley series in Math., 1970

References

AGRONSKY, S. J.
(1982) A generalization of a theorem of Maximoff and applications
Trans. Amer. Math. Soc. 273 (1982), 767-779 / MR 83h: 26003

ALEXANDROV, A. D.
(1940) Additive set-functions in abstract spaces
Mat. Sbornik 8(50) (1940), 307-348 / MR 2, 315

ALEXANDROFF, P. S., and HOPF, H.
(1935) Topologie I
Berlin, 1935

ANISZCZYK, B., and FRANKIEWICZ, R.
(1985) Non-homeomorphic density topologies / preprint

AVERSA, V., and LACZKOVICH, M.
(1982) Extension theorems on derivatives of additive interval functions
Acta Math. Acad. Sci. Hungar. 39 (1982), 267-277 / MR 84m: 26015 b

BAGEMIHL, F., and McMILLAN, J. E.
(1966) Characterization of the sets of angular and global convergence, and of the sets of angular and global limits, of functions in a half-plane
Fund. Math. 59 (1966), 177-187 / MR 35, 1782

BAGEMIHL, F., and PIRANIAN, G.
(1961) Boundary functions for functions defined in a disk
Michigan Math. J. 8 (1961), 201-207 / MR 26, 1467

BAIRE, R.
(1899) Sur les fonctions de variables réelles
Thése, Ann. di Math. 3 (1899), 1-123

BAISNAB, A. P., and PETERSEN, G. M.
(1971) Metric density and Lusin's theorem
Quart. J. Math. Oxford 22 (1971), 457-464 / MR 45, 488

BANACH, S.
(1930) Théorème sur les ensembles de première catégorie
Fund. Math. 16 (1930), 395-398

BAUER, H.
(1960) Minimalstellen von Funktionen und Extremalpunkte II
Arch. der. Math. 11 (1960), 200-205 / MR 24, A251

(1962) Axiomatische Behandlung des Dirichletschen Problems für elliptische und parabolische Differentialgleichungen
Math. Ann. 146 (1962), 1-59 / MR 26, 1612

(1963) Weiterführung einer axiomatischen Potentialtheorie ohne Kern (Existenz von Potentialen) / MR 27, 5926
Z. Wahrscheinlichkeitstheorie und Verw. Gebiete 1 (1963), 197-229

(1965) Propriétés fines des fonctions hyperharmoniques dans une théorie axiomatique du potentiel
Ann. Inst. Fourier 15 (1965), 137-154 / MR 32, 7772

(1975) Aspects of modern potential theory
Proc. Intern. Congress of Math., Vol. 1, Vancouver 1974, (1975), 41-51

(1981) Harmonische Räume / MR 84b: 31007
Jahrbuch Überblicke Mathematik 1981, Bibliograph. Institut (1981), 9-35

(1984a) Zum heutigen Bild der Potentialtheorie
Published in: Zum Werk Leonhard Eulers, Verlag Birkhäuser (1984), 3-20

(1984b) Harmonic spaces - a survey
Conferenze del Seminario di Matematica dell' Università di Bari, N° 197 (1984)

BAUERMANN, U.

(1977) Balayage-Operatoren in der Potentialtheorie
Math. Ann. 231 (1977), 181-186 / MR 58, 11472

BERG, Ch.

(1971) Quelques propriétés de la topologie fine, dans la théorie du potentiel et des processus standard
Bull. Sci. Math. 95 (1971), 27-31

BERTIN, E. M. J.

(1978) A. F. Monna on Dirichlet operators
In: Two decades of mathematics in the Netherlands, Mathematical Centre, Amsterdam, 1978, Part II, 351-360 / MR 82h: 01040

BESICOVITCH, A. S.

(1928) On the fundamental geometrical properties of linearly measurable plane sets of points
Math. Ann. 98 (1928), 422-464

(1945) A general form of the covering principle and relative differentiation of additive functions I
Proc. Cambridge Phil. Soc. 41 (1945), 103-110 / MR 7, 10

BICHTELER, K.

(1972) On the strong lifting property
Illinois Journ. Math. 16 (1972), 370-380 / MR 45, 5311

BLATTER, J., and SEEVER, G. L.

(1975) Interposition of semi-continuous functions by continuous functions
Analyse fonctionnelle et applications, Paris, 1975 / MR 55, 8749

(1976) Interposition and lattice cones of functions
Trans. Amer. Math. Soc. 222 (1976), 65-96 / MR 55, 11013

BLAŽEK, J., BORÁK, E., and MALÝ, J.
(1978) On Köpcke and Pompeiu functions
Časopis pěst. mat. 103 (1978), 53-61 / MR 58, 6101

BLIEDTNER, J.
(1980) Axiomatic foundation of potential theory
Published in: Functional Analysis, Surveys and Recent Results II, North-Holland Publ. Comp. (1980) / MR 81a: 46004

BLIEDTNER, J., and HANSEN, W.
(1975) Simplicial cones in potential theory
Inventiones Math. 29 (1975), 83-110 / MR 52, 8470
(1976) Cones of hyperharmonic functions
Math. Z. 151 (1976), 71-87 / MR 57, 16643
(1978) Simplicial cones in potential theory II (Approximation theorems)
Inventiones Math. 46 (1978), 255-275 / MR 58, 11473
(1978) Markov processes and harmonic spaces / MR 80a:60098
Z. Wahrscheinlichkeitstheorie verw. Gebiete 42 (1978), 309-325
(1980) Bases in standard balayage spaces
Lecture Notes in Mathematics 787 (1980), 55-63 / MR 82b: 31020
(1984) The weak Dirichlet problem
J. Reine Angew. Math. 348 (1984), 34-39 / MR 85h: 31012

BLIEDTNER, J., and JANSSEN, K.
(1974) A generalization of H. Bauer's minimum principle
Arch. Math. 25 (1974), 505-510 / MR 51, 912

BLUMBERG, H.
(1922) New properties of all real functions
Trans. Amer. Math. Soc. 24 (1922), 113-128
(1930) A theorem on arbitrary functions of two variables with applications
Fund. Math. 16 (1930), 17-24

BOBOC, N., and BUCUR, Gh.
(1985a) Natural localization and natural sheaf property in standard H-cones of functions. I
Rev. Roumaine Math. Pures Appl. 30 (1985), 1-21
(1985b) Natural localization and natural sheaf property in standard H-cones of functions. II
Rev. Roumaine Math. Pures Appl. 30 (1985), 193-213
(1985c) Potentials and supermedian functions on fine open sets in standard H-cones
Preprint

BOBOC, N., BUCUR, Gh., and CORNEA, A.
(1975) H-cones and potential theory
Ann. Inst. Fourier 25 (1975), 71-108 / MR 53, 8464

BOBOC, N., CONSTANTINESCU, C., and CORNEA, A.
(1965) Axiomatic theory of harmonic spaces. Balayage.
Ann. Inst. Fourier 15 (1965), 37-70 / MR 33, 1476
BOBOC, N., and CORNEA, A.
(1967) Comportement des balayées des mesures ponctuelles. Comportement des solutions du problème de Dirichlet aux points irréguliers
C. R. Acad. Sci. Paris, Sér. A 264 (1967), 995-997 / MR 37, 445
(1970a) Cônes convexes ordonnés. H-cônes et adjoints de H-cônes.
C. R. Acad. Sci. Paris 270 (1970), 596-599 / MR 42, 6273
(1970b) Cônes convexes ordonnés. H-cônes et biadjoints de H-cônes.
C. R. Acad. Sci. Paris 270 (1970), 1679-1682 / MR 42, 7929
(1970c) Cônes convexes ordonnés. Représentations integrales.
C. R. Acad. Sci. Paris 271 (1970), 880-883 / MR 49, 7469
BÔCHER, M.
(1903) Singular points of fractions which satisfy partial differential equations of the elliptic type
Bull. Amer. Math. Soc. 9 (1903), 455-465
BOGOMOLOVA, V. S.
(1924) Sur une classe des fonctions asymptotiquement continues
Rec. Math. XXXII, 1924
BOYTE, J. M., and LANE, E. P.
(1975) An insertion theorem for real functions
Fund. Math. 87 (1975), 29-30 / MR 50, 14643
BRADFORD, J. C., and GOFFMAN, C.
(1960) Metric spaces in which Blumberg's theorem holds
Proc. Amer. Math. Soc. 11 (1960), 667-670 / MR 26, 3832
BRELOT, M.
(1938a) Familles de Perron et problème de Dirichlet
Acta Sci. Math. (Szeged) 9 (1938-40), 133-153 / MR 1, 121
(1938b) Sur le potentiel et les suites de fonctions sous-harmoniques
C. R. Acad. Sci. Paris 207 (1938), 836-838
(1939) Sur la théorie moderne du potentiel
C. R. Acad. Sci. Paris 209 (1939), 828-830 / MR 1, 121
(1941) Sur la théorie autonome des fonctions sousharmoniques
Bull. Sci. Math. France 65 (1941), 78-98 / MR 3, 47
(1944) Sur les ensembles effilés
Bull. Sci. Math. 68 (1944), 12-36 / MR 7, 15
(1945) Minorantes sous-harmoniques, extremales et capacités
J. Math. Pures Appl. 24 (1945), 1-32 / MR 7, 521
(1950) Sur l'allure des fonctions harmoniques et surharmoniques à la frontière
Math. Nachr. 4 (1950/51), 298-307, 1-122 / MR 13, 35

(1952) La théorie moderne du potentiel
Ann. Inst. Fourier 4 (1952), 113-133 / MR 15, 527
(1958) Extension axiomatique des fonctions sous-harmoniques II
C. R. Acad. Sci. Paris 246 (1958), 2334-2337 / MR 21, 5094b
(1959) Axiomatique des fonctions harmoniques et surharmoniques dans un espace localement compact
Séminaire de Théorie du Potentiel, 2 (1959), 1.01-1.40 / MR 22, 4887
(1960) Lectures on potential theory
Tata Institute of Fundamental Research, Bombay, 1960 / MR 22, 9749
(1961) Sur un théorème de prolongement fonctionnel de Keldych concernant le problème de Dirichlet
J. Analyse Math. 8 (1960/61), 273-288 / MR 23, A2549
(1962) Quelques propriétés et applications nouvelles de l'effilement
Sém. Théorie du Potentiel 6 (1962), 1.27-1.40 / MR 26, 3560
(1965) Aspect statistique et comparé de deux types d'effilement
Anais da Ac. Brasilieira de ciências 37 (1965) / MR 33, 4303
(1966) Axiomatique des fonctions harmoniques
Montréal: Les Presses de l'Université de Montréal, 1966
(1967) Recherches axiomatiques sur un théorème de Choquet concernant l'effilemen
Nagoya Math. J. 30 (1967), 33-46 / MR 35, 5650
(1971) On topologies and boundaries in potential theory
Lecture Notes in Mathematics 175, Springer 1971 / MR 43, 7654
(1972) Les étapes et les aspects multiples de la théorie du potentiel
L'Enseignement math. 18 (1972), 1-36 / MR 51, 8436

BROWN, J. B.
(1983) Variations on Blumberg's theorem
Real Analysis Exchange 9 (1983-84), 123-137

BRUCKNER, A. M.
(1971) Differentiation of integrals
Amer. Math. Monthly, 78 No 9 Part II (1971), 1-51 / MR 45, 2124
(1977) Some observations about Denjoy's preponderant derivative
Bull. Math. Soc. Sci. Math. R.S.R. 21 (69) (1977), 1-10 / MR 57, 9922
(1983) Some new simple proofs of old difficult theorems
Real Analysis Exchange 9 (1983-84), 63-78

BRUCKNER, A., CEDER, J., and KESTON, R.
(1968) Representations and approximations by Darboux functions in the first class of Baire
Rev. Roum. Math. Pures et Appl. 13 (1968), 1247-1254 / MR 39, 4333

BRUCKNER, A., and LEONARD, J.
(1966) Derivatives
Amer. Math. Monthly 73 (1966), 24-56 / MR 33, 5797

BRUCKNER, A. M., O'MALLEY, R. J., and THOMSON, B. S.
(1984) Path derivatives: A unified view of certain generalized derivatives
Trans. Amer. Math. Soc. 283 (1984), 97-126

BUCUR, Gh.
(1969) Couples normaux de quasi-topologies. Théorèmes de prolongement.
Rev. Roum. Math. Pures Appl. 14 (1969), 1395-1422 / MR 41, 4470

BUCUR, Gh., and HANSEN, W.
(1984) Balayage, quasi-balayage, and fine decomposition properties in standard H-cones of functions
Rev. Roum. Math. Pures et Appl. 29 (1984), 19-41

CARTAN, H.
(1942) Capacité extérieure et suites convergentes de potentiels
C. R. Acad. Sci. Paris 214 (1942), 944-946 / MR 5, 146
(1945) Théorie du potentiel Newtonien: énergie, capacité, suites de potentiels
Bull. Soc. Math. France 73 (1945), 74-106 / MR 7, 447
(1946) Théorie générale du balayage en potentiel newtonien
Ann. Univ. Grenoble, Math. Phys. 22 (1946), 221-280 / MR 8, 581

ČECH, E.
(1937) On bicompact spaces
Ann. of Math. 38 (1937), 823-844

CEDER, J., and PEARSON, T. L.
(1981) On typical bounded Darboux Baire one functions
Acta Math. Acad. Sci. Hung. 37 (1981), 339-348 / MR 82j: 26005

ČERMÁKOVÁ, E.
(1978) The insertion of regular sets in potential theory
Časopis pěst. mat. 103 (1978), 356-362 / MR 80a: 31009

CHAIKA, M.
(1971) The Lusin-Menchoff theorem in metric space
Indiana U. Math. J. 21 (1971), 351-354 / MR 45, 489

CHAKRABARTI, S., and LAHIRI, B. K.
(1984) Density topology in a topological group
Indian J. pure appl. Math. 15 (1984), 753-764

CHITTENDEN, E. W.
(1919) On the limit functions of sequences of continuous functions converging relatively uniformly
Trans. Amer. Math. Soc. 20 (1919), 179-184

CHOQUET, G.
(1947) Application des propriétés descriptives de la fonction contingent à la théorie des fonctions de variable réelle et à la géometrie différentielle des variétés cartésiennes (Thèse)
J. Math. Pures Appl. 26 (1947), 115-226 / MR 9, 419

COLLINGWOOD, E. F., and LOHWATER, A. J.
(1966) The theory of cluster sets
Cambridge tracts in mathematics, 56 (1966)
COMFORT, W. W., and NEGREPONTIS, S.
(1982) Chain conditions in topology
Cambridge University Press, Cambridge-New York, 1982 / MR 84k: 04002
CONSTANTINESCU, C.
(1966) Die heutige Lage der Theorie der harmonischen Räume
Rev. Roumaine Math. Pures Appl. 11 (1966), 1041-1056 / MR 35, 4459
CONSTANTINESCU, C., and CORNEA, A.
(1963) On the axiomatic of harmonic functions II
Ann. Inst. Fourier 13 (1963), 389-394 / MR 29, 2416
CORNEA, A., and HÖLLEIN, H.
(1980) Bases and essential bases in H-cones
In: Functional Analysis: Surveys and recent results II, 69-86.
Amsterdam-New York-Oxford: North Holland 1980 / MR 81c: 31018
CSÁSZÁR, A., and LACZKOVICH, M.
(1975) Discrete and equal convergence
Studia Sci. Math. Hungar. 10 (1975), 463-472 / MR 81e: 54013
(1979) Some remarks on discrete Baire classes
Acta Math. Acad. Sci. Hung. 33 (1979), 51-70 / MR 80d: 26005
DAVIES, R. O.
(1973) Separate approximate continuity implies measurability
Proc. Camb. Phil. Soc. 73 (1973), 461-465 / MR 48, 4216
DAY, M. M.
(1944) Convergence, closure and neighborhoods
Duke Math. J. 11 (1944), 181-189 / MR 5, 212
DENJOY, A.
(1915) Sur les fonctions dérivées sommables
Bull. Soc. Math. France 43 (1915), 161-248
DIEUDONNÉ, J.
(1944) Une généralisation des espaces compacts
Journ. de Math. Pures et Appl. 23 (1944), 65-76 / MR 7, 134
DOOB, J. L.
(1956) Probability methods applied to the first boundary value problem
In: Proc. 3rd Berkeley Sym. on Math.Stat. and Prob., 1954-55, 49-80
Berkeley, University of California Press, 1956 / MR 18, 941
(1966) Applications to analysis of a topological definition of smallness of a set
Bull. Amer. Math. Soc. 72 (1966), 579-600 / MR 34, 3514
van DOUWEN, E. K., TALL, F.D., and WEISS, W. A. R.
(1977) Nonmetrizable hereditarily Lindelöf spaces with point-countable bases
from CH

Proc. Amer. Math. Soc. 64 (1977), 139-145 / MR 58, 24187

DOWKER, C.

(1951) On countably paracompact spaces
Canadian J. Math. 3 (1951), 219-224 / MR 13, 264

EAMES, W.

(1971) On a topology generated by measurable covers
Canad. Math. Bull. 14 (1971), 499-502 / MR 46, 9265

EILENBERG, S., and SAKS, S.

(1935) Sur la dérivation des fonctions dans des ensembles dénombrables
Fund. Math. 25 (1935), 264-266

ELLIS, H. W.

(1951) Darboux properties and applications to nonabsolutely convergent integrals
Canad. J. Math. 3 (1951), 471-484 / MR 13, 332

FEYEL, D.

(1978) Espaces de Banach fonctionnels adaptés. Quasitopologie et balayage.
Lecture Notes in Math. 681, Springer 1978, 81-102 / MR 81e: 31006

(1981a) Une mesure de l'épaisseur d'un ensemble presque borélien
C. R. Acad. Sci. Paris Sér. I Math. 292 (1981), 957-958 / MR 83b: 31007

(1981b) Ensembles singuliers associés aux espaces de Banach réticules
Ann. Inst. Fourier 31 (1981), 195-223 / MR 84i: 28016

FILLMORE, P. A.

(1966) On topology induced by measure
Proc. Amer. Math. Soc. 17 (1966), 854-857 / MR 33, 5830

FREUD, G.

(1958) Ein Beitrag zu dem Satze von Cantor und Bendixson
Acta Math. Hung. 9 (1958), 333-336 / MR 21, 7494

FROSTMAN, O.

(1938) Sur le balayage des mesures
Acta Sci. Math. Szeged 3 (1938-40), 43-51

(1939) Les points irréguliers dans la théorie du potentiel et le critère de Wiener
Medd. Lunds Univ. Mat. Sem. 4 (1939), 1-10

FUCHSSTEINER, B.

(1971) Bemerkungen zu den Minimumsätzen von H. Bauer
Arch. Math. 22 (1971), 287-290 / MR 45, 5777

FUGLEDE, B.

(1970) Fine connectivity and finely harmonic functions
C. R. Congr. Internat. Math., Nice (1970) / MR 54, 10637

(1971a) The quasi topology associated with a countably subadditive set function
Ann. Inst. Fourier 21 (1971), 123-169 / MR 44, 391

(1971b) Connexion en topologie fine et balayage des mesures
Ann. Inst. Fourier 21 (1971), 227-244 / MR 49, 9241

(1972) Finely harmonic functions
In: Lecture Notes in Math., Vol. 289, Springer 1972 / MR 56, 8883
(1974) Boundary minimum principles in potential theory
Math. Ann. 210 (1974), 213-226 / MR 50, 10293b
(1975a) Fonctions harmoniques et fonctions finement harmoniques
Ann. Inst. Fourier 24 (1975), 77-91 / MR 51, 3490
(1975b) Asymptotic paths for subharmonic functions
Math. Ann. 213 (1975), 261-274 / MR 52, 775
(1980) Asymptotic paths for subharmonic functions and polygonal connectedness of fine domains
Séminaire de théorie du potentiel, Paris, No. 5
In: Lecture Notes in Math., 814, Springer 1980, 97-116 / MR 82b: 31006
(1982) Localization in fine potential theory and uniform approximation by subharmonic functions
J. Functional Analysis 49 (1982), 57-72 / MR 84c: 31009

GAMELIN, T. W., and LYONS, T. J.
(1983) Jensen measures for R(K)
J. London Math. Soc. 27 (1983), 317-330 / MR 84d: 46079

GÅRDING, L.
(1979) The Dirichlet problem
Math. Intelligencer 2 (1979), 43-53 / MR 81d: 31001

GLEYZAL, A.
(1941) Interval functions
Duke Math. J. 8 (1941), 223-230 / MR 3, 226

GOFFMAN, C.
(1950) On Lebesgue's density theorem
Proc. Amer. Math. Soc. 1 (1950), 384-387 / MR 12, 167
(1975) Everywhere differentiable functions and the density topology
Proc. Amer. Math. Soc. 51 (1975), 250 / MR 51, 3366

GOFFMAN, C., and NEUGEBAUER, C. J.
(1960) On approximate derivatives
Proc. Amer. Math. Soc. 11 (1960), 962-966 / MR 22, 9562

GOFFMAN, C., NEUGEBAUER, C., and NISHIURA, T.
(1961) Density topology and approximate continuity
Duke Math. J. 28 (1961), 497-505 / MR 25, 1254

GOFFMAN, C., and WATERMAN, D.
(1961) Approximately continuous transformations
Proc. Amer. Math. Soc. 12 (1961), 116-121 / MR 22, 11082

GRANDE, Z.
(1978) Sur la r-continuité des fonctions de deux variables
Demonstratio Math. 11 (1978), 937-945 / MR 80g: 26005

(1979) Sur le prolongement des fonctions
Acta Math. Acad. Sci. Hung. 34 (1979), 43-45 / MR 80i: 26005

GRANDE, Z., and TOPOLEWSKA, M.

(1982) Sur les fonctions vectorielles approximativement continues
Časopis pěst. mat. 107 (1982), 333-340, 428 / MR 84k: 26005

de GUZMÁN, M.

(1970) A covering lemma with applications to differentiability of measures and singular integral operators
Studia Math. 34 (1970), 299-317 / MR 41, 8621

(1981) Real variable methods in Fourier analysis
Amsterdam, North-Holland, 1980 / MR 83j: 42019

HAHN, H.

(1917) Über halbstetige und unstetige Funktionen
Sitzungsberichte Akad. Wiss. Wien Abt. IIa 126 (1917), 91-110

HAMMER, P. C.

(1964) Extended topology: structure of isotonic functions
J. Reine Angew. Math. 213 (1964), 174-186 / MR 29, 576

HANSEN, W.

(1971) Fegen und Dünnheit mit Anwendungen auf die Laplace und Wärmeleitungsgleichung
Ann. Inst. Fourier, Grenoble, 21 (1971), 79-121

(1981) Semipolar sets and quasi-balayage
Math. Ann. 257 (1981), 495-517 / MR 83m: 31010

(1983) Convergence of balayage measures
Math. Ann. 264 (1983), 437-446 / MR 85c: 31012

(1985) On the identity of Keldych solutions
Czech. Math. J. 35 (110) (1985), 632-638

HARVEY, R., and POLKING, J. C.

(1972) A notion of capacity which characterizes removable singularities
Trans. Amer. Math. Soc. 169 (1972), 183-195 / MR 46, 5862

HASHIMOTO, H.

(1952) On some local properties on spaces
Math. Japonicae 2 (1952), 127-134 / MR 14, 396

(1954) On the resemblance of point sets
Math. Japonicae 3 (1954), 53-56 / MR 17, 179

(1976) On the *-topology and its applications
Fund. Math. 91 (1976), 5-10 / MR 54, 1179

HAUPT, O., and PAUC, Ch.

(1952) La topologie de Denjoy envisagée comme vraie topologie
C. R. Acad. Sci. Paris 234 (1952), 390-392 / MR 13, 728

(1954) Über die durch allgemeine Ableitungsbasen bestimmten Topologien
Ann. Mat. Pura Appl., Ser. 4, t. 36 (1954), 247-271 / MR 16, 388

HAUSDORFF, F.

(1914) Grundzüge der Mengenlehre / Leipzig, 1914

(1919) Über halbstetige Funktionen und deren Verallgemeinerung
Math. Z. 5 (1919), 292-309

HAYASHI, E.

(1964) Topologies defined by local properties
Math. Ann. 156 (1964), 205-215 / MR 29, 4024

(1979) Topologies defined by ideals of sets
Bull. Nagoya Inst. Tech. 31 (1979), 111-116 / MR 81j: 54003

HERVÉ, R.-M.

(1959) Développements sur une théorie axiomatique des fonctions sur-harmoniques
C. R. Acad. Sci. Paris 248 (1959), 179-181 / MR 21, 5097

(1962) Recherches axiomatiques sur la théorie des fonctions surharmoniques et du potentiel
Ann. Inst. Fourier 12 (1962), 415-571 / MR 25, 3186

HIGGS, D.

(1983) Iterating the derived set function
Amer. Math. Monthly 90 (1983), 693-697 / MR 85b: 26001

HOFFMANN-JØRGENSEN, J.

(1982) A general "in-between theorem"
Math. Scand. 50 (1982), 55-65 / MR 83h: 54020

HUREWICZ, W.

(1928) Relativ perfekte Teile von Punktemengen und Mengen (A)
Fund. Math. 12 (1928), 78-109

HYVÖNEN, J.

(1979) On the harmonic continuation of bounded harmonic functions
Math. Ann. 245 (1979), 151-157 / MR 80k: 31008

IKEGAMI, T.

(1981) On a generalization of Lukeš' theorem
Osaka J. Math. 18 (1981), 699-702 / MR 82m: 31012

(1983) On the simplicial cone of superharmonic functions in a resolutive compactification of a harmonic space
Osaka J. Math. 20 (1983), 881-898 / MR 85c: 31013

(1984) On the boundary behavior of the Dirichlet solutions at an irregular boundary point
Osaka J. Math. 21 (1984), 851-858

IONESCU-TULCEA, A., and IONESCU-TULCEA, C.

(1969) Topics in the theory of lifting
Springer, New York, 1969 / MR 43, 2185

ISEKI, K.

(1960) On the covering theorem of Vitali
Proc. Japan. Acad. 36 (1960), 539-542

JANSSEN, K., and HEDI BEN SAAD
(*) Topologie fine et mesure de référence / J.Reine Angew.Math.360(1985),153-159

JARNÍK, V.
(1926) Sur les fonctions de la première classe de Baire
Bull. Internat. Acad. Sci. Boheme (1926), 1-11

JEDRZEJEWSKI, J.
(1974) On limit numbers of real functions
Fund. Math. 83 (1974), 269-284 / MR 48, 8702

JESURAJ, R.
(1985) Continuous functions on polar sets
Proc. Amer. Math. Soc. 93 (1985), 262-266

JONES, F. B.
(1937) Concerning normal and completely normal spaces
Bull. Amer. Math. Soc. 43 (1937), 671-677

KAMKE, E.
(1927) Zur Definition der approximativ stetigen Funktionen
Fund. Math. 10 (1927), 431-433

KATĚTOV, M.
(1950) On nearly discrete spaces
Časopis pěst. mat. fys. 75 (1950), 69-78 / MR 12, 195
(1951) On real-valued functions in topological spaces
Fund. Math. 38 (1951), 85-91 / MR 14, 304
(1953) Correction to "On real-valued functions in topological spaces"
Fund. Math. 40 (1953), 203-205 / MR 15, 640

KATZNELSON, Y., and STROMBERG, K.
(1974) Everywhere differentiable, nowhere monotone functions
Amer. Math. Monthly 81 (1974), 349-354 / MR 49, 481

KELAR, V.
(1980) On strict local extrema of differentiable functions
Real Anal. Exchange 6 (1980-81), 242-244
(1983) On the first and the fifth class of Zahorski
Real Anal. Exchange 9 (1983-84), 233-250 / MR 85h: 28003

KELDYCH, M. V.
(1941a) On the resolutivity and the stability of Dirichlet problem (Russian)
Uspechi Mat. Nauk 8 (1941), 172-231 / MR 3, 123
(1941b) On the Dirichlet problem (Russian)
Dokl. Akad. Nauk SSSR 32 (1941), 308-309 / MR 6, 155

KELLOG, O. D.
(1926) Recent progress with the Dirichlet problem
Bull. Amer. Math. Soc. 32 (1926), 601-625

KELLY, J. C.
(1963) Bitopological spaces

Proc. London Math. Soc. 13 (1963), 71-89 / MR 26, 729

KEMPISTY, S.

(1921) Sur l'approximation des fonctions de première classe
Fund. Math. 2 (1921), 131-135

KNASTER, B.

(1945) Sur une propriété caractéristique de l'ensemble des nombres réels
Mat. Sb. 16 (1945), 281-288 / MR 7, 277

KÖHN, J.

(1968) Harmonische Räume mit einer Basis semiregulärer Mengen
In: Seminar über Potentialtheorie, Lecture Notes in Math. 69 (1968), 1-12, Springer Verlag, Berlin-Heidelberg-New York / MR 38, 6092

KÖHN, J., and SIEVEKING, M.

(1967) Reguläre und extremale Randpunkte in der Potentialtheorie
Rev. Roum. Math. Pures Appl. 12 (1967), 1489-1502 / MR 38, 1283

KRÁL, J., and NETUKA, I.

(1985) Fine topology in potential theory and strict maxima of functions
to appear

KUPKA, J.

(1983) Strong liftings with application to measurable cross sections in locally compact groups
Israel J. Math. 44 (1983), 243-261 / MR 84g: 28006

LACZKOVICH, M.

(1975) Separation properties of some subclasses of Baire 1 functions
Acta Math. Acad. Sci. Hungar. 26 (1975), 405-412 / MR 53, 8343

(1983) Baire 1 functions
Real Anal. Exchange 9 (1983-84), 15-28

LACZKOVICH, M., and PETRUSKA, G.

(1973) A theorem on approximately continuous functions
Acta Math. Acad. Sci. Hung. 24 (1973), 383-387 / MR 48, 4217

LAHIRI, B. K.

(1977) Density and approximate continuity in topological groups
Journal Indian Math. Soc. 41 (1977), 129-141 / MR 58, 28441

LANDIS, E. M.

(1969) Necessary and sufficient conditions for the regularity of a boundary point for the Dirichlet problem for the heat equation (Russian)
Dokl. Akad. Nauk SSSR 185 (1969), 517-520 / MR 41, 7308

(1971) Equations of the second order of elliptic and parabolic types (Russian)
Nauka, Moscow 1971 / MR 47, 9044

LANDKOF, N. S.

(1972) Foundations of modern potential theory
Grundlehren d. math. Wiss. 180 (1972), Springer-Verlag (Russian edition, Moscow 1966) / MR 50, 2520

LANE, E. P.

(1967) Bitopological spaces and quasi-uniform spaces
Proc. Lond. Math. Soc. 17 (1967), 241-256 / MR 34, 5054

(1971) Insertion of continuous functions
Glasnik Mat. Sov. III 6 (25) (1971), 165-171 / MR 49, 6145

(1975) A sufficient condition for the insertion of a continuous function
Proc. Amer. Math. Soc. 49 (1975), 90-94 / MR 50, 14589

(1976) Insertion of continuous functions
Pacific J. Math. 66 (1976), 181-190 / MR 57, 13833

(1979) PM-normality and the insertion of continuous functions
Pacific J. Math. 82 (1979), 155-162 / MR 81c: 54033

(1981) Insertion of a continuous function and X x I
Topology proceedings, 6 (1981), 329-334 / MR 84b: 54031

(1983) Lebesgue sets and insertion of continuous functions
Proc. Amer. Math. Soc. 87 (1983), 539-542 / MR 84g: 54013

LARSON, L.

(1983) The symmetric derivative
Trans. Amer. Math. Soc. 277 (2) (1983), 589-599 / MR 84j: 26009

LEBESGUE, H.

(1905) Sur les fonctions représentables analytiquement
Journ. de Math. Pures et Appl. 1 (1905), 139-216

(1907) Sur le problème de Dirichlet
Rend. Circ. Mat. Palermo 24 (1907), 371-402

(1913) Sur des cas d'impossibilité du problème de Dirichlet
C. R. Soc. Math. France 1913 (1913), 17

LEDERER, G.

(1960) Two theorems on Baire functions in separable metric spaces
Quart. J. Math. Oxford 11 (1960), 269-274 / MR 25, 3847

LE JAN, Y.

(1983) Quasi-continuous functions and Hunt processes
J. Math. Soc. Japan 35 (1983), 37-42 / MR 84c: 60116

LEVY, R.

(1973) A totally ordered Baire space for which Blumberg's theorem fails
Proc. Amer. Math. Soc. 41 (1973), 304 (Erratum, PAMS 45 (1974), 469)

(1974) Strongly non-Blumberg spaces
General Topology and Appl. 4 (1974), 173-177 / MR 49, 7976

LUKEŠ, J.

(1973) Principal solution of the Dirichlet problem in potential theory
Comment. Math. Univ. Carolinae 14 (1973), 773-778 / MR 49, 9243

(1974) Théorème de Keldych dans la théorie axiomatique de Bauer des fonctions harmoniques
Czech. Math. J. 24 (1974), 114-125 / MR 50, 2539

(1977a) The Lusin-Menchoff property of fine topologies
Comment. Math. Univ. Carolinae 18 (1977), 515-530 / MR 57, 4106

(1977b) Functional approach to the Brelot-Keldych theorem
Czech. Math. J. 27 (1977), 609-616 / MR 58, 28575

(1978) A topological proof of Denjoy-Stepanoff theorem
Časopis pěst. mat. 103 (1978), 95-96 / MR 57, 16527

(1984) Some old and new applications of Choquet theory in potential theory
Proc. of the 12th Winter School on abstract analysis, Srní, 1984
Suppl. Rend. Circ. Mat. Palermo (1984), 55-62

LUKEŠ, J., and MALÝ, J.

(1981) On the boundary behaviour of the Perron generalized solution
Math. Ann. 257 (1981), 355-366 / MR 83b: 31008

(1982) Fine hyperharmonicity without axiom D
Math. Ann. 261 (1982), 299-306 / MR 84k: 31008b

LUKEŠ, J., and NETUKA, I.

(1976) The Wiener type solution of the Dirichlet problem in potential theory
Math. Ann. 224 (1976), 173-178 / MR 54, 10638

(1979) What is the right solution of the Dirichlet problem?
In: Lecture Notes in Math. 743 (1979), 564-572 / MR 80, 31015

LUKEŠ, J., and ZAJÍČEK, L.

(1976) Fine topologies as examples of non-Blumberg Baire spaces
Comment. Math. Univ. Carolinae 17 (1976), 683-688 / MR 56, 9491

(1977a) The insertion of G_δ sets and fine topologies
Comment. Math. Univ. Carolinae 18 (1977), 101-104 / MR 56, 5808

(1977b) When finely continuous functions are of the first class of Baire
Comment. Math. Univ. Carolinae 18 (1977), 647-657 / MR 56, 15851

(1977c) Connectivity properties of fine topologies
Rev. Roum. Math. Pures et Appl. 22 (1977), 679-684 / MR 55, 10712

LUSIN, N. N.

(1951) Integral and trigonometric series (Russian)
Moskva-Leningrad, 1951

LYONS, T. J.

(1980) Finely holomorphic functions
J. Funct. Anal. 37 (1980), 1-18 / MR 82d: 31011a

(1982) Cones of lower semicontinuous functions and a characterisation of finely hyperharmonic functions
Math. Ann. 261 (1982), 293-297 / MR 84k: 31008a

MALÝ, J.

(1979a) A note on separation of sets by approximately continuous functions
Comment. Math. Univ. Carolinae 20 (1979), 207-212 / MR 80h: 26004

(1979b) The Peano curve and the density topology
Real Anal. Exchange 5 (1979-80), 326-330 / MR 81e: 54037

(1980) Dirichlet problem in the fine topology
Thesis, Charles University, Prague, 1980

(1983) Concerning the fine topology from the band $\mathcal{M}$
Comment. Math. Univ. Carolinae 24 (1983), 388-389

(1984) Quasitopological methods in fine potential theory / preprint

MARCUS, S.

(1953) Sur la dérivée approximative qualitative
Com. Acad. R. P. Romine 3 (1953), 361-364

(1963) Sur les dérivées dont les zéros forment un ensemble frontière partout dense
Rend. Circ. Mat. Palermo 12 (1963), 1-36 / MR 29, 4844

MARTIN, N. F. G.

(1960) A note on metric density of sets of real numbers
Proc. Amer. Math. Soc. 11 (1960), 344-347 / MR 25, 5152

(1961) Generalized condensation points
Duke Math. J. 28 (1961), 507-514 / MR 24, A3624

(1964) A topology for certain measure spaces
Trans. Amer. Math. Soc. 112 (1964), 1-18 / MR 28, 5157

MAULDIN, R. D.

(1974) Baire functions, Borel sets and ordinary function systems
Advances in Math. 12 (1974), 418-450 / MR 51, 4153

MAXIMOFF, I.

(1940) On density points and approximately continuous functions
Tôhoku Math. J. 47 (1940), 237-250 / MR 2, 352

MAZURKIEWICZ, S.

(1921) Sur les fonctions de classe 1
Fund. Math. 2 (1921), 28-36

MICHAEL, E. A.

(1956) Continuous selections
Ann. of Math. 63 (1956), 361-382 / MR 17, 990

MILLER, H. I.

(1981) Baire outer kernels of sets
Publ. Inst. Math. Beograd 30 (1981), 117-122 / MR 83j: 26004

MIŠÍK, L.

(1966) Über die Ableitung der Additiven Intervallfunktionen
Časopis pěst. mat. 91 (1966), 394-410 / MR 38, 291

(1969) A note concerning a paper by L. E. Snyder
Matematický časopis 19 (1969), 188-191 / MR 46, 1972

(1972) Notes on an approximate derivative
Mat. časopis Sloven. akad. vied 22 (1972), 108-114 / MR 47, 3606

MONNA, A. F.

(1938) Het probleem van Dirichlet
Nieuw. Arch. Wisk. 19 (1938), 249-256

(1939) On the Dirichlet problem and the method of sweeping-out
Nederl. Akad. Wetensch. Proc. Series A 42 (1939), 491-498

(1971) Note sur le problème de Dirichlet
Nieuw Arch. Wisk. 19 (1971), 58-64 / MR 46, 9374

(1975) Dirichlet's principle. A mathematical comedy of errors and its influence on the development of analysis
Oosthoek, Scheltema & Holkema, Utrecht, 1975

MOORE, R. L.

(1924) An extension of the theorem that no countable point set is perfect
Proc. Nat. Acad. Sci. 10 (1924), 168-170

MORSE, A. P.

(1947) Perfect blankets
Trans. Amer. Math. Soc. 61 (1947), 418-442 / MR 8, 571

MOTCHANE, L.

(1957) Sur la notion d'espace bitopologique et sur les espaces de Baire
C. R. Acad. Sci. Paris, Ser. A 244 (1957), 3121-3124 / MR 19, 1069

NAGAMI, K.

(1954) Baire sets, Borel sets and some typical semi-continuous functions
Nagoya Math. J. 7 (1954), 85-93 / MR 16, 1092

NETUKA, I.

(1974) Thinness and the heat equation
Čas. pěst. mat. 99 (1974), 293-299 / MR 50, 4991

(1975) Harmonic functions and mean value theorems
Čas. pěst. mat. 100 (1975), 391-409 / MR 57, 3411

(1980a) The Dirichlet problem for harmonic functions
Amer. Math. Monthly 87 (1980), 621-628 / MR 82c: 31005

(1980b) The classical Dirichlet problem and its generalizations
In: Potential theory Copenhagen 1979. Lecture Notes in Math. 787 (1980), 212-218 / MR 82e: 31018

(1982a) La représentation de la solution généralisée à l'aide des solutions classiques du problème de Dirichlet
Séminaire de Théorie du Potentiel, Paris, No 6
Lecture Notes in Math. 906 (1982), 261-268 / MR 83j: 31004

(1982b) L'unicité du problème de Dirichlet géneralisé pour un compact
Séminaire de Théorie du Potentiel, Paris, No 6
Lecture Notes in Math. 906 (1982), 269-281 / MR 83k: 31012

(1982c) Monotone extensions of operators and the first boundary value problem
In: Proceedings of the Fifth Czechoslovak Conference on Diff. equations and their applications, Bratislava 1981
Teubner-Texte zur Mathematik 47 (1982), Leipzig, 268-271 / MR 84i: 00016

NETUKA, I., and VESELÝ, J.

(1978) Harmonic continuation and removable singularities in the axiomatic

potential theory
Math. Ann. 234 (1978), 117-123 / MR 58, 1210

NETUKA, I., and ZAJÍČEK, L.
(1974) Functions continuous in the fine topology for the heat equation
Čas. pěst. mat. 99 (1974), 300-306 / MR 50, 4992

NISHIURA, T.
(1981) The topology of almost everywhere continuous, approximately continuous functions
Acta Math. Acad. Sci. Hungar. 37 (1981), No. 4, 317-328 / MR 82f: 26014

O'MALLEY, R. J.
(1976a) Baire 1*, Darboux functions
Proc. Amer. Math. Soc. 60 (1976), 187-192 / MR 54, 5405
(1976b) Approximately differentiable functions: The r topology
Real Analysis Exchange 2 (1976), 62-66
(1977) Approximately differentiable functions: the r-topology
Pacific J. Math. 72 (1977), 207-222 / MR 56, 5810
(1979a) Approximately continuous functions which are continuous almost everywhere
Acta Math. Acad. Sci. Hungar. 33 (1979), 395-402 / MR 80g: 26006
(1979b) Insertion of Baire * 1 functions
Rev. Roumaine Math. Pures Appl. 24 (1979), 1445-1448 / MR 81c: 26004

OSGOOD, W.
(1897) Non-uniform convergence and the integration of series term by term
Amer. Journ. of Math. 19 (1897), 155-190

OSTASZEWSKI, K.
(1981) Continuity in the density topology
Real Anal. Exchange 7 (1981-82), 259-270 / MR 83e: 26007
(1983a) Continuity in the density topology II
Rend. Circ. Mat. Palermo 32 (1983), 398-414
(1983b) Density topology and the Luzin (N) condition
Real Anal. Exchange 9 (1983-84), 390-393

OXTOBY, J. C.
(1961) Spaces that admit a category measure
J. Reine Angew. Math. 205 (1961), 156-170 / MR 25, 4054
(1971) Measure and category
Springer-Verlag, New York-Heidelberg-Berlin, 1971 / MR 52, 14213
(1975) The kernel operation on subsets of a T_1-space
Fund. Math. 90 (1975/76), 275-284 / MR 55, 4092

PATTY, C. W.
(1967) Bitopological spaces
Duke Math. J. 34 (1967), 387-392 / MR 36, 3310

PEKÁR, K.
(1981) Preponderant continuity and preponderant derivatives

Thesis, Charles University, Prague, 1981

PERRON, O.

(1923) Eine neue Behandlung der ersten Randwertaufgabe für $\Delta u = 0$
Math. Z. 18 (1923), 42-54

PETERSEN, G. M., and TO, T. O.

(1976) An extension of metric density
Quart. J. Math. Oxford 27 (1976), 463-466 / MR 54, 10538

PETRUSKA, G., and LACZKOVICH, M.

(1973) A theorem on approximately continuous functions
Acta Math. Acad. Sci. Hung. 24 (1973), 383-387 / MR 48, 4217

(1974) Baire 1 functions, approximately continuous functions and derivatives
Acta Math. Acad. Sci. Hung. 25 (1974), 189-212 / MR 52, 671

POINCARÉ, H.

(1894) Sur les équations de la physique mathématique
Rend. Circ. Mat. Palermo 8 (1894), 57-186

POMPEIU, D.

(1906) Sur les fonctions dérivées
Math. Ann. 63 (1906), 326-332

POREDA, W., WAGNER-BOJAKOWSKA, E., and WILCZYŃSKI, W.

(1985) A category analogue of the density topology
Fund. Math. 125 (1985), 167-173

POWDERLY, M.

(1981) On insertion of a continuous function
Proc. Amer. Math. Soc. 81 (1981), 119-120 / MR 81i: 54009

PREISS, D.

(1969) Limits of derivatives and Darboux-Baire functions
Rev. Roum. Math. Pures et Appl. 14 (1969), 1201-1206 / MR 40, 4397

(1971a) Limits of approximately continuous functions
Czechoslovak Math. J. 21 (1971), 371-372 / MR 45, 4154

(1971b) Approximate derivative and Baire classes
Czechoslovak Math. J. 21 (1971), 373-382 / MR 44, 4158

(1983) Dimension of metrics and differentiation of measures
In: General topology and its relations to modern analysis and algebra V, Proc. Fifth Prague Topol. Symp. 1981, Sigma series in pure mathematics 3, Berlin: Heldermann 1983, pp. 565-568 / MR 85b: 28014

PREISS, D., and VILÍMOVSKÝ, J.

(1980) In-between theorems in uniform spaces
Trans. Amer. Math. Soc. 261 (1980), No. 2, 483-501 / MR 82b: 54022

PRIESTLEY, H. A.

(1971) Separation theorems for semi-continuous functions on normally ordered topological spaces
J. London Math. Soc. 3 (1971), 371-377 / MR 43, 3999

PU, H. W., and PU, H. H.

(1982) Associated sets of Baire* 1 functions
Real Anal. Exchange 8 (1982-83), 479-485

RADO, T.

(1937) Subharmonic functions
Ergebnisse der Mathematik und Ihrer Grenzgebiete, Berlin, 1937

REMAK, R.

(1924) Über potentialkonvexe Funktionen
Math. Z. 20 (1924), 126-130

RIDDER, J.

(1929) Über approximativ stetigen Funktionen
Fund. Math. 13 (1929), 201-209

RIESZ, F.

(1926) Sur les fonctions subharmoniques et leur rapport à la théorie du potentiel
Acta Math. 48 (1926), 329-343

(1930) Sur les fonctions subharmoniques et leur rapport à la théorie du potentiel
Acta Math. 54 (1930), 321-360

SAKS, S.

(1932) On the generalized derivatives
J. London Math. Soc. 7 (1932), 247-251

SAMUELS, P.

(1975) A topology formed from a given topology and ideal
J. London Math. Soc. 10 (1975), 409-416 / MR 51, 11396

SCHEINBERG, S.

(1971) Topologies which generate a complete measure algebra
Adv. in Math. 7 (1971), 231-239 / MR 44, 4172

SCHIRMEIER, H., and SCHIRMEIER, U.

(1978) Einige Bemerkungen über den Satz von Keldych
Math. Ann. 236 (1978), No. 3, 245-254 / MR 81h: 31024

SCHWARZ, H. A.

(1872) Zur Integration der partiellen Differentialgleichungen
J. Reine Angew. Math. 74 (1872), 218-253

SEMADENI, Z.

(1963) Functions with sets of points of discontinuity belonging to a fixed ideal
Fund. Math. 52 (1963), 25-39 / MR 26, 6749

SIERPIŃSKI, W.

(1920) Sur un problème concernant les ensembles mesurables superficiellment
Fund. Math. 1 (1920), 112-115

(1921a) Sur les fonctions développables en séries absolument convergentes de fonctions continues
Fund. Math. 2 (1921), 15-27

(1921b) Demonstration d'un théorème sur les fonctions de première classe

Fund. Math. 2 (1921), 37-40
(1922) Démonstration de quelques théorèmes sur les fonctions mesurables
Fund. Math. 3 (1922), 314-321
(1923) Démonstration élémentaire du théorème sur la densité des ensembles
Fund. Math. 4 (1923), 167-171
(1924) Sur une propriété des ensembles ambigus
Fund. Math. 6 (1924), 1-5
(1927) Sur la densité linéaire des ensembles plans
Fund. Math. 9 (1927), 172-185
(1932) Sur les anneaux de fonctions
Fund. Math. 18 (1932), 1-22
(1934) Sur la séparabilité multiple des ensembles mesurables, B
Fund. Math. 23 (1934), 292-303

SLOBODNIK, S. G.
(1976) Expanding system of linearly closed sets (Russian)
Mat. Zametki 19 (1976), 67-84 / MR 53, 13494

SMALLWOOD, C. V.
(1972) Approximate upper and lower limits
Journ. of Math. Ann. Appl. 37 (1972), 223-227 / MR 45, 2103

SNYDER, L. E.
(1965) Continuous Stolz extensions and boundary functions
Trans. Amer. Math. Soc. 119 (1965), 417-427 / MR 31, 4865
(1966a) Approximate Stolz angle limits
Proc. Amer. Math. Soc. 17 (1966), 416-422 / MR 32, 5822
(1966b) The Baire classification of ordinary and approximate partial derivatives
Proc. Amer. Math. Soc. 17 (1966), 115-123 / MR 32, 4224
(1967a) Bi-arc boundary functions
Proc. Amer. Math. Soc. 18 (1967), 808-811 / MR 36, 2804
(1967b) Stolz angle convergence in metric spaces
Pacific J. Math. 22 (1967), 515-522 / MR 36, 848

STEPANOFF, W.
(1924) Sur une propriété caractéristique des fonctions mesurables
Rec. Math. Soc. Math. Moscow 30 (1924), 487-489

STERNBERG, W.
(1929) Über die Gleichung der Wärmeleitung
Math. Ann. 101 (1929), 394-398

STOICA, L.
(1983) On finely supermean valued functions
Expo. Math. 4 (1983), 361-364

STONE, M. H.
(1949) Boundedness properties in function-lattices
Canad. J. Math. 1 (1949), 176-186 / MR 10, 546

ŠVERÁK, V.

(1982) Two connected topologies on the real line
Real Anal. Exchange 8 (1982-83), 217-222 / MR 85a: 26005

TALL, F. D.

(1976) The density topology
Pacific J. Math. 62 (1976), 275-284 / MR 54, 7727

(1978) Normal subspaces of the density topology
Pacific J. Math. 75 (1978), 579-588 / MR 58, 18346

TAUTZ, G.

(1943) Zur Theorie der elliptischen Differentialgleichungen II
Math. Ann. 118 (1941-43), 733-770 / MR 6, 3

THOMSON, B. S.

(1982) Derivation bases on the real line, II
Real Anal. Exchange 8 (1982-83), 278-442 / MR 84i: 26008b

TODD, A. R.

(1981) Quasiregular, pseudocomplete, and Baire spaces
Pacific J. Math. 95 (1981), 233-250 / MR 83a: 54034

TOLSTOV, G.

(1938) Sur la dérivée approximative exacte
Rec. Math. (Mat. Sbornik) N. S. 4 (1938), 499-504

(1939) Sur quelques propriétés des fonctions approximativement continues
Rec. Math. (Mat. Sbornik) N. S. 5 (1939), 637-645 / MR 1, 206

TONG, H.

(1948) Some characterizations of normal and perfectly normal spaces (announcement)
Bull. Amer. Math. Soc. 54 (1948), 65

(1952) Some characterizations of normal and perfectly normal spaces
Duke Math. J. 19 (1952), 289-292 / MR 14, 304

TROYER, R. S., and ZIEMER, W. P.

(1963) Topologies generated by outer measures
J. Math. Mech. 12 (1963), 485-494 / MR 26, 5115

TUCKER, Ch.

(1968) Limits of a sequence of functions with only countably many points of discontinuity
Proc. Amer. Math. Soc. 19 (1968), 118-122 / MR 36, 2112

VANĚČEK, S.

(1985) On uniform approximation of bounded approximately continuous functions by differences of lower semicontinuous and approximately continuous ones
Czech. Math. J. 35(110) (1985), 28-30

VEDENISOV, N. B.

(1936) Sur les fonctions continues dans les espaces topologiques
Fund. Math. 27 (1936), 234-238

(1940) Généralisation de quelques théorèmes sur la dimension

Comp. Math. 7 (1940), 194-200 / MR 1, 107

VERBLUNSKY, S.

(1930) The generalized third derivative and its application to the theory of trigonometric series
Proc. London Math. Soc. 31 (1930), 387-406

(1933) On the theory of trigonometric series V
Fund. Math. 21 (1933), 168-210

VESELÝ, J.

(1983) Some aspects of modern potential theory
In: Recent trends in Math., Teubner Texte zur Math., Band 50, Teubner, Leipzig 1983, 284-293 / MR 84a: 00012

VETRO, P.

(1983) Ordinarily approximately continuous selections
Rend. Circ. Mat. Palermo 32 (1983), 415-420

VLACH, M.

(1982) On nonerlaging closures in affine spaces
General topology and its relations to modern analysis and algebra V, Proc. Fifth Prague Topol. Symp. 1981, Heldermann Verlag, Berlin 1982, 653-656 / MR 84c: 54003

(1984) Closures and neighborhoods induced by tangential approximations
In: Selected topics in operations research and mathematical economics, Lecture Notes in Economics and Math. Systems, Springer-Verlag, Berlin-Heidelberg-New York-Tokyo 226 (1984), 119-127

VUORINEN, M.

(1982) On the Harnack constant and the boundary behavior of Harnack functions
Annal. Acad. Sci. Fennicae AI, 7 (1982), 259-277 / MR 84e: 31008

WALLIN, H.

(1963) Continuous functions and potential theory
Ark. Mat. 5 (1963), 55-84 / MR 29, 2425

WALSH, J. B.

(1971) Some topologies connected with Lebesgue measure
Lecture Notes in Math. 191 (1971), 290-310 / MR 51, 11638

WARD, A. J.

(1933) On the points where $AD_+ < AD^-$
J. London Math. Soc. 8 (1933), 293-299

WEIL, C. E.

(1976) On nowhere monotone functions
Proc. Amer. Math. Soc. 56 (1976), 388-389 / MR 53, 730

WEISS, W. A. R.

(1975) A solution to the Blumberg problem
Bull. Amer. Math. Soc. 81 (1975), 957-958 / MR 52, 11825

(1977) The Blumberg problem

Trans. Amer. Math. Soc. 230 (1977), 71-85 / MR 55, 11198

WHITE, Jr. H. E.

(1974) Topological spaces in which Blumberg's theorem holds
Proc. Amer. Math. Soc. 44 (1974), 454-462 / MR 49, 6130

(1975a) An example involving Baire spaces
Proc. Amer. Math. Soc. 48 (1975), 228-230 / MR 50, 14691

(1975b) Some Baire spaces for which Blumberg's theorem does not hold
Proc. Amer. Math. Soc. 51 (1975), 477-482 / MR 53, 14438

WIENER, N.

(1924) Certain notions in potential theory
J. Math. Massachussetts 3 (1924), 24-51

(1925) Note on a paper of O. Perron
J. Math. Massachussetts 4 (1925), 21-32

WILCZYŃSKI, W.

(1982) A generalization of density topology
Real Anal. Exchange 8 (1982-83), 16-20

(1984) Remarks on density topology and its category analogue
Proc. 12th Winter School on Abstract Analysis, Srní, 1984
In: Suppl. Rend. Circ. Mat. Palermo II, N° 5 (1984)

(1985) A category analogue of the density topology, approximate continuity and the approximate derivative
Real Anal. Exchange 10 (1984-85), 241-265

WILCZYŃSKI, W., and AVERSA, V.

(1984) Some remarks on $\mathcal{I}$-approximately continuous functions
Ricerche di Matematica 33 (1984), 63-79

WILDENHAIN, G.

(1983) Die feine Topologie der Potentialtheorie und ihre Anwendung in der komplexen Funktionentheorie
14th Styrian mathematical symposium (Stift Rein, 1983), Exp. No. 5, Bericht, 208-213, Forschungszentrum Graz, Graz, 1983

WITTMANN, R.

(1982) Kacsche Potentialtheorie für Resolventen, Markoffsche Prozesse und harmonische Räume
Abhandlungen der Bayerische Akademie der Wissenschaften, München 1982, Heft 161 / MR 84g: 60126

(1983) On the existence of Shilov boundaries
Proc. Amer. Math. Soc. 89 (1983), 62-64 / MR 84m: 31012

WROŃSKI, S.

(1978) Remarks on the derivative of a set in the density topology
Bull. Acad. Pol. Sci. 26 (1978), 385-388 / MR 58, 17022

XUAN LOC NGUYEN, and WATANABE, T.

(1972) Characterization of fine domains for a certain class of Markov processes

with application to Brelot harmonic spaces
Z. Wahrsch. Verw. Gebiete 21 (1972), 167-178 / MR 47, 1136

ZAHORSKI, Z.

(1941) Über die Menge der Punkte in welchen die Ableitung unendlich ist
Tôhoku Math. J. 48 (1941), 321-330 / MR 10, 359

(1947) Sur l'ensemble des points singuliers d'une fonction d'une variable réelle admettant les derivées de tous les ordres
Fund. Math. 34 (1947), 183-245 / MR 10, 23

(1948) Sur la classe de Baire des dérivées approximatives d'une fonction quelconque
Ann. Soc. Polon. Math. 21 (1948), 306-323 / MR 11, 89

(1950) Sur la première dérivée
Trans. Amer. Math. Soc. 69 (1950), 1-54 / MR 12, 247

ZAJÍČEK, L.

(1974) On cluster sets of arbitrary functions
Fund. Math. 83 (1974), 197-217 / MR 49, 3060

(1979) An elementary proof of the one-dimensional density theorem
Amer. Math. Monthly 86 (1979), 297-298 / MR 80c: 28004

(1981) On approximate Dini derivatives and one-sided approximate derivatives of arbitrary functions
Comment. Math. Univ. Carolinae 22 (1981), No. 3, 549-560 / MR 82m: 26004

ZALCWASSER, Z.

(1927) Sur les fonctions de Köpcke
Prace Mat. Fiz. 35 (1927-28), 57-99

ZAREMBA, S.

(1911) Sur le principe de Dirichlet
Acta Math. 34 (1911), 293-316

ZARYCKI, M.

(1930) Über den Kern einer Menge
Jahresber. d. D. Math. Ver. 39 (1930)

LIST OF SPECIAL SYMBOLS

Basic Notations

$\mathbb{N}$ - set of all natural numbers

$\mathbb{Z}$ - set of all integers

$\mathbb{Q}$ - set of all rational numbers

$\mathbb{R}$ - set of all real numbers

$\overline{\mathbb{R}}$ - $\mathbb{R} \cup \{-\infty, +\infty\}$

$\mathbb{R}^n$ - Euclidean n-space

e, e_n - Euclidean topology on $\mathbb{R}$, $\mathbb{R}^n$

λ_n - Lebesgue measure on $\mathbb{R}^n$

λ - λ_1

λ_n^* - outer Lebesgue measure on $\mathbb{R}^n$

$\mathcal{C}(X) = \mathcal{C}(X,t)$ - set of all real continuous functions on a topological space $X = (X,t)$

$\mathcal{C}^{(n)}$ - set of all n-times continuously differentiable functions

$\mathcal{K}(X)$ - set of all real continuous functions on X with compact support

$\exp X$ - family of all subsets of X

CA or $X \setminus A$ - complement of A

$\overline{A}$ or $\mathrm{cl}\, A$, $\mathrm{int}\, A$, ∂A - closure, interior, boundary of the set A

$\mathrm{der}\, A$ - set of all accumulation points of A

$\mathcal{V}(x)$ - set of all neighborhoods of x

$U(x,r)$ or $U^r(x)$ - open ball with center x and radius r

$\mathfrak{F}_\delta$ - family of all countable intersections of sets from $\mathfrak{F}$

$\mathfrak{F}_\sigma$ - family of all countable unions of sets from $\mathfrak{F}$

$[f \geqq g]$ - $\{x \in X: f(x) \geqq g(x)\}$

$\mathrm{supt}\, f$ or $\mathrm{supp}\, f$ - support of f, i.e. $\overline{[f \neq 0]}$

c_A or χ_A - characteristic function of a set A

$f \upharpoonright A$ - restriction of f to A

$\mathrm{Coz}(\Phi)$ - family $\{[f \neq 0]: f \in \Phi\}$ of all Φ - cozero sets

$Z(\Phi)$ - family $\{[f = 0]: f \in \Phi\}$ of all Φ - zero sets

$\mathfrak{F}_+$ - nonnegative functions in $\mathfrak{F}$

f^+ - $\max(f,0)$

f^- - $\max(-f,0)$

$\mathfrak{F}^\uparrow$ - $\{f:$ there is $\{f_n\} \subset \mathfrak{F}$, $f_n \nearrow f\}$, analogously $\mathfrak{F}^\downarrow$

$f * g$ - superposition of f and g

w-lim - weak limit

$\hat{f}$ - lower semicontinuous regularization (i.e. the greatest lower semicontinuous minorant) of f

$B_1(M) = B_1(\rho)$ - system of all Baire one functions on M w.r.t. ρ.

$B_2(M) = B_2(\varrho)$ - system of all Baire two functions on M w.r.t. ϱ.

ε_x - Dirac measure at x (i.e. $\varepsilon_x(A) = c_A(x)$)

supt μ or supp μ - support of a measure μ

CH - continuum hypothesis

c - cardinal number of $\mathbb{R}$

$\mathrm{contg}_x(M)$ or cont (x,M) - contingent of M at x (p.98)

residual set - complement of a first category set

function on X - mapping of X into $\overline{\mathbb{R}}$

Riemann function - the function f on an interval defined by $f(x) = 0$ for x irrational and by $f(x) = \max\{1/q : q \in \mathbb{N},\ qx \in \mathbb{Z}\}$ for x rational

Dirichlet function - $\chi_{\mathbb{Q}}$

positive function f - $f > 0$ (it means everywhere)

nonnegative function f - $f \geqq 0$

increasing function f - $f(y) \geqq f(x)$ for $y > x$

strictly increasing function f - $f(y) > f(x)$ for $y > x$

Baire functions on X - the smallest family of functions on X which contains all real continuous functions and which is closed under pointwise limits

Borel functions on X - Borel measurable functions on X

discrete set - a set M for which $M \cap \mathrm{der}\, M = \emptyset$

isolated set - a set M for which $\mathrm{der}\, M = \emptyset$

$\mathcal{F}$ is linearly separating - $\mathcal{F}$ is a family of functions on X such that for every couple of points $x \neq y$ of X and for every $\lambda \geqq 0$ there is $f \in \mathcal{F}$ such that $f(x) \neq \lambda f(y)$

completely regular or normal space - need not be T_1

perfectly normal space - normal space in which each closed set is a G_δ-set

Symbol index

INDEX